T0201513

identify several candidate genes presumed to be important in causing strokes in these animals. These genes include the atrial naturetic peptide and the brain naturetic peptide loci. One of the challenges is to determine if these advances in animal models have any relevance to strokes in humans. Another challenge is that stroke and genetics in humans are such broad fields, extending the subject even further to include animals was too cumbersome for this particular volume. Perhaps future editions or other publications will focus on genetic aspects of animal stroke models.

In choosing the topics for each chapter, I wanted to include some of the important risk factors for stroke. One could rightly question why I included a chapter on lipid metabolism, while omitting a chapter on diabetes. The data supporting a role for lipids as an independent risk factor for stroke are not clear, particularly when one includes patients with mildly or modestly elevated cholesterol levels. Yet there are data showing that patients with markedly elevated lipids clearly have an increased occurrence of stroke. With regard to the genetics of diabetes, this is clearly an area of very active and productive research. Unfortunately, I was unable to identify an individual with expertise in this area who could author a chapter within the allotted time frame.

Most of the clinical chapters include sections on specific stroke etiologies, mechanisms, and syndromes. Chapter 9, "Inherited Cardiac Diseases that Cause Stroke," proved to be a challenging chapter to author due to the marked progress and ongoing research in this area. Likewise, I had to be selective in organizing Chapter 14. "Inherited Systemic Disorders that Cause Stroke," since one could include hundreds of rare disorders. In general, I feel that this book offers a nice balance between common inherited disorders and less common, but scientifically important, genetic diseases that cause stroke. Examples include sickle cell disease, disorders of homocysteine metabolism, various vasculopathies, and coagulation disorders.

When I was approached several years ago to author a book dealing with the genetics of cerebrovascular disease, I was initially delighted. Such an effort would combine the two medical and scientific areas that most interested me—stroke and genetics. I was also thrilled that a publisher had shown an interest in producing a book on such an important, yet overlooked topic.

As the process of soliciting coauthors, writing chapters, editing, and proofreading got underway, my initial feeling of being thrilled with the prospect of being a book author turned to one of apprehension as the scope of this project came into focus. The field of cerebrovascular disease is large enough, but adding the discipline of genetics to the complex clinical landscape of stroke was a daunting task. Another significant challenge was the rapid increase in the knowledge base in both areas, but particularly genetics. With breakthroughs occurring almost on a weekly basis, our ability to publish an up-to-date book was limited severely. Several chapters have undergone revisions prior to publication in an effort to provide updated information.

I fully realize that the widespread access to computerized databases and literature search programs makes traditional books somewhat less important in a fast moving area such as genetics. The publisher and I envision this book to be a reference tool, providing in some detail the key clinical features of some familial stroke types and syndromes, as well as the key genetic advances for a number of stroke mechanisms and etiologies. In this regard, most of the chapters have very extensive reference sections. This was done for several reasons. As the first (and hopefully definitive) publication in this area, I felt that well-referenced chapters would be important and valuable to our audience. Another consideration was that many of the important works in this area were published prior to the establishment of computerized literature databases; therefore, including extensive references would be helpful for those readers who desire a broad or historical perspective. Lastly, many of the topics covered in each chapter are quite complex and diverse, necessitating a large number of references.

As with most medical publications, every effort has been made to ensure the accuracy and correctness of clinical information with regard to tests and therapies. However, the rapid advances and changing knowledge in the medical and genetics arenas can make these recommendations outdated fairly rapidly. Therefore, before using information in this book for medical decision making, it is strongly suggested that the reader consult more current sources of medical information, as well as colleagues and experts in the subject area.

With these various caveats and qualifications, I hope that you, the reader, will find this book useful and helpful in your daily practice, as well as in your research endeavors that focus on stroke, genetics, or both. Please contact me with suggestions for improvements as well as areas that you feel are particularly well covered.

Mark J. Alberts, MD
Durham, North Carolina

Contents

Part I. Genetic Principles and Techniques

Part II. Genetics of Risk Factors for Cerebrovascular Disease

Part III. Genetics of Specific Stroke Etiologies and Syndromes

Part IV. Clinical Applications

Part 1

Genetic Principles and Techniques

Chapter 1

Genetic Concepts and Techniques

Mark J. Alberts, MD

This chapter will review the basic genetic concepts and molecular genetic techniques that are used for the study of human diseases. Subsequent chapters will demonstrate how these concepts and techniques are used in the investigation of inherited diseases, namely genetic epidemiology, genetic linkage analysis, and the isolation of pathogenic genes.

As with any new and rapidly changing field, mastering the terminology is important yet often intimidating. In discussing the basic genetic concepts, only those terms and procedures that are important and used widely will be defined. Specific clone designations, probe names, and the like will be specifically cited only when they are important for understanding the study at hand.

Molecular Genetic Concepts

Deoxyribonucleic acid (DNA) is the basis of the genetic code for all human cells. DNA is composed of four different nucleotides that are arranged in a linear polymeric structure. A nucleotide has three distinct parts: a sugar moiety, a phosphate group, and an organic base. The bases can be classified into purines (adenine [A] and guanine [G]) and pyrimidines (thymine [T], cytosine [C], and uracil [found only in RNA]).[1] DNA is normally arranged in a double helix with two antiparallel chains. One strand is oriented in the 5′→3′ direction, and the other in a 3′→5′ direction (Figure 1).[2] The directionality of DNA (5′, 3′) denotes the position of the carbon atoms on the sugar moiety that link each nucleotide. By convention, DNA sequences are written in a 5′→3′ orientation. Bases on opposite strands are bound to each other via hydrogen bonds. This bond formation occurs due to the specific size and chemical characteristics of the nucleotides. Due to these limitations, an adenine can form a complementary bond with a thymine (or uracil for a DNA:RNA hybrid), and a cytosine with a guanine. This base pair complementarity forms the basis for many of the hybridization reactions discussed below.[3] DNA is replicated when cells undergo mitosis. The replication of DNA is semiconservative, since one strand is used as a template for the synthesis of a new complementary strand.

Each human cell contains approximately 3 billion bases of DNA, organized into 23 pairs of chromosomes (22 pairs of

From Alberts MJ (ed). *Genetics of Cerebrovascular Disease.* Armonk, NY: Futura Publishing Company, Inc., © 1999.

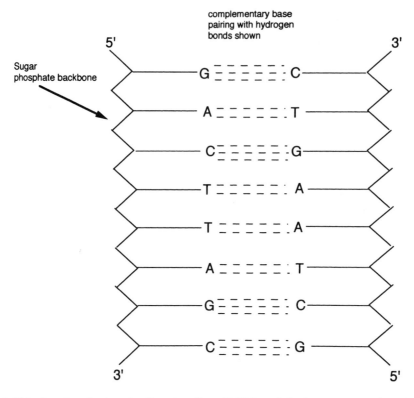

complementary base
pairing with hydrogen
bonds shown

5'   3'

Sugar
phosphate backbone

G ====== : C

A ====== : T

C ====== : G

T ====== : A

T ====== : A

A ====== : T

G ====== : C

C ====== : G

3'   5'

Figure 1. This drawing depicts the directionality of DNA and the base pairing of nucleotides. Note the two complementary strands with 5'→3' and 3'→5' orientations. Adenine-thymine bases are held together by two hydrogen bonds, and cytosine-guanine bases are held together by three hydrogen bonds.

autosomes and 1 pair of sex chromosomes) located within the cell nucleus. Each pair of chromosomes is composed of one chromosome inherited from each parent. It is estimated that the normal complement of DNA encodes between 50,000 and 100,000 genes. There is another important pool of cellular DNA, mitochondrial DNA (mtDNA). Human mtDNA is a circular molecule of 16,569 bases. It encodes 13 proteins that are important for mitochondrial metabolism, including cytochrome oxidase subunits and subunits of NADH-CoQ reductase.[4] Another unique feature of mtDNA is its maternal inheritance. Since only the ovum has mitochondria, essentially all mtDNA (and any mtDNA mutations) will follow maternal lineage.[5]

The DNA of a particular gene is transcribed into ribonucleic acid (RNA) within the cell nucleus. RNA is a single stranded molecule with ribose as its sugar backbone.[6]

The RNA molecule will reflect the base sequence of the gene (DNA) with the exception of uracil replacing thymine. There are three types of RNA: messenger RNA (mRNA), which is an intermediary in protein synthesis; ribosomal RNA (rRNA), which is a key component in RNA translation; and transfer RNA (tRNA), a series of molecules that transport amino acids for peptide synthesis (see below).

Messenger RNA is often modified in several ways before it is transported out of the nucleus. A poly A tail is added to the 3' end of the molecule. The poly A tail is present in most eukaryotic mRNAs and may play a role in directing transportation of the mRNA. Another modification is the addition of of methylguanosine cap at the 5' end of the RNA. This cap may be important in guiding ribosomal translation.

Following these modifications, the mRNA is transported into the cytoplasm, modified in various ways, then translated on ribosomes into a string of amino acids (a peptide).[7] Translation involves the "reading" of the mRNA beginning with the start codon AUG, which codes for methionine. The mRNA sequence is read in contiguous groups of three nucleotides, termed a codon. Transfer RNA (tRNA) carries amino acids to the ribosomes.[8] There is a specific tRNA for each of the 20 different amino acids. Each tRNA has a specific anticodon that will pair with its complementary codon on the mRNA, thereby translating the nucleotide sequence into a peptide sequence. Interestingly, human mtDNA codes for 22 tRNAs that are used in the translation of mRNAs derived from mitochondrial genes. There are several differences in the codon assignments between mtDNA and genomic DNA. Several stop codons exist that terminate translation. The peptide may be posttranslationally modified by enzymatic

cleavage into a smaller peptide, or sugar moieties may be added (for production of a glycoprotein), or it may be phosphorylated. These events are summarized in Figure 2.

Most genes are organized in a noncontiguous manner, with the exception of mtDNA. A typical gene is composed of exons, introns, a promoter region (in some cases) and other specialized regions (Figure 3).[9] The exon is that portion of a gene that encodes a mRNA. Introns are stretches of DNA between exons that are eventually spliced out of the RNA transcript. The vast majority (estimated to be 90–95%) of DNA is intronic, meaning that it does not code for specific proteins. However, intronic DNA may be important in directing the splicing and arrangement of exons.[10] Intronic DNA also has other properties (described below) that make it quite useful in the study of genetic diseases.

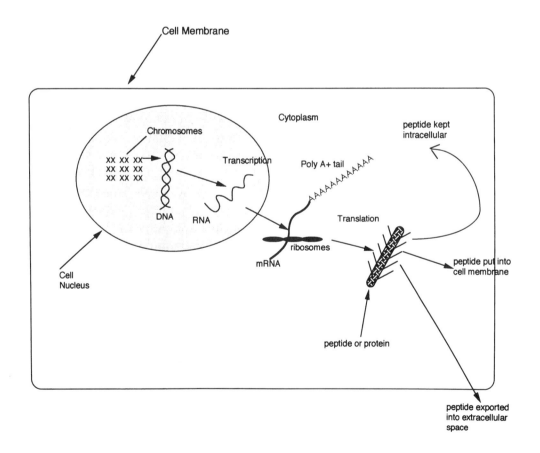

Figure 2. The cellular processes involved in protein synthesis are shown. DNA is initially transcribed into RNA, and the RNA is translated into peptides. (See text for more detail.)

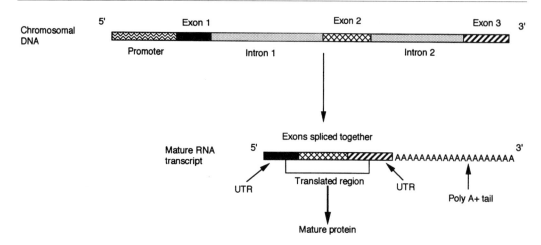

Figure 3. In the human genome most genes are noncontiguous due to the separation of coding regions (exons) by noncoding regions (introns). During DNA and RNA processing the noncoding regions are spliced out, leaving a mature messenger RNA. Note the 3' poly A+ tail, which is important for proper transportation of the mRNA. Also note that even a mature transcript may have untranslated regions (UTRs) that are not included in the mature protein.

Every cell does not express every gene, but every cell does express some common genes, the so-called housekeeping genes. These genes code for proteins that are vital for the survival of each cell (e.g., membrane proteins, structural proteins, metabolic proteins, enzymes, etc.). However, the unique genes that a specific cell or organ expresses are important in determining the special functions of that tissue. For example, some neurons express the gene for choline acetyl transferase, which is an important enzyme for the production of acetylcholine, a major neurotransmitter. There are several factors that control gene expression. Promoters, enhancers, transcription factors, and other mechanisms such as alternative splicing control the level of expression of most genes. A complete discussion of these complex molecules and processes is beyond the scope of this chapter. The interested reader is referred to one of several excellent reviews on this subject.[11,12]

When DNA was studied initially, it was thought that the overall sequence of DNA from one individual to another would be essentially the same, with some obvious exceptions. Most of these differences in the DNA sequences, termed genotypic polymorphisms, produce easily identifiable physical or biochemical changes, termed phenotypic polymorphisms. Examples of these phenotypic polymorphisms include the different blood types among individuals, differences in hair and eye color, and variations in some enzyme isoforms. We now know that there are many other significant variations in the DNA sequences among individuals. Many of these differences occur in the intronic regions of DNA, and probably do not affect genes directly. These differences in the sequence of DNA, termed DNA polymorphisms, have proven critical in the study of genetic diseases (see below).

Laboratory Techniques for the Analysis and Cloning of DNA

Extraction, Blotting, and Hybridization

Although most cells in the body (except red blood cells) have ample amounts of DNA, most of the DNA used in human research is extracted from either white blood cells obtained via venipuncture, fibroblasts obtained by a skin biopsy, or epithelial cells scrapped from the inside wall of the cheek (a so-called "buccal smear"). Although easy to obtain, buccal smears provide relatively small amounts of DNA suitable for only a few PCR experiments. Lymphocytes and fibroblasts can be immortalized and supply large amounts of DNA long after a patient has died. Although DNA can be extracted from other tissues, the ease of obtaining whole blood via venipuncture has made white blood cells the tissue of choice for the study of genomic DNA. Whole blood is centrifuged, and the buffy coat (containing the white cells) is separated from the red cells and plasma. The white cells are then lysed to release the intracellular contents. These contents include proteins, lipids, DNA, and RNA. The proteins and lipids can be removed from the nucleic acids by phenol/chloroform extractions, leaving the genomic DNA. It is often very thick and stringy, owing to the large size of the DNA

molecules. Several hundred micrograms of DNA can be obtained from 15–30 cc of whole blood. Automated systems are available now that can isolate and prepare DNA from white cells in a matter of hours. Benchtop systems are available for the rapid (several minutes) isolation of small amounts of DNA from several cell types.

DNA is often analyzed by electrophoresis through gels made of agarose or acrylamide. Since DNA has a negatively charged phosphate backbone, it will be drawn towards a positive charge. The drag and mobility of a particular fragment of DNA is dependent on its size, therefore small DNA fragments will migrate further than larger fragments when separated in this manner. The DNA can be stained with a fluorescent dye, ethidium bromide, allowing it to be photographed and analyzed (Figure 4).[13]

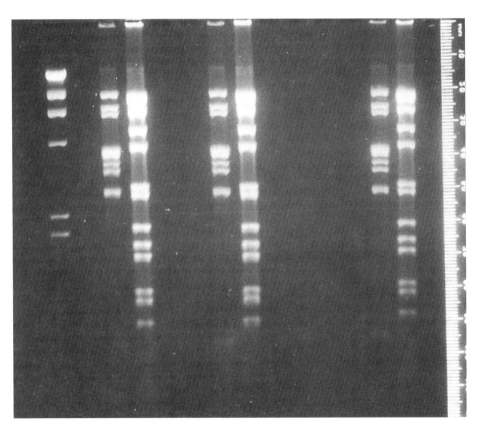

Figure 4. DNA from various vectors (cosmids and phage) was isolated and cut with various restriction endonucleases, then separated in a 1% agarose gel. The gel was then stained with ethidium bromide, which causes DNA to fluoresce when exposed to ultraviolet light. Each bright band represents a specific fragment of DNA. The first lane on the left is a DNA size standard.

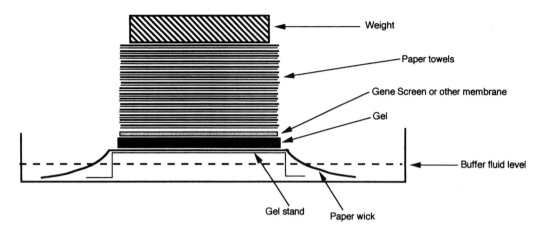

Figure 5. This diagram shows the construction of a Southern blot. The buffer is drawn through the gel by capillary action, transferring the DNA from the gel onto the Gene Screen or other membrane. Paper towels provide the capillary action. Newer techniques are electrical currents or vacuum to speed the blotting process.

Once DNA is electrophoresed through a gel, it is difficult to manipulate. This problem can be overcome by transferring the DNA to a solid membrane, usually made of nylon or a related substance. Such a transfer is usually performed using simple capillary action, although it can be done by vacuum or with electrical charges. This process is termed "Southern blotting," and the membrane with the bound DNA is called a Southern blot (named after its creator, Dr. Southern).[14] The transfer of RNA is done in a similar manner, and it is called a "Northern blot" (sorry, there was no Dr. Northern!).[15] The blot can then be used for various hybridizations (see below). Figure 5 illustrates how such blots are made. Once such a blot is made, the DNA is immobilized and can be used in subsequent experiments.

The development of Southern and Northern blots paved the way for the detection of specific DNA or RNA molecules through the use of hybridization experiments. The initial step is to create a radioactive probe (usually a fragment of DNA) by one of two techniques. First, a radioactive molecule of ^{32}P can be enzymatically transferred to the end of a DNA molecule by using T4 polynucleotide kinase. Another common technique is to use a DNA polymerase and random primers to synthesize a radioactive DNA probe using a nonradioactive DNA as a template. If cold adenine, guanine, and thymine triphosphates (ATP, GTP, TTP) are used along with radioactive cytosine triphosphate (usually ^{32}P-CTP), then every cytosine of the synthesized DNA probe will be radioactive. This technique produces probes of very high specific activity.

The radiolabeled probes are placed in a plastic bag with the blot of interest and allowed to incubate with gentle shaking. During the hybridization the DNA probe will preferentially bind to any DNA molecules with a sequence that is complementary to the probe. Following hybridization, the blot is washed under increasing conditions of stringency, which usually means decreasing salt concentrations and increasing temperatures (both of which will tend to reduce nonspecific binding). The blot is then exposed to a piece of X-ray film for several hours to days, in a process called autoradiography.[16] The autoradiograph is then developed, and the positive hybridization signals can be seen as dark bands (Figure 6). New nonradioactive labels are now being used extensively. These labels use either a fluorescent or colorometric technique to visualize that band of interest. In addition, phophoimagers and fluorescent imagers as well as other nonradiologic imaging techniques and apparati are being used to more rapidly visualize the DNA on a gel or membrane.

Cloning, PCR, and Sequencing

Current techniques allow researchers to manipulate DNA in several ways. The primary manipulations include cleavage (cutting), ligation (joining), and amplification.[12,17] A major advance in our ability to work with DNA was the discovery of various enzymes (mostly of bacterial origin) that can recognize and cut DNA at specific base sequences.[18] These enzymes are called restriction endonucleases. For example, the enzyme *Hind*III recognizes the sequence "AAGCTT" and cuts between the two "As". Another restriction enzyme, *Eco*RI, recognizes the sequence GAATTC, and cuts between the "G" and "A". Most of these enzymes cut only double-stranded DNA, and in many cases produce overhanging ends that are prone to rejoining other complementary overhanging ends. Fragments of DNA can be joined to other pieces of DNA in a process called ligation, using specific enzymes (ligases).[17] The process of cutting and ligating various DNA fragments is loosely termed DNA cloning, although cloning also includes other manipulations such as transformation and transfection (see below).

A major problem in studying DNA is the need to amplify specific DNA fragments. Two general approaches have been developed to solve this problem. In one case, a specific fragment of DNA can be isolated and ligated into a bacterial plasmid, which is a small extrachromosomal circular piece of DNA. When the plasmid DNA, which usually contains a gene for antibiotic resistance, is introduced into bacteria (in a process called transformation), the bacteria will grow in the presence of the specific antibiotic and produce millions of copies of the recombinant plasmid containing the cloned DNA fragment.[19,20] The recombinant plasmid DNA containing the fragment of interest can be isolated in milligram quantities from a large bacterial culture and further studied.[21] Fragments of DNA can also be cloned into bacteriophage vectors (bacteriophage are viruses that infect only bacteria) or other viral vectors. Such vectors infect host cells in a process termed transfection, and eventually replicate to produce large quantities of the recombinant virus. This technique is rarely used, due to the ease and wide availability of PCR technology (see below).

A very important recent development has been the polymerase chain reaction (PCR). PCR is a procedure used to enzymatically amplify a specific fragment of DNA. PCR makes use of the Taq DNA polymerase

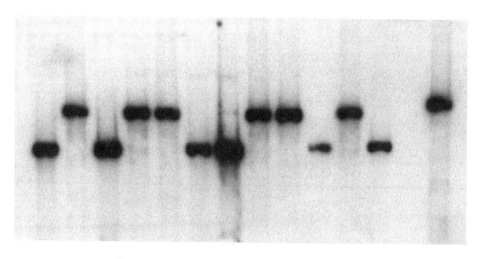

Figure 6. This is an example of an autoradiograph.

enzyme and specific DNA primers.[22,23] The template DNA (containing the DNA to be amplified) is denatured (made single stranded), then cooled, allowing DNA primers (fragments of DNA usually 16–30 bases in length) to attach to complementary sequences flanking the target sequence. The mixture is then heated to 70–72°C for several minutes, during which time the Taq polymerase will attach to the primer-template complex and synthesize a new strand of DNA using the template as a guide (Figure 7). This sequence of events is repeated for 25–35 cycles, thereby producing millions of copies of the target DNA. PCR has made it possible to produce microgram quantities of a specific DNA fragment in only a few hours, as opposed to the plasmid amplification method

which can require several days. Newer PCR based techniques have made it possible to amplify trace amounts of DNA from forensic samples and archival material. Even single molecules of DNA in single cells can be amplified using PCR techniques. In addition, PCR is now routinely used for the detection of DNA mutations, DNA sequencing, amplification of mRNA, construction of cDNA libraries, and many other applications.[24–25b]

DNA can be cloned into many different types of vectors depending on the size of the DNA and the purpose of the experiment. Plasmid vectors (described above) are easy to use and can reliably accept and amplify inserts up to 7–8 kb. They are amplified by growing within bacteria. Bacteriophage lambda is a virus that infects bacterial cells.

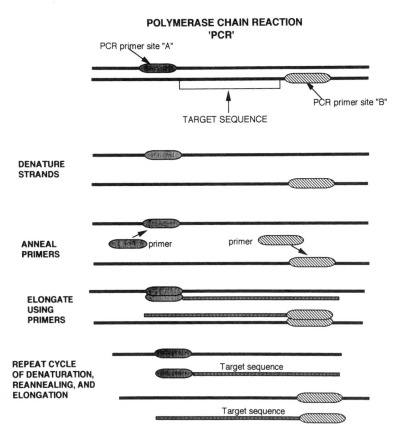

Figure 7. The polymerase chain reaction (PCR) makes use of specific primers that flank the DNA sequence of interest. By going through mutliple cycles of denaturation, reannealing, and elongation, several million copies of the target sequence can be synthesized within a few hours. The enzyme *Taq* polymerase is relatively stable at high temperatures and can function through several dozen cycles.

Such phage vectors can accept DNA inserts of about 15 kb.[26] The phage transfects bacterial cells, replicates, and lyses the cells, producing millions of new phage. The cycle of transfection and lysis is repeated as long as there are viable phage and new cells to transfect.

A cosmid vector has some of the characteristics of a plasmid (it can infect and grow in bacteria), but it can accept larger inserts, usually 30–40 kb.[27] Yeast artificial chromosomes (YACs) can accept DNA inserts of several hundred thousand bases up to 1–2 million bases.[28] Other vectors exist that grow well in eukaryotic (nonbacterial) cells as opposed to prokaryotic (bacterial) cells. Some vectors have been modified to enhance certain features such as transcription of an insert, selection of recombinants by a color marker system, and ease of amplification by PCR.

The base sequence of a fragment of DNA can be determined with high accuracy using various methods. Most rely on sequentially cutting DNA after specific bases (e.g., each adenine, guanine, cytosine, thymidine), or synthesizing a new DNA strand using a mixture of deoxynucleotides and dideoxynucleotides. The dideoxynucleotides will be incorporated randomly, and will terminate the further synthesis of a new DNA strand at various points, producing a population of products of varying lengths. The products of these reactions are then electrophoresed on high resolution acrylamide gels, and the sequence can be read directly after the gel is autoradiographed (Figure 8).

Laboratory Techniques for Studying RNA and Gene Expression

As described above, there are three major types of RNA: messenger, transfer, and ribosomal. Our major focus is messenger RNA, often termed mRNA or polyA+ RNA, since mRNA is a key intermediary between DNA and a peptide. In general, RNA is much more difficult to work with than DNA due to the presence of enzymes called ribonucleases (RNAses). RNAses rapidly degrade most types of RNA, but are extremely difficult to destroy or deactivate. Most RNAses are relatively resistant to heating, boiling, and other decontamination techniques. Glassware used for RNA work must be baked in an oven for several hours, or washed with a base (sodium hydroxide) to remove any RNAses. Total RNA can be isolated from cells and organs using various homogenization techniques. Most types of mRNA have a long polyA+ tail, meaning that 10–20 adenines have been attached to the transcribed RNA molecule. By using a synthetic molecule containing many deoxythymidine residues (poly-dT) bound to a resin and packed in a column, the poly A+ RNA can be separated from the other RNA and studied. The long poly A+ tail of mRNA will cause these molecules to bind to the poly-dT in the column, and the poly A+ mRNA can then be eluted and collected. RNA can also be electrophoresed on agarose gels and blotted, producing a Northern blot.

Since the collection of mRNAs from a specific organ defines all the genes expressed in that organ, it is often of great interest to study the population of mRNAs in various circumstances. However, as noted above, RNA is extremely difficult to manipulate without causing degradation. These problems can be overcome by using RNA as a template to produce complementary DNA (cDNA), which is much more stable. A retroviral enzyme, reverse transcriptase, uses mRNA as a template for the synthesis of DNA (complementary DNA, or cDNA). Another enzyme can be added to the newly synthesized single-stranded cDNA to produce double-stranded cDNA. This double-stranded cDNA can then be cloned into various vectors (such as a plasmid or phage) for amplification and further study.[29,30] An assortment of such cDNA clones made from mRNA of a specific organ is called a cDNA library. Such libraries are extremely useful when studying gene expression or trying to isolate disease specific genes.

Gene expression (as judged by mRNA production) can be determined at a cellular

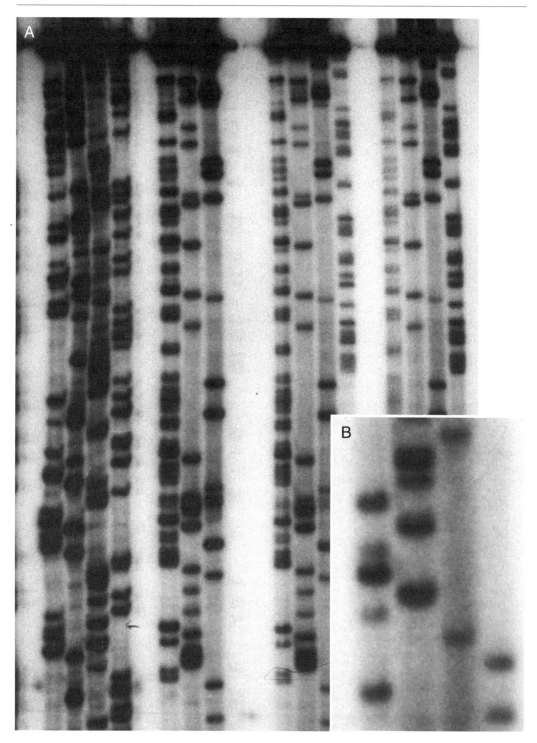

Figure 8. Panel A: An autoradiograph of a large sequencing gel is shown. Each group of four lanes represents one sample. Each lane within a group is used for one base (adenine, cytosine, guanine, or thymine). Panel B is a close-up of a portion of one sample. The sequence is read from the bottom to the top. The four lanes represent (left to right) adenine, cytosine, guanine, and thymine. The base sequence of the DNA in this picture would be: "T A T G A C A A C A C C G".

level by using *in situ* hybridization techniques. This involves sectioning and fixing the tissue of interest, typically on a glass slides. By pretreating the tissue section with a DNAse, the probe will not hybridize to the cellular genomic DNA. The fixed section is then hybridized to a labeled fragment of either RNA or DNA which corresponds to the gene of interest. Following washing, the labeling in the section can be either directly visualized or autoradiographed, depending on the type of probe label which was used. The results will show if specific cells in the section express the gene of interest, and can provide semiquantitative estimates of how much a particular gene is expressed in the tissue or organ under study. This technique is very useful for distinguishing gene expression among populations of cells that may appear quite similar by morphology or location, yet have very different genetic properties.

At times it is important to analyze the protein product of a particular gene. While these proteins can sometimes be obtained after production *in vivo*, at other times this is impractical (as in many brain proteins), or the protein is produced in such minute quantities *in vivo* that analysis is impossible. However, *in vitro* expression systems exist that allow researchers to clone a gene into an expression vector that will produce large quantities of the desired protein in a test tube or other easily manipulated system.[31] The target protein can then be isolated in large quantities and studied for biological activity and other characteristics.

Gene Localization

The localization of genes to a particular chromosome or chromosomal region can be accomplished using several techniques. Genetic linkage analysis (discussed in detail in *Chapter 2*) allows gene localization based on a statistical analysis of crossover events during meiosis. Another technique to localize genes on a chromosome is fluorescent *in situ* hybridization (FISH). A gene probe (such as a cDNA or other clone) is made fluorescent by attaching special chemical moieties. The probe is then hybridized to a spread of metaphase chromosomes. The preparation is washed and visualized using special microscopes. Areas of probe hybridization appear as bright yellow specks against a red background. Since each chromosome has a different morphologic appearance when using special dyes, it is possible to localize the probe's signal to a single chromosome. In some cases, subregional localization (e.g., band q22) can be determined using FISH.[32] Using different fluorescent labels, it is possible to study the relative positions and localizations of two or more different probes in one FISH experiment. FISH can also be used to detect deletions or duplications involving regions of chromosomes.[32a]

Another useful resource for mapping genes is somatic cell panels. These are hybrid cell lines (usually from rodents) that contain parts of one or more human chromosomes. DNA from these cell lines is isolated, cut, separated by gel electrophoresis, then immobilized by Southern blotting. The radiolabeled probe of interest is hybridized to the Southern blot, and following autoradiography a pattern of hybridization is seen (Figure 9). By comparing the hybridizaton results of several cell lines that contain different human chromosomes (with appropriate controls) it is usually possible to map the clone of interest to a particular chromosome or a subchromosomal region.[33]

The techniques described above provide a brief outline of some of the important and most commonly used tools in molecular genetics. Many variations of these techniques are used in specific circumstances. There are several other laboratory techniques that will not be described here but are discussed in subsequent chapters. Some techniques related mostly to the study of disease causing genes, such as transgenic models, will be discussed in *Chapter 3*. The interested reader can refer to one of the excellent laboratory manuals or molecular biology textbooks for more information about these techniques and concepts.[12,21]

SOMATIC CELL HYBRID PANEL

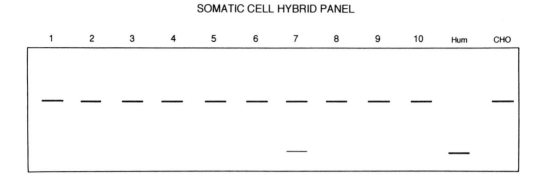

Figure 9. This drawing represents a typical hybrid somatic cell panel. The numbers represent the human chromosome in each cell line. Hum indicates only human DNA. CHO is Chinese hamster ovary, which is a common cell line in which to make somatic cell hybrids. The probe used hybridizes to both human and CHO DNA, meaning it is present in both species. Note that a unique human band is seen in lane 7, meaning that the probe maps to that human chromosome.

Genotypes and Phenotypes

The definition of a gene has changed since the days of Mendel's breeding experiments with peas. In molecular terms, a gene consists of all nucleic acid sequences (i.e., exonic and intronic DNA, promoters, etc.) necessary for the production of a specific RNA or peptide. A gene should be distinguished from an allele, which is an alternative form of a gene encoding or producing alternative forms of a specific trait. These traits may be clinically apparent (such as hair or eye color) or they may consist of a variation in the sequence of DNA without obvious clinical changes.

The understanding of how a gene product can produce a specific physical attribute is one area of human genetics that is rapidly expanding as new genes and gene products are discovered. Some basic terms should be defined. A genotype refers to the genetic makeup of an organism, and a phenotype is the detectable manifestation of a specific genotype. Phenotypes can include obvious physical attributes such as hair and eye color, biochemical characteristics such as blood type or isoenzyme patterns, and diseases.

In general terms, inherited diseases can be classified as monogenic, polygenic, or multifactorial. A monogenic disease is caused by a mutation in one specific gene,

while a polygenic disease requires mutations in more than one gene. Multifactorial diseases, also called complex diseases, are due to interactions among several genes and environmental influences.[34] Examples of such diseases may include hypertension, diabetes, multiple sclerosis, and atherosclerosis. Finding pathogenic genes and studying complex genetic diseases are discussed in *Chapter 3*.

Another concept is dominant and recessive traits. Dominant traits are those produced by only one allele at a specific gene locus, and a recessive trait requires two similar alleles at a gene locus. Each classic autosomal genetic locus has two alleles. Both alleles can carry the dominant trait (homozygous dominant), the recessive trait (homozygous recessive), or one of each (heterozygous). By definition, the dominant trait will be expressed over the recessive trait if a person is heterozygous. However, genotypes involving the sex (X and Y) chromosomes are a little more complex, especially in males. Such traits are referred to as sex-linked disorders. For example Duchenne's muscular dystrophy (DMD) is due to mutations of the dystrophin gene which is on the X chromosome. The disease is inherited as an X-linked recessive. So, a mother (genotype XX) may carry a dystrophin mutation on one of her X chromosomes, but since it is a recessive trait it will not be expressed due to the presence of a normal allele on the other X chromosome. (This is somewhat of an oversimplification since such carriers may have subtle biochemical abnormalities due to partial inactivation in the female of the normal X chromosome in some tissues.) However, if such a carrier mother has a male (XY) offspring, there is a 50% chance that he will have the DMD mutation and develop clinical DMD. (Since he has to get the Y chromosome from his father, there is a 50% chance that he will inherit the defective X chromosome from his mother.) Since he does not have a normal X chromosome to counteract the recessive DMD mutation inherited from his mother, he will develop DMD. These various modes of transmission are illustrated in the sample pedigrees in Figure 10.

Another type of inheritance involves mitochondrial transmission. As discussed above, mtDNA is carried only in the ovum, therefore transmission of mitochondrial traits is determined by the mother's mtDNA. This type of transmission is termed maternal inheritance, since it is solely the mother's mtDNA genotype that will be transmitted to all of her offspring.[5] One unusual characteristic of diseases caused by mutations in mtDNA is the marked degree of phenotypic heterogeneity among individuals with the same mutation.[35] This issue is discussed in more detail in *Chapter 3*.

If an individual carries a genotype for a specific genetic disease, it does not necessarily mean that the individual will develop that disease. Many factors can influence the expression of that genotype, including environment, age, and interactions with other genes. For example, the gene for Huntington's disease (HD) has a peak expression at age 40–45 years. If an individual carrying the HD gene dies in a car accident at age 20,

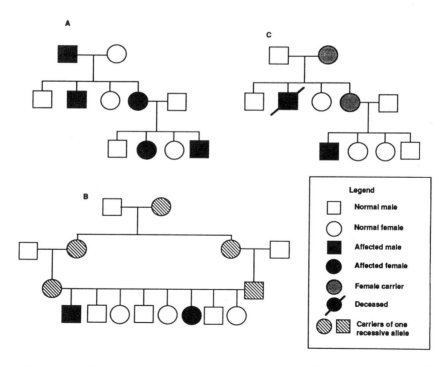

Figure 10. Typical pedigrees depicting some common forms of disease transmission. Pedigree A shows autosomal dominant transmission. Note the male-to-male transmission, which rules out an X-linked disorder. Pedigree B shows an autosomal recessive trait that is expressed in a family with some inbreeding. Pedigree C depicts X-linked transmission. Note the absence of male-to-male transmission, and the presence of female carriers.

the phenotype of HD will never be expressed, although the individual carries the HD genotype. This issue is particularly problematic for late-onset diseases such as cerebrovascular disease and Alzheimer's disease (AD). In the case of late-onset AD, it was first thought that genetic factors were unimportant since in most cases a clear family history of AD was not apparent. However, several groups performed a detailed analysis of late-onset AD families and found evidence for an autosomal dominant pattern of inheritance when at-risk relatives lived to the maximum age of onset (which is probably around 90 years).[36,37] This example illustrates the importance of understanding the age-of-onset pattern for a genetic disorder and selecting the study population accordingly.

Environmental factors can have important roles in determining the expression of a disease genotype. For example, children with the gene defect for phenylketonuria may not develop any major clinical manifestations if they are fed a diet low in phenylalanine. Perhaps other environmental factors such as radiation exposure or ingested toxins may trigger a genetic disease in patients with a susceptible genetic makeup. It is well known that certain genes contain "hot spots" that have a high predilection for recombination and mutation when exposed to exogenous factors such as radiation.[38]

When studying genetic diseases, one must be concerned about phenocopies. A phenocopy occurs when an abnormal phenotype, which is identical or very similar to a genetically produced phenotype, is produced by environmental factors.[34] For example, certain types of motor neuron disease may be caused by toxin exposure, but other types are due to genetic defects. If the history of toxin exposure was not clear, an affected individual could be mistakenly diagnosed as having an inherited motor neuropathy, especially if other members of the family have the disease. The issue of phenocopies complicates the study of those diseases that are likely to have both genetic and environmental etiologies.

Other terms that are often used incorrectly are penetrance and expression or expressivity. Genetic diseases do not have variable penetrance. Either everyone with the disease genotype expresses the disease phenotype (complete penetrance), or only some individuals with the disease genotype develop the disease phenotype (incomplete penetrance). Therefore, penetrance is either complete or incomplete, but not variable. However, genetic diseases may have variable expressivity, meaning that individuals with the same genotype will have somewhat different phenotypes.[39,40] Examples of neurogenetic disease with variable expressivity include myotonic muscular dystrophy and neurofibromatosis type I. Variable expressivity may be due to gene-gene interactions, environmental effects, or a combination of factors. As more detailed studies into some of these diseases have progressed, it has become clear that genetic heterogeneity (in the form of different mutations in a single gene, or mutations in different genes producing similar phenotypes) may be very important in defining the variable clinical expression of a disease. Examples of such correlations include mutations in the dystrophin gene in Duchenne's muscular dystrophy and mutations in the iodyoronate sulfatase gene in Hunter's syndrome.[41,42]

Summary

This chapter has provided a review of some of the basic genetic concepts needed to understand modern molecular and clinical genetic research. The overall message is that modern techniques have made the manipulation and study of DNA and RNA significantly easier and more productive. However, since the human genome has 3 billion bases of DNA, we have studied in detail only a very small percentage of the entire genome. Subsequent chapters will focus on the techniques that allow us to prioritize the areas of the genome that require more intense study to locate disease-causing genes. These chapters will emphasize population genetic techniques in the form of genetic linkage studies, and cloning genes and identifying disease-causing mutations.

It was not possible in this chapter to review every molecular genetic technique and concept. The interested reader is referred to several excellent publications in the reference section that can provide supplemental material about other laboratory methods and genetic concepts.

Acknowledgements: The author would like to thank Rohan deSilva, PhD, and Ms. Susan Howard Alberts for their review, advice, and assistance with this chapter.

References

1. Wyatt G. The purine and pyrimidine composition of deoxypentose nucleic acids. Biochem J 1951;48:584–590.
2. Watson J, Crick F. Molecular structure of nucleic acids: A structure for deoxyribose nucleic acid. Nature 1953;171:737–738.
3. Crick F, Watson J. The complementary structure of deoxyribonucleic acid. Proc Roy Soc A 1954;223:80–96.
4. Shoffner J, Wallace D. Oxidative phosphorylation diseases: Disorders of two genomes. Adv Hum Genet 1990;19:267–330.
5. Giles R, Blanc H, HM C, et al. Maternal inheritance of mitochondrial DNA. Proc Natl Acad Sci USA 1980;77:6715–6719.
6. Hoagland M, Stephenson M, Scott J, et al. A soluble ribonucleic acid intermediate in protein synthesis. J Biol Chem 1958;231:241–257.
7. Dintzis H. Assembly of the peptide chain of hemoglobin. Proc Natl Acad Sci 1961;47:247–261.
8. Leder P, Nirenberg M. RNA code words and protein synthesis II: Nucleotide sequence of a valine RNA code word. Proc Natl Acad Sci 1964;52:420–427.
9. Chambon P. Split genes. Sci Am 1981;244:60–71.
10. Witkowski J. The discovery of split genes: A scientific revolution. Trends Biochem Sci 1988;13:110–113.
11. Johnson P, McKnight S. Eukaryotic transcriptional regulatory proteins. Ann Rev Biochem 1989;58:797–839.
12. Watson J, Gilman M, Witkowski J, et al. Recombinant DNA. New York: WH Freeman, 1992.
13. Sharp P, Sugden B, Sambrook J. Detection of two restriction endonucleases activities in *Haemophilus parainfluenzae* using analytical agarose-ethidium bromide electrophoresis. Biochem 1973;12:3005.
14. Southern E. Detection of specific sequences among DNA fragments separated by gel electrophoresis. J Mol Biol 1975;98:503–517.
15. Alwine J, Kemp D, Stark G. Method for detection of specific RNAs in agarose gels by transfer to diazobenzloxymethyl-paper and hybridization with DNA probes. Proc Natl Acad Sci 1977;74:5350–5354.
16. Laskey R, Mills A. Enhanced autoradioactive detection of 32P and 125I using intensifying screens and hypersensitized film. FEBS Lett 1977;82:314–316.
17. Lobban P, Kaiser A. Enzymatic end-to-end joining of DNA molecules. J Mol Biol 1973;79:453–471.
18. Meselson M, Yuan R. DNA restriction enzyme from *E Coli*. Nature 1968;217:1110–1114.
19. Cohen S, Chang H, Boyer H, et al. Construction of biologically functional bacterial plasmids in vitro. Proc Natl Acad Sci 1973;70:3240–3244.
20. Hanahan D. Studies on transformation of *Escherichia coli* with plasmids. J Mol Biol 1983;166:557–580.
21. Sambrook J, Fritsch E, Maniatis T. Molecular cloning: A laboratory manual. Cold Spring Harbor, NY: Cold Spring Harbor Press, 1989.
22. Mullis K, Faloona F. Specific synthesis of DNA in vitro via a polymerase catalyzed chain reaction. Meth Enzymol 1987;55:335–350.
23. Saiki R, Scharf S, Faloona F, et al. Enzymatic amplification of beta-globin sequences and restriction site analysis for diagnosis of sickle cell anemia. Science 1985;230:1350–1354.
24. Innis M, Myambo K, Gelfand D, et al. DNA sequencing with Thermus aquaticus DNA polymerase, and direct sequencing of PCR-amplified DNA. Proc Natl Acad Sci 1988;85:9436–9440.
25. Erlich H, Gelfand D, Sninsky J. Recent advances in the polymerase chain reaction. Science 1991;252:1643–1651.
25a. Farrell RE. DNA amplification. Immunol Invest 1997;26:3–7.
25b. White TJ. The future of PCR technology: Diversification of technologies and applications. Trends Biotechnol 1996;14:478–483.
26. Blattner F, Williams B, Blechl E, et al. Charon phages: Safer derivatives of bacteriopahge lambda for DNA cloning. Science 1977;196:161–169.
27. Collins J, Hohn B. Cosmids: A type of plasmid gene cloning vector that is packageable in vitro in bacteriophage heads. Proc Natl Acad Sci 1978;75:4242–4246.

28. Burke D, Carle G, Olsen M. Cloning of large segments of exogenous DNA into yeasy by means of artificial chromosome vectors. Science 1987;236:806–812.
29. Gubler U, Hoffman B. A simple and very effective method for generating cDNA libraries. Gene 1983;25:263–269.
30. Huynh T, Young R, Davis R. Constructing and screening cDNA libraries in lambda gt 10 and lambda gt 11, In Glover D (ed): DNA cloning: A practical approach. Oxford, IRL Press, 1985.
31. Rosenberg A, Lade B, Chui D, et al. Vectors for selective expression of cloned DNAs by T7 RNA polymerase. Gene 1987;56:125–135.
32. Lebo R, Lynch E, Bird T, et al. Multicolor in situ hybridization and linkage analysis order Charcot-Marie-Tooth type 1 (CMT1A) gene-region markers. Am J Hum Genet 1992;50:42–55.
32a. Heng HH, Spyropoulos B, Moens PB. FISH technology in chromosome and genome research. Bioessays 1997;19:75–84.
33. Witkowski J. Somatic cell hybrids: A fusion of biochemistry, cell biology, and genetics. Trends Biochem Sci 1986;11:149–152.
34. Lander E. Mapping complex genetic traits in humans. In Davies K (ed): Genome analysis: A practical approach. Oxford, IRL Press, 1988, 171–187.
35. Shoffner J, Wallace D. Mitochondrial genetics: Priniciples and practice. Am J Hum Genet 1992;51:1179–1186.
36. Huff F, Auerbach J, Chakravarti A, et al. Risk of dementia in relatives of patients with Alzheimer's disease. Neurol 1988;38:786–790.
37. Breitner J, Silverman J, Mohs R, et al. Familial aggregation in Alzheimer's disease: Comparison of risk among relatives of early- and late-onset cases, and among male and female relatives in successive generations. Neurol 1988;38:207–212.
38. Coulondre C, Miller J, Farabaugh P, et al. Molecular basis of base substitution hotspots in Escherichia coli. Nature 1978;274:775–780.
39. Risch N. Linkage strategies for genetically complex traits. I. Multilocus models. Am J Hum Genet 1990;46:222–228.
40. Risch N. Genetic linkage and complex diseases, with special reference to psychiatric disorders. Genet Epidemiol 1990;7:17–45.
41. Passos-Bueno M, Bakker E, Kneppers A, et al. Different mosaicism frequencies for proximal and distal Duchenne muscular dystrophy (DMD) mutations indicate difference in etiology and recurrence risk. Am J Hum Genet 1992;51:1150–1155.
42. Flomen R, Green P, Bentley D, et al. Detection of point mutations and a gross deletion in six Hunter syndrome patients. Genomics 1992;13:543–550.

Chapter 2

Genetic Linkage:

Concepts and Methods

Marcy C. Speer, PhD

Introduction

Genetic linkage analysis is one of several methods for determining the chromosomal localization of genes. Other approaches to gene localization include *in situ* hybridization and the co-occurrence of gross chromosomal rearrangements (i.e., deletions or translocations) and/ or polymorphisms with trait phenotypes (discussed in detail in *Chapters 1 and 3*). Linkage analysis, especially in recent years, has proven to be a tremendously powerful tool in the primary localization of genes, particularly when the causative defect for a disease phenotype is unknown. Linkage analysis has been of critical import ance in the eventual identification of the genes responsible for several genetic diseases including Duchenne/Becker muscular dystrophy,[1] cystic fibrosis,[2] and neurofibromatosis type I,[3] and was instrumental in determining the relationship of a defect in the superoxide dismutase (SOD1) gene to familial amyotrophic lateral sclerosis.[4]

The most frequent successes to date in the localization of disease-causing genes with linkage analysis have been with diseases with a known mode of inheritance (as discussed above). These disorders are often highly or completely penetrant and are due to a defect in a single gene, yet these Mendelian disorders are often relatively rare in the population. Some of the most common and deadly diseases, such as atherosclerosis and cancer, have significant genetic components. These diseases are termed "complex" because they are likely to be due to the interaction of multiple factors, both environmental and genetic.

Apparent Mendelian transmission of a "complex" trait may be in a subset of families, such as in Alzheimer's disease,[5,6] amyotrophic lateral sclerosis,[4] and breast cancer.[7,8] The study of a complex disorder presents special yet surmountable challenges in linkage analysis for gene localization. Elucidation of the genetic defect in a subset of families may provide insights into the pathogenesis of the non-Mendelian form of such a complex disease.

This chapter will review some of the basic concepts of linkage analysis and demonstrate how linkage techniques can be applied in various situations.

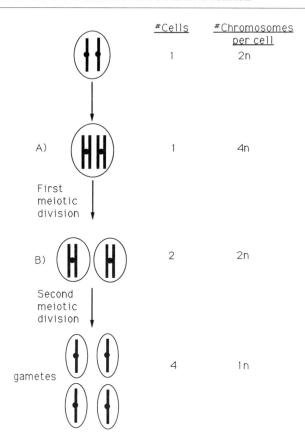

Figure 1. Simplified view of meiosis. The purpose of meiosis is to generate gametes with one complete complement of the chromosomal material. Each gamete is haploid and combines with another gamete to produce a diploid cell (e.g., zygote). In this figure (where n represents the number of chromosomes), only one pair of chromosomes is represented; in humans, 23 pairs of chromosomes are present in each diploid cell. **A:** The $4n$ cell enters meiosis and cell division leads to two $2n$ cells. **B:** The two $2n$ cells divide, giving rise to four $1n$ cells. These cells are either male (sperm) or female (egg) gametes.

Basic Concepts

Genetic Linkage and Recombination

The ability to detect linkage between two loci (genes or other markers in the genome) is dependent on the scoring of recombination events. Genetic recombination occurs during meiosis (Figure 1), the process by which gametes are formed, and is caused by physical crossing over between homologous chromosomes during meiosis (Figure 2). A "genetic recombination" is "scored" or counted when the physical crossover occurs between two loci whose alleles can be differentiated. By measuring the frequency with which two genes recombine, an estimate of the genetic distance between the genes is made. Linkage analysis is fundamentally based on Mendel's second law of independent assortment which states that unlinked pairs of genetic loci will be transmitted independently at meiosis. A pair of loci can be unlinked in one of two respects. First, two loci may be on different chromosomes. Second, two loci may be *syntenic* (on the same chromosome), yet far enough apart that transmission of one during meiosis is independent of transmission of the second (Figure 3). Loci that are transmitted to gametes in a nonrandom fashion are said to be *linked*.

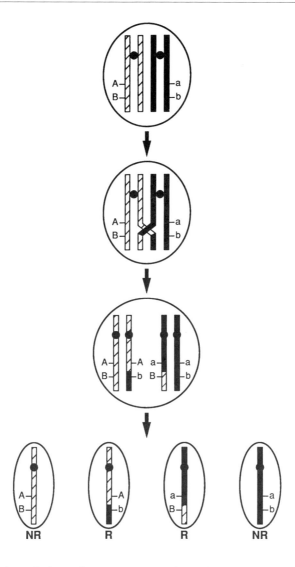

Figure 2. Simplified view of physical crossing over. Physical crossing over between homologous chromosomes occurs during meiosis. This crossing over leads to genetic recombination (the observation of the physical crossover by scoring of offspring for two traits). **A:** Alleles (A,a and B,b) are shown for two markers linked to one another on a chromosome. **B:** Physical crossing over between a pair of homologous chromosomes before the first cell division in meiosis. **C:** Two subsequent cell divisions lead to four gametes, two of which are recombinant (A,b and B,a) when compared to the parental cell and two of which are nonrecombinant (A,B and a,b).

When two genes are unlinked, Mendel's second law of independent assortment predicts that approximately 50 of the 100 gametes will be recombinant for the two genes. Consequently, the upper limit of expected recombination between two unlinked genes is 50%. As another example, if 100 gametes are scored and 10 show evidence of recombination between two genes, two conclusions can be drawn: (1) the genes are linked because there is a nonrandom assortment; and (2) the estimated distance between the genes is 10 units (10/100).

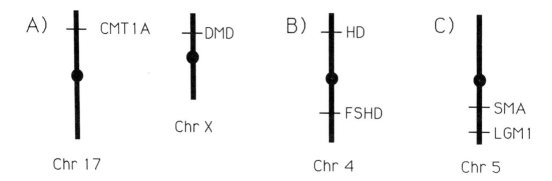

Figure 3. **A:** The genes for Charcot-Marie-Tooth disease type 1A (CMT1A) on chromosome 17 and Duchenne muscular dystrophy (DMD) on the X chromosome are unlinked because they are on different chromosomes. **B:** The genes for Huntington disease (HD) on chromosome 4p and facioscapulohumeral muscular dystrophy (FSHD) on chromosome 4q are *syntenic*, located on the same chromosome, but *unlinked* because they are far enough apart that their transmission to gametes is independent. **C:** The genes for spinal muscular atrophy (SMA) on chromosome 5q and the autosomal dominant form of limb-girdle muscular dystrophy (LGM1) are *syntenic* and *linked*, because they are close enough together that their transmission to gametes is not independent.

Genetic Markers

The loci used in genetic linkage studies need not be genes of functional significance; anonymous segments of DNA (stretches of DNA with no known function), known as genetic markers, have been of tremendous value in genetic linkage analysis. In order for genetic markers to be of benefit in a linkage analysis, the chromosomal localization of the marker must be known. More importantly, *variation* in the sequence or length of these markers among individuals is a necessity (i.e., it must be polymorphic). Molecular genetic markers are classified into three broad categories: restriction fragment length polymorphisms (RFLPs), variable number of tandem repeat markers (VNTRs), and short tandem repeat polymorphisms (STRPs).

The technical details that describe each type of marker system are beyond the scope of this chapter. The interested reader is referred to excellent reviews and papers.[9–11] Although the technical details for generating genotypes using these three types of markers vary from marker system to marker system, the underlying purpose is the same: to determine genotypes at a genetic marker locus for a series of individuals

and calculate the frequency with which *genetic recombination* between a pair of traits occurs. An STRP marker system is described here because it is currently the most frequently used marker system in genetic linkage studies.

STRPs[12–14] represent strings of short sequences of base pairs (usually 2, 3, or 4 base pairs representing dinucleotide, trinucleotide, and tetranucleotide repeats) that are repeated in tandem at a marker locus. For instance, the two base pair sequence CA (for the bases cytosine and adenosine) could be repeated at a locus three times (CA-CACA), four times (CACACACA), or five times (CACACACACA). The number of tandem repeats in a typical $(CA)_n$ STRP usually ranges from 11 to several dozen. By extracting DNA and performing PCR reactions (described in *Chapter 1*), the size of the repeat sequence at a marker locus can be scored. The difference in the number of repeats per chromosomes between individuals provides the variation necessary for scoring alleles, as shown in Figure 4.

STRP loci are ubiquitous throughout the genome, with an estimate of about 1 per 30,000–60,000 base pairs of DNA.[14] Assuming that the human genome is comprised of

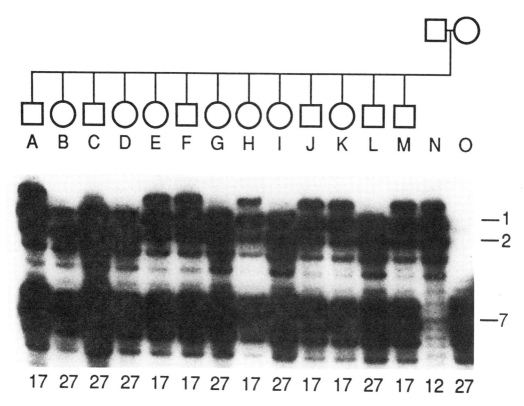

Figure 4. Genotypes were obtained using PCR amplification of a dinucleotide repeat marker [(CA)$_n$]. Pedigree symbols are indicated above the diagram, with individuals N and O being parents of individuals A-M. Although shadow bands are present, genotypes can be scored by assessing the most intense bands. For instance, individual O can be scored as a "77" and individual N as a "12".

approximately 3.3×10^9 base pairs of DNA, about 50,000–100,000 of these loci are expected, although these numbers are likely to overestimate the number of useful (informative) STRPs. STRPs are invaluable in linkage analysis for two reasons: (1) they are common throughout the genome; and (2) they tend to be extremely polymorphic. In general, STRPs have been thought to have no function. However, recent studies have shown that the expansion of certain triplet repeats can cause several genetic diseases including fragile X syndrome,[15,16] X-linked spinal bulbar atrophy,[17] myotonic dystrophy,[18,19] and Huntington's disease.[20] For instance, the normal range of repeats in myotonic dystrophy ranges from 5–30; individuals with > 50 repeats are considered to have an expansion.

Example

For illustrative purposes, consider a family pedigree in which an autosomal dominant, fully penetrant disease trait is segregating (Figure 5). Visual inspection of pedigree data can frequently aid in assessing whether a marker and trait are linked. This process involves counting the number of recombinant and nonrecombinant individuals in a pedigree. If a marker is linked to the disease trait, cosegregation of the disease trait with one of the alleles at the marker locus should be apparent; in other words, a particular marker allele is inherited more frequently with the disease allele than with any of the other alleles within a pedigree. In Figure 5, individual II-1 has inherited the disease trait together with marker allele 2 from his affected mother.

Thus, linkage *phase* (the disease allele segregates with marker allele 2) has been established and the offspring can be scored as either recombinant or nonrecombinant. All *affected* offspring of II-1 and II-2 who have inherited marker allele 2 from their father will be scored as nonrecombinant for the disease and marker; *affected* offspring who have inherited the 1 allele will be scored as recombinant for the disease and marker. Similar reasoning applies to the unaffected offspring, except that the *unaffected* offspring who have inherited allele 1 from their father are nonrecombinant and those who have inherited allele 2 are recombinant. All affected children in this pedigree have received the disease and marker allele 2 from their affected father and are thus nonrecombinant for the disease and marker. Four of the five unaffected children have inherited marker allele 1 from their affected father and are thus also nonrecombinant with respect to the disease and marker loci. Individual III-6, however, is unaffected and has inherited marker allele 2 from his af-

fected father. This individual is thus scored as a recombinant individual (a recombinant meiosis), while the remaining nine individuals are scored as nonrecombinant individuals (nonrecombinant meioses). Thus, visual inspection of this pedigree suggests that the marker and trait are linked because there is evidence of nonrandom segregation of the disease and marker.

Alternatively, in Figure 6, out of 10 scorable meioses, 5 are nonrecombinant and 5 are recombinant, suggesting that the disease and marker are unlinked to one another. When a disease and marker locus (or two marker loci) are unlinked to one another, the recombination fraction (θ) is considered to be 0.50. The upper limit for observed recombination between two unlinked loci is set at 50% because the frequency with which an odd number of recombination events between a pair of loci occurs should equal the frequency with which an even number of recombination events occurs. When an even number of recombination events occurs between two loci,

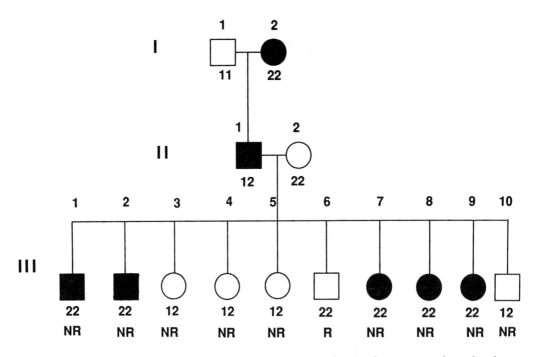

Figure 5. Pedigree demonstrating nonrandom segregation of trait phenotype and marker locus. Shaded individuals are affected; nonshaded individuals are unaffected. Marker genotypes are indicated beneath each individual. NR = nonrecombinant; R = recombinant. (See text for details.)

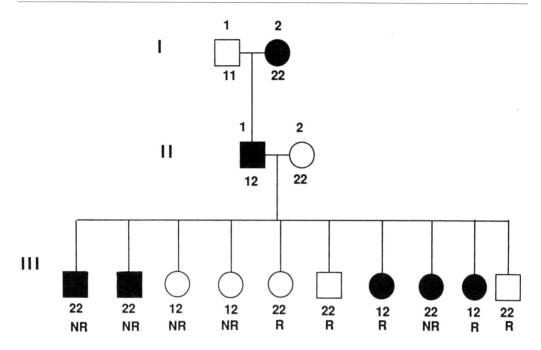

Figure 6. Pedigree demonstrating nonlinkage of trait phenotype and marker locus.

the resultant gametes *appear* to be nonrecombinant for the two loci and hence these recombination events are unobserved.

Measures of Linkage Information

Several series of nuclear families are shown in Figure 7, some of which are *informative* for linkage analysis and some of which are *uninformative*. A general "rule of thumb" is that at least one parent must be doubly heterozygous for the pedigree to be useful in linkage analysis. Thus, in an autosomal dominant disease, an affected parent must be heterozygous at a marker locus for the offspring to contribute information to the linkage analysis. The higher the number of alleles at a marker locus, the more likely an indivdiual is to be heterozygous. This rule of thumb is an oversimplification to some extent, as methods for measuring allele sharing between relatives can also be applied successfully.

The amount of variation inherent within a genetic marker system is quantified by a heterozygosity (H) value, which mea-

sures the probability that an individual will be heterozygous (as opposed to homozygous) at a marker locus. H is calculated as:

$$H = 1.0 - \sum_{i=1}^{n} p_i^2$$

where n is the number of alleles and p_i is the frequency of the ith allele. For instance, in a marker system with two equally frequent alleles, the heterozygosity is calculated as:

$$H = 1.0 - [(0.5^2) + (0.5^2)] = 0.50$$

In general, genetic marker systems with high heterozygosities are desirable because there is a high probability that an individual in a genetic linkage analysis will be heterozygous for that marker, and thus likely to contribute information to a genetic linkage study. Many STRP markers have high heterozygosity values.

The PIC (Polymorphism Information Content) is a modification of the heterozygosity measure that subtracts from the H value an additional probability that an individual in a linkage analysis does not contribute information to the study. Formally, the PIC value excludes from the H value the probability of obtaining a heterozygous offspring from an intercross mating. For example, when a mother whose marker genotype is 13 and a father whose marker genotype is 13 have an offspring whose marker genotype is also 13, it is impossible to tell whether the 1 allele came from the mother or from the father; thus, this combination contributes essentially no information to the linkage analysis.

The PIC is calculated as:

$$PIC = 1 - (\sum_{i=1}^{n} p_i^2) - \sum_{i=1}^{n-1} \sum_{j=i+1}^{n} 2p_i^2 p_j^2$$

where n is the number of alleles at the marker locus and p_i is the frequency of the ith allele.

AUTOSOMAL DOMINANT DISEASES

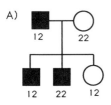

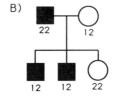

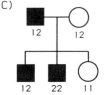

AUTOSOMAL RECESSIVE DISEASES

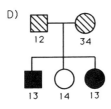

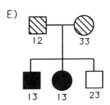

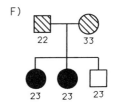

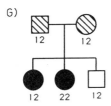

Figure 7. Pedigree examples demonstrating families which are informative and noninformative for linkage analysis. Both the autosomal dominant and autosomal recessive disease are assumed to be fully penetrant. Shaded individuals are affected with the disease, unshaded individuals are unaffected, and stippled individuals are asymptomatic gene carriers (in the recessive disease). Marker results are indicated beneath each individual. *Autosomal dominant diseases:* The affected parent, who is by definition homozygous at the disease locus, must also be heterozygous at the marker locus. **A.** Fully informative pedigree: all three offspring can be scored for linkage analysis. **B.** Uninformative pedigree: cannot tell which 2 allele is transmitted with the disease allele. **C.** Partially informative pedigree (an "intercross"): homozygous offspring of this mating contribute significantly to linkage analysis. *Autosomal recessive diseases:* For pedigrees to be fully informative, both parents, who are heterozygous at the disease locus by virtue of the fact that they have at least one affected child with a recessive disease, must be also heterozygous *for different alleles* at the marker locus. **D.** Fully informative pedigree: both paternal and maternal gametes can be scored with respect to disease and marker locus. **E.** Partially informative pedigree: only paternally contributed gametes contribute to linkage analysis. **F.** Uninformative pedigree: both parents are homozygous at the marker locus. **G.** Partially informative pedigree: homozygous offspring contribute significantly to the linkage analysis.

The Risch ratio (λ): In a series of seminal papers, Risch[20a–c] defined a measure of the potential genetic contribution to a complex disease λ as the ratio of recurrence risk of disease to a relative of an affected individual divided by the population prevalence of disease. For instance, in neural tube defects (NTDs) the recurrence risk for a sibling of an affected individual is 2% and the population prevalence is about 0.01%. Thus, the λ_s (where the subscript s indicates that the relationship quantified is siblings) for NTDs is $0.02/0.001 = 20$. In general, the larger the λ, the stronger the genetic component for the disease although this value is highly dependent on the population prevalence of disease; the higher the prevalence, the lower the λ will tend to be. And, although in most contexts high λ values are felt to represent strong genetic components, high λ values can be due to nongenetic causes of familial aggregation such as shared environment.

The following sections are intended to be a practical introduction to the difficulties inherent in human linkage analysis. For a more detailed review, the interested reader is referred to several excellent monographs.[21–24]

Approaches to
Linkage Analysis

The goal of linkage analysis is to determine the chromosomal location of the gene(s) (also called genetic mapping) for a trait in a set of families. Typically, the family material (individuals within pedigrees collected for a study) will be genotyped for a series of genetic markers whose chromosomal location is known. These markers are usually selected from available genetic maps such as those recently published by the NIH/CEPH (National Institutes of Health/Centre d'Etude du Polymorphisme Humain) Collaborative Group[25] and Weissenbach et al.[26] Markers throughout the genome will then be tested in what is termed a "genomic screen". In diseases with some obvious candidate loci, polymorphic markers located near such genes may be appropriate for immediate study. If the marker and disease trait are un-

linked to one another, then the alleles at the marker locus will segregate independently from the disease allele; if the marker and disease trait are linked to one another, then the alleles at the marker locus will show a nonrandom assortment.

Linkage analysis has traditionally been performed in one of two manners: a parametric approach or a nonparametric approach. Parametric approaches require the assumption of a genetic model and the specification of various parameters such as disease gene frequency and disease penetrance. In contrast, nonparametric methods do not require specification of a genetic model and thus do not have the potentially hazardous effects of model misspecification. When the genetic model for a disease is clearly known, parametric (likelihood) based methods are more powerful than the nonparametric approaches.

The LOD Score Approach
to Linkage Analysis

A conclusion that the disease and marker appear to be linked requires a statistical quantification: How far apart are the disease and marker, and how certain is the conclusion of linkage? An intuitive, direct estimate of the recombination frequency (usually designated as θ), which is related to the genetic distance between the disease and marker locus, is the proportion of recombinant meioses counted among total meioses scored. In the earlier example of Figure 5, the direct estimate of the recombination frequency, based on the observation of one recombinant out of ten scorable meioses, is $1/10$ or 10%. This direct approach is possible because these are phase known events allowing recombinant and nonrecombinant individuals to be counted.

Not all pedigree structures are as tractable as the one in Figure 5. Morton[27] suggested a likelihood ratio approach for use in general pedigrees in which the likelihood of the pedigree and marker data is calculated under the null hypothesis which assumes free recombination ($\theta = 0.50$)

between the disease and trait, where θ represents the recombination fraction. This likelihood is compared to the likelihood calculated at various increments of $\theta < 1/2$ within the range of allowable values (0.00–0.49). The likelihood ratio is constructed as:

$$\frac{L(pedigree|\theta = x)}{L(pedigree|\theta = 0.50)}$$

where x is some value of θ.

In the pedigree of Figure 5, the likelihood of observing the pedigree data is θ^R $(1-\theta)^{NR}$ where R is the number of recombinants, NR is the number of nonrecombinants, and N is the total number of scored offspring (R + NR = N); this likelihood is the numerator of the likelihood ratio. The denominator is the likelihood when the two loci are unlinked (i.e., when $\theta = 0.5$). Thus, the likelihood ratio of the two hypotheses is constructed as:

$$L.R. = \frac{\theta^R(1-\theta)^{NR}}{(0.5)^R(0.5)^{NR}}$$

and reduces to:

$$L.R. = \frac{\theta^R(1-\theta)^{NR}}{0.5^N}$$

Typically, the base 10 logarithm of this ratio is taken to obtain an LOD (for Logarithm of the Odds of linkage) score; $z(\theta)$ is sometimes used to denote an LOD score. An LOD score of 3 indicates odds of 10^3:1 (1000:1) in favor of linkage and is considered in most cases to be conclusive evidence *for* linkage between two markers (or a marker locus and disease locus). An LOD score of -2 indicates odds of 10^{-2}:1 (0.01:1) in favor of linkage, or more commonly, odds of 100:1 *against* linkage for the two markers. An LOD score of -2 or less is considered to be conclusive evidence that the two markers under study are unlinked at the specified recombination fraction. LOD scores between the values of -2 and 3 are considered inconclusive and warrant additional study. A study is "expanded" by rendering the currently available family data more informative (i.e., testing the family with different or more informative markers) or by increasing the number of families under study.

An LOD score should always be considered in conjunction with its respective estimate of θ.

It is important to note that using an LOD score of 3.0 (odds 1000:1 in favor of linkage) as a test statistic does not equate with a type 1 error rate (α) of 0.001, which is much more stringent than the standard type 1 error rate of 0.05 in most statistical analyses. Because of the prior probability of linkage between two traits, the true p-value associated with an LOD score of 3.0 is approximately 0.04, consistent with a false-positive rate of 1 in 25. In other words, when evidence for linkage between two loci is declared significant at an LOD score of 3.0, there is a 4% chance that it is a spurious positive result (for more detail, see Ott[21] and Conneally and Rivas[24]).

LOD scores can be calculated by hand in some circumstances and are generally determined for a variety of θ values spanning the range of allowable values, as recommended by the Human Gene Mapping Linkage and Gene Order Committee.[28] For example, in order to calculate the LOD score when $\theta = 0.01$ for the data observed in Figure 5, recall that 9 nonrecombinant meioses and 1 recombinant meiosis were scored. The numerator for the likelihood ratio, which is the likelihood of the pedigree under the hypothesis that the trait and marker are linked to one another at $\theta = 0.01$, is constructed as $L = (0.01)^1(0.99)^9 = 0.009135$; the denominator of the likelihood ratio, which is the null hypothesis that the trait and marker are unlinked to one another (i.e., $\theta = 0.50$), is constructed as $L = (0.50)^{10} = 0.000977$. The LOD score is then calculated as the $\log_{10}$ of the ratio of the two likelihoods and is: $\log_{10}$ (0.009135/0.000977) = 0.97. The interpretation of this result is that the pedigree data provides odds of 9.35:1 ($10^{0.97}$:1) in favor of linkage between the disease and the marker at $\theta = 0.01$. Table 1 shows the construction of the likelihoods for LOD scores calculated for the pedigree data in Figure 5 at additional recombination fractions. The results of such a linkage analysis are often displayed graphically, as shown in Figure 8. When reporting results of a linkage analysis, it is also necessary to find the maximum

Table 1
Numerical Calculation of LOD Scores

Θ	Calculation of Numerator of Likelihood Ratio*	LOD Score
0.01	$(0.01)^1(0.99)^9 = 0.009135$	0.97
0.05	$(0.05)^1(0.95)^9 = 0.031512$	1.51
0.10	$(0.10)^1(0.90)^9 = 0.038742$	1.60
0.15	$(0.15)^1(0.85)^9 = 0.034742$	1.55
0.20	$(0.20)^1(0.80)^9 = 0.026844$	1.44
0.30	$(0.30)^1(0.70)^9 = 0.012106$	1.09
0.40	$(0.40)^1(0.60)^9 = 0.004031$	0.62

*Recall that the denominator is $\log_{10}(0.50)^{10} = 0.000977$. See text for details.

likelihood estimate (MLE) of the recombination fraction, or that value of θ at which the LOD score is largest. It is at this value of the LOD score (the highest LOD score) that the determination of statistical significance is made. In this case, the maximum LOD score is 1.60 which occurs at $\theta = 0.10$; therefore, 0.10 is the MLE for θ in this example (note that this result is identical to the direct estimate for θ obtained earlier). This result does not meet the established criteria of odds $\geq$ 1000:1 (an LOD score $\geq$ 3) for concluding evidence for linkage, yet it is interesting and perhaps merits further investigation by the genotyping of additional tightly linked markers or additional families.

The family studied in this example is large and relatively atypical for a human pedigree. How much information is required to obtain significant results from a linkage analysis? The answer to this question can be complicated and highly dependent on the inheritance pattern of the disease (i.e., dominant, recessive, or sex-linked), the penetrance of the disease, whether sporadic cases of the disease (phenocopies) may be present, and other factors. As a guideline, each scorable phase known meiotic event (with no recombination between the trait and marker locus) contributes about 0.30 to the LOD score when there is full penetrance and no phenocopies. Thus, at a minimum, 10 phase

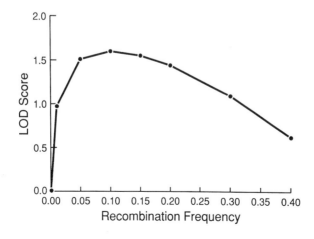

Figure 8. Graph of LOD score curve.

Table 2

Number of Phase Known Meiotic Events
Required to Obtain a LOD Score of at
Least 3.0 with Power = 0.90*

Penetrance	Θ 0.01	0.05	0.10
1.00	16	22	31
0.90	21	28	39
0.80	26	35	48

*Adapted from Ott, 1992.

known meiotic events demonstrating no recombination between the trait and marker locus are required to obtain an LOD score of 3.0. Table 2 shows approximate numbers of phase known meiotic events required to obtain a LOD score of 3.0 with 90% power for various recombination fractions. In general, human pedigrees are complex and do not have a multitude of "phase known meiotic events". Therefore, computer simulation studies should be performed to assess the power of an available data set to detect linkage under an assumed genetic model prior to initiating a screen of the entire genome to detect linkage. Several programs including SIMLINK and SLINK are available for performing these power studies.[29–32]

When significant evidence for linkage is found, a formal presentation of results of a linkage analysis is incomplete without an indication of a support interval (similar to a confidence limit) for the MLE of θ. A description of the calculation of these support intervals, usually performed using the "one LOD score down" method, can be found in either Ott[21] or Conneally and Rivas.[24]

As described above, it is quite tedious to calculate LOD scores by hand. The calculation of LOD scores by hand is practical for a limited number of pedigree structures. Several well-tested computer programs including LIPED,[33] the LINKAGE computer package,[34] and newly developed faster programs (e.g., FASTLINK[34a,b]) or novel algorithms (VITESSE[34c]) are often used for cal-

culation of these scores. Likelihoods and LOD scores in subsequent parts of this chapter were calculated using LIPED and are included so the reader can compare results with those presented here.

Complications in Linkage Analysis

Reduced Penetrance

The examples just described are relatively straightforward because phenotypes with respect to disease status and marker genotypes were available on all family members. We will now review some complications encountered frequently in linkage analysis, some of which are outlined in Table 3. First, assume that the disease is not fully penetrant. In other words, there are individuals who carry the disease gene but who fail to show signs or symptoms of the disease trait. Assume for this analysis that the penetrance, the probability of showing signs of the disease given that an individual carries the disease gene, is 60% and thus 40% of gene carriers fail to show symptoms.

In considering the pedigree with an incompletely penetrant gene, it is still certain that the affected individuals are truly gene carriers, but the unaffected individuals in Figure 5 can no longer be confidently scored: each has a 40% chance of carrying

Table 3

Factors that Can Confound
Linkage Analysis

Misspecification of parameters
Disease allele frequency
Disease penetrance
Disease mode of inheritance
Marker allele frequency
Frequency of sporadic cases

Scoring errors
Incorrect disease phenotype
Incorrect marker phenotype
Incorrect family relationships

Other factors
Linkage heterogeneity

the disease gene (and be currently asympto-matic) and a 60% chance to be a nongene carrier. Individual III-6 who was scored as a recombinant meiosis in the example assum-ing complete penetrance can no longer be scored absolutely as a recombinant meiosis. LOD scores for various θ values for the pedigree in Figure 5, assuming 60% pene-trance of the disease gene, are listed in Table 4, item number 2. The maximum value of the LOD score is 1.86 which occurs when θ = 0.0. These results are considerably differ-ent from those obtained in the previous ex-ample and the differences are entirely due to the *assumptions* regarding penetrance of the trait. The MLE for θ in this example is 0.0 because there are no *absolute* recombinants.

Missing Genotypes

Rarely are research families as large as the family in Figure 5 where all members are available and willing to participate. Often, some individuals are deceased and others refuse to cooperate in the studies. Assume that individuals I-2 and II-2 are unavailable for genotyping (Figure 9). For this example, assume a 60% penetrance for the disease gene. Because individual I-2 is a founder and her marker genotype cannot be inferred from her offspring, population estimates for the

marker alleles must be used. Assume here that the true frequency for marker allele 1 is 0.99. In the calculation of the likelihood, all possible genotypes of the missing individu-als must be considered and weighted by pop-ulation allele frequencies. For each missing individual in the pedigree, the set of all pos-sible genotypes includes genotypes 11, 12, and 22. These genotypes have population frequencies (based on allele frequencies and assuming Hardy-Weinberg equilibrium) of 0.9801, 0.0198, and 0.0001, respectively. Based on the available family information, genotype 11 is impossible as her offspring II-1 must have inherited a 2 allele from his af-fected mother. Of the remaining possible genotypes, the most likely genotype is 12 (based on population genotypic frequen-cies). For individual II-2, genotypes 12 and 22 are possible (genotype 11 is impossible since she has offspring who are genotype 22 and thus must have inherited a 2 allele from her). Of the possible genotypes for II-2, genotype 12 is more likely than genotype 22, again be-cause of population frequencies. Valuable in-formation on linkage phase is lost in this sit-uation. LOD scores and associated θ values are shown in Table 4.3. The maximum LOD score is 1.49 when θ = 0.0. When compared to the reduced penetrance model when all in-dividuals were genotyped, 20% of the avail-able information in the pedigree is lost be-

Table 4
LOD Scores Calcuated for Pedigree in Figure 5 under Various Genetic Models

	0.01	0.05	0.10	0.15	0.20	0.30	0.40
1. Complete penetrance.	0.97	1.51	1.60	1.55	1.44	1.09	0.62
2. Penetrance = 0.60.	1.86	1.75	1.61	1.45	1.29	0.92	0.50
3. Model 2 and frequency of marker allele 1 = 0.99; individuals I-2 and II-2 not available for genotyping.	1.49	1.40	1.28	1.16	1.03	0.73	0.40
4. Model 3 and disease allele frequency = 0.50; individuals I-2 and II-2 without information in disease status.	0.63	0.55	0.46	0.37	0.29	0.16	0.06

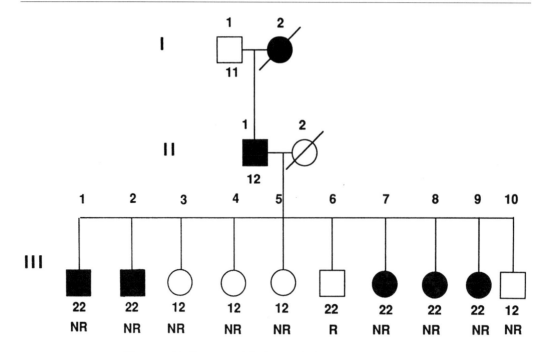

Figure 9. Pedigree modified to include missing genotypes.

cause of the ungenotyped family members. It is important to note that the correct specification of marker allele frequencies is critically important in situations analogous to the one just considered. When marker genotypes on pedigree members are missing, results of linkage analysis can vary dramatically depending on specification of marker allele frequencies. Incorrect specification of the allele frequencies can lead to a large bias in estimates of θ and incorrect conclusions of linkage. (See Ott[35] and Knowles et al.[36] for a more thorough discussion of the problems of allele frequencies and LOD scores with the highly polymorphic microsatellite markers.)

Missing Phenotypes

Other parameters of the genetic model can be altered or incorrectly specified in linkage analysis. As an example, consider that in addition to the restrictions placed above (penetrance of the disease gene = 0.60, no genotyping on individuals I-2 and II-2), the disease status of I-2 is uncertain (Figure 10). Thus, because of the reduced penetrance, the disease gene could have come from I-1 or from I-2. Assume further that the disease is frequent in the population$enso common, in fact that the gene frequency is 0.50 (this is an unrealistic situation, but is used for illustrative purposes). Here, both linkage phases, either that the disease is passed with the 1 allele from the father (phase 1) or with the 2 allele from the mother (phase 2), must be considered in the formal likelihood calculations. Because the linkage phase that assumes the fewest recombinant individuals has a higher likelihood (in this case, linkage phase 2), it contributes more to the LOD score. Note the marked decrease in the LOD score once the two disease phenotypes become unknown (Table 4.4).

Locus Heterogeneity

The reader should note that all analyses thus far have dealt with only a single family.

The advantage of working with a single large pedigree in a rare disease is that the possibility of locus heterogeneity is essentially eliminated. This advantage comes at the cost of a loss of generalizability of the results: a disease gene in a single family may be unique to that family. This possibility should always be considered when two or more families with similar clinical phenotypes are studied. Genetic heterogeneity (i.e., two or more loci causing the same phenotype) has been particularly frequent in neurogenetic diseases, including Charcot-Marie-Tooth disease type 1,[37] Alzheimer's disease,[5,6,38] facioscapulohumeral muscular dystrophy,[39] and many inherited mitochondrial disorders.

In a typical linkage analysis, many families with a disease trait that is clinically similar between families are available and willing to participate in the linkage studies. Before initiating a genomic screen to detect linkage of the disease trait to a genetic marker, the investigator will not be certain whether the disease is genetically homogeneous or heterogeneous. Clinical differences between families may suggest genetic heterogeneity, and this clinical information should be carefully catalogued for future reference. In some cases, clinical differences between families may represent solely differences in expression of a single disease gene. If LOD scores are calculated and summed across families, the overall results may not be significant in the presence of locus heterogeneity. Thus, genomic screening of a series of clinically similar families should be conducted with careful regard for individual family LOD scores so as not to miss a linkage in a subset of families.

Although it is tempting to inspect LOD scores visually and include or exclude families based on whether their LOD scores are positive or negative, this practice can lead to extreme biases. When genetic heterogeneity is suspected, formal tests of heterogeneity should be performed.[21,40–44]

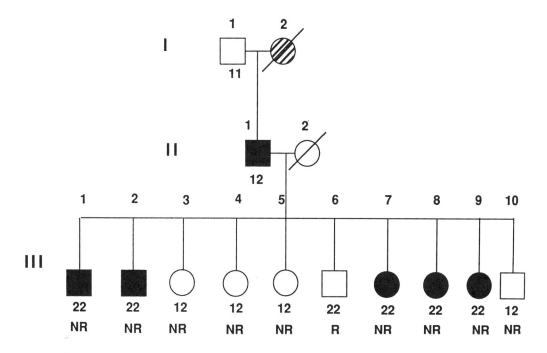

Figure 10. Pedigree modified to include missing phenotypes. Stippled individual is of unknown disease phenotype.

Mode of Inheritance

The specified mode of inheritance is the most fundamental assumption in linkage analysis. The likelihood-based approach to linkage analysis requires specification of a mode of inheritance. Often, the mode of inheritance of a disease is uncertain. For instance, in the neural tube defects, there is evidence for both dominant and recessive models[45,46] based on segregation analysis. How can the investigator deal with this situation which is frequently encountered in many complex diseases? One solution is to analyze the data using a series of models, including both dominant and recessive, and spanning a range of penetrances and gene frequencies. This approach can be quite problematic: a positive result for linkage attained under a specified model is valid only under that model. Whether or not a significant LOD score generated under a specific model (which was chosen with no underlying purpose) is conclusive evidence of the "correctness" of the model has been a source of significant controversy.[47,48] What is certain is that the maximization of results over genetic models must be considered in the interpretation of results.[49] Furthermore, extreme care must be exercised when using this approach so that chromosomal regions are not excluded from further consideration. A chromosomal region can be excluded only for the specified model and not for all possible models.

Nonparametric (Model-Free) Methods of Linkage Analysis

Other approaches to linkage analysis employ model-free methods, in which mode of inheritance or other parameters of the genetic model do not need to be specified. Two well-recognized approaches using nonparametric methods of linkage analysis are sib-pair analysis and affected pedigree member (APM) analysis.

The sib-pair approach to linkage analysis was first described by Penrose.[50] The premise is straightforward: if pairs of siblings are selected at random from a series of families in which two traits (usually a disease and a marker) are under consideration, then four classes of sibling pairs can be identified. Let L_1 represent sibling pairs that are alike for trait 1, and L_2 represent sibling pairs that are alike for trait 2. Similarly, U_1 and U_2 will represent sibling pairs that are unalike for traits 1 and 2 respectively. In the test defined by Penrose,[50] if the traits are linked then there should be an increase in the frequency of pairs that are alike (class L_1L_2) or unalike (U_1U_2) with respect to the traits, relative to the frequency of pairs that are alike for one trait and unalike for the other (L_1U_2 and U_1L_2). If the traits are unlinked, these four classes should occur with approximately equal frequency. Such scorings can also be tested with a 2×2 chi square analysis (see, for example, Elston and Johnson[51]). The chi square test is a test for an association (as opposed to a linkage) between two traits, and should also be significant in the presence of linkage when data is tabulated as above. However, the chi square test does not take into account the fact that under linkage, classes L_1L_2 or U_1U_2 should be more frequent than the L_1U_2 and U_1L_2 classes. The difference between an association and a linkage is subtle but important. Association implies that two traits occur together more frequently than expected by chance alone, which could be caused by many factors. Genetic linkage of two loci presumes that two loci are located sufficiently near one another on the same chromosome so that they cosegregate within a family with a frequency proportional to the recombination frequency between the loci. If two traits show an association, they may or may not be linked to one another. By definition, two traits that are linked to one another will show an association *within* a family, but the associated allele may vary from family to family.

Penrose[50] illustrates the sib-pair approach to linkage using the traits red hair, agglutinogen A, and blue eyes. Two examples from his original manuscript are shown

Table 5
Agglutinogen A and Red Hair

Red hair	Agglutinogen A		
	Like	**Unalike**	**Total**
Like	40	17	57
Unalike	0	3	3
Total	40	20	60

in Tables 5 and 6. Table 5 shows results of scoring agglutinogen A and red hair in 60 pairs of siblings. Here, there is an excess in the L_1L_2 class, consistent with linkage of the trait for agglutinogen A and red hair (p = 0.03). Table 6 shows results of scoring agglutinogen A and blue eyes in 50 pairs of siblings; the test is nonsignificant, indicating no evidence for linkage between the traits agglutinogen A and blue eyes.

Sib-pair approaches to linkage analysis are based on assessing the number of alleles that a pair of sibs shares that are identical by descent (IBD). Figure 11 shows a nuclear family with four offspring. Siblings 1 and 2 share no alleles IBD, siblings 1 and 3 share 1 allele IBD, and siblings 1 and 4 share 2 alleles IBD. Under the null hypothesis of no linkage, the number of alleles that a pair of siblings shares that are identical by descent are determined by Mendelian ratios: the probability that a pair of sibs shares 0 alleles IBD is 1/4, 1 allele IBD is 1/2, and 2 alleles IBD is 1/4. Various test statistics have been proposed[52,53] each of which is generally more powerful than Penrose's original design, and are based on testing whether the deviations from these expected ratios under the null hypothesis of no linkage are significant.

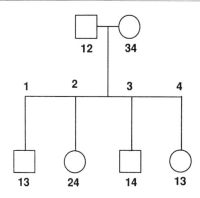

Figure 11. A nuclear family with four offspring. Marker genotypes are indicated beneath each individual. Note that siblings 1 and 3 share one allele identical by descent (allele 1, inherited from the father, is common to these sibs, while the maternally inherited allele differs. Sibling 1 inherited maternal allele 3 and sibling 2 inherited maternal allele 4).

Rough sample size guidelines for affected sib-pair studies under a variety of assumptions about λ_s are shown in Table 7. Generally, a sampling scheme whereby only affected sib pairs are studied is more powerful than a scheme including a mixture of affected/affected and unaffected/affected sib pairs or all unaffected/affected sib pairs. However, if the penetrance of the disease is essentially complete or the disease population prevalence is high, the power of a sampling scheme that includes a mixture of affected/unaffected pairs or all unaffected/affected pairs is also high. An excellent example of a successful study utilizing a sib-pair approach to linkage analysis is the determination of linkage between hypertension and the angiotensin gene.[54] A very thorough and readable example of the use of the sib-pair approach to screening multiple loci for linkage to a trait is found in Wilson et al.[55] Recently, a series of genomic screens in a variety of complex diseases using this approach have been published.[55a–d]

Penrose's initial sib-pair method was based on the use of dichotomous traits, but he[56] and later Haseman and Elston[57] extended this approach to include continuous traits. As proposed by Haseman and Elston,[57] the number of alleles IBD is regressed

Table 6
Agglutinogen A and Blue Eyes

Blue eyes	Agglutinogen A		
	Like	**Unalike**	**Total**
Like	22	11	33
Unalike	11	6	17
Total	33	17	50

Table 7

Approximate Numbers of Affected Sibling Pairs Required to Detect Linkage for a Trait Linked at 0% Recombination with a Fully Informative Marker and 80% Power*

λ_s	Approximate Number of Affected Sibling Pairs
1.7	300
2.0	200
3.0	100

*Adapted from reference 20b.

on the squared trait difference (STD) between sibs for continuous traits. Under linkage, as the STD decreases, the number of alleles IBD increases. Various extensions of this approach, including incorporation of age of onset[58] and the allowance for multiple trait measurements,[59] have been proposed. When families larger than nuclear families are available, other relative pairs can be utilized in these approaches if the IBD relationships can be established by genotyping a sufficient number of family members. Test statistics are based on whether the deviations from the expected ratios under the null hypothesis of no linkage are significant, where the expected ratios are based on the relative pair under study. For instance, the probabilities that a half-sib pair (a half-sib pair shares only one parent in common) shares 0, 1, and 2 alleles IBD are 1/2, 1/2, and 0 respectively.

Sib-pair tests may not be practical in rare diseases or in diseases with late age of onset because attaining a sufficient sample size of sib pairs may be difficult. In late-onset diseases, the parental generation is often unavailable and at-risk family members may die before reaching typical-onset age. In these cases, IBD relationships for a pair of alleles cannot be established with certainty. Here, statistical tests have been developed and are based on shared alleles that are identical by state (IBS).[60] Although these tests are generally less powerful than tests based on alleles IBD, they may be more practical in

some situations. The Affected Pedigree Member (APM) method, based on IBS relationships, was successfully implemented in the primary detection of association of late-onset Alzheimer's disease with chromosome 19.[38] This novel approach was recently modified to approximate IBD relationships in a program called Sim IBD.[38a] Figure 12 shows a typical pedigree demonstrating an affected relative pair (here, uncle-nephew) which shares one allele IBS, the 1 allele. The 1 allele may in fact be IBD, although given the pedigree structure, this can never be proven. If a sib-pair analysis (as opposed to an APM test) had been implemented in the study of this disease, no information would be available from this pedigree.

The power of these IBS tests is highly dependent upon many factors including marker allele frequency and type of relative pair used (e.g., Bishop and Williamson[61]). Additionally, for all of the nonparametric linkage tests, the power of the study to detect linkage (or association) is highly dependent upon distance between marker and trait loci and is drastically reduced as the distance between the marker and trait loci increases.

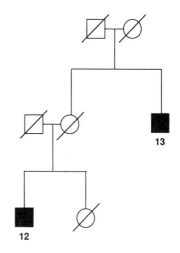

Figure 12. The pair of affected relatives (an uncle and a nephew) share one allele identical by state (the 1 allele). The 1 allele may also be IBD, although given the pedigree structure, this cannot be confirmed.

Planning a Linkage Analysis

A study to screen the genome for loci that are causative in human diseases often requires a multidisciplinary approach, including physicians, statistical geneticists, and molecular biologists. Typically, and this is particularly true in the study of complex diseases, the critical and most difficult components of the study will be in phenotype definition and assignment, and in data analysis. A general outline of important considerations in planning a linkage analysis is presented in Table 8.

Concluding Remarks

Linkage analysis is a powerful tool for the primary detection of genes leading to human disease. This chapter is meant to serve as a general introduction and brief overview of the methods and problems encountered when planning and performing studies to locate genes that lead to human disease.

Methods for localizing genes in diseases that are clearly Mendelian have been highly successful. Two-point linkage analysis, described in this chapter, provides a primary approximation to disease localization.

Table 8
Considerations for Initiating Studies to Search for Genes Leading to Human Disease

A. **Define Phenotype of Interest**
- What evidence is there that the trait has a hereditary component?
- How reliable are diagnostic tests? Are there differences in scoring between clinicians? (e.g., consider schizophrenia where inter-examiner variability in scoring is high).
- Are there clinical differences among individuals from different families which might suggest genetic heterogeneity?
- Given the phenotype of interest, what method of linkage analysis will be used and how will this affect sampling scheme?
- Given disease phenotype information and available sample size, is a successful study feasible?

B. **Family Ascertainment and Collection**
- Determination of sampling scheme (e.g., only affected relatives, only affected sibs, all sibs, extended families).
- Search for high density disease populations which can be studied.
- Careful and thorough documentation of phenotype with maintainence of records for future re-evaluation, if necessary.
- Pedigree and family history documented with names, phone numbers, and addresses of contact persons.
- Maintain communication with study participants (i.e., phone calls, newsletters).

C. **Initiation of Genomic Screening**
- Consideration of candidate gene loci, if available.
- Development of rational approach to screening the genome (i.e., select highly heterozygous markers, select inter-marker spacing, test by chromosome or more randomized approach).
- Linkage analysis to assess significance of results.
- Maintainence of records of genomic areas screened and/or excluded to influence future selection of markers.

The localization of a disease or trait locus can be refined using multipoint linkage analysis (e.g., Hellsten et al.,[62] Gilbert et al.,[63] and for methodological descriptions, Ott[21]; Lathrop et al.[34]), which should be supplemented by visual inspection of actual crossover data via haplotype analysis.

Extreme care and caution must be exercised when planning a linkage analysis, particularly with complex diseases, since errors in the specification of the model can cause significant errors in the final results. Although linkage analysis of more complex diseases, in which the mode of inheritance is unclear, is less straight-forward, methods are available to address such situations. The potential rewards of localizing and cloning genes causing the complex and most common diseases are abundant, especially with respect to public health and policy.

Acknowledgements: The author would like to thank Mark Alberts, Elizabeth Corder, Pete Gaskell, Suzanne Leal, Margaret Pericak-Vance, and Larry Yamaoka for scientific and editorial critique, and Caren Hurwitz for assistance with the graphic design of the figures.

References

1. Monaco AO, Bertelson CJ, Colletti-Feener C, et al. Localization and cloning of Xp21 deletion breakpoints involved in muscular dystrophy. Hum Genet 1987;75:221–227.
2. Riordan JR, Rommens JM, Kerem B, et al. Identification of the cystic fibrosis gene: Cloning and characterization of complementary DNA. Science 1989;245:1066–1073.
3. Wallace MR, Andersen LG, Saulino AM, et al. A de novo Alu insertion results in neurofibromatosis 1. Nature 1991;353:864–866.
4. Rosen D, Siddique T, Patterson D, et al. Mutations in Cu/Zn superoxide dismutase gene are associated with familial amyotrophic lateral sclerosis. Nature 1993;362:59–62.
5. St. George-Hyslop PH, Tanzi RE, Polinsky RJ, et al. The genetic defect causing familial Alzheimer's disease maps on chromosome 21. Science 1987;235:885–890.
6. Schellenberg GD, Bird TD, Wijsman EM, et al. Genetic linkage evidence for a familial Alzheimer's disease locus on chromosome 14. Science 1992;258:668–671.
7. Hall JM, Lee MK, Newman B, et al. Linkage of early onset familial breast cancer to chromosome 17q21. Science 1991;250:1684–1689.
8. Hall JM, Friedman L, Guenther C, et al. Closing in on a breast cancer gene on chromosome 17q. Am J Hum Genet 1992;50:1235–1242.
9. Botstein D, White RL, Skolnick M, et al. Construction of a genetic linkage map in man using restriction fragment length polymorphisms. Am J Hum Genet 1980;32:314–331.
10. Emery AEH. An Introduction to Recombinant DNA. New York: John Wiley and Sons, Ltd, 1984.
11. Nakamura Y, Leppert M, O'Connell P, et al. Variation number of tandem repeat (VNTR) markers for human gene mapping. Science 1987;235:1616–1622.
12. Litt M, Luty JA. A hypervariable microsatellite revealed by in vitro amplification of a dinucleotide repeat within the cardiac muscle action gene. Am J Hum Genet 1989;44:397–402.
13. Smeets HJM, Brunner HG, Ropers HH, et al. Use of variable simple sequence motifs as genetic markers: Application to study of myotonic dystrophy. Hum Genet 1989;83:245–251.
14. Weber JL. Abundant class of human DNA polymorphisms which can be typed using the polymerase chain reaction. Am J Hum Genet 1989;44:388–396.
15. Sherman SL, Jacobs JA, Morton NE, et al. Further segregation analysis of the fragile X syndrome with special reference to transmitting males. Hum Genet 1985;69:289–299.
16. Fu YH, Kuhl DP, Pizzuti A, et al. Variation of the CGG repeat at the fragile X site results in genetic instability: Resolution of the Sherman paradox. Cell 1991;67:1047–1058.
17. LaSpada AR, Wilson EM, Lubahn DB, et al. Androgen receptor gene mutations in X-linked spinal and bulbar muscular atrophy. Nature 1991;352:77–79.
18. Harley HG, Brook JD, Rundle SA, et al. Expansion of an unstable DNA region and phenotypic variation in myotonic dystrophy. Nature 1992;355:545–546.
19. Mahadevan M, Tsilfidis C, Sabourin L, et al. Myotonic dystrophy mutation: An unstable CTG repeat in the 3' untranslated region of the gene. Science 1992;255:1253–1255.
20. The Huntington's Disease Collaborative Research Group. A novel gene containing a trinucleotide repeat that is expanded and unstable on Huntington's disease chromosomes. Cell 1993;72:971–983.

20a. Risch N. Linkage strategies for genetically complex traits I: Multilocus models. Am J Hum Gen 1990;46:222–228.

20b. Risch N. Linkage strategies for genetically complex traits II: The power of relative pairs. Am J Hum Gen 1990;46:229–241.

20c. Risch N. Linkage strategies for genetically complex traits III: The effect of marker polymorphism on analysis of affected pairs. Am J Hum Gen 1990;46:242–253

21. Ott J. Analysis of Human Genetic Linkage. Johns Hopkins University Press, 1991.

22. Lander ES.Mapping complex genetic traits in humans. In Davies KE (ed). Genome Analysis. Oxford: IRL Press, 1988, 171–190.

23. Ott J. A short guide to linkage analysis. In Davies KE (ed). Human Genetic Diseases. Oxford: IRL Press, 1986, 19–32.

24. Conneally PM, Rivas ML. Linkage analysis in man. Adv Hum Genet 1980;10:209–263.

25. NIH/CEPH Collaborative Mapping Group. A comprehensive genetic linkage map of the human genome. Science 1992;258:67–86.

26. Weissenbach J, Gyapay G, Dib C, et al. A second generation linkage map of the human genome. Nature 1992;359:794–801.

27. Morton NE. Sequential tests for the detection of linkage. Am J Hum Genet 1956;8:80–96.

28. Conneally PM, Edwards JH, Kidd KK. Report of the committee on methods in linkage analysis reporting. Eighth International Workshop on Human Gene Mapping. Cytogenet Cell Genet 1985;40:356–359.

29. Boehnke M. Estimating the power of a proposed linkage study: A practical computer simulation approach. Am J Hum Genet 1986;39:513–527.

30. Ploughman LM, Boehnke M. Estimating the power of a proposed linkage study for a complex genetic trait. Am J Hum Genet 1989;44:543–551.

31. Ott J. Computer simulation methods in human linkage analysis. Proc Natl Acad Sci USA 1989;86:4175–4178.

32. Weeks DE, Ott J, Lathrop GM. SLINK: A general simulation program for linkage analysis. Am J Hum Genet 1990;47:A204.

33. Ott J. Estimation of the recombination fraction in human pedigrees: Efficient computation of the likelihood for human linkage studies. Am J Hum Genet 1974;26:588–597.

34. Lathrop GM, Lalouel JM, Julier C, et al. Strategies for multilocus linkage analysis in humans. Proc Natl Acad Sci USA 1984; 81:3443–3446.

34a. Cottingham RW Jr, Idury RM, Schaffer AA. Faster sequential genetic linkage computations. Am J Hum Gen 1993;53:252–263.

34b. Schaffer AA, Gupta SK, Shriram K, Cottingham RW. Avoiding recomputation in linkage analysis. Hum Hered 1994;44:225–237.

34c. O'Connell JR, Weeks DE. The VITESSE algorithm for rapid exact multilocus linkage analysis data via genotype set-recording and fuzzy inheritance. Nat Gen 1995;11:402–408.

35. Ott J. Strategies for characterizing highly polymorphic markers in human gene mapping. Am J Hum Genet 1992;41:283–290.

36. Knowles JA, Vieland VJ, Gilliam TC. Perils of gene mapping with microsatellite markers. Am J Hum Genet 1992;51:905–909.

37. Vance JM, Nicholson GA, Yamaoka LH, et al. Linkage of Charcot-Marie-Tooth disease type 1a to chromosome 17. Exp Neurol 1989;104:186–189.

38. Pericak-Vance MA, Bebout JL, Gaskell PC, et al. Linkage studies in familial Alzheimer's disease: Evidence for chromosome 19 linkage. Am J Hum Genet 1991;48:1034–1050.

38a. Davis S, Schroeder M, Goldin LR, Weeks DE. Nonparametric simulation-based statistics for detecting linkage in general pedigrees. Am J Hum Gen 1996;58:867–880.

39. Gilbert JR, Stajich JM, Wall S, et al. Evidence for heterogeneity in facioscapulohumeral muscular dystrophy. Am J Hum Genet 1993;53:401–408.

40. Ott J, Bhattacharya S, Chen JD, et al. Localizing multiple X chromosome-linked retinitis pigmentosa loci using multiple locus homogeneity tests. Proc Natl Acad Sci USA 1990;87:701–704.

41. Risch N. A new statistical test for linkage heterogeneity. Am J Hum Genet 1988;42:353–364.

42. Cavalli-Sforza LL, King M-C. Detecting linkage for genetically heterogeneous diseases and detecting heterogeneity with linkage data. Am J Hum Genet 1986;38:599–616.

43. Ott J. The number of families required to detect or exclude linkage heterogeneity. Am J Hum Genet 1986;39:159–165.

44. Ott J. Linkage analysis and family classification under heterogeneity. Ann Hum Genet 1983;47:311–320.

45. Demenais F, le Merrer M, Briard ML, et al. Neural tube defects in France: Segregation analysis. Am J Med Genet 1992;11:287–298.

46. Fineman RM, Jorde LB, Martin RA, et al. Spinal dysraphia as an autosomal dominant defect in four families. Am J Med Genet 1982;12:457–464.

47. Elston RC. Man bites dog? The validity of maximixing LOD scores to determine the mode of inheritance (editorial). Am J Med Genet 1989;34:487–488.

48. Greenberg DA. Inferring mode of inheritance by comparison of LOD scores. Am J Med Genet 1989;34:480–486.

49. Weeks DE, Lehner T, Squires-Wheeler E, et al. Measuring the inflation of the LOD score due to its maximization over model parameter values in human linkage analysis. Genet Epidemiol 1990;7:237–243.

50. Penrose LS. The detection of autosomal linkage in data which consists of pairs of brothers and sisters of unspecified parentage. Ann Eugen 1935;5:133–148.

51. Elston RC, Johnson WD. Essentials of Biostatistics. Philadelphia: FA Davis Company, 1987.

52. Suarez BK, van Eerdewegh P. A comparison of three affected sib pair scoring methods to detect HLA-linked disease susceptibility genes. Am J Med Genet 1984;18:135–146.

53. Weitkamp LR, Stancer HC, Persad E, et al. Depressive disorders and HLA: A gene on chromosome 6 that can affect behavior. New Engl J Med 1981;305:1301–1306.

54. Jeunemaitre X, Soubrier F, Kotelevtsev YV, et al. Molecular basis of human hypertension: Role of angiotensin. Cell 1992;71:169–180.

55. Wilson AF, Elston RC, Tran LD, et al. Use of the robust sib-pair method to screen for single-locus, multiple-locu, and pleiotropic effects: Application to traits related to hypertension. Am J Hum Genet 1991;48:862–872.

55a. Hanis CL, Boerwinkle E, Chakraborty R, et al. A genome-wide search for human noninsulin-dependent (type 2) diabetes genes reveales a major susceptibility locus on chromosome 2. Nat Gen 1996;13:161–166.

55b. Davies JL, Kawaguchi Y, Bennett ST, et al. Genome-wide search for human type 1 diabetes susceptibility genes. Nature 1994;371:130–136.

55c. Sawcer S, Jones HB, Feakes R, et al. A genome screen in multiple sclerosis reveals susceptibility loci on chromosome 6p21 and 17q22. Nat Gen 1996;13:464–476.

55d. The Multiple Sclerosis Genetics Group. A complete genomic screen for multiple sclerosis underscores a role for the major histocompatibility complex. Nat Gen 1996;13;469–471.

56. Penrose LS. Genetic linkage in graded human characters. Ann Eugen 1938;8:233–237.

57. Haseman JK, Elston RC. The investigation of linkage between a quantitative trait and marker locus. Behav Genet 1972;2:3–19.

58. Dawson DV, Kaplan EB, Elston RC. Extensions to sib-pair linkage tests applicable to disorders characterized by delayed onset. Genet Epidemiol 1990;7:453–466.

59. Amos CI, Elston RC, Bonney GE et al. A multivariate method for detecting genetic linkage with application to a pedigree with an adverse lipoprotein phenotype. Am J Hum Genet 1990;47:247–254.

60. Weeks DE, Lange K. The affected-pedigree member method of linkage analysis. Am J Hum Genet 1988;42:315–326.

61. Bishop DT, Williamson JA. The power of identity by state methods for linkage analysis. Am J Hum Genet 1990;46:254–265.

62. Hellsten E, Vesa J, Speer MC, et al. Refined assignment of the infantile neuronal ceroid lipofuscinosis (INCL, CLN1) locus at 1p32: Incorporation of linkage disequilibrium in multipoint linkage analysis. Genomics 1993;16:720–725.

63. Gilbert JR, Stajich JM, Speer MC, et al. Linkage studies in facioscapulohumeral muscular dystrophy (FSHD). Am J Hum Genet 1991;51:424–427.

Chapter 3

Advanced Genetics:

Finding Genes that Cause Disease

Mark J. Alberts, MD

The current molecular genetic revolution has made available the biological tools to isolate and study genes and mutations that cause diseases. There are several different approaches that can be used to isolate such pathogenic genes. The particular aspects of a specific disease will determine which approach is most useful and efficient. This chapter will describe these various approaches by building upon the concepts and techniques covered in *Chapters 1 and 2*.

When studying an inherited disease, a key initial step is to determine if the defective protein is known, or if a promising candidate protein exists. For example, sickle cell anemia was found to be due to a single amino acid substitution in the globin moiety of hemoglobin.[1] The protein mutation was known many years prior to the isolation and cloning of the globin gene. This scenario can be contrasted with Duchennne's muscular dystrophy (DMD), for which the pathogenic protein was unknown until intense genetic studies were undertaken and the dystrophin gene was cloned.[2] The identification of dystrophin depended on prior experiments which showed that the mutant gene was on the X chromosome, and a variety of other cloning studies focused on isolating the pathogenic gene. In the case of Alzheimer's disease (AD), β-amyloid was an excellent candidate gene and protein.[3] After intensive study, however, it has become apparent that mutations in the β-amyloid gene are responsible for only a very small minority of familial AD cases.[4] Further genetic linkage studies were used to help isolate the presenilin genes responsible for most cases of early-onset familial AD.[4a,4b]

These three examples highlight some of the most common general approaches to finding pathogenic genes and the limitations of each approach. Sickle cell anemia was due to an obvious, known protein, and isolation of the globin gene was not necessary for the isolation of the mutated protein. DMD was due to a mutation in a previously unknown protein that required discovery of the gene before the protein could be isolated and analyzed. In the case of familial AD, β-amyloid was an excellent candidate gene/protein, but it clearly did not account for all inherited cases of AD. Additional studies were needed to identify the genes responsible for other cases of AD. The case of familial AD illustrates genetic heterogeneity, where mutations in more than one gene can produce very similar clinical phenotypes. The different approaches are explained in more detail below and in Figure 1. In some cases, other

From Alberts MJ (ed). *Genetics of Cerebrovascular Disease.* Armonk, NY: Futura Publishing Company, Inc., © 1999.

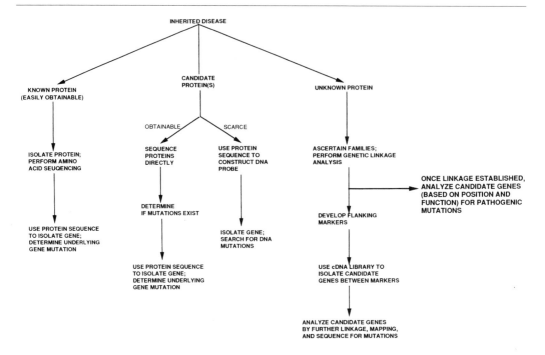

Figure 1. Common approaches for isolating pathogenic genes in various types of inherited diseases. A relatively new approach, positional cloning, is shown in the far right side of the figure.

less common approaches and/or serendipity might be used to isolate pathogenic genes. For example, the co-occurrence of two diseases in one patient might suggest a genetic locus, particularly if one of the disorders had a known locus. In other cases, a gross chromosomal abnormality (i.e., deletion, translocation) might be detected, thereby suggesting an obvious locus for the pathogenic gene.

Protein Analysis

When the pathogenic protein is known or suspected, it can be isolated and sequenced directly to determine its amino acid composition. By comparing the amino acid sequence of the protein from normal and affected individuals, it is possible to determine the nature of the abnormality. In some cases, the amount of synthesized protein is altered, or the size of the protein is abnormal. In other cases, a single amino acid change is sufficient to cause a devastating disease (as is the case with sickle cell anemia). Using this information, the pathophysiology of the disease can often be elucidated. However, this approach will not be useful for all diseases. If the disease is due to a total or near total absence of a protein, it will not be possible to isolate and study the protein from affected individuals. While it could be present in normals, researchers may not know which protein to isolate without performing comparative protein fingerprinting experiments. Even using this technique, very rare proteins may be exceedingly difficult to identify and sequence.

In sickle cell anemia, it was found that the hemoglobin differed from normal by a single amino acid substitution. (In fact, many different mutant hemoglobins have been described.)[5] Since the protein defect was established, it was not of great importance to clone the defective gene, although this was eventually done. However, most neurogenetic diseases differ from sickle cell anemia in that the defective protein is unknown.

A variation of this approach can be seen in the case of β-amyloid. β-amyloid protein has long been considered a candidate gene for AD, since it is deposited in AD brains. In the 1980s, the β-amyloid protein was isolated from an AD brain and partially sequenced.[6] This amino acid sequence was used to make an oligonucleotide probe in an attempt to isolate the β-amyloid gene.[3] When the gene was isolated and localized, it was found that β-amyloid is part of a much larger gene, the amyloid precursor protein (APP).[7] APP has been mapped to chromosome 21 (CH21), which is not unexpected considering that Down's syndrome (trisomy 21) is associated with the extensive deposition of β-amyloid in the brain.[3] Once this genetic marker was available, it was used to perform genetic linkage analysis in families with AD. It was found that some early-onset AD families showed genetic linkage to the APP locus on CH21.[8] Exons 16 and 17 of APP, which form β-amyloid, were sequenced and several mutations were identified in patients with early-onset familial AD.[9,10] Other mutations in the APP gene have been identified in Dutch patients with hereditary cerebral hemorrhage (see *Chapter 11* for details).[11] Therefore, β-amyloid is an example of a candidate protein that led to the identification of several pathogenic mutations in its gene.

There are several important differences between the sickle cell anemia/globin and β-amyloid/AD cases. In the case of sickle cell disease, the candidate protein was very abundant and easy to obtain. Essentially all of the clinical cases can be attributed to different mutations in the same gene or gene family. AD is quite different in several respects. β-amyloid is difficult to obtain from human brain during life (for obvious reasons), and even when it is obtained, it is difficult to solubilize and study. It was only through genetic linkage studies that several families were found to be linked to the APP locus.[12] This led to intense sequencing of the APP gene in those linked familles, with the discovery of several different mutations.[4,9,10] Therefore the identificaiton of the APP gene via the β-amyloid protein has led to important studies and findings that would have

been very difficult, if not impossible, using just the protein. However, the vast majority of patients with early-onset familial AD, and essentially all of the patients with late-onset familial AD, have not shown linkage to the APP gene, nor have they been found to have mutations in this gene.[13–15] A whole genome search using new linkage techniques was used to identify and clone the two presenilin genes responsible for most cases of early-onset familial AD.[4a,4b] Therefore, familial AD displays genetic heterogeneity.[16]

The situation with respect to the APP gene is even more complex since there is clear evidence of allelic heterogeneity. This means that different mutations in the same gene can produce different phenotypes. Depending on the location of the mutations in exon 17 of APP, patients may develop AD or a form of intracerebral hemorrhage (see *Chapter 11* for more detail).[10,11]

Reverse Genetics

Most inherited diseases, particularly the neurogenetic and neurodegenerative diseases, are caused by defects in unknown proteins. With the advent and widespread use of molecular genetic techniques, these diseases can be studied through a process inaccurately called reverse genetics. The term reverse genetics refers to a process where genetic techniques are used to identify the gene responsible for a disease when the gene's protein product is unknown. In purely scientific terms, this is hardly reverse genetics, since genes precede proteins in a functional sense. However, until the early 1980s, the technology did not exist to identify genes for unknown proteins; therefore, the approach outlined below was considered "reverse" genetics. It is now relatively easy to design experiments to identify novel genes that code for previously unknown proteins. Some researchers have objected to the term reverse genetics, preferring to use "positional cloning" instead.[17,18]

Assume we wish to study an inherited neurologic disease that causes the formation of vascular malformations. The gene and protein responsible for this disease are unknown. The first step would be to collect a number of families with the disease. These efforts might yield a few very large pedigrees or many small pedigrees with the disease, depending upon the disease under study and our resources. The study of several very large families could simplify genetic linkage studies. However, the results obtained from a few large families may not be generalizable to other families with the disease. The study of many smaller families makes linkage studies more difficult, and if there is marked genetic heterogeneity such an approach may produce equivocal results. Despite these potential problems, the ascertainment step has several goals including: (1) confirming that the disease is inherited with some type of Mendelian pattern; (2) determining if there is any significant clinical heterogeneity (which could suggest genetic heterogeneity); and (3) providing a source of DNA (and perhaps other tissue) for further studies.

Once DNA was available from many affected individuals (and normals if possible) in several families, we could begin genetic linkage studies. The concepts of genetic linkage are discussed in *Chapter 2*. For the initial study of a complex disease, where the inheritance pattern and the number of involved genes may be unknown, the use of newer linkage techniques such as the Affected Pedigree Member method (APM) or sib-pair analysis could be used to identify regions of chromosome warranting more intense study.[19-21] Using the APM or sib-pair techniques combined with highly informative microsatellite repeat polymorphisms, complex diseases can be studied even if the family structure of the available pedigrees is limited.[22,23] Since genetic markers exist for most regions of each chromosome, a whole genome search for linkage can be performed.[24] The goal is to find one (or more) markers with a statistically significant association to the disease locus. Once a positive marker is identified, several surrounding markers would be tested to con-

firm or disprove the initial linkage by using standard likelihood analysis. (These genetic linkage techniques, as well as other, are described in more detail in *Chapter 2*.) By studying multiple markers in a small, defined region, it is possible to develop flanking markers. The identification of flanking markers means that the disease locus (gene) is between the two flanking markers.

An alternative or complementary approach to performing genetic linkage studies with probes from throughout the genome is to study several promising candidate genes or probes near those genes. Such candidate genes are defined as being likely to play a role in the pathogenesis of the disease due to some obvious association. For example, Ehlers-Danlos syndrome can be caused by mutations in a collagen gene, and is also associated with aneurysm formation.[25] Therefore, it is reasonable to study collagen as a candidate gene in patients with aneurysms but without Ehlers-Danlos syndrome. Multiple sclerosis (MS) has long been associated with the abnormal destruction of myelin basic protein (MBP). It seems reasonable to study MBP as a candidate gene in families with MS (in fact, this was done). [26]

The Human Genome Project (HGP) has been tremendously helpful for identifying disease genes, due to two main contributions. First, the HGP has identified hundreds of additional polymorphic markers that can be utilized in whole genome linkage studies. Second, the HGP (and other laboratories and companies) have identified and characterized humdreds of cDNAs and genes throughout the genome. Partial sequences of such cDNAs (also termed expressed sequence tags, ESTs) are now available for thousands of genes throughout the genome.[26a] These newly discovered genes provide a rich resource of candidate genes that can be useful for mapping a disease (see below).

Using either a whole genome search with many DNA probes, or a candidate gene approach, it should be possible to map the disease locus to a chromosomal region. If a probe produces a very high LOD score without any crossovers, it is possible that the pathogenic gene is part of the probe or very near the probe. Linkage studies with additional markers near and around the positive marker would be used to confirm linkage.

If only a modestly positive LOD score (i.e., 3–4) is generated and there are a few crossovers, additional markers near this locus should be studied. The search for such markers is facilitated by using vectors capable of cloning large fragments of DNA. Vectors such as cosmids and yeast artificial chromosomes (YAC) can be used to clone DNA fragments of 40 kb to 1 Mb, respectively.[27,28] Techniques such as chromosome walking and chromosome jumping can be used to determine which cosmid or YAC clone is contiguous to the linked marker. These clones can then be studied to determine if they contain an informative polymorphism that is suitable for further linkage analysis. By repeating this approach it is possible to close in on a specific genetic locus, especially if a marker flanking the locus is available.

Some of the approaches described above use genetic linkage to close in on a pathogenic gene. However, several complicating factors may limit the usefulness of this approach. Polymorphic markers or promising candidate genes may not exist within the region of interest, and additional linkage studies can be time consuming and may not help in defining the disease locus. Also, some genes are quite large (> 1–2 million bases) or may have "hot spots" for recombination. Therefore, further linkage studies that use markers located within a large gene or unstable segment of DNA may produce misleading crossovers, further complicating the search for the pathogenic gene.

This type of approach using genetic markers to establish progressively tighter linkage is rapidly being replaced (or complemented) by a newer paradigm, the so-called positional candidate approach. The concept of positional candidate cloning makes use of the increasingly powerful genetic linkage techniques and the proliferation of cloned cDNAs throughout the genome.[26a] Several examples of positional candidate cloning

will illustrate the fidelity of this technique. Marfan's syndrome was genetically linked to long arm of CH15. In unrelated experiments, an excellent candidate gene, fibrillin, was cloned and localized to the region showing linkage to Marfan's syndrome. Researchers then investigated fibrillin as the causative gene for Marfan's syndrome based on both its location and protein characteristics. Such studies did in fact show that mutations in fibrillin were the cause of Marfan's syndrome. As another example, the gene for fibroblast growth factor receptor 3 had been cloned and localized to the end of chromosome 4. There were no known clinical implications for this gene, until familial achondroplasia was genetically linked to the same region on CH4. Further studies showed that mutations in the FGFR3 gene are the cause of achondroplasia. Many experts believe that the positional candidate approach will become the dominant technique for finding causative genes for many inherited diseases.[26a]

At this point, some researchers might suggest a different approach for identifying the pathogenic gene. If one or several cosmid or YAC clones from within the flanking markers can be identified, these clones can be studied intensely for specific genes. For example, cDNA from the tissue or organ of interest could be used to probe the cosmid or YAC clones to determine which clones possibly contain expressed genes. Smaller subclones of the positive cosmids and YACs could then be used to screen the cDNA libraries to identify and isolate the specific cDNAs that are coded for by these clones. Approaches similar to this were used to identify the cystic fibrosis gene from a sweat gland cDNA library.[29,30]

Another new approach to detect and isolate new genes is termed "exon-trapping". This technique uses the gene processing apparatus of bacteria and helper phages to identify, excise, and amplify DNA fragments likely to represent exonic regions of genes.[30a,30b] Although these techniques can be technically demanding and somewhat capricious, they can be quite effective and time-saving when they work properly.

Once a potential candidate gene is identified, it can be analyzed in several ways to determine if it has a role in causing the disease. The cDNA and genomic clones could be hybridizied against Southern blots made from other species (mouse, dog, etc.) to determine if the gene is a conserved sequence. The cDNAs could be hybridized against a Northern blot containing mRNA from diseased and control tissue, and mRNA from multiple organs. If differences in a transcript were identified (i.e., absent, different size, different intensity) this would be strong evidence that the cDNA did code for a gene that was altered in some way in the diseased tissue. The cDNA should be mapped to a specific chromosome to confirm that it is localized to the same region that showed positive linkage.

The next step would be to sequence the cDNA, identify open reading frames and determine the amino acid sequence of the putative protein. By comparing the DNA sequence and the amino acid sequence through Genbank with the sequences of other known genes and proteins, it is possible to infer some characteristics and functions of the gene.[31] For example, a protein with extensive hydrophobic regions is unlikely to pass through the cell membrane. Even if only a small portion of the amino acid sequence has homology with other known proteins, it might provide important clues about the protein's function. The neurofibromatosis 1 gene was found to have some homology to proteins that interreacted with the *ras* oncogene, suggesting it could be involved in signal regulation.[32]

The most promising cDNAs from normals and affecteds should be sequenced to determine if they contain any mutations. Several types of mutations have been identified, including insertions, deletions, inversions, duplications, and repeats (Figure 2). Since many mutations are point mutations, affecting only 1 or 2 bases, there may not be any gross differences in the size of the transcripts between normal and affect-

eds on a Northern blot. However, if such a point mutation produced a premature stop codon, than it could produce an abnormally small and dysfunctional protein. A single deletion or insertion would cause a shift in the triplet codon reading frame, resulting in a dramatically altered protein (or a premature stop codon). In other cases, a single base change (substitution) could cause a dramatic functional change in a normally sized protein (i.e., sickle cell disease), thereby producing a disease. This can occur because of the triplet codons used to translate the mRNA into a peptide. A change of one nucleotide in the DNA sequence will cause an mRNA codon to change, resulting in an entirely different amino acid being used in the peptide. The new amino acid may have dramatically different properties, affecting the folding of the protein, or its transportation through the cell, or its active site if it is an enzyme. Such drastic changes can obviously result in physiologic consequences.

The polymorphic trinucleotide repeats are a new class of mutations. A segment of DNA containing several triplet repeats (i.e., GTC, CAG) sometimes expands in successive generations and produces disease. This type of mutation was totally unexpected before it was discovered. Repeat mutations have been described in the fragile X syndrome and myotonic muscular dystrophy, and may explain the phenomena of anticipation (where successive generations exhibit earlier onset of the disease).[33] It is not clear exactly how these repeats produce the disease phenotype, but they may alter gene expression or function, since they may occur in the exonic region of the gene. A recent study found 40 human genes with polymorphic trinucleotide repeats, although not all were considered pathogenic.[34] Another unusual mutation type is a duplication, which occurs when a specific section of chromosomal DNA is present in three copies per cell (as opposed to the usual diploid or two copies per cell allotment).[35]

Mutations can also exist in intronic regions that are important for controlling the splicing of an initial RNA transcript into mature mRNA. Such intronic mutations can cause skips (where an exon is inadvertently spliced out of the mature mRNA), or can change a reading frame if an exon is not spliced correctly from an intron.[36] Even silent mutations (ones that do not change the amino acid sequence) in exons can cause exon skipping that markedly disrupts the final gene and protein product.[36a] In some cases, it appears that a mutation's effect can be determined by other DNA polymorphisms that occur nearby.[37] For example, a single mutation in the prion gene can produce Creutzfeldt-Jakob disease. However, if that mutation occurs in an individual with a specific DNA polymorphism nearby, it will produce fatal familial hypersomnia.[38] The full extent and meaning of such allelic heterogeneity is still being studied, but it may provide an important explanation for the occurrence of some complex or heterogeneous genetic diseases.

In the case of genes expressed only in the brain, it is usually not possible to obtain mRNA from large numbers of living individuals for the construction of a cDNA library and analysis of cDNA clones. However, cDNA libraries can be constructed using postmortem brain. Such libraries may lack some rare genes due to postmortem mRNA degradation, but they are usually sufficient to permit isolation and sequencing of some genes.[39] Once the coding regions for a gene are known, this information can be used to determine the gene's organization. Since

Types of mutations

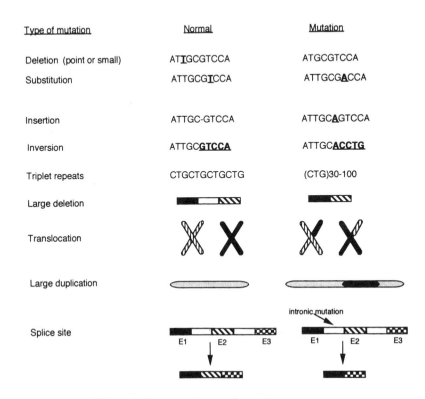

Figure 2. Common types of mutations.

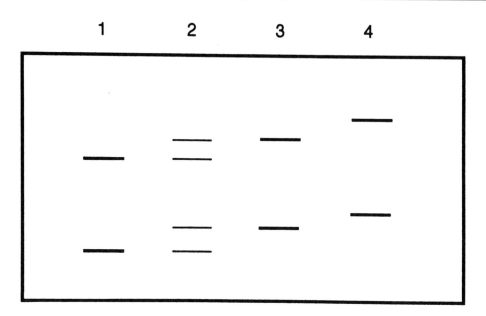

Figure 3. The concept of single-strand conformational polymorphisms (SSCP) is depicted. Specific fragments of DNA (typically exons) are PCR amplified using radiolabeled primers, then electrophoresed on special gels. The conformation of each DNA fragment is dependent on the DNA sequence, and the conformation determines the mobility of each fragment. Therefore, single mutations in a DNA fragment may produce differences in mobility that can be detected. Lane 1 shows 2 bands representing the normal alleles. Lane 2 has 4 bands representing 2 normal alleles and 2 mutated alleles (each band is one single-strand DNA fragment). Lane 3 is homozygous for the mutated allele, and lane 4 shows a different homozygous mutation. Note the ability of SSCP to distinguish normal from mutated alleles, and its ability to detect different mutations.

most genes are composed of several small exons separated by large intronic regions, it is time consuming and expensive to sequence every exon looking for a mutation. However, once the sequence of an exon and the adjacent intronic regions is known, primers can be synthesized and used to PCR amplify (see *Chapter 1*) the exons from genomic DNA (which can be isolated easily from lymphocytes). The PCR-amplified exons can be electrophoresed in denaturing acrylamide gels. Using special techniques such as denaturing gradient gel electrophoresis (DGGE) and single-strand conformational polymorphisms (SSCP), it is possible in some cases to determine which exons are likely to contain mutations (Figure 3).[40–41a] These techniques can also be used to detect mutations in cDNAs. A single-point mutation often produces a different electrophoretic pattern due to the presence of normal and abnormal DNA (assuming the mutation was present in a heterozygous state).[42] If the mutation was present in a homozygous state, the PCR-amplified exonic DNA could still have a different migration pattern compared to control DNA. By studying each exon or various cDNA fragments, it is possible to determine which ones contain mutations. These exons can then be PCR-amplified from many patients and sequenced to determine the exact nature of the mutations. Unfortunately, DGGE and SSCP do not detect every mutation in every exon. In addition, they work best on relatively small fragments of DNA (200–300 base-pairs). Another limitation of DGGE and SSCP is that they may detect polymorphisms that have no pathologic significance.[41a] In addition to these techniques, the use of automated DNA sequencing techniques and devices has simplified the screening and identification of some mutations.

Analysis of Mutations

Once a mutation is identified, how does one prove that it causes the disease under study? This can be a challenging task in many cases. Three general rules can be applied in proving a causal relationship between a mutation and a disease:

1. All affected individuals must have a mutation (not necessarily the same mutation, since different mutations can produce the same disease, as is the case with cystic fibrosis and DMD). Usually all affected members of a given family will have the same mutation.
2. The function of the normal and mutated protein must account for the pathophysiology of the disease (this can be proven in some cases by using transgenic animals, which are described below).
3. Use of the normal gene can reverse the effects of the mutated gene (this is termed complementation).

These conditions do not imply that all persons carrying the mutation must have the disease, since late-onset diseases will not be expressed in childhood. Also, these points do not exclude the existence of other mutations that can cause the same or a similar phenotype. In the case of complex diseases, mutations at more than one locus may have to be present to produce the disease phenotype, or certain environmental factors may be present that will accelerate or stifle expression of the disease.[42a]

In general, mutations can produce a disease state via three different mechanisms: (1) loss of function, (2) a dominant-negative effect, whereby a mutation alters the subunit assembly or interaction of a protein, and (3) a mutation produces a new function that has an overall deleterious effect.[42b] The experiments described below can be used to determine the most likely mechanism linking the mutation to the disease phenotype.

Gene Expression

Once a suspected pathogenic gene is identified and cloned, and mutations are detected, it is necessary to expand the study of the protein product and prove that the mutation is responsible for the disease. A first step is to clone the cDNA into an expression vector that will produce sufficient quantities of the protein for further analysis.[18,43] In addition to classic protein analysis, the protein can be used to produce monoclonal antibodies. These antibodies can be used for immunohistochemical staining of tissue sections to determine the localization of the protein. The protein could be radiolabeled and studied in various cell systems to analyze its metabolism and interaction with other proteins.[44] Using these techniques, the molecular and cellular biology of the protein can be studied.

One of the most useful tools for studying mutated and pathogenic genes is transgenic animals (Figure 4). In these experiments, the defective gene is directly injected into a fertilized gamete removed from a pregnant mouse.[45] The injected gamete is then reintroduced into another fertile mouse, producing several affected offspring in the litter. Up to 40% of the offspring will have the foreign DNA incorporated into their genome in both somatic and germ cell lines. Therefore, these transgenic mice can be used to establish whole colonies of transgenic mice. It must be proven that the transgene is expressed in the tissue(s) of interest. This can be done by looking for gene and protein expression in those tissues. Normal and affected offspring can be studied for the genetic, biochemical, and clinical effects of the mutated gene.[46] The production of a transgenic animal is an important step in studying the pathogenicity of a specific mutation and the effects of various treatments.

The study of transgenic animals has several advantages compared to *in vitro* gene expression systems. Transgenics provide an opportunity to assess cell-cell interactions and whole organism effects of a particular gene construct. This can greatly aid in

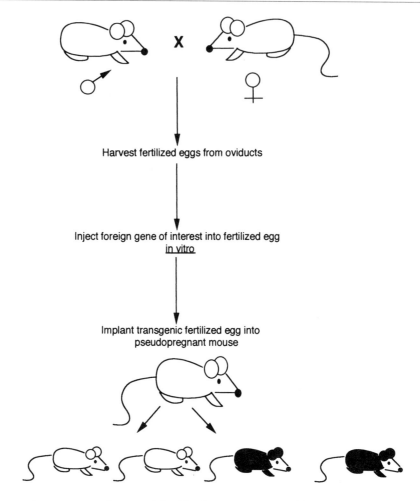

Figure 4. Transgenic mouse experiments. A fertilized ovum is removed from the oviducts of a pregnant mouse. The foreign gene is introduced *in vitro*, and the transformed ovum is implanted into a pseudopregnant mouse. Some of the offspring (shaded) will express the foreign gene. Such offspring can then be used to start a colony of transgenic mice.

understanding the pathophysiology of a mutation. But there are some limitations to the use of transgenics. The metabolism of certain proteins may differ between humans and rodents, therefore a transgenic mouse may not produce the disease of interest even if the correct gene, or a mutated gene, is introduced. Some gene sequences may not be stable when inserted into the mouse genome, making establishment of a stable transgenic colony or line difficult. Despite these problems, the use of transgenic animals will continue to be an important tool in the study of genetic diseases and treatments.

Complex Diseases

Complex diseases refer to those conditions that are not inherited or expressed as simple Mendelian traits. Most of the deadly and common diseases of our time, such as cancer, atherosclerosis, and diabetes are likely to fall under the definition of a complex disease. These diseases appear to have genetic components, but do not appear to be expressed as the result of a single gene acting alone. Several factors may contribute to a "complex" phenotype or complicate the study of these complex traits. These factors include the effects of multiple genes, environmental factors, complex age-of-onset curves, primary and secondary risk factors, phenocopies, genetic heterogeneity, and allelic heterogeneity.[18,42a,47,48]

One of the most common and clinically important complex diseases is atherosclerosis. Atherosclerosis is clearly not due to the actions of a single gene. Perhaps it is more accurate to view atheroscleorsis as a synthetic trait, in that several other traits or risk factors have to act together (in many cases) to produce the larger trait. This further complicates things, since each individual trait, such as hypertension or hyperlipidemia, is in itself determined by the interaction of one or more genes and environmental factors. Examples of such multigenic systems include lipid and lipoprotein metabolism, fibrinogen synthesis, thrombosis factors, blood sugar control, and probably several undiscovered factors. Environmental factors such as diet, smoking, and exercise are likely to play an important role in modifying certain genetic traits. Other noninherited factors may also precipitate or accelerate the formation of atherosclerotic lesions. These issues are discussed in more detail in *Chapter 6*.

Several approaches exist for the study of complex genetic diseases. One methodology uses advanced linkage techniques such as sib-pair analysis or APM analysis.[20,21] These techniques have the advantage that a mode of inheritance does not have to be known, and only known affecteds need to be studied. By looking for common alleles among affecteds, it is possible to calculate the probability that a specific allele is close to a locus that has a role in determining disease susceptibility. However, diseases caused by several multigenic traits with prominent environmental factors may be too complex for this type of analysis.

Another technique is to look for a locus that is responsible for a significant proportion of the risk for a well-defined trait. The identification of such a major susceptibility locus for a complex disease would be a significant accomplishment. In the case of some cancers, several genes and proteins

have been identified that appear to play a significant role in defining cancer risk and even in producing some types of tumors.[49,50] The identification of similar genes for various subtypes of stroke, or traits determined to be risk factors for stroke, would be highly significant.

Another approach for studying complex diseases is to identify families in which there are many affected individuals who display the disease at a very young age. The study of such families may simplify the confounding effects of other risk factors, while highlighting the effects of a major susceptibility locus.[51] The early onset of the disease could reduce the influence of other metabolic and environmental risk factors, especially for a disease that typically has a late age of onset. If large numbers of affected individuals could be studied from such families, they might indicate that the major susceptibility loci behave as Mendelian traits, which could make linkage studies easier. One drawback to this approach is that the findings in such studies may not reflect the abnormalities present in all affected individuals (as is the case for early-onset familial AD).

Yet another line of investigation that is sometimes fruitful with genetic diseases (complex and otherwise) is to identify families or individuals that have the disease of interest plus other obvious phenotypic abnormalities. In many cases, this clustering of several obvious abnormalities indicates the presence of a gross chromosomal rearrangement such as a translocation or deletion. If confirmed by karyotyping, this finding can provide important information as to which chromosome and subchromosomal region is likely to contain the gene of interest or a major susceptibility gene.[52] Diseases with high rates of spontaneous mutations (such as DMD) are sometimes associated with large chromosomal abnormalities such as large deletions.

The identification of a pathogenic gene along with an understanding of the pathophysiology of the disease is a critical step in planning and approaching therapy for inherited diseases. In addition, once a pathogenic gene is found, it makes accurate genetic counseling possible. The issues of genetic counseling and gene therapy are discussed in more detail in *Chapters 16 and 17*, respectively.

Summary

This chapter has summarized and applied the concepts and techniques of *Chapters 1 and 2* to demonstrate how disease-causing genes can be defined, isolated, and studied. Since many of the most common and deadly human diseases are likely to be due to the interactions of several genes and environmental factors, such complex diseases require a somewhat different approach for the identification of major susceptibility genes. However, the tools and techniques exist to make the study of such diseases timely and productive.

Acknowledgements: The author would like to thank Rohan deSilva, PhD, and Susan H. Alberts for their assistance in reviewing and editing this chapter.

References

1. Ingram V. Gene mutations in human hemoglobin: The chemical difference between normal and sickle cell hemoglobin. Nature 1957;180:326–328.
2. Koenig M, Hoffman E, Bertelson C, et al. Complete cloning of the Duchenne muscular dystrophy (DMD) cDNA and preliminary genomic organization of the DMD gene in normal and affected individuals. Cell 1987;50:509–517.
3. Goldgaber D, Lerman M, McBride O, et al. Characterization and chromosomal localization of a cDNA encoding brain amyloid in Alzheimer's disease. Science 1987;235:877–880.
4. Kamino K, Orr H, Payami H, et al. Linkage and mutational analysis of familial Alzheimer disease kindreds for the APP gene region. Am J Hum Genet 1992;51:998–1014.
4a. Sherrington R, Rogaev EI, Liang Y, et al. Cloning of a gene bearing missense mutations in early-onset familial Alzheimer's disease. Nature 1995;375:754–760.
4b. Sherrington R, Froelich S, Sorbi S, et al. Alzheimer's disease associated with mutations in presenilin 2 is rare and variably penetrant. Hum Mol Genet 1996;5(7):985–988.
5. Weatherall DJ, Clegg JB, Higgs DR, et al. The hemoglobinopathies. In Scriver C, Beaudet A, Sly W, et al (eds). The Metabolic Basis of Inherited Disease. New York: McGraw-Hill, 1995, 3417–3484.
6. Glenner G, Wong C. Initial report of the purification and characterization of a novel cerebrovascular amyloid protein. Biochem Biophys Res Comm 1984;120:885–890.
7. Kitaguchi N, Takahashi Y, Tokushima Y, et al. Novel precursor of Alzheimer's disease amyloid protein shows protease inhibitory activity. Nature 1988;331:530–532.
8. Tanzi R, Gusella J, Watkins P, et al. Amyloid protein gene: cDNA, mRNA distribution, and genetic linkage near the Alzheimer locus. Science 1987;235:880–882.
9. Murrell J, Farlow M, Ghetti B, et al. A mutation in the amyloid precursor protein associated with hereditary ALzheimer's disease. Science 1991;254:97–99.
10. Goate A, Chartier-Harlin C, Mullan M, et al. Segregation of a missense mutation in the amyloid precursor protein gene with familial Alzheimer's disease. Nature 1991;349:704–706.
11. Levy E, Carman M, Fernandez-Madrid I. Mutation of the Alzheimer's disease amyloid gene in hereditary cerebral hemorrhage Dutch type. Science 1990;248:1124–1126.
12. St George-Hyslop P, Tanzi R, Polinsky R, et al. The genetic defect causing familial Alzheimer's disease maps on chromosome 21. Science 1987;235:885–890.
13. Pericak-Vance M, Yamaoka L, Haynes C, et al. Genetic linkage studies in Alzheimer's disease families. Exp Neurol 1988;102:271–279.
14. Pericak-Vance M, Bebout J, Gaskell P, et al. Linkage studies in familial Alzheimer's disease: Evidence for chromosome 19 linkage. Am J Hum Genet 1991;48:1034–1050.
15. Schellenberg G, Bird T, Wijsman E, et al. Absence of linkage of chromosome 21 q21 markers to familial Alzheimer's disease in autopsy-documented pedigrees. Science 1988;241:1507–1510.
16. St George-Hyslop P, et al. Genetic linkage studies suggest that Alzheimer's disease is not a single homogeneous disorder. Nature 1990;347:194–197.
17. Gelehrter T, Collins F. Principles of Medical Genetics. Baltimore, MD: Williams and Wilkins, 1990.

18. Watson J, Gilman M, Witkowski J, et al. Recombinant DNA. New York: WH Freeman, 1992.

19. Weeks D, Lange K. The affected-pedigree-member method of linkage analysis. Am J Hum Genet 1988;42:315–326.

20. Risch N. Linkage strategies for genetically complex traits. II. The power of affected relative pairs. Am J Hum Genet 1990;46:229–241.

21. Dawson D, Kaplan E, Elston R. Extensions of sib-pair linkage tests applicable to disorders characterized by delayed onset. Genet Epidemiol 1990;7:453–466.

22. Weber J, May P. Abundant class of human DNA polymorphisms which can be typed using the polymerase chain reaction. Am J Hum Genet 1989;44:388–396.

23. Weber J. Informativeness of human (dC-dA)n,(dT-dG)n polymorphisms. Genomics 1990;7:524–530.

24. Hudson T, Elgelstein M, Lee M, et al. Isolation and chromosomal assignment of 100 highly informative human simple sequence repeat polymorphisms. Genomics 1992;13:622–629.

25. Rubinstein M, Cohen N. Ehlers-Danlos syndrome associated with multiple intracranial aneurysms. Neurol 1964;14:125–132.

26. Tienari P, Wikstrom J, Sajantila A, et al. Genetic susceptibility to multiple sclerosis linked to myelin basic protein. Lancet 1992;340:987–991.

26a. Collins FS. Positional cloning moves from perditional to traditional. Nat Genet 1995;9:347–350.

27. Collins J, Hohn B. Cosmids: A type of plasmid gene cloning vector that is packageable in vitro in bacteriophage heads. Proc Natl Acad Sci 1978;75:4242–4246.

28. Burke D, Carle G, Olsen M. Cloning of large segments of exogenous DNA into yeast by means of artificial chromosome vectors. Science 1987;236:806–812.

29. Rommens J, et al. Identification of the cystic fibrosis gene: Chromosome walking and jumping. Science 1989;245:1059–1065.

30. Riordan J, Rommens J, Kerem B-S, et al. Identification of the cystic fibrosis gene: Cloning and characterization of complementary DNA. Science 1989;245:1066–1073.

30a. Chen H, Chrast R, Rossier C, et al. Cloning of 559 potential exons of genes of human chromosome 21 by exon trapping. Genome Res 1996;6:747–760.

30b. Krizman DB. Gene isolation by exon trapping. Methods Mol Biol 1997;68:167–182.

31. Smith T. The history of the genetic sequence databases. Genomics 1990;6:701–707.

32. Xu G, O'Connell P, Viskochil D, et al. The neurofibromatosis type 1 gene encodes a protein related to GAP. Cell 1990;62:599–608.

33. Caskey C, Pizzuti A, Fu Y-H, et al. Triplet repeat mutations in human disease. Science 1992;256:784–789.

34. Riggins G, Lokey L, Chastain J, et al. Human genes containing polymorphic trinucleotide repeats. Nature Genetics 1992;2:186–191.

35. Lupski J, Montes de Oca-Luna R, Slaugenhaupt S, et al. DNA duplication associated with Charcot-Marie-Tooth disease type 1A. Cell 1991;66:219–232.

36. Kontusaari S, Tromp G, Kuivaniemi H, et al. Inheritance of an RNA splicing mutation (G^{+1} $_{ivs20}$) in the type III procollagen gene (COL3A1) in a family having aortic aneurysms and easy bruisability: Phenotypic overlap between familial arterial aneurysms and Ehlers-Danlos syndrome type IV. Am J Hum Genet 1990;47:112–120.

36a. Liu W, Qian C, Francke U. Silent mutation induces exon skipping of fibrillin-1 gene in Marfan syndrome. Nat Genet 1997;16:327–328.

37. Hsaio K, Meiner Z, Kahana E, et al. Mutation of the prion protein in Libyan Jews with Creutzfeldt-Jakob disease. N Eng J Med 1991;324:1091–1097.

38. Goldfarb L, Petersen R, Tabaton M, et al. Fatal familial insomnia and familial Creutzfeldt-Jakob disease: Disease phenotype determined by a DNA polymorphism. Science 1992;258:806–808.

39. Alberts M, Ioannou P, Deucher R, et al. Isolation of a cytochrome oxidase gene overexpressed in Alzheimer's disease brain. Mol Cell Neurosci 1991;3:461–470.

40. Gray M. Detection of DNA sequence polymorphisms in human genomic DNA by using denaturing gradient gel blots. Am J Hum Genet 1992;50:331–346.

41. Orita M, Suzuki Y, Sekiya T, et al. Rapid and sensitive detection of point mutations and DNA polymorphisms using the polymerase chain reaction. Genomics 1989;5:874–879.

41a. Grompe M. The rapid detection of unknown mutations in nucleic acids. Nat Genet 1993;5:111–117.

42. Myers R, Maniatis T, Lerman L. Detection and localization of single-base changes by denaturing gradient gel electrophoresis. Methods Enzymol 1987;155:501–527.

42a. Lander ES, Schork NJ. Genetic dissection of complex traits. Science 1994;265:2037–2048.

42b. Willems PJ. Dynamic mutations hit double figures. Nat Genet 1994;8:213–215.

43. Rosenberg A, Lade B, Chui D, et al. Vectors for selective expression of cloned DNAs by T7 RNA polymerase. Gene 1987;56:125–135.

44. Milstein C. Monoclonal antibodies. Sci Am. 1980;243:66–74.

45. Gordon J, Scangos G, Plotkin D, et al. Genetic transformation of mouse embryos by

microinjection of purified DNA. Proc Natl Acad Sci 1980;77:7380–7384.

46. Jaenisch R. Transgenic animals. Science 1989;240:1468–1474.

47. Lander E. Mapping complex genetic traits in humans. In Davies K (ed). Genome Analysis: A Practical Approach. Oxford: IRL Press, 1988, 171–187.

48. Risch N. Linkage strategies for genetically complex traits. I. Multilocus models. Am J Hum Genet 1990;46:222–228.

49. Weinberg R. Oncogenes, anti-oncogenes, and the molecular bases of multistep carcinogenesis. Cancer Res 1989;49:3713–3721.

50. Levine A, Momand J, Finlay C. The p53 tumour suppressor gene. Nature 1991;351:453–456.

51. Alberts M. Genetic aspects of cerebrovascular disease. Curr Concepts of Cerebrovascular Disease and Stroke 1990;25:25–29.

52. Kunkel L, Monaco A, Middlesworth W, et al. Specific cloning of DNA fragments absent from the DNA of a male patient with an X-chromosome deletion. Proc Natl Acad Sci 1985;82:4778–4782.

Part II

Genetics of Risk Factors for Cerebrovascular Disease

Chapter 4

Genetics of Hypertension

*Laura P. Svetkey, MD, Edmond O'Riordan, MRCPI,
Peter J. Conlon, MB, Osemwegie Emovon, MD*

Hypertension is one of the most powerful risk factors for ischemic and hemorrhagic stroke. The age-adjusted risk of stroke in persons with definite hypertension (BP>160/95 mmHg) is more than four times that of persons with normal blood pressure. Even among individuals with "borderline hypertension" or mildly elevated blood pressure, stroke risk is twice that of normotensive individuals. More than half of all strokes can be directly attributed to hypertension.[1]

Fortunately, effective treatment of hypertension substantially reduces the risk of stroke. Over the past 20 years stroke deaths have declined by 57%, in part due to improved detection and treatment of hypertension.[2] Hypertension-related strokes, however, remain prevalent, suggesting that primary prevention of hypertension might be an effective strategy for primary prevention of cerebrovascular disease. A preventive strategy requires a better understanding of the pathophysiology of essential hypertension, including its genetic basis. Therefore, this chapter will review the role of genetics in determining the trait of essential hypertension.

Genetic Epidemiology

Blood pressure is a continuous trait that is normally distributed. As blood pressure increases, so does the risk of cardiovascular and cerebrovascular morbidity and mortality, with no clear threshold between normal and elevated levels.[2] By convention, however, we define a threshold in order to label some portion of the population "hypertensive". Regardless of where we set this somewhat arbitrary threshold, hypertension aggregates in families. In Table 1, for instance, hypertension is defined as a blood pressure greater than 140/80 mmHg.[3] If neither parent exceeds this threshold, 3.1% of offspring exceed it in adulthood. In contrast, if both parents have hypertension, almost half of the adult offspring also develop this condition. These findings persist regardless of the threshold selected to distinguish hypertensive from normotensive individuals.

When no threshold is imposed and blood pressure is considered a continuous variable, it is also highly correlated among family members. This familial aggregation is found throughout the entire range of blood pressure and in virtually all popula-

From Alberts MJ (ed). *Genetics of Cerebrovascular Disease*. Armonk, NY: Futura Publishing Company, Inc., © 1999.

Table 1
Risk of hypertension (BP > 140/80 mmHg)
in 1524 Individuals in 227 Families

Number of Parents with HTN	Proportion of Offspring with HTN
0	3.1%
1	28.3%
2	45.5%

HTN, hypertension.

tions that have been studied.[4] Of course, familial aggregation does not necessarily imply genetic influence. Blood pressure aggregation may be attributed to the fact that family members share a common environment and develop similar dietary and health habits. However, environmental influences are not likely to entirely explain the familial aggregation of blood pressure since this phenomenon is independent of sodium consumption, body weight, alcohol consumption, and smoking.[5] It is clear that biologically related individuals also share influences on blood pressure that are genetic.

There are two types of studies that substantiate this concept: adoption studies[6] and twin studies.[7-9] Adoption studies help to determine the degree to which a trait is genetically determined because, within a family, individuals related by adoption are less similar to each other genetically than are biologically related individuals. Therefore, if a trait is in part genetically determined, we expect less correlation for this trait in adoptive than in biological relationships. Table 2 demonstrates that biologically unrelated individuals (parents compared to each other and parents compared to adopted children) are less similar with respect to blood pressure than biologically related individuals (parents compared to biological children and biological siblings compared to each other).

The results of adoption studies, however, could be explained by environmental effects that occur in utero or very early in life (i.e., before adoption). In contrast, twins share environmental influences beginning with conception. Twin studies, therefore, may be more helpful in determining the degree to which blood pressure is genetically determined. Twin methods rely on the fact that monozygotic twins are genetically identical while dizygotic twins on average are homologous at 50% of their genetic loci. Presuming that the twins being studied were raised together and therefore share the same environment, correlation between monozygotic twin members should be high for a genetic trait. Because of the possibility of genetic differences between members of dizygotic twin pairs, however, a genetically determined trait will be less highly correlated between members of these pairs. Table 3 demonstrates that both systolic and diastolic blood pressure are more highly correlated in monozygotic twin pairs than in dizygotic twins. Among children, the correlation coefficient for systolic blood pressure among monozygotic twin pairs is as high as 0.85, but only 0.50 for dizygotic twins.[8] Similarly, diastolic blood pressure is more similar in pairs of monozygotic than dizygotic twins. Interestingly, in this study heritability of blood pressure is higher in females than in males. Lower correlation for systolic blood pressure among adult twins suggests that the increase in systolic blood pressure with age may be independent of earlier genetic influences on blood pressure.[9] These results lead to estimates of heritability of blood pressure of 0.60–0.70 for systolic blood pressure and 0.52–0.61 for diastolic blood pressure, suggesting that more than half of the variability of both systolic and diastolic blood pressure in Western populations is genetically determined.

Having established that there is a genetic component to essential hypertension, many investigators have sought to identify the hypertension gene(s). As mentioned previously, however, hypertension represents the upper end of the distribution of blood pressure, which is a quantitative trait. Thus, the genetic basis of hypertension can only be elucidated by studying the genes that influence this distribution (i.e., genes that affect the variability in blood pressure within a popu-

Table 2
Aggregation of Blood Pressure (BP) within Adoptive families[6]

Correlation	Systolic BP	Diastolic BP
Between parents	0.16	0.17
Between mother and biological children	0.27	0.26
Between mother and adopted children	0.08	0.10
Between biological siblings	0.38	0.53
Between adoptive siblings	0.16	0.29

lation). The quantitative (as opposed to qualitative) nature of blood pressure as well as the large number of biological factors that influence it make it unlikely that there is a single "hypertension gene". Indeed, in animal models of genetic hypertension, the mode of inheritance is consistent with six to ten genes having major effects.[10] If human essential hypertension is analogous, there is the possibility that we can identify a limited number of genes that account for most of the genetic basis of blood pressure variability.

Candidate Genes for Essential Hypertension

Hypertension may represent the common endpoint of several pathophysiologic pathways, and therefore may be due to an interacting family of genetic abnormalities. The search for the genetic basis of hypertension may be best served by the identification of hypertension subtypes that are likely to have a common etiology (see "Hypertension Subtypes" later in the chapter), or by the search for pedigrees with Mendelian

forms of hypertension. Alternatively, many investigators have focused on genes that regulate blood pressure and therefore may contribute to the broad phenotype, hypertension. Several candidate genes may play an etiologic role in the development of hypertension. These candidates are related to the major determinants of blood pressure, namely intravascular volume, cardiac output, and peripheral vascular resistance. These determinants are influenced by environmental factors, such as dietary salt intake, but are also biologically regulated. Each point of biological regulation is determined by genes encoding a particular protein through which regulation occurs, and thus each of these genes is a potential candidate in the pathophysiology of hypertension.

Many of the candidate genes that have been evaluated in hypertensive animals and/or humans are summarized in Table 4. In this discussion we will focus on the two pathways that have received the most attention: the renin-angiotensin-aldosterone axis and the cell membrane transport systems that regulate intracellular cation concentration.

Table 3
Blood Pressure Correlations among Monozygotic and Dizygotic Twins

Population	Systolic			Diastolic		
	MZ	DZ	h^2	MZ	DZ	h^2
Children (n=200)[8]	.85	.50	.70	.80	.54	.52
Adults (n=514)[9]	.55	.25	.60	.58	.27	.61

Renin-Angiotensin-Aldosterone (RAA) System

This hormone system regulates blood pressure through several interrelated mechanisms (Figure 1). Angiotensinogen substrate is cleaved to angiotensin I by renin. Angiotensin I is converted to angiotensin II (AII) by angiotensin converting enzyme (ACE). The action of AII on vascular smooth muscle cell receptors promotes increased blood pressure via vasoconstriction. In addition, AII stimulates aldosterone release, which elevates blood pressure via renal sodium retention.[11] Overactivity at any of these steps (or inhibition of the negative feedback loop by which elevated blood pressure reduces renin release) would promote elevation of blood pressure. Interest in the renin-angiotensin-aldosterone system as a potential genetic determinant of blood pressure is justified by evidence that this system is altered not only in essential hypertension but also in normotensive individuals with a parental history of hypertension.[12,13] Thus, the genes encoding angiotensinogen, renin, angiotensin converting enzyme and the angiotensin II receptors are all candidate genes for hypertension.

Angiotensinogen Gene

Increased activity of angiotensinogen would lead to increased substrate (AI) for angiotensin converting enzyme, the rate-limiting step in the activation of the vasoconstrictor angiotensin II.[14] The gene encoding angiotensinogen is a reasonable candidate in the etiology of essential hypertension both because this gene is expressed in tissues directly involved in blood pressure regulation,[15] and because plasma concentrations of angiotensinogen are elevated in subjects with hypertension or a family history of hypertension.[16] Thus, hypertension may be due to genetic alterations leading to increased transcription of angiotensinogen.

In order to develop a genetic model of overexpressed angiotensinogen, Kimura et al.[17] created transgenic mice, which contain the entire rat angiotensinogen (ANG) gene. Overexpression of the ANG gene in these transgenic mice was associated with elevated circulating angiotensin II and hypertension.[17] In another transgenic model, mice carrying the rat angiotensinogen gene developed hypertension only when the transgenic animals also carried the rat renin gene.[18]

Despite these apparent inconsistencies in transgenic models, the angiotensinogen gene was implicated in a human linkage study.[19] Taking advantage of a highly polymorphic dinucleotide repeat element,[20] the angiotensinogen gene was studied in 347 affected sib pairs in two centers. In subjects studied in Paris, but not in those studied in Utah, linkage was statistically significant, with an excess of 7.7% of alleles shared by hypertensive siblings. When data from both populations were pooled, evidence for linkage was stronger. Furthermore, in a subset with severe hypertension, there was a 15–18% excess of shared alleles.

Two of the 15 allelic variants identified in these sib pairs, both involving amino acid substitutions, were associated with the presence of hypertension, and with higher plasma levels of angiotensinogen. Since the pioneering work of Jeaunemaitre et al.,[19] published studies have produced conflicting results. These have shown marked ethnic variation in the frequency of the 235T allele.[20a,20b] Even in studies limited to Caucasian populations, results have been evenly divided between those that have suggested[20c,20d] an association and those that have rejected it.[20e,20f] A recent meta analysis[20g] highlighted major methodological differences which, although making interpretation of meta analysis difficult, possibly explained some of the discrepancies observed. The principal problems included inadequate definitions of cases or controls, large anthropometric differences between the subject groups, and possible selection bias. Significantly, however, sample size was similar between the positive and negative studies, and analyses of the negative studies suggest they were powered to detect differences. The meta analysis did find a weak but

Table 4
Candidate Genes for Essential Hypertension (Partial List),
with Location on the Human Genome

Gene Product	Chromosomal Location	Model	Method	Results
Renin-angiotensin system				
Renin[27]	1p25–q32	Dahl-S	Segreg	Segregation of BP with Dahl-S allele
Renin[32]		Human	Sib-pairs	No linkage
ACE[34]	17q23	SHR	Mapping	Linkage on chromosome 10
ACE[36]		Human	Assoc	RFLP associated w/EHT
Angiotensinogen	1q42–43			
Angiotensin II receptor	3q21–25			
Cell membrane transport systems				
Na$^+$, K$^+$-ATPase				
alpha$_1$ subunit[57]	1p13–11	Dahl-S	Assoc	Mutation unique to Dahl-S
alpha$_2$ subunit[105]	1cen-q32	SHR	Assoc	RFLP unique to SHR
alpha$_2$ subunit[106]		Human	"Segreg"	Discordant segregation
beta$_1$ subunit[106]	1q23–q25	Human	"Segreg"	Discordant segregation
Na$^+$/H$^+$ antiporter[107,108]	1p36–p35	Human	Sib-pairs	No linkage
Other systems				
Calmodulin[109]	14q24–31 2p21, 19q13	SHR	Sequencing	No changes
Calcium ATPase	1q25–32 12q21–23			
ANP[107]	1p36	Human	Sib-pairs	No linkage
ANP[31]		Human	Assoc	No assoc of RFLP w/EHT
Phospholipase C[110]	20q12–13	SHR	Assoc	No RFLP unique to SHR
Kallikrein[111]	19q13	SHR	Segreg	Segreg of BP with SHR allele
Insulin receptor[112]	19p13	Human	Assoc	Possible assoc of RFLP w/EHT
Insulin[112]	11p15	Human	Assoc	No assoc of RFLP w/EHT
Endothelin[113]	6p24–23	Human	Assoc	No assoc of RFLP with BP
Endothelin[113]		Human	Twin	No assoc of RFLP with within pair BP variability
Adrenergic receptors				
alpha$_1$	5q31–32			
alpha$_2$[114]	2, 4,10q24–26	Human	Assoc	No assoc of RFLP w/EHT
beta$_1$	10q23–25			
beta$_2$	5q31–32			

Dahl-S, Dahl salt sensitive rat; segreg, segregation analysis; BP, blood pressure; ACE, angiotensin converting enzyme; SHR, spontaneously hypertensive rat; assoc, association study; RFLP, restriction fragment length polymorphism; EHT, essential hypertension; Na$^+$, K$^+$-ATPase, sodium-potassium-adenosine triphosphatase; Na$^+$/H$^+$ antiporter, sodium-proton antiporter; ANP, atrial natriuretic protein.

positive association between 235T polymorphism and hypertension.[20h] A recent study showed almost complete linkage between a guanine and adenosine transition −6 bp upstream from the initial transcription site. This substitution affects the gene transcription rate and may explain the elevated angiotensinogen levels seen in those with the 235T polymorphism. Nonmodulation[20i] has also been demonstrated in this population, which may cause up-regulation of other mediators in the renin-angiotensin-aldosterone axis and hence become implicated in the genesis of hypertension. Other work[20g] is showing a more definite association between 235T polymorphism and those with more severe hypertension and earlier onset of disease. Additional investigators[20j,21] have suggested that the angiotensinogen gene will only be associated with hypertension in males as estrogens may modify gene expression.

Furthermore, the significance of the elevated levels of circulating angiotensinogen is unclear since the presence of this hormone is affected by dietary sodium intake[22] and by secondary hypertension.[23] The role of angiotensinogen gene in human blood pressure regulation may be further elucidated by studies of transgenic mice harboring the human angiotensinogen gene.[24]

Renin Gene

In rat models of hypertension it is clear that aberrant expression of renin can cause severe hypertension. This phenomenon has been demonstrated in transgenic rats harboring the mouse renin gene (Ren-2), a model that develops severe hypertension. The exact pathophysiology of hypertension in this model is unclear since renin content was not elevated in either the plasma or the kidneys.

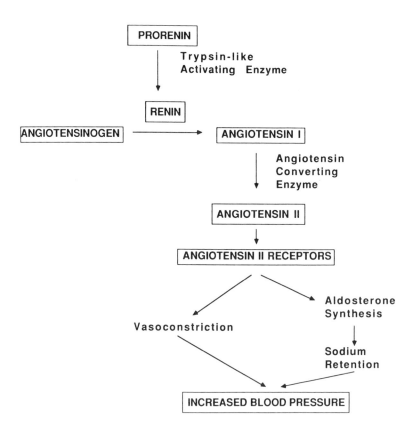

Figure 1. Renin-angiotensin-aldosterone axis.

Nonetheless, these transgenic rats are extremely sensitive to the blood pressure lowering effects of an angiotensin converting enzyme inhibitor, suggesting that the transgene is causing hypertension through alterations in the renin-angiotensin-aldosterone (RAA) system at a local level.[25] In support of this hypothesis, there was increased angiotensin II release from vascular smooth muscle of these animals, apparently due to effects of local renin production.[26]

The role of the renin gene has also been investigated in genetically hypertensive rat strains. In inbred Dahl salt-sensitive rats, Rapp and colleagues[27] demonstrated that a Bgl-II polymorphism in the first intron of the renin gene cosegregates with blood pressure levels. When salt-sensitive rats were bred with salt-resistant rats and the offspring were exposed to a high sodium diet, the renin allele from the salt-sensitive animals (the S-allele) was associated with a dose-dependent increment in blood pressure. In other words, rats in the F2 generation with one "dose" of the S-allele (i.e., rats with one renin allele from the salt-sensitive rat and one from the salt-resistant rat) had blood pressure 10 mmHg higher than rats with no S-alleles (i.e., homozygous for renin alleles derived from the salt-resistant rat), and 10 mmHg lower than those with two "doses" of the S-allele (i.e., homozygous for renin alleles derived from the salt-sensitive rat).

In the Dahl rat, as in the transgenic model mentioned above, the etiologic role of the renin gene is unclear since this model of essential hypertension is a low renin model, the polymorphism is in a noncoding region of the renin gene, and the polymorphism is not associated with a change in the protein sequence of the renin molecule.[27] Nonetheless, the renin gene has also been implicated in another genetic model that differs from the Dahl salt-sensitive rat in that elevated blood pressure does not depend on high sodium consumption, namely the spontaneously hypertensive rat (SHR).[28] In this experiment, inbred SHRs crossed with the normotensive Lewis rat resulted in an F2 generation in which homozygotes for

the SHR renin gene had higher blood pressure than rats homozygous for the renin gene from the Lewis rat.

Like the Dahl rat mentioned previously, the SHR is not generally considered a high renin form of hypertension, but the authors speculated that the expression of the renin gene in the kidney of the SHR leads to abnormal function of the RAA system locally. If renin is responsible for hypertension in both the Dahl salt-sensitive rat and the SHR, (models of genetic hypertension that have quite different natural histories and hemodynamic profiles), it must act by different mechanisms. This point is highlighted by the fact that, unlike the Dahl-S × Dahl-R offspring, the SHR × Lewis F2 rats that were heterozygous at the renin locus had blood pressure that was as high as that of rats that were homozygous for the SHR renin gene.[28]

In addition to the possibility that the renin gene is responsible for elevated blood pressure in hypertensive rats, these data are also consistent with a hypertension gene closely linked to the renin gene. In any case, regardless of what truths ultimately emerge concerning the renin gene in hypertensive rats, it is unclear to what extent findings in these animals reflect pathophysiology in human hypertension. The animal studies suggesting a role for the renin gene are not borne out in human studies. For instance, Soubrier and colleagues studied restriction fragment length polymorphisms (RFLPs) of the renin gene in an association study comparing 102 hypertensives with hypertensive parents to 120 normotensives with normotensive parents. Allelic frequencies did not differ between these two groups of French subjects.[29] These findings were subsequently replicated in an Australian population,[30] and in a population survey in Great Britain.[31] In addition, linkage analysis in a single multigenerational pedigree with multiple hypertensive members was negative,[30] as was an affected sib-pair linkage analysis.[32]

These studies argue against a role for the renin gene in the pathogenesis of essential hypertension, but these conclusions may be limited by four methodological problems. First, none of the human studies cited above employed statistical methods accommodating the variable age of expression of hypertension. Second, each of these studies considered hypertension as a single clinical entity. It is possible that the renin gene is important in some forms of essential hypertension but not in others. Third, subjects were not evaluated with respect to renin profile, which might have identified a subset in whom renin genotype is associated with hypertension. Lastly, two of the three negative association studies and both linkage studies discussed above included only Caucasian subjects. Renin genes may be more important in the pathogenesis of hypertension in nonwhites. In support of this hypothesis, the British study mentioned above[31] was negative when Afro-Caribbean and white European subjects were analyzed together. But when Afro-Caribbeans were considered separately, there was a significant association between blood pressure quintile and a Bgl-I polymorphism of the renin gene.[31] These data suggest that the renin gene may be pathogenetic in a subgroup of hypertensives, perhaps those who are salt sensitive. Therefore, the issue of genetic heterogeneity may be of key importance in studying the etiology of hypertension. Further evidence that ethnicity may be important in the effect of the renin gene, is provided by a recent study[32a] showing a significant association between Hind III polymorphisms and essential hypertension. The same study however, failed to show a difference between the frequency of Bg1 I polymorphism in hypertensives and normotensives. Currently the overall results in this area are conflicting.

Angiotensin Converting Enzyme (ACE) Gene

In stroke-prone spontaneously hypertensive rats (SHRSP), Jacob and colleagues[33] identified a gene that has a major effect on blood pressure. Using genetic mapping, they identified a locus (RD17) on rat chromosome 10 that was tightly linked to sodium-loaded blood pressure (LOD score 5.10). Genotype at this locus accounted for nearly 10% of the variance in baseline diastolic blood pressure and for nearly 20% of the variance in both systolic and diastolic blood pressure in sodium-loaded animals. This gene demonstrated close linkage with the gene encoding ACE (0% recombinations out of 224 crossovers).[33] These findings were confirmed independently in another laboratory studying SHRSP rats,[34] and in a cross between normotensive rats and Dahl salt-sensitive (Dahl-S) rats.[35] Subsequently however, a further cross-breeding experiment[35a] involving SHRSP and Wistar Kyoto rats(WKY-HD), using a 6 cM long homologous segment introgressed in chromosome 10, demonstrated that although plasma ACE activity was linked to genetic effect at the ACE gene locus the ACE locus did not cosegregate with blood pressure.[35b]

In humans, Zee et al. compared ACE genotype in 80 hypertensives with hypertensive parents and 93 normotensives with normotensive parents. Hypertensives were significantly more likely than normotensives to have an insertion allele in an insertion/deletion polymorphism in intron 16 of the ACE gene (56% versus 41%, $p<0.01$).[36]

Subsequently, other studies have had conflicting results, one of the most extensive failing to show linkage among hypertensive sib pairs in Utah.[36a] Several hypotheses have been considered, including ethnicity, to explain the difference in these results. Other investigators have attempted to link ACE polymorphisms with diabetic nephropathy,[36b] ischemic heart disease,[36c] left ventricular hypertrophy,[36d] and renal artery stenosis.[36e] Although these associations appeared promising initially, further larger studies have been unable to repeat their results.[36f,36g,36h] Unfortunately, studies of this kind are subject to type I errors and reporting bias.

There have also been attempts to determine if combinations of gene polymorphisms in the renin angiotensin system could explain differences, particularly those attributed to ethnicity. Two separate investigators recently looked for associations between hypertension and polymorphisms of both angiotensinogen and ACE-1 gene in Czech[36i] and German[36j] populations. Unfortunately, conflicting answers were obtained. These might be explained by some methodological differences and further studies may give a more definitive picture.

Currently, the results in this area of research are confusing and the lack of 'intermediate phenotypes' is the principal difficulty. This will need to be resolved before these results can be properly interpreted.[36k]

Other Genes in the RAA System

The renin-angiotensin system exerts its influence through stimulation of angiotensin II receptors. The genes encoding these receptors, therefore, are rational candidates for essential hypertension. There are at least three receptors (Ia, Ib, and II). Not surprisingly given the findings in the other areas, promising results in animal experiments have proved to be difficult to transfer to the human population. Initial associations shown by Hingorami et al.[36l] were not reproducible in a more homogenous Japanese population in a study which failed to find an association between AT1 and AT2 receptor gene variants and hypertension.[36m] With further characterization of these genes, the role of angiotensin II receptor genes in the development of hypertension will be elucidated.

Transmembrane Transport Systems

In an attempt to explain the role of sodium in the pathogenesis of essential hypertension, many investigators have focused on the transmembrane transport systems that regulate intracellular sodium and other cations. Active transport of sodium out of cells creates a concentration gradient that provides energy for movement of calcium out of the cell. In fact, changes in transmembrane sodium transport have major effects on intracellular calcium concentration. Since cytosolic calcium is the major determinant of vascular smooth muscle tone, the higher the cytosolic calcium concentration, the greater the tonic contraction. Thus, any abnormality that increases the intracellular sodium concentration can potentially lead to increased peripheral resistance and hypertension. This connection between intracellular sodium and vascular tone has led to hypotheses concerning the role of sodium transport in the genesis of hypertension.

The possibility that hypertension is associated with and perhaps due to increased intracellular sodium was first proposed by Tobian and Binion in 1952, who reported that the sodium content of renal arteries obtained at autopsy is higher in hypertensive than in normotensive individuals.[37] These findings have since been confirmed in more accessible human tissue. Based on the assumption that abnormalities in transport across blood cell membranes will reflect transport abnormalities elsewhere, several investigators have demonstrated increased sodium concentrations in red[38,39] and white[40–43] blood cells of hypertensive individuals. Furthermore, when blood pressure is considered a continuous variable, it is positively associated with intracellular sodium concentration.[42,44]

Although these findings are not consistently reported,[39] a genetic basis is suggested by the fact that increased intracellular sodium is found in circulating cells of normotensive offspring of hypertensives.[38,45] Furthermore, increased intracellular sodium has been described in blacks compared with whites, men compared with women, and obese compared with lean individuals, conditions which are each associated with hypertension on a presumed genetic basis.[44]

Several transmembrane transport systems affect intracellular sodium concentration (Figure 2), and the genes encoding them are candidates for essential hypertension. These transport systems include sodium-potassium-ATPase, sodium-lithium countertransport, and sodium-potassium cotransport, among others (see also Table 4). The studies reviewed below attempt to establish a role for each of these systems in the pathogenesis of essential hypertension.

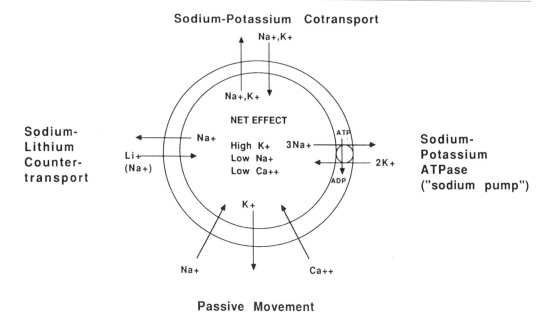

Figure 2. Transmembrane transport systems.

Sodium-Potassium-Adenosine Triphosphatase (Na-K-ATPase)

Several active and passive transport mechanisms regulate the intracellular concentration of sodium. However, the transmembrane sodium concentration gradient is primarily due to the extrusion of sodium by Na-K-ATPase, the so-called "sodium pump". Decreased pump function will decrease the energy available for the transport of calcium out of the cell, and will potentially lead to vasoconstriction. Although many investigators have focused on a search for a circulating factor that would inhibit the sodium pump (the natriuretic factor proposed by Blaustein[46]), the genes encoding both the alpha and beta subunits of sodium-potassium-ATPase are logical candidates for essential hypertension.

There is some evidence that essential hypertension is associated with heritable influences on sodium pump activity. Several investigators have demonstrated decreased sodium pump activity in blood cells of normotensive individuals with a family history of hypertension.[44,47] Although others have

demonstrated strong environmental effects on pump activity,[48] race association studies provide additional evidence of a genetic component. When red blood cells from black individuals were compared to red cells obtained from whites, the blacks had a lower V_{max} for pump activity,[49,50] decreased pump density,[51,52] and higher intracellular sodium concentration.[44,49,52–54] These studies performed in circulating cells may be representative of sodium pumps in other tissues. For instance, intracellular sodium concentration in cultured fibroblasts obtained from black subjects was higher than in whites. This cell culture study also suggested that the change in sodium pump activity is inherent to the cells (i.e., genetically determined) rather than secondary to circulating factors.[55] Although most studies have been performed in normotensive individuals, pump activity is decreased in those groups most prone to hypertension: those with a family history of hypertension, blacks, and men.

These association studies in humans suggest that the genes encoding Na-K-ATPase may be involved in the development of essential hypertension, but the data are not entirely consistent.[39] For instance, in cultured fibroblasts, one investigator reported that intracellular sodium was increased in association with an *increase* in pump activity.[55] Another well-designed study that contrasts with the data summarized above showed that hypertensives had normal numbers of ouabain binding sites,[56] conflicting with findings of decreased pump density.[51,52] There are no genetic linkage studies in humans to confirm these association studies.

In animal models of genetic hypertension, the Na-K-ATPase genes may be more strongly implicated. The role of Na-K-ATPase genes is particularly intriguing in Dahl rats. The Dahl rat is bred to be either salt sensitive (Dahl-S), in which case it develops elevated blood pressure if fed a high (8%) sodium diet, or salt resistant (Dahl-R), in which case it does not develop hypertension despite high sodium intake. The Dahl-S rat is a model of genetically determined salt sensitivity that may be analogous to human salt sensitivity, and the Dahl-R rat is genetically similar but salt resistant. In these two strains, Herrera and Ruiz-Opazo found a RFLP of the gene encoding the alpha$_1$ subunit of Na-K-ATPase. The Dahl-S rats contained a gene that differed from the Dahl-R rat due to three nucleotide substitutions, one of which resulted in an amino acid substitution in the protein that changed the pump's hydropathy profile. This conformational change was associated with reduced ouabain-sensitive potassium influx, indicating a decrease in sodium pump function that could lead to increased intracellular sodium concentration.[57]

These investigators identified a single amino acid substitution in a gene encoding a functional portion of Na-K-ATPase that was associated with abnormal pump function in Dahl-S rats. These findings do not, however, confirm that this genetic polymorphism is responsible for the difference in salt-induced changes in blood pressure in Dahl-S and -R rats. When the etiologic role of this polymorphism was tested in a segregation analysis in inbred Dahl-S and -R rats, the Dahl-S form of the gene did not segregate with high blood pressure.[58] Furthermore, direct sequencing of the gene for the alpha$_1$ subunit in Dahl-S rats was not consistent with the findings reported by Herrera and Ruiz-Opazo,[59] although methodological differences might explain this inconsistency.[57,59] Thus, the significance of this polymorphism is unclear in the pathogenesis of hypertension in Dahl-S rats. It is also unclear whether hypertension in this model is analogous to salt-sensitive hypertension in humans. Linkage studies in humans are needed to resolve this issue.

Sodium-Lithium (Na-Li) Countertransport

Red blood cells in culture medium can exchange intracellular lithium for extracellular sodium. The significance of this exchange is unknown since in physiologic medium, i.e., in the absence of lithium, there appears to be an exchange of sodium for sodium without affecting intracellular ion concentrations. Nonetheless, several studies establish that erythrocyte Na-Li countertransport activity is higher in hypertensive than in normotensive individuals.[56,60–63] A genetic basis for this finding is suggested by the fact that countertransport is also elevated in normotensive, first-degree relatives of hypertensives.[56,60,61,63,64] In fact, Na-Li countertransport is highest for hypertensives, lowest for normotensives with normotensive parents, and intermediate for normotensives with hypertensive parents.[60] These differences are independent of the potentially transport-modulating effects of age, body size, gender, or ethnic origin.[63] Additional evidence that Na-Li countertransport may contribute to the heritability of hypertension includes the fact that it is not altered in secondary hypertension[61] (although this is not a consistent finding[63]). In both normotensives and hypertensives, the level of countertransport activity aggregates in families.[56,60]

In a large Caucasian cohort, a pedigree analysis suggested a model of polygenic inheritance for Na-Li countertransport.[65] Overall, more than 80% of the variation in countertransport was explained by genetic factors.[66] In a subset of families that were ascertained for early coronary heart disease, there was statistical evidence of a major gene effect. In this subset, individuals with a high probability of homozygosity for the high Na-Li countertransport allele were twice as likely to have hypertension as those without the high countertransport genotype.[65] In addition, high likelihood of the high countertransport genotype in normotensives was associated with the subsequent development of hypertension.[67] More recent research has linked left ventricular hypertrophy to elevated Na-Li transport.[67a] It has also shown that this phenomenon is associated with recognized risk factors for cardiovascular disease, such as high alcohol intake, dyslipidemia, and obesity.[67b] A possible link between insulin resistance and high Na-Li countertransport[67c] could provide a biological mechanism for the association. A molecular basis for this remains to be discovered.

The studies cited above for the most part were performed in Caucasian populations. In contrast, black hypertensives have normal Na-Li countertransport activity.[68,69] This ethnic difference suggests that there is heterogeneity in the importance of countertransport disturbances in the development of essential hypertension, with greater importance in whites than blacks. Alternatively, the inconsistency of findings in different groups of hypertensives could mean that countertransport is not important in the development of hypertension, and that reported associations are spurious.

Some researchers believe that the Na-Li countertransport system has the strongest claim as a candidate gene (or at least a marker gene) for essential hypertension,[70] but it remains difficult to explain a mechanism by which increased sodium-sodium exchange would lead to increased intracellular sodium, with its attendant changes in cytosolic calcium. Proposed interactions between Na-Li countertransport and the exchange of sodium for hydrogen ion[70] remains an area of controversy with evidence both for[70a] and against[70b] a common link between a combined abnormality of Na-Li and Na-H transporters in these patients. Current evidence suggests it is unlikely that the same membrane protein is involved.[70c] The case for the Na-Li transporter as a marker or indeed as a mechanism for heritable hypertension remains unproven despite extensive research. Difficulties with methodology for measuring Na-Li transport could explain some of the conflicting results.[70d]

Sodium-Potassium (Na-K) Cotransport

This system, inhibited by furosemide, simultaneously transports sodium and potassium either into or out of the cell, depending on the concentration gradients of the two ions. Na-K cotransport apparently plays an important role in maintaining cell volume. It exists in erythrocytes,[71] which makes this system accessible to investigation in human cells. It also exists in vascular smooth muscle[72,73] and renal tubular epithelium[74,75] making it a reasonable system to consider in the pathogenesis of hypertension.

Unfortunately, such considerations have yielded inconsistent results. Hypertension has been associated with both increased[76,77] and decreased[68,78,79] cotransport activity. These inconsistencies may be the result of methodological variables, since different laboratories use different methods for measuring cotransport.[39] Alternatively, they may be explained by differences in the populations being studied.[80] Subpopulations of hypertensives may exist in which cotransport plays a genetic role. This concept is supported by evidence of ethnic differences in cotransport activity.[68,78] However, the role of Na-K cotransport in the inheritance of hypertension remains undefined; to date there are no animal models or human linkage studies that clearly demonstrate alterations in Na-K cotransport genes.

Na-H Cotransport and G proteins

G proteins are a group of proteins widely distributed in the body and are involved in signal transduction of a number of hormones. In a recent report by Siffert,[80a] the role of G proteins as an intermediary pathway in heritable hypertension was reviewed. The Na-H exchanger isoform 1 (NHE-1) has been shown in multiple studies to be enhanced in a group of patients with essential hypertension.[80b] However, the activity of this pump is modulated by many factors such as oral glucose,[80c] metabolic acidosis,[80d] and chronic NaCl loading.[80e] This has been circumvented by immortalizing cells, using Epstein-Barr virus, from hypertensive and normotensive patients and then demonstrating that the cells from hypertensives had preservation of high NHE-1 activity.[80f]

Subsequent stimulation of these cells with platelet-activating factor (PAF) showed a marked increase in intracellular Ca^{++} in hypertensive cell lines when compared to normotensive[80g] derived cells. This effect was then blocked by pertussis toxin (PTX), suggesting the involvement of Gi and Gq/11 type G protein pathways. This suggested signal transduction was enhanced via PTX-sensitive G proteins in hypertensive cell lines.[80h] These findings were further supported by experiments in which cells from hypertensive and normotensive patients, in whom NHE-1 activity was not determined, were immortalized and again marked PAF evoked Ca^{++} signals were seen in the hypertensive group.[80i] Currently, a C825T polymorphism in the gene encoding for the G protein subunit Gβ3 has been shown to be associated with enhanced intracellular signal transduction and hypertension.[80j]

α-Adducin

Adducin is a cytoskeletal protein that forms the backbone of the cell and is important in the regulation of cell membrane transport proteins. α-Adducin has also been implicated as a mechanism of heritable hypertension. The elucidation of this has followed the classical pathway from rat to hu-

man. Hypertension in the Milan hypertensive strain (MHS) is associated with an increased ion transport rate which may cause a rise in renal tubular reabsorption of sodium. This elevated tubular sodium reabsorption is not found in the Milan normotensive strain (MNS) and results from an increase in all sodium transport systems.[80k] This was then found to be abolished by removing the cytoskeleton.[80l] The difference between the two strains was shown to be mediated by a cytoskeletal protein known as adducin.

Adducin, which is composed of α, β, and γ subunits, directly binds to actin and modulates its assembly into filaments.[80m] Polymorphism of the α and β subunits were then shown to be responsible for 50% of the differences in blood pressure between the two strains of Milan rat. These polymorphisms have been shown to increase the number and the V_{max} of the Na-K pumps in the MHS rat. These results were shown to be applicable to humans with hypertension when an association was shown between the α adducin G460W polymorphism in both a case control[80n] and sib-pair linkage study[80o] in a Caucasian population. Although the evidence for the effect of this polymorphism in humans is still being assembled, it has been shown that patients with this polymorphism are 8 times as likely to be salt sensitive and 5 times as likely to have a good clinical response to diuretic therapy.[80p] There is also further indirect evidence that the pressure nateuresis curves are altered in patients with this G460W polymorphism but localization of this effect to the kidney has yet to be shown.[80q] This elegant research is a further link in the chain but will explain only a further subset of patients with inherited hypertension.

Other Transport Systems

The studies reviewed above suggest that hypertension may be due, in some individuals, to genetic alterations leading to increased sodium transport, presumably leading to increased intracellular calcium. Alternatively, cytosolic calcium can accumulate if there is reduced activity of calcium adenosine triphosphatase (calcium-ATPase), a "calcium pump" that is analogous to Na-K-ATPase and that actively transports calcium out of the cell. However, little data exists implicating Ca-ATPase in essential hypertension, and the data that does exist are inconsistent. For instance, hypertensives have a lower V_{max} for basal and calmodulin-stimulated Ca-ATPase and reduced affinity for calcium in erythrocytes,[81] but a higher V_{max} in platelets.[82] Platelets share many properties with vascular smooth muscle, including the presence of calcium-dependent contractile proteins, so we would expect the V_{max} in vascular tissue to be similarly increased. This would lead to decreased cellular calcium which would result in decreased vascular tone. Shaw et al. have reported some studies that look at the sensitivity of myofiloaments to Ca^{++} in both SHR and WKY rats. They could find no differences between the hypertensive rats and the controls.[82a] Therefore, it is difficult at this time to implicate genetic alterations of Ca-ATPase in the pathogenesis of hypertension.

In summary, several genes encoding components of the renin-angiotensin-aldosterone axis, and others encoding cell membrane transport proteins, have been evaluated as candidate genes for hypertension. Other reasonable candidate genes are summarized in Table 4. These data suggest that each hypertensive animal model may have its own pattern of variability involving these candidate genes, but this variability may or may not be pathogenetic in the development of high blood pressure. Similarly, in humans there is the possibility that these genes may be involved in the pathogenesis of essential hypertension. However, much work remains to be done to substantiate these hypotheses. Formal genetic linkage studies must confirm observed associations between genotype and phenotype, and biochemical studies (including studies in transgenic animals) must demonstrate alterations of gene products that would lead to elevated blood pressure.

Hypertension Subtypes

In our opinion, the search for hypertension genes has been hampered by the definition of phenotype. In all the studies discussed so far, the phenotype is defined as the presence or absence of hypertension (making a dichotomy out of a continuous trait), or it is defined as blood pressure itself (making a single entity out of a complexly regulated phenomenon). This is likely to be a gross oversimplification. In some individuals, hypertension is associated with elevated renin activity, in others with salt sensitivity or vascular hyperreactivity. Each of these groups of individuals may have a distinct derangement of blood pressure regulation that is genetically determined. Looking for genetic etiologies in a population that includes individuals with several different pathophysiologic profiles will dilute even strong genetic influences on some or all of the subtypes. Thus, the identification and characterization of hypertension subtypes (often referred to as "intermediate phenotypes"), each with its own candidate genes, will improve our chances of understanding the genetic basis of hypertension. Some subtypes to consider are indicated in Table 5, and discussed below.

Salt Sensitivity

Population studies have shown that there is a direct relationship between dietary salt and the prevalence of hyperten-

---

Table 5
Hypertension Subtypes

Salt sensitivity
Vascular hyperreactivity
Familial hyperlipidemic hypertension
Insulin resistance

---

sion. However, even in populations with habitually high dietary salt consumption, only a fraction of people develop hypertension, suggesting individual variability in susceptibility to dietary salt.[83] In an attempt to define and characterize susceptibility to salt intake, investigators have introduced the concept of "salt sensitivity," a term which simply describes blood pressure response to manipulations of sodium balance. Several manipulations have been used including both dietary and parenteral increases in sodium intake.[84]

Weinberger et al.[85] defined salt sensitivity by a volume expansion/volume depletion protocol. The volume expansion phase of the experiment was achieved with the infusion of two liters of normal saline (0.9% sodium chloride) intravenously over 4 hours, followed by the volume depletion phase, which was achieved with dietary salt restriction (10 mEq per day) and three oral doses of a diuretic (furosemide, 120 mg total dose). The mean arterial blood pressure (MABP) at the end of the saline infusion was then compared with that at the end of the volume depletion maneuver. Individuals with at least 10 mmHg decrease in blood pressure were characterized as salt sensitive, and those with less than 5 mmHg decrease (including those with a blood pressure increase) were considered salt resistant. Individuals with an intermediate change in blood pressure (a decrease of between 5 and 10 mmHg) were described as "indeterminate" with respect to salt sensitivity. Several other methods have been used to characterize patients with regard to salt responsivity, leading to variable estimates of the prevalence of salt sensitivity.[84] Nevertheless, these studies suggest that 15–30% of white normotensives, 50% of white hypertensives, 30–40% of African-American normotensives, and as many as 70% of African-American hypertensives are salt sensitive.

The possibility that this subtype of essential hypertension is genetically determined is suggested by similarities between human salt sensitivity and the analogous Dahl salt-sensitive rat model of genetic hypertension, described above. Further evidence is derived from four types of human studies. First, the fractional excretion of sodium is similar among family members.[86] Second, blood pressure response to a 12-week sodium restricted diet is more closely correlated in monozygotic twins than in non-twin siblings.[87] Third, there is an increased frequency of salt sensitivity in individuals who have hypertensive first-degree relatives.[88] Lastly, salt sensitivity has been associated with haptoglobin phenotype.

These associations suggest an hereditary component to salt sensitivity. In addition, an evolutionary hypothesis has been advanced to explain the high prevalence of salt sensitivity among African-Americans.[90,91] According to this hypothesis, salt sensitivity in African-Americans relates to selection pressure in favor of salt and water retention in the tropical west African climate. It is further suggested that events during the Trans-Atlantic slave trade, during which 50% of incarcerated Africans died of dehydration-related illness, may have resulted in an intense selection pressure for salt conservation. This hypothesis could be tested systematically by comparing African-Americans with native Africans, but these studies have not yet been performed.

Despite the limited data defining the heritability of salt sensitivity, characterizing individuals according to this aspect of blood pressure regulation, rather than characterizing them by blood pressure itself, may permit genetic studies in a more homogeneous group. Salt sensitivity itself suggests a long list of candidate genes. The blood pressure response to sodium intake is presumably regulated by several factors including the renin-angiotensin-aldosterone axis, autonomic nervous system, renal kinin-kallikrein system, and by all the other local and circulating influences on vascular tone and renal sodium excretion. The genes regulating each of these functions are thus candidates for salt sensitivity. Given the multitude of factors that can potentially lead to salt sensitivity, this trait, like blood pressure, may be multifactorial and polygenic.

Very little data are available investigating these candidate genes for salt sensitivity in humans. Recently, there has been a significant advance with the linking of adrenergic receptors to the process of salt sensitivity. In an association study[91a] and a linkage study,[91b] a relationship was demonstrated with the β-adrenergic receptors in hypertensive black populations. An association was also demonstrated between the α receptor and hypertensive whites. In another human study, Weinberger et al.[89] report an association between haptoglobin 1–1 phenotype and salt sensitivity in both normotensive and hypertensive individuals. In this study, both adults and children with the haptoglobin 1–1 phenotype were significantly more salt sensitive than those with the 2–2 phenotype.[89] This data has been confirmed in Japanese populations[91c] in whom the frequency of haptoglobin 1–1 genotype is much less than in American populations. A gene close to this locus is thus a possible candidate gene for salt sensitivity. Since there is no apparent relationship between haptoglobin phenotype and sodium excretion or blood pressure regulation, these data, if confirmed, suggest that the haptoglobin gene may be a marker for a salt-sensitivity-related gene.

There is a clear need for further investigation of candidate genes for salt sensitivity. In addition, there is a need to develop standardized criteria for characterizing individuals as salt sensitive, so that independent investigators can be sure they are studying the same phenomenon. Despite these methodological limitations, we believe that the stratification of patients based on their salt responsivity may be useful in genetic studies, narrowing the focus from the broad heterogenous phenotype, hypertension, to patient populations sharing similar characteristics and perhaps representing a more genetically homogeneous group.

Vascular Hyperreactivity

Vascular hyperreactivity may define another subtype of essential hypertension that may be useful in the study of the genetic basis of hypertension. Vascular hyperreactivity refers to increased vascular responsiveness to environmental stimuli, such as exposure to cold or a mental arithmetic task. These stimuli are thought to be mediated by the adrenergic nervous system with relative specificity for particular classes of adrenergic receptors. For instance, exposure to cold is thought to stimulate alpha-adrenergic receptors, while a mental arithmetic task is thought to stimulate beta receptors. Some investigators have suggested that a chronically exaggerated pressor response to environmental stimuli eventually leads to sustained hypertension.

There is some evidence that vascular hyperreactivity, like salt sensitivity, is a subtype of hypertension that is genetically determined. Twin studies suggest heritability of pressor response to both mental arithmetic (h = 0.72 to 0.84)[92] and cold stimulus (h = 0.62–0.72).[93] In addition, parental hypertension is associated with increased vascular reactivity in normotensive offspring.[94,95]

A genetic influence is further suggested by racial/ethnic differences in patterns of reactivity. The pressor response to mental arithmetic is lower in blacks than in whites,[96] but the response to cold stimulus is greater.[97] In addition, family history of hypertension affects reactivity in blacks and whites differently. Black normotensives with hypertensive parents show a greater response to both cold and mental arithmetic than blacks without this family history, but the response to cold is significantly more vigorous than the response to the mental task.[94] In contrast, whites with parental hypertension are more reactive to the mental arithmetic task.[95]

In a recent study which looked at a white male population, hyperreactivity to stressors that evoke an active behavioral response was shown to be mediated by an increase in cardiac output. This was due to elevation of plasma epinephrine and norepinephrine levels on a background of reduced β_2 receptor responsiveness.[97a] This down-regulation of β_2 receptors could explain the reduced cardiac output and resultant increased peripheral vascular resistance that is felt to be the physiological milieu from which hypertension emerges.[97b] Similar alterations in β receptor responsiveness have not been found in black populations.[97c]

While these data suggest that vascular hyperreactivity is an inherited subtype of hypertension, other investigators have been unable to demonstrate that this response is associated with familial hypertension,[98,99] and no candidate genes have been investigated. In addition, although muscle sympathetic nerve activity does correlate with pressor response to cold,[100] the receptor specificity of these responses is unclear since there have been no studies employing pharmacologic blockade of the receptor subtypes. Nonetheless, it appears that a high degree of pressor response to these stimuli is associated with the presence of essential hypertension in some subjects, and further investigation of the genetic basis of this hypertension subtype is warranted.

Familial Dyslipidemic Hypertension

As an alternative method of defining hypertension subtypes, some investigators have attempted to identify unique forms of hypertension in large pedigrees. Along these lines, researchers in Utah have identified adult sibships with a syndrome labeled "familial dyslipidemic hypertension" (FDH).[101] These siblings have hypertension that was diagnosed before age 60 along with various lipid abnormalities (fasting triglyceride and/or LDL cholesterol above the 90th percentile, and/or fasting HDL cholesterol below the 10th percentile). In the large Utah population that was screened, 12% of the adults with hypertension and 1% to 2% of all the adults screened fulfilled the criteria for FDH. These authors have further subdivided FDH by the presence or absence of familial combined hyperlipidemia (FCHL), which refers to the absolute levels of apolipoprotein A-1 and B and very low density lipoprotein (VLDL) cholesterol.[102] Further studies are certainly needed to independently confirm FDH as a true genetic subtype of hypertension. Without attempting to underplay the importance of this approach to the genetics of hypertension, and with full appreciation of the need to identify and treat lipid abnormalities in the hypertensive population, it is important to point out that FDH may represent two unrelated predispositions (i.e., for dyslipidemia and hypertension), each with its unique genetic determinants.

Insulin Resistance

Similar to FDH, some investigators have proposed that insulin resistance associated with hypertension represents a unique familial syndrome.[103,104] Attempts are currently underway to refine possible genetic mechanisms that may show the precise relationship between essential hypertension and the causes of insulin resistance. A recent paper published by Chiang et al.,[104a] links essential hypertension to a molecular variant arising from a G to A transition at nucleotide 258 of the liver glucokinase promoter region which would give an attractive explanation for the simultaneous occurrence of these two common conditions. Other groups[104b] found no association with polymorphisms of the insulin receptor, the insulin responsive glucose transporter, and glycogen synthase. Some investigators have suggested a different mechanism which would result in the coexistence of insulin resistance and hypertension. In this construct a mechanism exists which controls total body sodium. This pathway is mediated via the insulin receptor. A genetic defect in the insulin receptor in SHR would result in these rats being both hypertensive and resistant to insulin. There appears to be an association between insulin resistance and hypertension, but the exact genetic or reactive mechanism remains to be elucidated. Again, the possibility of the coexistence among family members of two very common conditions has not been excluded.

In summary, each of these hypertension subtypes may represent multifactorial and/or polygenic traits. In addition, they may not be completely distinct entities regulated by different genes. In fact, there is some evidence that vascular hyperreactivity and salt sensitivity are closely related phenomena. However, investigation of these syndromes is likely to be more fruitful in determining the genetic influences on hypertension than investigation of blood pressure itself.

Conclusions

Hypertension is a major risk factor for stroke. The risk of developing hypertension is strongly influenced by genetic background. Numerous candidate genes have been proposed and investigated, with some intriguing results. However, the identification of genetic influences on blood pressure are complicated by the quantitative, multifactorial, and polygenic nature of blood pressure, the variable age of onset of hypertension, and the phenotypic heterogeneity underlying this diagnosis. New genetic linkage techniques such as the affected pedigree member method and sib-pair analysis may be useful in the study of hypertension.

References

1. Wolf PA, Kannel WB, McGee DL. Prevention of ischemic stroke: Risk factors. In Barnett HJM, Stein BM, Mohr JP, Yatsu FM (eds). Stroke: Pathophysiology, Diagnosis, and Management. New York: Churchill Livingstone Press, 1986, 967–988.
2. Fifth Report of the Joint National Committee on Detection, Evaluation, and Treatment of High Blood Pressure (JNC V). National High Blood Pressure Education Program, National Heart, Lung, and Blood Institute, National Institutes of Health. October, 1992.
3. Ayman D. Heredity in arteriolar (essential) hypertension: A clinical study of blood pressure of 1,524 members of 277 families. Arch Int Med 1934;53:792–803.
4. Ward R. Familial aggregation and genetic epidemiology of blood pressure. In Laragh JH, Brenner BM (eds). Hypertension: Pathophysiology, Diagnosis and Management. New York: Raven Press, 1990.
5. Muldoon MF, Terrell DF, Bunker CH, Manuck SB. Family history studies in hypertension research: Review of the literature. Am J Hypertens 1993;6:76–88.
6. Annest JL, Sing CF, Biron P, Mongeau J-G. Familial aggregation of blood pressure and weight in adoptive families. II. Estimation of the relative contributions of genetic and common environmental factors to blood pressure correlations between family members. Am J Epidemiol 1979;110:492–503.
7. Stocks P. A biometric investigation of twins and their brothers and sisters. Ann Eugen 1930;4:49–62.
8. McIlhany ML, Shaffer JW, Hines EA. The heritability of blood pressure: An investigation of 200 twin pairs using the cold pressure test. Johns Hopkins Med J 1975;136:57–74.
9. Feinleib M, Garrison RJ, Fabsitz R, et al. The NHLBI twin study of cardiovascular disease risk factors: Methodology and summary of results. Am J Epidemiol 1977;106:284–295.
10. Rapp JP, Wang S-M. Genetic hypertension: Classical and molecular genetic concepts as applied to the renin gene. In Laragh JH, Brenner BM (eds). Hypertension: Pathophysiology, Diagnosis, and Management. New York: Raven Press, 1990, 955–964.
11. Dzau VJ. Significance of the vascular renin-angiotensin pathway. Hypertension 1968;8:553–559.
12. van Hooft IMS, Grobbee DE, Derkx FHM, et al. Renal hemodynamics and the renin-angiotensin-aldosterone system in normotensive subjects with hypertensive and normotensive parents. New Engl J Med 1991;324:1305–1311.
13. Widgren BR, Herlitz H, Aurell M, et al. Increased systemic and renal vascular sensitivity to angiotension II in normotensive men with positive family histories of hypertension. Am J Hypertens 1992;5:167–174.
14. Sealey JE, Laragh JH. The renin-angiotensin-aldosterone system for normal regulation of blood pressure and sodium and potassium homeostasis. In Laragh JH, Brenner BM (eds). Hypertension: Pathophysiology, Diagnosis, and Management. New York: Raven Press, 1990, 1287–1317.

15. Campbell DJ, Habener JF. Angiotensinogen is expressed and differentially regulated in multiple tissues of the rat. J Clin Invest 1986; 78:1427–1431.

16. Fasola AF, Martz BL, Helmer OM. Plasma renin activity during supine exercise in offspring of hypertensive parents. J Appl Physiol 1968;25:410–415.

17. Kimura S, Mullins JJ, Bunnemann B, et al. High blood pressure in transgenic mice carrying the rat angiotensinogen gene. EMBO J 1992;11:821–827.

18. Ohkubo H, Kawakami H, Kahehi Y, et al. Generation of transgenic mice with elevated blood pressure by introduction of the rat renin and angiotensinogen genes. Proc Natl Acad Sci USA 1990;87:5153–5157.

19. Jeunemaitre X, Soubrier F, Kotelevtsev YV, et al. Molecular basis of human hypertension: Role of angiotensinogen. Cell 1992;71:169–180.

20. Kotelevtsev YV, Clauser E, Corvol P, Soubrier F. Dinucleotide repeat polymorphism in the human angiotensinogen gene. Nucleic Acids Res 1991;19:6978.

20a. Kamitami R, Rakugi H, Hikagi J, et al. Association analysis of a polymorphism of the angiotensinogen gene with essential hypertension in the Japanese. J Hypertens 1994;8:521–524.

20b. Rotimi C, Puras A, Cooper R, et al. Polymorphisms of the renin angiotensin genes among Nigerians, Jamaicans and African Americans. Hypertension 1996;27:558–563.

20c. Jeunemaitre X, Charry A, Chatellier G, et al. M235T variant of the human angiotensinogen gene in unselected hypertensive patients. J Hypertens 1993;11(suppl 5): S80–S81.

20d. Jeunemaitre X, Inoue I, Williams C, et al. Haplotypes of angiotensinogen in essential hypertension. Am J Hum Genet 1997;60:1448–1460.

20e. Hingorami AD, Sharma P, Haiyan J, et al. Blood pressure and the M235T Polymorphism of the angiotensinogen gene. Hypertension 1996;28:907–911.

20f. Tiret L, Richard S, Poirer O, et al. Genetic variation at the angiotensinogen locus in relation to high blood pressure and myocardial infarction: The ECTIM study. J Hypertens 1995;13:311–317.

20g. Kunz R, Kreutz R, Beige J, et al. Association between the angiotensinogen 235T-variant and essential hypertension in whites: A systematic and methodological appraisal. Hypertension 1997;30:1331–1337.

20h. Inoue I, Naajima T, Williams CS, et al. A nucleotide substitution in the promoter of human angiotensinogen is associated with essential hypertension and effects basal transcription. J Clin Invest 1997;99:1786–1797.

20i. Hopkins PN, Lifton RP, Hollenberg NK, et al. Blunted renovascular response to angiotensin II is associated with a common variant of the angiotensinogen gene and obesity. J Hypertens 1996;14:199–207.

20j. Fischer NDC, Ferri C, Bellinno C, et al. Age, gender and non modulation: A sexual dimorphism in essential hypertension 1997;29:980–985.

21. Bachmann J, Seldmer M, Ganten U, et al. Sexual dimorphism of blood pressure: Possible role of the renin-angiotensin system. J Steroid Biochem Mol Biol 1991;40:511–515.

22. Naftlan AJ, Zuo WM, Inglefinger J, et al. Localization and differential regulation of angiotensinogen mRNA expression in the vessel wall. J Clin Invest 1992;89:1695.

23. Shiota N, Miyazaki M, Okunishis H. Increase of angiotensin converting enzyme gene expression in the hypertensive aorta. Hypertension 1992;20:168–174.

24. Takahashi S, Fukamizu A, Hasegawa T, et al. Expression of the human angiotensinogen gene in transgenic mice and transfected cells. Biochem Biophys Res Commun 1991;180:1103–1109.

25. Mullins JJ, Peters J, Ganten D. Fulminant hypertension in transgenic rats harboring the mouse Ren-2 gene. Nature 1990;344:541–544.

26. Hilgers KF, Peters J, Veelken R, et al. Increased vascular angiotensin formation in female rats harboring the mouse Ren-2 gene. Hypertension 1992;19:687–691.

27. Rapp JP, Wang S-M, Dene H. A genetic polymorphism in the renin gene of Dahl rats cosegregates with blood pressure. Science 1989;243:542–544.

28. Kurtz TW, Simonet L, Kabra PM, et al. Cosegregation of the renin allele of the spontaneously hypertensive rat with an increase in blood pressure. J Clin Invest 1990;85:1328–1332.

29. Soubrier F, Jeunemaitre X, Rigat B, et al. Similar frequencies of renin gene restriction fragment length polymorphisms in hypertensive and normotensive subjects. Hypertension 1990;16:712–717.

30. Zee RYL, Ying L-H, Morris BJ, Griffiths LR. Association and linkage analyses of restriction fragment length polymorphisms for the human renin and antithrombin III genes in essential hypertension. J Hypertens 1991;9:825–830.

31. Barley J, Carter ND, Cruickshank JK, et al. Renin and atrial natriuretic peptide restriction fragment length polymorphisms: Association with ethnicity and blood pressure. J Hypertens 1991;9:993–996.

32. Jeunemaitre X, Rigat B, Charru A, et al. Sib

pair linkage analysis of renin gene haplotypes in human essential hypertension. Hum Genet 1992;88:301–306.

32a.Chiang FT, Hsu KL, Tseng CD, et al. Association of the renin gene polymorphism with essential hypertension. Clin Genet 1997; 51 (6):370–374.

33. Jacob HJ, Lindpainter K, Lincoln SE, et al. Genetic mapping of a gene causing hypertension in stroke-prone spontaneously hypertensive rat. Cell 1991;67:213–224.

34. Hilbert P, Lindpaintner K, Beckman SS, et al. Chromosomal mapping of two genetic loci associated with blood-pressure regulation in hereditary hypertensive rats. Nature 1991; 353:521–529.

35. Deng Y, Rapp JP. Cosegregation of blood pressure with angiotensin converting enzyme and atrial natriuretic peptide receptor genes using Dahl salt-sensitive rats. Nature Genet 1992;1:267–272.

35a.Kreutz R, Hubner N, James MR, et al. Dissection of a quantitative trait locus for genetic hypertension on rat chromosome 10. Proc Natl Acad Sci USA 1995;92:8778–8782.

35b.Kreutz R, Hubner N, Ganten D, Lindpaintner K. Genetic linkage of the ACE gene to plasma angiotensin-converting enzyme activity but not to blood pressure: A quantitative trait locus confers identical complex phenotypes in human and rat hypertension. Circulation 1995;92:2381–2384.

36. Zee RYL, Lou Y-K, Griffiths LR, Morris BJ. Association of a polymorphism of the angiotension I-converting enzyme gene with essential hypertension. Biochem Biophys Res Commun 1992;184:9–15.

36a.Jeunemaitre X, Lifton RP, Hunt SC, et al. Absence of linkage between the angiotensin converting enzyme and human essential hypertension. Nat Genet 1992;1:72–75.

36b.Marre M, Bernadet P, Gallois Y, et al. Relationship between angiotensin 1 converting enzyme gene polymorphism, plasma levels and diabetic renal complication. Diabetes 1994;43:384–388.

36c. Cambien F, Poirer O, Lecoef L, et al. Deletion polymorphism in the gene foe ACE is a potent risk factor for MI. Nature 1992;359: 641–644.

36d.Schunkert A, Hense H-W, Holmer SR, et al. Association between a deletion polymorphism of the angiotensin converting enzyme gene and LVH. N Engl J Med 1994;330: 1634–1638.

36e. Missouris CG, Barley J, Jeffrey S, et al. Genetic risk for renal artery stenosis: Association with deletion polymorphism in angiotensin 1 converting enzyme gene. Kidney Int 1996;44: 375–380.

36f. Schmodt S, Schone N, Ritz ED. Association of ACE gene polymorphism and diabetic nephropathy: The Diabetic Nephropathy Study Group. Kidney Int 1995;47:1176–1181.

36g.Lindpaintner K, Pfeffer MA, Kreuntz R, et al. A prospective evaluation of the angiotensin-converting enzyme gene polymorphism and the risk of ischemic heart disease. N Engl J Med 1995;332:706–711.

36h.Lindpaintner K, Lee M, Larson MG, et al. Absence of association or genetic linkage between the ACE gene and left ventricular mass N Engl J Med 1996;334:1023–1028.

36i. Vasku A, Soucek M, Znojil V, et al. Angiotensin 1 converting enzyme and angiotensinogen gene interaction and prediction of essential hypertension. Kidney Int 1998;53:1479–1482.

36j. Mondorf U, Russ A, Wiesemwa A, et al. Contribution of angiotensin 1 converting enzyme gene polymorphism and angiotensinogen gene polymorphism to blood pressure regulation in essential hypertension. Am J Hypertens 1998;11:174–183.

36k. Hollenberg NK. Genes, hypertension and intermediate phenotypes. Curr Opin Cardiol 1996;11:457–563.

36l. Hingorami AD, Jia MJ, Stevens PA, et al. Renin angiotensin system gene polymorphisms influence blood pressure and the response to ace inhibitors. J Hypertens 1995; 13:1602–1609.

36m.Takami S, Katsuya T, Rakugi H, et al. Angiotensin II type 1 receptor gene polymorphism is associated with an increase in left ventricle mass but not with hypertension. Am J Hypertens 1998;11:316–321.

37. Tobian L, Binion JT. Tissue cations and water in essential hypertension. Circulation 1952;5: 754–758.

38. Zidek W, Vetter H, Dorst K-G, et al. Intracellular Na$^+$ and Ca^{2+} activities in essential hypertension. Clin Sci 1982;63:41s–43s.

39. Hilton PJ. Cellular sodium transport in essential hypertension. N Engl J Med 1986; 314:222–229.

40. Edmondson RPS, Thomas RD, Hilton PJ, et al. Abnormal leucocyte composition and sodium transport in essential hypertension. Lancet 1975;1:1003–1005.

41. Araoye MA, Khatri IM, Yao LL, Freis ED. Leukocyte intracellular cations in hypertension: Effect of antihypertensive drugs. Am Heart J 1978;96:731–738.

42. Ambrosioni E, Tartagni F, Montebugnoli L, Magnani B. Intralymphocytic sodium in hypertensive patients: A significant correlation. Clin Sci 1979;57:325s–327s.

43. Poston L, Sewell RB, Wilkinson SP, et al. Evidence for a circulating sodium transport in-

hibitor in essential hypertension. Br Med J 1981;282:847–849.

44. Rygielski D, Reddi A, Kuriyama S, et al. Erythrocyte ghost Na$^+$-K$^+$-ATPase and blood pressure. Hypertension 1987;10:259–266.

45. Chien Y-W, Zhao G-S. Abnormal leucocyte sodium transport in Chinese patients with essential hypertension and their normotensive offsprings. Clin Exp Hypertens 1984; 2279–2296.

46. Blaustein MP. Sodium ions, calcium ions, blood pressure regulation, and hypertension: A reassessment and a hypothesis. Am J Physiol 1977;232:C165–C173.

47. Aviv A, Lasker N. Proposed defects in membrane transport and intracellular ions as pathogenic factors in essential hypertension. In Laragh JH, Brenner BM (eds). Hypertension: Pathophysiology, Diagnosis, and Management. New York: Raven Press, 1990, 923–937.

48. Cusi D, Tripodi G, Alberghini E, et al. Heritability of sodium transport systems and hypertension. Ann NY Acad Sci 1986;488: 576–578.

49. Lasker N, Hopp L, Grossman S, et al. Race and sex differences in erythrocyte Na$^+$, K$^+$ and Na$^+$-K$^+$ adenosine triphosphatase. J Clin Invest 1985;75:1813–1820.

50. Beutler E, Kuhl W, Sacks P. Sodium-potassium-ATPase activity is influenced by ethnic origin and not by obesity. N Engl J Med 1983;309:756–760.

51. Izumo H, Lear S, Williams M, et al. Sodium-potassium pump, ion fluxes, and cellular dehydration in sickle cell anemia. J Clin Invest 1987;79:1621–1628.

52. Hopp L, Lasker N, Grossman S, et al. [^{3}H]-ouabain binding of red blood cells in whites and blacks. Hypertension 1986;8:1050–1057.

53. Tuck ML, Corry DB, Maxwell M, Stern N. Kinetic analysis of erythrocyte Na$^+$-K$^+$ pump and cotransport in essential hypertension. Hypertension 1987;10:204–211.

54. Munro-Faure AD, Hill DM, Anderson J. Ethnic differences in human blood cell sodium concentration. Nature 1971;231:457–458.

55. Kuriyama S, Hopp L, Tamura H, et al. A higher cellular Na$^+$ turnover rate in cultured skin fibroblasts from blacks. Hypertension 1988;11:301–307.

56. Smith JB, Ash KO, Hunt SC, et al. Three red cell sodium transport systems in hypertensive and normotensive Utah adults. Hypertension 1984;6:159–166.

57. Herrera VLM, Ruiz-Opazo N. Alteration of alpha1 Na$^+$, K$^+$-ATPase^{86}Rb$^+$ influx by a single amino acid substitution. Science 1990; 249:1023–1026.

58. Rapp JP, Dene H. Failure of alleles at the Na$^+$, K$^+$-ATPase alpha$_1$ locus to cosegregate with blood pressure in Dahl rats. J Hypertension 1990;8:457–462.

59. Simonet L, St Lezin E, Kurtz TW. Sequence analysis of the alpha$_1$ Na$^+$, K$^+$-ATPase gene in the Dahl salt-sensitive rat. Hypertension 1991;18:689–693.

60. Cusi D, Barlassina C, Ferrandi M, et al. Relationship between altered Na$^+$-K$^+$ contransport and Na$^+$-Li$^+$ countertransport in the erythrocytes of essential hypertensive patients. Clin Sci 1981;61:33s–36s.

61. Canessa M, Adragna N, Solomon HS, et al. Increased sodium-lithium countertransport in red cells of patients with essential hypertension. N Engl J Med 1980;302:772–776.

62. Clegg G, Morgan DB, Davidson C. The heterogeneity of essential hypertension: Relation between lithium efflux and sodium content of erythrocytes and a family history of hypertension. Lancet 1982;2:891–894.

63. Turner ST, Boerwinkle E, Johnson M, et al. Sodium-lithium countertransport in ambulatory hypertensive and normotensive patients. Hypertension 1987;9:24–34.

64. Woods JW, Falk RJ, Pittman AW, et al. Increased red-cell sodium-lithium countertransport in normotensive sons of hypertensive parents. N Engl J Med 1982;306:593–595.

65. Dadone MM, Hasstedt SJ, Hunt SC, et al. Genetic analysis of sodium-lithium countertransport in 10 hypertension-prone kindreds. Am J Med Genet 1984;17:565–577.

66. Hasstedt SJ, Wu LL, Ash KO, et al. Hypertension and sodium-lithium countertransport in Utah pedigrees: Evidence for major-locus inheritance. Am J Hum Genet 1988; 43:14–22.

67. Hunt SC, Stephenson SH, Hopkins PN, et al. A prospective study of sodium-lithium countertransport and hypertension in Utah. Hypertension 1991;17:1–7.

67a. Yap L, Arrazola A, Soria F, Diez J. Is there increased cardiovascular risk in essential hypertensive patients with abnormal kinetics of red blood cell sodium-lithium countertransport. J Hypertens 1989;7: 667– 673.

67b. Hardman TC, Dubrey S, Soni S, Lant AF. Relationship of sodium-lithium counter-transport activity to markers of cardiovascular risk in normotensive subjects. J Hum Hypertens 1995;9:589–596.

67c. Doria A, Fioretta P, Avogaro A, et al. Insulin resistance is associated with high sodium-lithium counter-transport in essential hypertension. Am J Physiol 1991;261: E684–E691.

68. Canessa M, Spalvins A, Adragna N, Falkner B. Red cell sodium countertransport and cotransport in normotensive and hypertensive blacks. Hypertension 1984;5:344–351.

69. Weder AB, Torretti BA, Julius S. Racial dif-

ferences in erythrocyte cation transport. Hypertension 1984;6:115–123.

70. Hilton PJ. Sodium transport in hypertension. Diabetes Care 1991;14:233–239.

70a.Morgan K, Canessa ML. Interactions of external and internal H and Na with Na\Na Na\h exchange of rabbit red cells: Evidence for a common pathway. J Membr Biol 1990;118:193–214.

70b.Kahn AM. Difference between human red blood cell Na\Li counter-transport and renal Na-H exchange. Hypertension 1985;9:7–12.

70c.Lifton RP, Hunt SC, Williams RR, et al. Exclusion of the Na-H antiporter as a candidate gene in human essential hypertension. 1985;9:7–12.

70d.Hardman TC, Lant AF. Controversies surrounding erythrocyte sodium-lithium countertransport. J Hypertens 1995;14:695–703.

71. Wiley JS, Cooper RA. A furosemide-sensitive contransport of sodium plus potassium in the human red cell. J Clin Invest 1974;53: 745–755.

72. O'Donnel ME, Owen NE. Atrial natriuretic factor stimulates Na/K/Cl cotransport in vascular smooth muscle cells. Proc Natl Acad Sci USA 1986;83:6132–6136.

73. Tokushige A, Kino M, Tamura H, et al. Bumetanide-sensitive sodium-22 transport in vascular smooth muscle cell of the spontaneously hypertensive rat. Hypertension 1986;8:379–385.

74. Herbert SC, Friedman PA, Culpepper RM, Andreoli TE. Salt absorption in the thick ascending limb of Henle's loop: NaCl cotransport mechanisms. Semin Nephrol 1983;2: 316–327.

75. Warnock DG, Greger R, Dunham, PB, et al. Ion transport processes in apical membranes of epithelia. Fed Proc 1984;43: 2473–2487.

76. Adragna NC, Canessa ML, Solomon H, et al. Red cell lithium-sodium countertransport and sodium-potassium cotransport in patients with essential hypertension. Hypertension 1982;4:795–804.

77. Cusi D, Fossali E, Piazza A, et al. Heritability estimate of erythrocyte Na-K-Cl cotransport in normotensive and hypertensive families. Am J Hypertens 1991;4:725–734.

78. Tuck ML, Gross C, Maxwell MH, et al. Erythrocyte Na^+, K^+-cotransport and Na^+, K^+-pump in black and caucasian hypertensive patients. Hypertension 1984;6:536–544.

79. Garay RP, Dagher G, Pernollet M-G, et al. Inherited defect in a Na-K cotransport system in erythrocytes from essential hypertensive patients. Nature 1980;284:281.

80. Muldoon MF, Terrell DF, Bunker CH, Manuck SB. Family history studies in hypertension research: Review of the literature. Am J Hypertens 1993;6:76–88.

80a.Siffert W. G proteins and hypertension: An alternative candidate gene approach. Kidney Int 1998;53:1466–1470.

80b.Rosskopf D, Dusing R, Siffert W. Membrane sodium-proton exchange and primary hypertension. Hypertension 1993;21: 607–617.

80c.Tepel M, Schlotmann R, Barenbrock M, et al. Lymphocytic Na^+-H^+ exchange increases after an oral glucose challenge. Circ Res 1995; 77:1024–1029.

80d.Reusch HP, Reusch R, Rosskopf D, et al. Na^+/H^+ exchange in human lymphocytes and platelets in chronic and subacute metabolic acidosis. J Clin Invest 1993;92:858–865.

80e.Gobel BO, Hoffman G, Ruppert M, et al. The lymphocyte NA^+/H^+ antiport: Activation in primary hypertension and during chronic NaCl loading. Eur J Clin Invest 1994;24: 529–539.

80f.Ng LL, Sweeney FP, Siczkowskib M, et al. Na^+/H^+ antiporter phenotype, abundance and phosphorylation of immortalised lymphoblasts from humans with hypertension. Hypertension 1995;25:971–977.

80g.Siffert W, Rosskopf D, Moritz A, et al. Enhanced G protein activation in immortalised lymphoblasts from patients with essential hypertension. J Clin Invest 1995;96:759–766.

80h.Mazer BD, Sawami H, Tordai A, Gelfand EW. Platelet-activating factor mediated transmembrane signaling in human B lymphocytes is regulated through a pertussis and cholera toxin sensitive pathway. J Clin Invest 1992;90:759–765.

80i. Gruska S, Ihrke R, Stolper S, et al. Prevalence of increased intracellular signal transduction in immortalised lymphoblasts from patients with essential hypertension and normotensive subjects. J Hypertens 1997;15: 29–33.

80j. Siffert W, Rosskopf D, Siffert G, et al. A C825T polymorphism in the gene encoding for the G protein subunit Gβ3 was shown to be associated with alternative splicing of the gene, enhanced intracellular signal transduction and hypertension. Nat Genet 1998; 18:45–48.

80k.Hanozet G, Parenti P, Salveti P. Presence of a potential-sensitive Na transport across renal brush border membrane vesicles from rat of the Milan hypertensive strain. Biochim Biophys Acta 1985;819:179–186.

80l. Ferrari P, Torielli L, Salardi S, et al. Na-K-Cl cotransport in released ghosts from erythrocytes of the Milan hypertensive rats. Biochim Biophys Acta 1992;111:111–119.

80m.Kuhlman PA, Hughes CA, Bennet V, Fowler VM. A new function for adducin. J Biol Chem 1996;271:7986–7991.

80n.Casari G, Barlassina C, Cusi D, et al. Association of the α adducin locus with essential hypertension. Hypertension 1995;25: 320–326.

80o.Barlassina C, Citterio L, Casari G, et al. Linkage of α adducin polymorphism to human essential hypertension [abstract]. Hypertension 1996;28:694.

80p.Cusi D, Barlassina C, Glorioso N, et al. α Adducin polymorphism in primary hypertension: linkage and association studies; relationship to salt sensitivity. Lancet 1997; 349(9062):1353–1357.

80q.Manunta P, Cusi D, Barlassina C, et al. α Adducin polymorphisms and renal sodium handling in essential hypertensive patients. Kidney Int. 1998;53:1471–1478.

81. Vincenzi FF, Morris CD, Kinsel LB, et al. Decreased calcium pump adenosine triphosphatase in red blood cells of hypertensive subjects. Hypertension 1986;8:1058–1066.

82. Resink TJ, Tkachuck VA, Erne P, Buhler FR. Platelet membrane calmodulin-stimulated calcium-adenosine triphosphatase: Altered activity in essential hypertension. Hypertension 1986;8:159–166.

82a.Shaw LM, Ohanian J, Heagerty AM. Calcium sensitivity and agonist induced calcium sensitisation in small arteries of young and adult spontaneously hypertensive rats. Hypertension 1997;Sep 3 (Pt 10) 442–448.

83. Sasaki, N. The relationship of salt intake to hypertension in the Japanese. Geriatrics 1964;19:735–774.

84. Sullivan JM. Salt sensitivity: Definition, concept, methodology, and long-term issues. Hypertension 1991;17 (suppl I):I61–I68.

85. Weinberger MH, Miller JZ, Luft FC, et al. Definitions and characteristics of sodium sensitivity and blood pressure resistance. Hypertension 1986;8:(suppl II):II127–II134.

86. Whitfield JB, Martin NG. Genetic and environmental causes of variation in renal tubular handling of sodium and potassium: A twin study. Gen Epidemiol 1985;2:17–27.

87. Miller JZ, Weinberger MH, Christian JC, Daugherty SA. Familial resemblance in blood pressure response to dietary sodium restriction in normotensive adults. J Chronic Dis 1987;40:245–250.

88. Oshima T, Matsuura H, Ishibashi K, et al: Familial influence upon NaCl sensitivity in patients with essential hypertension. J Hypertension 1992;10:1089–1094.

89. Weinberger MH, Miller JZ, Fineberg NS, et al. Association of haptoglobin with sodium sensitivity and resistance of blood pressure. Hypertension 1987;10:443–446.

90. Wilson TW, Grim CE. Biohistory of slavery and blood pressure differences in blacks today: A hypothesis. Hypertension 1991;17 (suppl II):I122–I128.

91. Jackson FLC. An evolutionary perspective on salt, hypertension, and human genetic variability. Hypertension 1991;17(suppl I): I129–I132.

91a.Svetkey LP, Chen Y-T, McKeown SP, et al. Preliminary evidence of linkage salt sensitivity in black Americans at the β adrenergic receptor locus. Hypertension 1997;29:918–922.

91b.Svetkey LP, Timmons PZ, Emovon O, et al. Association of hypertension with β2 and α2c10-adrenergic receptor genotype. Hypertension 1996;27:1210–1215.

91c.Kofima S, Inenaga T, Matsuoka H, et al. The association between salt sensitivity of blood pressure and some polymorphic factors. J Hypertens 1994;12:797–801.

92. Carmelli D, Ward MM, Reed T, et al. Genetic effects on cardiovascular responses to cold and mental activity in late adulthood. Am J Hypertens 1991;4:239–244.

93. McIlhany ML, Shaffer JW, Hines EA Jr. The heritability of blood pressure: An investigation of 200 pairs of twins using the cold pressor test. Johns Hopkins Med J 1975;136: 57–64.

94. Johnson EH, Nazzaro P, Gilbert DC. Cardiovascular reactivity to stress in black male offspring of hypertensive parents. Psychosom Med 1991;53:420–432.

95. Somova L. Parental history of hypertension in Zimbabwe and cardiovascular reactivity to psychological experimental stress. Central African J Med 1992;38:214–220.

96. Anderson NB. Racial differences in stress-induced cardiovascular reactivity and hypertension: Current status and substantive issues. Psychol Bulletin 1989;105:89–105.

97. Anderson NB, Lane JD, Muranaka M, et al. Racial differences in blood pressure and forearm vascular responses to the cold face stimulus. Psychosom Med 1988;50:57–63.

97a.Sherwood A, Hinderliter AL, Light KC. Physiological determinants of hyperreactivity to stress in borderline hypertension. Hypertension 1995;25:384–390.

97b.Lund-Johansen P. Hemodynamic patterns in the natural history of borderline hypertension. J Cardiovasc Pharmacol 1986;8(suppl 5):S8–S14.

97c.Sherwood A, Hindlighter AL. Responsiveness to α and β-adrenergic receptor agonists: Effects of race in borderline hypertensive compared to normotensive men. Am J Hypertens 1993;6:630–635.

98. Anderson NB, Lane JD, Taguchi F, Williams RB Jr. Patterns of cardiovascular responses to stress as a function of race and parental hy

pertension in men. Health Psychol 1989; 8:525–540.

99. Anderson NB, Lane JD, Taguchi F, Williams RB Jr. Race, parental history of hypertension, and patterns of cardiovascular reactivity in women. Psychophysiology. 1989;26:39–47.

100. Calhoun DA, Mutinga M, Collins AS, et al. Normotensive blacks have larger sympathetic response to cold pressor test than normotensive whites. Clin Res 1993;40:843A.

101. Williams RR, Hunt SC, Hopkins PN, et al. Familial dyslipidemic hypertension. JAMA 1988;24:3579–3585.

102. Hunt SC, WU LL, Hopkins PN, et al. Apolipoprotein, low density lipoprotein subfraction, and insulin associations with familial combined hyperlipidemia. Arteriosclerosis 1989;9:335–344.

103. Ferrannini E, Buzzigoli G, Bonadonna R, et al. Insulin resistance in essential hypertension. N Engl J Med 1987;317:350–357.

104. Brands MW, Hall JE. Insulin resistance, hyperinsulinemia and obesity-associated hypertension. J Am Soc Nephrol 1992;3: 1063–1077.

104a. Chiang FT, Chiu KC, Tseng YZ, et al. Nucleotide(-258) G-to-A transition variant of the liver glucokinase gene is associated with essential hypertension. Am J Hypertens 1997;Sep 10(9 pt 1): 1049–1052.

104b. Ikegami H, Yamato E, Fujisawa T, et al. Analysis of candidate genes for insulin resistance in essential hypertension. Hypertens Res 1996;(suppl 1): S31–S34.

105. Nojima H, Yagawa Y, Kawakami K. The Na, K-ATPase alpha 2 subunit gene displays restriction fragment length polymorphisms between the genomes of normotensive and hypertensive rats. J Hypertens 1989;7:937–940.

106. Shull MM, Hassenbein D, Loggie J, et al. Discordant segregation of Na^+, K^+-adenosine triphosphatase alleles and essential hypertension. J Hypertens 1992;10: 1005–1010.

107. Dudley CRK, Giuffra L, Reeders ST. The contribution of candidate genes to the heritable component of blood pressure. J Am Soc Neph 1990;1:488. Abstract.

108. Lifton RP, Hunt SC, Williams RR, Pouyssegur J, Lalouel J-M. Exclusion of the Na^+-H^+ antiporter as a candidate gene in human essential hypertension. Hypertension 1991; 17:8–14.

109. Nojima H, Sokabe H. Genes and pseudogenes for calmodulin in the spontaneously hypertensive rat. J Hypertens 1988;6:s231–s233.

110. Yagisawa H, Emor Y, Nojima H. Phospholipase C genes display restriction fragment length polymorphisms between the genomes of normotensive and hypertensive rats. J Hypertens 1991;9:303–307.

111. Pravenec M, Kren V, Kunes J, et al. Cosegregation of blood pressure with a kallikrein gene family polymorphism. Hypertension 1991;17:242–246.

112. Ying L-H, Zee RYL, Griffiths LR, Morris BJ. Association of a RFLP for the insulin receptor gene, but not insulin, with essential hypertension. Biochem Biophys Res Commun 1991;181:486–492.

113. Berge KE, Berg K. No effect of a Taql polymorphism in DNA at the endothelin I (EDNI) locus on normal blood pressure level or variability Clin Genet 1992;41:90–95.

114. Sun L, Schulte N, Pettinger P, et al. The frequency of alpha$_2$-adrenoceptor restriction fragment length polymorphisms in normotensive and hypertensive humans. J Hypertens 1992;10:1011–1015.

Chapter 5

Molecular Genetics of Lipid Metabolism

Carmelo Graffagnino, MD, FRCP(C)

Introduction

Atherosclerotic vascular disease is one of the most significant health problems of western society. Coronary artery disease (CAD) and cerebrovascular disease account for one-third of the deaths in the United States.[1] In 1991 about 500,000 Americans had a stroke and more than 143,000 died.[1] Coronary artery disease results almost exclusively from atherosclerosis, while a significant percent of strokes are atherothrombotic in origin and are secondary to lesions in large vessels (mostly extracranial) or occlusive intracranial small vessel disease.[2]

Atherosclerosis results from the interaction of a number of complex processes leading to vascular injury (reviewed in *Chapter 6*). Dysfunctional lipoprotein matabolism contributes significantly towards the genesis of atherosclerosis. While dyslipoproteinemia has unequivocally been shown to be associated in the genesis of ischemic heart disease[3–12] and atherosclerotic peripheral vascular disease,[13,14] the role of dyslipidemia in stroke has been controversial.

Unlike myocardial infarction which almost exclusively results from atherosclerotic coronary artery disease, stroke is a syndrome in which a focal neurologic deficit results either from brain infarction secondary to occlusive vascular disease or from brain hem-

orrhage secondary to a ruptured cerebral vessel. Atherosclerosis is rarely responsible for the latter, therefore inclusion of such patients in a study of the role of lipids in stroke is inappropriate. Some studies have in fact shown an inverse relationship between plasma cholesterol concentration and cerebral hemorrhage.[2,15,16] Most studies assessing the role of lipids in stroke have not accounted for the hypolidemic effect of acute stroke or TIA,[17,18] thereby missing hyperlipidemia by measuring lipids too early. While some earlier studies have failed to show an association between dyslipidemia and stroke,[19–21] recent reports have shown an association between dyslipidemia and ischemic cerebrovascular disease similar to that for CAD.[22,23] Total cholesterol, triglycerides and LDL-C were found to be higher in patients with stroke or TIA compared to controls.[22]

Abnormalities in lipid metabolism result from the interaction of environmental and genetic factors. A number of mutations affecting genes whose products are involved in lipid metabolism has been reported. The majority of patients with dyslipidemias however, do not harbor these mutations but rather are likely to have genetic polymorphisms which result in structurally variable gene products. These variations may thus account for the differences in response to similar environmental factors

From Alberts MJ (ed). *Genetics of Cerebrovascular Disease.* Armonk, NY: Futura Publishing Company, Inc., © 1999.

such as dietary intake of fats among individuals and between populations.

Classification of Lipid Disorders

The hyperlipoproteinemias are a heterogeneous group of disorders. This has made their classification difficult. Hyperlipoproteinemias have traditionally been classified by descriptive terms derived by Fredrickson et al.[24] This system describes phenotypes of hyperlipoproteinemia (HLP) according to the lipids and lipoproteins that are elevated regardless of the etiology (phenotypic classification). Five phenotypic classes result:

- Type I hyperlipoproteinemia (HLP): increased chylomicrons.
- Type IIa HLP: increased low-density lipoprotein (LDL).
- Type IIb HLP: increased very-low-density lipoprotein (VLDL) and LDL.
- Type III HLP: increased VLDL of abnormal electrophoretic mobility (beta VLDL).
- Type IV HLP: increased VLDL.
- Type V HLP: increased chylomicrons and VLDL.

Although simple, this system of classification has limited utility, as phenotypic groupings do not always correspond to specific genotypes. An example of this is the case of familial combined hyperlipidemia (FCHL), an autosomal dominant disorder which may be expressed as type IIa, type IIb or type IV hyperlipoproteinemia.[25] A more scientifically sound classification of the dyslipidemias is based on the molecular genetics of lipoprotein metabolism and its various regulatory elements. An overview of lipoprotein metabolism and its regulation precedes a review of the molecular genetics of lipoprotein synthesis.

Lipoprotein Metabolism

Plasma lipids are water insoluble and therefore are transported in the blood as spherical macromolecular complexes of lipid and protein.[26] Lipoproteins are spherical particles whose lipid components are triglyc-erides, phospholipids, and esterified and unesterified cholesterol. Apolipoproteins are the protein components of lipoproteins.Lipoproteins transport cholesterol and triglycerides from sites of absorption and synthesis to sites of utilization. Lipoproteins are classified into five major groups by size, density, electrophoretic mobility, and lipid and protein composition (Table 1). Apolipoproteins act as cofactors for specific enzymes or as cofactors for specific receptors involved in lipid metabolism (Table 2). Conveniently, lipoprotein metabolism can be divided into two pathways.[27] The exogenous pathway transports dietary triglyceride and cholesterol from the gut to the tissues and the endogenous pathway transports lipids synthesized in the liver to the tissues (Figure 1).

Exogenous Pathway

Cholesterol and fatty acids, absorbed by intestinal cells after a meal, are the building blocks for the synthesis of chylomicrons by the intestinal endothelial cells. The newly synthesized chylomicrons are then secreted into the lymph and eventually enter the systemic circulation. The core of chylomicrons is composed of triglycerides (mainly) and cholesteryl ester. On the surface of the chylomicron particles are the apolipoproteins (apo B-48, A-I, A-II, A-IV).[28,29] Once chylomicrons enter the lymph and the blood they acquire apolipoproteins C-II, C-III and E from HDL.[30,31] Endothelial lipoprotein lipase is activated by apo C-II and the core triglycerides of the chylomicron are hydrolyzed.[32] This liberates fatty acids for use as an energy source by muscle and for storage by adipose.

The process of hydrolysis results in a smaller chylomicron particle called a chylomicron remnant. In the process of remnant formation, free cholesterol, phospholipid, apo C-II and apo C-III are transferred to high-density lipoprotein (HDL) producing a particle enriched in cholesteryl ester. Cholesterol esters are formed in the HDL particles from excess surface cholesterol and triglycerides by the enzyme lecithin cholesterol acyltransferase (LCAT).[33] The cholesterol esters are then transferred to the chylomicron rem-

Table 1
Human Lipoproteins[62,66,463]

Lipoprotein	Diameter (A)	Density (g/ml)	Electrophoretic Mobility	Major Apolipoproteins	Transport Function
Chylomicrons	800–5,000	<0.94	origin	apo A-I apo A-II apo A-IV apo B-48 apo C-I apo C-II apo C-III apo E	dietary triglycerides
VLDL	300–800	0.96–1.006	pre-beta	apo B-100 apo C-I apo C-II apo C-III apo E	endogenous triglycerides
IDL	250–350	1.006–1.019	pre-beta	apo B-100 apo C-III apo E	endogenous triglycerides and cholesterol esters
HDL$_2$/HDL$_3$	75–120	1.003–1.21	alpha	apo A-I apo A-II apo C-I apo C-II apo C-III apo D apo E	cholesterol esters from extrahepatic tissues
Lp(a)	300–400	1.061–1.11	pre-beta	apo B-100	unknown

nants in exchange for triglycerides in a reaction mediated by the plasma protein, cholesteryl ester: triglyceride exchange protein (CETP).[34] Apo B-48 and apo E are the remaining surface apoproteins.

The liver quickly removes these remnants by an active transport system in which apo E binds to a receptor different from the LDL receptor (B/E receptor).[35,36] The LDL-receptor-like protein (LRP) was once thought to be a possible candidate for the chylomicron remnant receptor; however, recent findings showing that mice deficient in this gene had defective embryogenesis, while those with the heterozygous state had normal development and normal cholesterol and triglyceride levels, make the role of LRP in lipoprotein metabolism unclear at present.[37] Once internalized, the hep-

atocytes further digest the particles and cholesterol is esterified for use as bile acid, for endogenous lipoprotein synthesis, or for the maintenance of cell membranes.

Endogenous Pathway

The endogenous pathway of lipid metabolism has many parallels to the exogenous pathway. The triglyceride-rich particle of the endogenous pathway is the VLDL which is synthesized and secreted by the liver.[38] Like the chylomicron particles which they resemble (functionally), VLDL contain apo C-I, C-II, C-III, and E.[39] The apo B molecule on VLDL (apo B-100), however, is larger than that on chylomicrons (apo B-48) and it is synthesized in the liver rather than the intestine.[39]

Table 2
Characteristics and Function of Major Apolipoproteins[62,66,463]

Apolipoprotein	Molecular Weight	Origin	Function
apo A-I	28,016	liver intestine	activates LCAT*; structural role in HDL
apo A-II	17,414	liver intestine	possibly has a structural role in HDL
apo A-IV	44,465	intestine	cofactor for LCAT
apo B-48	264,000	intestine	needed for assembly and secretion of chylomicrons, structural role
apo B-100	512,000	liver	structural role for VLDL, IDL, LDL; ligand for LDL receptor
apo C-I	6,630	liver	possibly a cofactor for LCAT
apo C-II	8,900	liver	activates LPL#
apo C-III	8,800	liver	inhibits uptake of chylomicron and VLDL remnants by liver; may inhibit LPL
apo D	22,000	liver	may be cofactor for CETP##
apo E	34,135	liver	ligand for LDL receptor and chylomicron remnant receptor

* LCAT, lecithin-cholesterol acyltransferase;
# LPL, lipoprotein lipase;
CETP, cholesterol ester transfer protein.

VLDL particles may be taken up directly into muscle and fat cells by a receptor homologous but different from LDL[40] or they may undergo lipolysis. Endothelial lipoprotein lipase activated by apo C-II on VLDL[41,42] hydrolyzes core triglycerides and releases free fatty acids for muscle energy metabolism and adipose storage. The removal of core triglycerides results in an excess of surface phospholipids and protein which are removed and become associated with HDL.[43] Following hydrolysis the VLDL particles are also referred to as remnants. Half of the remnant particles bind to LDL receptors on hepatocytes and are taken up while the other half become intermediate density lipoproteins (IDL).[44] IDL, like chylomicron remnants, are enriched in cholesteryl ester relative to triglyceride and contain apo B-100 (apo B-48 in chylomicron remnants) and apo E. Hepatic triglyceride lipase converts IDL to LDL[45] by removing triglycerides and phospholipids. The product is a particle rich in cholesteryl ester.[46] During the conversion to LDL, IDL loses apo E and apo C while retaining a single molecule of apo B-100.[28] LDL particles function to transport cholesterol to extrahepatic cells for steroid hormone synthesis and cell membrane synthesis. Nearly 75% of the circulating cholesterol is carried by LDL. Cellular LDL uptake is mediated by a glycoprotein receptor molecule that binds to apo B-100.[47] Hepatocytes containing LDL receptors remove most of the VLDL remnants and nearly 70% of the LDL.[48,49] As the free cholesterol accumulates within the hepatocytes, HMG-CoA reductase (rate-limiting enzyme in cholesterol synthesis) synthesis decreases,[50] acyl cholesterol acyltransferase (ACAT) (esterifies free cholesterol into cholesteryl ester) is activated,[51] and LDL receptor synthesis is inhibited.[52,53]

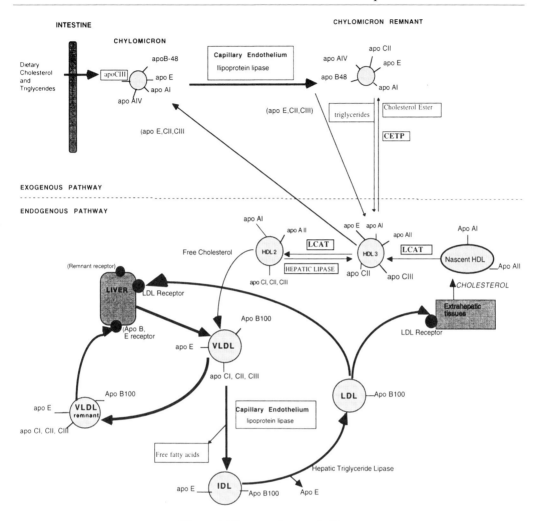

Figure 1. Lipoprotein metabolism.

Reverse Cholesterol Transport

High-density lipoprotein (HDL) removes cholesterol from extrahepatic sites and returns it to the liver. This occurs by two pathways, a direct one in which cholesterol is transferred to HDL and then carried back to the liver for direct uptake, and an indirect pathway. In the latter, cholesteryl ester is transferred first from HDL to VLDL and then taken to the liver either through IDL or later through LDL.[54]

HDL and their associated apoproteins are synthesized in liver and intestinal cells and start as disk-like structures containing apo A-I and apo A-II. The early forms of HDL are cholesterol poor and are called HDL$_3$. Initially, free cholesterol from endothelial cell membranes diffuses into the HDL$_3$ surface coat, after which circulating lecithin-cholesterol acyltransferase (LCAT) associates with it and is activated by apo A-I. Free cholesterol on the surface is then esterified and shifts to the core. As reviewed earlier, chylomicron and VLDL lipolysis results in phospholipid, free cholesterol, and apolipoprotein (A-I, A-II, C-I, C-II and C-III) transfer to HDL$_3$. The resulting particles, called HDL$_2$, are enriched in cholesteryl ester. The cholesteryl ester is transferred to triglyceride-rich lipoproteins (chylomicron remnants, VLDL remnants, IDL) in exchange for triglycerides.[55] The rate of HDL cholesterol esterification depends on the cholesteryl ester content of acceptor VLDL and LDL particles.[56] This reaction is mediated by cholesteryl ester transfer protein. The newly acquired triglycerides are hydrolyzed by hepatic triglyceride lipase, converting HDL_2 to HDL_3.

Molecular Biology of the Apolipoproteins

Apo A-I-C-III-A-IV Gene Cluster

Apolipoproteins A-I, C-III, and A-IV are considered part of the same gene family. They are clustered tightly in an area spanning approximately 15kb on the long arm of chromosome 11, region 11q13-qter.[57-60] Apo A-I is the major protein component of HDL as well as a cofactor for LCAT and possibly a ligand for an HDL receptor (Table 2). Apo C-III is a minor constituent of VLDL and HDL and inhibits lipoprotein lipase. This may result in the delay of triglyceride-rich lipoprotein clearance.[61] Apo A-IV, although a major component of chylomicrons, rapidly dissociates from these particles during lipolysis and thus its function is not clear.[62]

Apolipoprotein A-1

Apo A-I has been the most studied gene in the cluster. The apo A-I protein consists of a single polypeptide composed of 243 amino acids.[63,64] The apo A-I protein and the gene sequence contain repeated units.[63-65] Apo A-I contains eight 22-amino acid repeats and two 11-amino acid repeats.[66] It has been proposed that the lipid-binding domain is in the α-helical arrangement which is formed by the 22-mer.[67-69]

Apo A-I is synthesized in the liver and the intestine with mRNA levels which are controlled by hormonal and nutritional factors.[70-75] When apo A-I is newly synthesized it contains a six-residue amino-terminal prosegment which is removed by proteolysis after it leaves the cell.[76-78] Apo A-I functions both as a structural protein, contributing to

the formation of HDL particles, and as an effective cofactor for LCAT activity allowing cholesterol esterification on HDL particles.[79-82] Apo A-I also stimulates the efflux of cholesterol from peripheral cells, providing a substrate for the LCAT reaction and contributing to the process of reverse cholesterol transport.[83-85] The cholesteryl esters are then either transferred to VLDL and LDL[86] for removal by the liver, using the LDL receptor, or catabolized directly by peripheral tissues after acquisition of apo E.[87]

The apo A-I gene is 2 kb long and has four exons and three introns.[88] The first Apo A-I mutation described, apo A-I Milano, results from a 173Arg to Cys substitution[89] and is associated with low HDL cholesterol and apo A-I levels but not with premature atherosclerosis.[90,91] Several other apo A-I mutations have been described which are associated with abnormal lipid levels (apo A-I Giessen, Pro-143-Arg substitution[92] and apo A-I Marburg, Lys-107 deletion).[93,94] The apo A-I Marburg subjects have hypertriglyceridemia and reduced HDL cholesterol levels.[94] A mutation resulting from a frameshift starting at residue 202 and incorporation of 27 random carboxy-terminal residues has been associated with HDL deficiency, partial LCAT deficiency, and corneal opacities.[95] Familial amyloidotic polyneuropathy III is an autosomal dominant disorder resulting from a Gly-26 to Arg substitution which causes amyloid fibrils to be formed from the 83 amino-terminal residues of apo A-I.[96,97] A number of other rare mutations have been identified which affect the structure of the protein or its expression.[98] The majority of the mutations are the result of amino acid substitutions and most of the affected subjects are asymptomatic with normal lipids and lipoprotein profiles.[98,99]

In Tangier disease, levels of HDL, apo A-I, and apo A-II are low due to increased catabolism. Apo A-I however is structurally normal and the problem results from a defect in retroendocytosis of HDL with subsequent accumulation of cholesterol in macrophages and other types of cells.[100-102] Although these mutations are rare, there is convincing evidence of family clustering of plasma apo A-I levels.[103,104] There is evidence that both a single major locus as well as polygenic loci determine apo A-I levels.[105]

Apolipoprotein C-III

Apo C-III is a 79-amino acid protein which forms a major component of chylomicrons and VLDL and a minor component of HDL.[106] Apo C-III is synthesized mostly in the liver and to a small extent in the intestine.[107–109] The apo C-III mRNA encodes for a 99-amino acid protein which undergoes a number of modifications prior to and following secretion into the plasma.[110–112] Apo C-III contains a 33-, an 11-, and a 21-residue unit at nucleotides 8–40, 41–51, and 52–72, respectively.[66] It has been predicted that Apo C-III has α-helices at residues 1–39 and 54–69, β-turns at residues 39–42 and 72–75,[113] and an amphipathic α-helix at residues 40–67.[67] It has been proposed that the phospholipid-binding domain of apo C-III is localized in the 39 carboxy-terminal amino-acid residues 41–79 which contain the 11- and 22-residue repeated units.[66,114]

Apo C-III inhibits the function of lipoprotein lipase in vitro.[115–117] It also inhibits the binding of apo E-containing lipoproteins to the LDL receptor, but not to the LDL receptor-related protein (LRP).[118,119] Whereas apo E enhances the receptor-mediated binding and catabolism of lipoproteins, apo C-III has an inhibitory function[61,120,121] suggesting that apo C-III plays a role in the catabolism of triglyceride-rich lipoproteins.

The apo C-III gene is 3.2 kb long and contains four exons and three introns.[57,58] It is localized 2.5 kb downstream of the apo A-I gene and 5 kb upstream of the apo A-IV gene.[57,58,122] The apo C-III gene is transcribed in the opposite direction to that of the apo A-I and apo A-IV genes. Premature atherosclerosis has been described in individuals with a deletion of the apo A-I, apo C-III, apo A-IV loci.[123] These individuals had severe deficiencies in plasma apo C-III and HDL. A similar clinical presentation has been reported in

association with an inversion of the apo A-I, apo C-III loci.[124] Elevated apo C-III levels are also seen in individuals with type III, type IV, and type V hyperlipoproteinemia.[112,125]

Apolipoprotein A-IV

Apo A-IV has been found in human plasma, in the plasma fraction with a density > 1.21 as well as in lymph, chylomicrons, VLDL, and HDL.[126–130] Apo A-IV has twelve 22-amino acid repeats, one 11-amino acid repeat, and one 19-residue truncated repeat which lacks the first three amino-terminal residues.[66] Most of the apo A-IV is made in the intestine[122,131] and is controlled both by hormonal and nutritional factor.[74,132–134]

Apo A-IV potentiates the activation of lipoprotein lipase by apo C-III[32] as well as activating LCAT.[135] Cholesterol efflux has also been shown to be potentiated by apo A-IV in tissue culture experiments[136] suggesting a role in reverse cholesterol transport similar to that of apo A-I.

The apo A-IV gene is 2.8 kb long and contains three exons and two introns encoding for a protein that is 396 amino acids long including a 20-amino acid signal peptide.[122,137,138] As previously noted, apo A-IV is closely linked with the apo A-I and apo C-III genes[57,58] on the long arm of chromosome 11 at the 11q13-qter region.[59] There are several different phenotypes of apo A-IV encoded by two common and three rare alleles.[139–142] Five homozygous phenotypes designated apo A-IV-0 through apo A IV-4 result from structural mutations in the apo A-IV gene. The following amino acid substitutions have been described at the corresponding residues: Glu-Lys (165), Lys-Gln (167), Glu-Lys (230), Thr-Ser (347), and a four-residue insertion between amino acids 361 and 362.[143–145] Several reports have suggested that the apo A-IV-2/1 phenotype is associated with increased HDL cholesterol and decreased plasma triglyceride levels.[140,146,147] It has been shown that apo A-IV-2 activates LCAT and lipoprotein lipase better than apo A-IV-1[147,148] thus explaining the lipid and lipoprotein profile observed in people with the apo A-IV-2/1 phenotype.[140,146,147]

There have been reports of patients manifesting with premature atherosclerosis associated with deletions of the apo A-I, apo C-III, apo A-IV loci as well as others with an inversion of the apo A-I, apo C-III, apo A-IV locus.[123,149] In addition, a patient with deficient apo A-I synthesis resulting from a homozygous mutation at the codon which specifies residue 84 of apo A-I has also been reported to be associated with premature atherosclerosis.[150]

A large number of polymorphisms at the A-I-C-III-A-IV locus have been identified.[60] However, the correlation between plasma lipid levels and specific mutations appears to show a closer association within families than among populations of individuals.[151-153] Several studies suggest that polymorphisms of this locus are associated with various forms of hyperlipidemia. Although conflicting reports have been published regarding the association of the SstI and XmnI polymorphisms (found in the noncoding portion of apo C-III's last exon) and hypertriglyceridemia when all of the studies are taken together, the results support the conclusion that genetic variations of the A-I-C-III-A-IV locus are involved with certain hypertriglyceridemias.[154-159] Hypoalphalipoproteinemia has also been associated with polymorphisms of the same locus. An association has been shown between the PstI polymorphism, located 3′ to the apo A-I gene, and hypoalphalipoproteinemia as well as coronary artery disease.[160] Seventeen percent of individuals studied with coronary artery disease possessed the polymorphism compared to 3 percent of controls. The effect on HDL-cholesterol level is even more pronounced when studying kindreds with hypoalphalipoproteinemia (HDL cholesterol < 10th percentile). Ordovas et al. found that 8 of 12 index cases with such a diagnosis carried the rare allele.[160] A similar association between an SstI-MspI haplotype and HDL cholesterol levels less than the 10th percentile has been reported in an American population.[161]

Apolipoprotein A-II

Apo A-II is the second most abundant protein of HDL. It contains two identical 77-amino acid long polypeptide chains linked by disulfide bonds at residue 6.[162] Besides having a structural role, apo A-II activates hepatic lipase and may also affect LCAT activity.[163] Apo A-II is synthesized almost exclusively in the liver.[107,110] The apo A-II gene is on chromosome 1 and maps to the 1p21–1qter region.[164-166] It contains four exons and three introns. The apo A-II protein encoded by the mRNA is 101 amino acids in size including an 18-residue signal peptide.[110] There is a five-residue N-terminal prosegment which is cleaved both intracellularly and extracellularly.[110,167,168] After further molecular modification apo A-I is incorporated into lipoprotein particles[110] containing apo A-I and apo A-II (LpAI:AII).[169-173] Apo A-II has a 33-, an 11-, and a 21- residue unit at nucleotides 7–39, 40–50, and 51–71 respectively[66] forming α-helices at residues 24–30, 42–47, and 51–58, β-sheet at residues 10–21 and 60–71 and β-turns at residues 4–7, 20–23, and 31–34.[66] The phospholipid-binding domains have been localized to residues 12–31, 39–47, and 50–61.[174,175] Apo A-II and apo A-I occupy overlapping domains on the HDL surface and are thus capable of displacing each other from HDL particles.[176] HDL may therefore consist of LpAI or LpAI:AII particles. Animal studies have shown that altered apo A-II production results in a significant change in HDL structure.[177] The HDL particles in strains of mice with high levels of apo A-II are larger and have nearly twice the ratio of apo A-II to apo A-I as compared to strains with low levels of apo A-II.[60] It therefore seems that the stoichiometry for incorporation of apolipoproteins into HDL is determined in part by the relative production of apo A-II and apo A-I.

A hereditary variation for apo A-II levels in humans has been identified using an MspI polymorphism that occurs in an Alu repeat 3′ to the apo A-II gene.[178] Using this polymorphism, two different alleles of the gene were distinguished in a study population. Those individuals who were homozygous for the rare allele (allele frequency of .20) had significantly higher levels of serum apo A-II levels (P<0.005) and significantly higher apo A-II/apo A-II ratios (P=0.02) than individuals homozygous for the common allele.[178] This same polymorphism was also used to demonstrate an association with hypertriglyceridemia in a British population.[179]

Apolipoprotein B

Apolipoprotein B, the main protein particle of LDL, also plays a key function in VLDL and chylomicron particles.[180] Apo B mediates the recognition of LDL by LDL receptors. This interaction mediates the clearance of LDL from the plasma, therefore regulating cellular cholesterol biosynthesis.[180] Apo B is also needed for the assembly and secretion of LDL and VLDL.[181,182]

The gene for apo B has been mapped to the short arm of chromosome 2 between p23 and p24.[183,184] The apo B gene is composed of 29 exons and 28 introns extending over 43 kb.[185,186] Chylomicrons and VLDL each contain a single molecule of apo B-48 and apo B-100 respectively.[28,187] The nomenclature in use for description of the two apo B types was derived on the basis that apo B-48 is 48% the size of apo B-100 on SDS-PAGE.[188] Apo B-48 and apo B-100 are encoded by the same gene; however, the mRNA for apo B-48 is produced by converting nucleotide 6666 from a **C** to a **U** by means of a unique processing event.[189–192] As a result, a stop codon is inserted at position 2153 instead of glutamine, resulting in a truncated translation product (apo B-48). The introduction of the stop codon activates cryptic polyadenylation signals downstream and yields a 7 kb product.[193] Apo B takes about 10 minutes to translate on the ribosome.[194] It is synthesized as a single polypeptide chain which remains tightly associated with the original lipoprotein particle as it is transported from the endoplasmic reticulum through the Golgi, and then afterwards as the LDL particle matures

and is metabolized to LDL remnants. While it is associated with the membranes of the endoplasmic reticulum it undergoes a number of post-translational modifications such as glycosylation, phosphorylation, S-S bonding, and covalent binding to apo(a).[195] Although it is recognized that a significant amount of the apo B that is synthesized is degraded within the cell and is not secreted, the mechanism regulating its secretion is not currently understood.[195]

Apo B is susceptible to cleavage by proteolytic enzymes at a number of sites. There are two regions which are particularly vulnerable to proteolysis. One site is between residues 1280 and 1320 at the N-terminal region and the other is at the C-terminal region between residues 3180 and 3282.[196] Both kallikrein and thrombin are able to cleave apo B at lysines 1297 and 3294 yielding three fragments designated T4, T3, and T2 (in order from the N-terminal).[197] As a result, two other forms of apo B may be observed in human plasma, apo B-26(T4) and apo B-74.[188] Apo B-26 (T4) and apo B-74 are thus complementary fragments of apo B-100; apo B-26 corresponds to the N-terminal and apo B-74 the C-terminal fragment.

Several findings support the notion that the LDL-receptor-binding domain corresponds to a region centered at the T2/T3 junction at lysine 3249.[183] First, this region includes two basic sequences; residues 3147–3157 and 3359–3367. The basic sequence located between residues 3359 and 3367 has a high degree of homology to the sequence which is responsible for binding apo E to the LDL receptor. Second, LDL receptor binding has been inhibited using monoclonal antibodies which recognize epitopes flanking the region at the T2/T3 junction.[198] Further corroboration for this region being involved in LDL receptor binding comes from the finding that receptor binding can also be abolished by mutations at residue 3500 either by directly interfering with receptor binding or as a result of conformational changes which secondarily interfere with binding.[199–202] Direct protein sequencing has been used to assess the lipid-binding domains of apo B-100[203,204] suggesting that residues 1–1000 (amino-terminal end) and residues 3017–4100 (a domain near the carboxy terminal) are enriched in lipid-free peptides whereas residues 4101–4536 (the carboxy-terminal end) and residues 1701–3070 (a domain in the middle of apo B) are enriched in lipid-free peptides. The domains between residues 1001 and 1700 have both lipid-bound and lipid free peptides.[203, 204]

Early studies of apo B using alloantisera obtained from multiply transfused patients revealed ten Mendelian factors which when placed into five antithetical pairs composed the Ag system.[205,206] These five factors represent distinct allelic variations of the apo B-100 molecule. The cloning of the apo B gene has allowed the identification of a number of DNA polymorphisms, some of which have been associated with the Ag antigenic sites of apo B.[60] Butler and coworkers have reported at least 14 different haplotypes occurring in various populations.[207,208] As the study of alloantisera is likely to reveal only a fraction of genetic differences, it becomes apparent that apo B is highly polymorphic. Newly found genetic markers are now available for familial and population-based studies. Alleles of apo B may be associated with altered lipid levels and atherosclerosis. In a study analyzing data from 10 different populations, individuals with Ag(y) have been noted to have higher serum cholesterol and triglyceride levels than those with Ag(x).[209] An XbaI polymorphism near the middle of the apo B gene has been associated with variable cholesterol and triglyceride levels. Individuals homozygous for the absence of the restriction site have been found to have triglyceride and cholesterol levels lower than those heterozygous or heterozygous for the presence of the restriction site.[210,211] XbaI and EcoRI polymorphisms have also been associated with myocardial infarction.[212]

A number of rare disorders result in either underproduction or overproduction of apo B. Those disorders associated with deficient or underproduction of apo B are not associated with atherosclerotic vascular disease, but rather result in fat malabsorption and are associated with neurodegenerative disorders.

Abetalipoproteinemia is characterized by the absence of apo B (B-100 and B-48), low levels of plasma cholesterol and triglycerides, and the absence of VLDL, LDL, and chylomicrons.[213,214] Clinical manifestations include fat malabsorption, failure to thrive, neuropathy, ataxia, acanthocytosis, and retinitis pigmentosa.[213,215] It is an autosomal recessive disorder with heterozygotes carrying nearly normal levels of apo B.[216,217] Persons with this disorder have a normal-sized apo B mRNA in their liver as well as immunoreactive apo B protein in liver and intestine.[218,219] None of the patients examined to date show any deletions or gross rearrangements of the apo B gene.[218] These findings suggest the possibility that a post-translational defect may be responsible. It has recently been shown that patients with abetalipoproteinemia are deficient in the 88k-Da subunit of the microsomal triglyceride transfer protein suggesting that this activity, which normally mediates transport of cholesteryl ester triglycerides and phospholipids between membranes, may participate in VLDL and LDL assembly.[220]

Hypobetalipoproteinemia is a similar disorder characterized by an absence of both apo B-100 and apo B-48 in homozygotes; however, heterozygotes have only half of normal apo B plasma levels.[221,222] Heterozygotes are phenotypically normal whereas homozygotes lack apo B, VLDL, LDL and chylomicrons and are phenotypically identical to patients with abetalipoproteinemia.[223] It appears that a structural or regulatory mutation of the apo B gene may be involved. There is a great deal of heterogeneity in this disorder. Some patients exhibit impaired apo B synthesis and produce a truncated form of apo B,[224,225] while others have normal synthesis of apo B but increased removal of VLDL remnants.[226] A variant of apo B deficiency has been described in which only apo B-100 but not apo B-48 is absent.[227]

The hyperlipidemic states on the other hand promote atherosclerosis leading to cerebrovascular and cardiovascular disease.

Because apo B plays a central role in the metabolism of triglyceride-rich lipoproteins, genetic variations in this molecule may contribute to the various hyperlipidemias and thus to atherosclerosis. These may result either from overproduction of VLDL-apo B or from decreased clearance of LDL and lipoprotein remnants.[228,229]

Familial defective apo B-100 (FDB) is a dominantly inherited genetic disorder in which a G to A mutation at nucleotide 10708 in exon 26 of the apo B gene (causing a substitution of glutamine for arginine at codon 3500) results in a defective apo B-100 leading to decreased binding of LDL to the LDL receptors.[199] In North America and Europe, the frequency of this mutation is between 1/500 and 1/700.[230] The arginine$_{(3500)}$ →glutamine mutation changes the pK of the lysine residues which are known to be involved in the binding of apo B-100 to the LDL receptor, thus possibly changing the conformation of the binding domain.[201,202] The presence of this single mutation is associated with a 20-fold depression in receptor-binding activity.[231] All FDB subjects examined to date have the identical Arg3500 to Glu mutation.[200]

APOE-C-I-C-II Gene Complex

Apolipoprotein E

Apolipoprotein E (apo E) is normally found on HDL and LDL[232] where it serves as a ligand for recognition of lipoproteins by cellular receptors.[233] Apo E is a member of a multigene family consisting of APOI, APOII, APOIV, APOI, APOII, and APOIII. Apo E, C-I, and C-II are located on the long arm of human chromosome 19.[234] Apo E is synthesized primarily in the liver; however, it is also made by a variety of tissues whose function is not associated with lipid synthesis.[235–237] Apo E is a 299-amino acid protein with two folded structural domains.[238] Residues 1–191 make up the NH2-terminal region which binds to the LDL receptor. Several studies have localized the LDL receptor-binding domain to the region between residues 130 and 150.[239,240]

The gene for human apo E is polymorphic.[241] There are three major alleles, E2, E3, and E4 that code for three apo E isoforms, E2, E3, and E4.[242] These isoforms differ from each other by an amino acid substitution at residue 112 and/or residue 158. Apo E2 has a Cys and apo E4 has and Arg at both sites, whereas apo E3 has a Cys at site 112 and an Arg at site 158.[243] As the LDL receptor-binding domain is between residues 130 and 150, the Arg to Cys substitution at residue 158 in apo E2 produces a protein with defective receptor-binding capabilities.[244]

Lipoprotein remnants are poorly cleared and therefore accumulate, resulting in type III hyperlipidemia (elevated triglycerides, low cholesterol) in some individuals with the homozygous state.[245] Not all individuals with the E2/E2 homozygous state have type III hyperlipidemia, suggesting that other factors are also likely to be involved. What these other genetic or environmental factors are is not clear; however, familial combined hyperlipidemia, diabetes, and obesity often coexist in families with HLP III.[246] A common DNA polymorphism of the apo E gene detected with the enzyme HpaI has been shown to have a rare allele frequency of 0.97 in 39 individuals with HLP III and E2/E2 phenotype.[247] Individuals with E2/E2 phenotype and normolipidemia had the same allele frequency as the general population (0.38). This finding raises the possibility that there are genetic variations in or near the apo E gene that contribute to HLP III and are in linkage disequilibrium with the Hpa I polymorphism. In contrast to apo E2, the apo E4 allele has been associated with hypercholesterolemia and HLP 5.[248–250] Other rare alleles of apo E have also been reported in association with hyperlipidemia and atherosclerosis.[251]

Apo C-I and C-II

Apo C-I and apo C-II are peptides associated mostly with triglyceride-rich lipoproteins. Apo C-I is a 57-amino acid protein which forms a major component of VLDL and a minor component of HDL.[252,253] The apo C-I gene is located on the long arm of chromosome 19,[234] 5.5 kb downstream of the apo E gene, and the apo C-I pseudogene is located 7.5 kb 3′ to the apo C-I gene.[254] The apo C-II gene is 20 kb downstream from the apo C-I′ pseudogene[255] and the LDL receptor genes are closely linked with the apo E, apo C-I cluster.[234,256,257]

Apo C-I is synthesized almost exclusively in the liver[107,109] from an mRNA which encodes for an 83-amino acid protein with a 26-amino acid signal peptide.[254,258] Apo C-I activates LCAT(766) as well as lipoprotein lipase.[62] Apo C-I and to a lesser extent apo C-II inhibit the binding of apo E-containing lipoproteins such as β-VLDL to the LDL receptor-related protein (LRP) as well as to the LDL receptor.[118,119]

Apo C-II is a major component of chylomicrons and VLDL and a minor component of HDL. It is mostly synthesized in the liver and to a minor extent in the small intestine.[107–109] The human apo C-II gene is 3.3 kb long and contains four exons and three introns.[259] It is closely linked to the genes for apo E, apo C-I, and the LDL receptor on the long arm of chromosome 19. The mRNA encodes for a 101-amino acid protein with a 22-residue signal peptide although the mature protein is 79 amino acids long.[260] Apo C-II contains a 33- and a 22-residue unit at nucleotides 18–50 and 51–72, respectively.[66] Apo C-II activates lipoprotein lipase but not hepatic lipase.[223] Apo C-II deficiency is characterized by a marked elevation of plasma triglycerides and chylomicrons and decreased HDL and LDL (HLP 1). The apo C-II deficiency appears to result both from mutations resulting in regulatory as well a structural abnormalities.[261–264]

Apolipoprotein D

Apo D is a glycoprotein found in HDL particles.[265] It is a glycoprotein which contains 18% carbohydrate by weight and was previously known as apo A-III.[266,267] The apo D gene has been mapped to the distal long arm of chromosome 3 in the 3p14.2-qter region.[268] The apo D gene is 12kb long and is made up of six exons and five introns[268,269] encoding a 189-amino acid protein with a 20-amino acid signal peptide.[268] It is not structurally related to the other apolipoproteins but rather belongs to the α-2 microglobulin superfamily.[270] Apo D is synthesized by numerous tissues including brain, liver, intestine, heart, kidney, pancreas, adrenal gland, spleen, urinary bladder, and skin.[269,271]

Although apo D has been found in association with LCAT, CETP and apo A-I, suggesting a possible role in the esterification and transfer of cholesterol from HDL, VLDL and LDL, the precise role of apo D remains to be clarified.[272–274]

Lipoprotein (a) and Apo(a)

Lipoprotein (a) [Lp(a)] is a distinct type of lipoprotein first described by Berg in 1963.[275] It was initially thought to be a genetic variant of LDL which was inherited as a single autosomal dominant trait present in 30% of Northern Europeans[275]; however, later studies showed that Lp(a) was present in all people, with levels ranging from less than 1 mg/dl to greater than 100 mg/dl.[276,277] Lp(a) is composed of two very different components linked to each other covalently by a disulfide bridge.[221,278] One component has the structural and functional properties of LDL: it binds to LDL receptors, it contains one molecule of apo B-100, and its density is 1.04 grams per liter.[278] The other component, apo(a), is a highly glycosylated protein with molecular masses ranging from 200 to 700 kD.[279,280] The heterogeneity in size appears to be genetically controlled.[221]

There is a high degree of homology between apo(a) and plasminogen. This was first suspected based on immunological studies showing cross-reactivity of apo(a) and plasminogen[281] and then confirmed by sequencing the apo(a) at the protein and cDNA levels.[282,283] Plasminogen is a serine protease zymogen of the fibrinolytic system. It has a trypsin-like protease domain and five tandemly repeated homologous domains called kringles. Kringles are convoluted structures in which the amino acid chains are interconnected by three disulfide bonds. Each plasminogen kringle is distinctive, with some containing binding sites for fibrin while others bind regulatory enzymes involved in thrombolysis. The apo(a) gene contains a single serine protease domain with 94% homology to that of the plasminogen gene and two plasminogen-like kringle domains.[284] One of the kringle domains is present as a single copy and is homologous to kringle 5 of plasminogen. The other kringle domain is present in multiple copies and resembles plasminogen's kringle 4. There is, however, a single amino acid substitution (serine for arginine) in apo(a) which prevents it from being converted to an active protease by tPA, streptokinase, or urokinase as would occur in plasminogen.[283] Apo(a) binds to apo B-100 covalently by a disulfide bond as a result of one extra unpaired cysteine residue in one of the kringle-4 repeats. This high degree of homology to plasminogen suggests that apo(a) and plasminogen may have derived from a common ancestral gene. Linkage studies have been conducted between plasminogen protein variation and Lp(a) phenotype, Lp(a) lipoprotein level or various isoforms of the polypeptide chains with the Lp(a) antigens. LOD scores in excess of 10 have confirmed that the three Lp(a) phenomena mentioned above are closely linked to plasminogen and thus likely resulting from a single gene on chromosome 6.[285–287]

The apo(a) locus is the major locus that determines plasma Lp(a) concentrations.[221,288,289] Individuals may have several apo(a) isoforms ranging in size from 400 to 700 kD. An individual may have no major apo(a) isoforms, or one or two (never more than two).[284] Six different isoforms designated F, B, S1, S2, S3, and S4 (S stands for slow, F for fast, and B for the position of apo B-100) have been identified. The plasma concentration of Lp(a) is inversely related to the molecular mass of the apo(a) isoforms present and is related to the number of kringle-4

domains present.[221,288,289] The apo(a) alleles affect Lp(a) concentrations in an additive manner.[290] Heterozygotes therefore have plasma Lp(a) concentrations that nearly equal the sum of the respective single-band types. The variability in Lp(a) concentrations cannot however be completely explained by the variation at the apo(a) locus as it has been found that even within a single phenotype there is a heterogeneous distribution in Lp(a) concentration. It has been estimated that approximately 40% of the variability in the Lp(a) concentration can be explained by the apo(a) locus variability.[290] It is not known for certain what accounts for the remaining variability in concentrations of Lp(a); however, it has been shown that LDL receptor defects affect Lp(a) concentrations by a mechanism not directly related to impaired apo B-100 binding.[291]

The major site of synthesis of plasma apo(a) appears to be the liver.[284] It is postulated that LDL and apo(a) are secreted independently from each other and assembled in the plasma. The site and mechanism of catabolism of Lp(a) are not known.

It is currently not known what the function of Lp(a) is in human metabolism. The homology of apo(a) to plasminogen has led some to speculate that Lp(a), by its ability to bind to fibrin, provides a mechanism for the delivery of cholesterol to areas where injury has occurred thus aiding with wound healing.[292] Lp(a) also has been postulated to regulate thrombosis. In vitro studies have shown that Lp(a) prolongs fibrinolysis time as stimulated by streptokinase.[281] Lp(a) inhibits plasminogen activation by competitive inhibition at low concentration and by noncompetitive inhibition at high concentration. Lp(a) competes with plasminogen for binding to the plasminogen receptor.[293–295]

Although the physiologic function of Lp(a) is not known, it is recognized that elevated levels of Lp(a) have been associated with various forms of atherosclerotic vascular disease. Initial studies focused on coronary artery disease, however peripheral vascular disease and cerebrovascular disease have also been shown to be associated with Lp(a) levels in the top quartile.[23,296–302]

Molecular Biology of Lipoprotein Receptors

Low-Density Lipoprotein (LDL) Receptor

LDL receptors are located on the surface of fibroblasts, endothelial cells, hepatocytes and a number of other human cell types.[223] LDL, the major cholesterol transport protein in human plasma, binds with high affinity to LDL receptors located in clusters in coated pits which have clathirin as their main protein component.[303,304–306] Besides LDL which contain only apo B, the LDL receptor also binds apo E-containing lipoproteins such as IDL, β-VLDL, and HDL with apo E (HDLwE).[307–309] The coated pits, with the LDL receptor complex, are internalized in the form of endocytic vesicles called endosomes which carry LDL to lysosomes.[52,310] Within the endosomes, the lipoprotein receptor dissociates in the setting of an acidic pH, allowing the free receptor to recycle back to the cell surface.[311] As cholesterol accumulates within the cell, LDL receptor synthesis is decreased.[52,53] The number of LDL receptors synthesized is regulated at the level of transcription.[312]

The LDL receptor gene is located at the distal end of the short arm of chromosome 19 (p13.1-p13.3).[256,257] It is linked closely to apo E, apo C-I, and apo C-II.[234,255] It is 45 kb long and is composed of 18 exons and 17 introns.[313] The human LDL receptor cDNA and gene encode a protein with 860 amino acids with a 21-residue signal sequence and five domains.[313,314] Mutations in domain 1 affect the binding of LDL to the LDL receptor.[315] This Cys rich domain has seven homologous sequences which contain an average of 40 amino acids each and are homologous to the C8 component of complement.[223] Domain 2 has homology to a segment of the epidermal growth factor (EGF) precursor protein and consists of 400 amino acids. Domain 3 has 58

amino acids, of which 18 are Ser and are the sites of O-linked glycosylation. Domain 4 is the membrane spanning region and domain 5 is the cytoplasmic tail containing 50 amino acids.[223] Between residues 804 and 807 of the cytoplasmic domain is a conserved NPVY sequence required for the movement of the receptor to the coated pits. The EGF precursor homologous region has 2 adjacent repeats designated A and B.[313] Deletions of mutations within each of the repeat sequences has shown that repeats 2,3,6,7, and A are required for maximum binding of LDL but not β-VLDL, whereas repeat 5 is required for maximum binding of both LDL and β-VLDL.[315,316]

Familial hypercholesterolemia (FH) results from mutations that eliminate or almost eliminate the function of the LDL receptor. Four classes of mutations result which affect the LDL receptor. Class I mutants result from insertions or deletions in the receptor gene or nonsense mutations leading to premature chain termination and are thus characterized by the absence of receptor protein.[317–323] Class 2 mutants have defective modification of the precursor N- and O-linked carbohydrate resulting from amino acid substitutions or deletions in exons 2, 4, 6, 11, and 14.[223,324–331] The resulting mutant precursor protein is trapped in the endoplasmic reticulum.[332] Deletions or insertions in exons 2–8 in the cysteine-rich domain of the receptor result in class 3 mutations in which the receptors reach the surface but have reduced affinity for LDL.[331,333,334] Class 4 mutants all have alterations within the first 21 residues of the cytoplasmic tail and thus do not have the ability to cluster into the coated pits. Although these receptors bind normally to LDL, they can not internalize the complex.[335]

Individuals heterozygous for such mutations have twofold to threefold elevations of plasma LDL while homozygotes have sixfold to eightfold elevations.[47] The LDL accumulation in the plasma is due to two factors. Defects of the LDL receptors result in a decreased rate of LDL removal from the circulation (Figure 1) as well as an overproduction of LDL. Normally a significant fraction of

IDL is cleared by the LDL receptor before it is converted to LDL. This process is mediated by the interaction of the receptor with apo E, which is present on IDL and has a higher affinity for the receptor than apo B-100. Therefore, reduced levels of receptors result in the further shunting of IDL to LDL.[27]

Numerous heterogeneous mutations have been reported which affect the LDL receptor.[47,324,336] These mutations involve various aspects of receptor expression, including receptor synthesis, intracellular transport, binding to LDL and clustering in coated pits.

Chylomicron Remnant Receptor (Apolipoprotein E/α2 Macroglobulin [E/α2M] Receptor)

The presence of a high-affinity chylomicron remnant receptor was initially designated as the apo E receptor following work with rat liver plasma membranes and, later, dog hepatic membranes.[337,338] Chylomicron remnant catabolism is normal in patients lacking LDL receptors (homozygous for familial hypercholesterolemia), further demonstrating the existence of a unique and separate receptor for apo E.

cDNA clones coding for a 4224-amino acid protein with structural and biochemical similarities to LDL (B/E) receptor have been isolated.[339] The protein has been named lipoprotein receptor-related protein (LRP).[340] LRP mRNA and protein are found in a number of different tissues including brain, lung, and liver.[339] LRP and LDL are structurally similar at a number of sites including the cysteine-rich ligand-binding domain, the EGF precursor homologous region, and the 100-residue, carboxy-terminal cytoplasmic tail.[340] LRP does not have the O-linked sugar domain but rather replaces it with six repeated sequences found in the EGF. β-VLDL (containing apo E), after binding to the LRP in a calcium-dependent process, is internalized and degraded by lysosomes with the LRP recycled in a similar manner to the LDL receptor.[341] LRP has a number of other functions including uptake of activated α2 macroglobulin-pro-

tease complexes, plasminogen activator-inhibitor complexes, lactopherin, and lipoprotein lipase.[342–344] To date, there have been no mutations reported in LRP in humans or animals.

High-Density Lipoprotein-Binding Protein (HDLBP)

There are epidemiological studies showing that levels of high-density lipoprotein (HDL) are inversely related to the risk of coronary artery disease. This may be due to the role of HDL in reverse cholesterol transport.[345] There is evidence that cholesterol shifts from cells to HDL lipoproteins involve the binding of HDL to specific plasma membrane receptors.[346–349] The cDNA for a 110-kDa cellular membrane protein that binds HDL avidly has been cloned.[350,351] The gene for high-density lipoprotein-binding protein has been localized to human chromosome 2q37.[352] Specific mutations in the HDLBP have yet to be identified.

Molecular Biology of the Enzymes of Lipid Metabolism

Lipoprotein Lipase (LPL)

Lipoprotein lipase (LPL) functions in the hydrolysis of the 1,3 ester bonds of triglycerides in the core of circulating chylomicrons and VLDL, producing free fatty acids and mostly 2-monoglycerides.[353,354] It functions on the surface of vascular endothelium where it is anchored by a membrane-bound glycosaminoglycan chain and can be released into plasma by injection of heparin.[355–357] Endogenous triglycerides and free fatty acids are hydrolyzed by LPL in the presence of cofactor apo C-II[115,354,358,359]

Lipoprotein lipase is synthesized in high levels by a number of tissues throughout the human body including adrenal, brain, heart, muscle and adipose tissue.[360–363] Lipoprotein lipase mRNA levels and LPL enzymatic activity are regulated both by hormones as well as by nutritional factors.[364–378]

The gene for human LPL has been localized to chromosome 8, region 8p22.[379,380] It contains ten exons and nine introns which produce a 475-amino acid protein with a 27-residue signal peptide.[379, 380] Lipoprotein lipase, hepatic lipase and pancreatic lipase belong to a common lipase gene family in which all three enzymes have a triad of catalytic residues, Asp-His-Ser.[381]

Lipoprotein lipase deficiency is associated with type I hyperlipoproteinemia (HLP-I) in which there is accumulation of triglyceride-rich lipoproteins (VLDL, chylomicrons) in plasma as well as decreased LDL and HDL levels.[353,354,382] Pancreatitis, eruptive xanthomas, abdominal pain, and lipemia retinalis are other features of HLP-I.[383–385] LPL deficiency appears to be transmitted as an autosomal recessive trait with a frequency of $1/10^6$ for homozygotes and $1/500$ for heterozygotes.[382,386] Heterozygous patients may occasionally present with a lipid profile resembling that of familial combined hyperlipoproteinemia, depending on the influence of other genetic and environmental factors.[387,388] Patients with LPL deficiency show impaired catabolism of triglyceride-rich lipoproteins, which in turn results in decreased production of LDL and HDL. The low HDL levels in these patients may then increase their risk for atherosclerotic disease.[387]

Patients with LPL deficiency, however, have been shown to have enzyme levels at or above normal, suggesting that the enzyme is nonfunctional and thus results from a structural gene mutation.[389]

Three classes of LPL mutations have been described. Patients with class I mutations lack immunoreactive LPL mass in their postheparin plasma either as a result of nonsense mutations early in the protein sequence or from deletions or insertions in the LPL gene.[223] Class II mutations occur in exons 4–6, which contain the residues involved in the binding of lipids as well as the catalytic triad consisting of Ser-132, Asp-156, and His-

241.[390] These patients have immunoreactive LPL mass in the postheparin but not in the preheparin plasma. Class III mutations result in patients with immunoreactive LPL mass both in the preheparin and postheparin plasma but with conformational changes which alter the heparin-binding site of LPL encoded by exon 2.[391] There have been a number of mutations reported in the LPL gene. These have recently been reviewed[223,386,392] A number of other variants of LPL deficiency have been identified including individuals showing tissue-specific deficiency, combined LPL and hepatic lipase (HL) deficiency, combined LPL and apo C-II deficiency, and transient LPL deficiency.[389] In these cases it appears that the mutations affect pathways common to the final expression of all of these proteins as the gene for LPL is on chromosome 8, HL is on chromosome 15 and apo C-II is on chromosome 19.

Hepatic Lipase (HL)

Hepatic lipase (HL) functions primarily in the metabolism of cholesteryl ester-rich lipoproteins. It is a glycoprotein with subunit molecular mass of 62.5 kD[393] or 53kD.[394,395] Although it does not require Apo C-II for activation,[354] how HL functions still remains poorly understood.[396] HL is synthesized in the hepatocytes[392,397] and then modified with N-linked carbohydrate chains in order to allow its secretion.[398] HL is localized on the surface of sinusoidal endothelial cell.[399] As with LPL, it can be released from its binding site with heparin.[400]

The cDNA for HL has been isolated and sequenced and appears to be a member of a family of lipases that includes LPL and pancreatic lipase.[401,402] The human HL gene contains nine exons and eight introns encoding for a 499-amino acid protein with a 23-residual signal peptide.[403,404] The gene for HL is on chromosome 15, region 15q21.[380,403]

Deficiencies of HL have been described. These patients have abnormally triglyceride-rich LDL and elevated HDL and β-VLDL.[405–409] Two families with complete HL deficiency have been found to have a substitution of Met-383 for Thr; however, it has been postulated that a second mutation may also exist in the other allele as the subjects were heterozygous for the mutation.[410]

Lecithin:Cholesterol Acyltransferase (LCAT)

Lecithin:cholesterol acyltransferase (LCAT) is involved in the esterification of HDL-cholesterol.[84,411] LCAT is activated by apo A-I and to some extent apo C-I, apo A-IV, and apo E.[81,135,148,412,413] The enzyme is a glycoprotein which is synthesized mainly by the liver and after modification with N-linked carbohydrate chains, it is secreted into the plasma.[414] LCAT mRNA has been detected in a number of other tissues including brain and testes.[415]

Human LCAT cDNA has been cloned and sequenced.[416,417] It contains six exons and four introns which encode for a 440-amino acid protein with a 24-residue signal peptide.[417] The LCAT gene is located on the long arm of chromosome 16, region 16q22.[415]

Familial LCAT deficiency is a rare autosomal recessive disorder associated with corneal opacities and xanthomas. It is characterized by abnormalities of all lipoprotein subclasses.[33,418] The abnormalities include: large- and intermediate-size LDL_2 particles with increased concentrations of unesterified cholesterol and lecithin but depleted of apo B and cholesterol esters; (βVLDL enriched in unesterified cholesterol, VLDL particles with increased apo E and apo C-I, and decreased apo C-II and apo C-III concentrations[419,420]; and discoidal HDL particles with abnormal lipid and apoprotein composition.[419,421] The discoidal HDL particles contain either apo E or apo A-I and apo A-II and have increased concentration of unesterified cholesterol and lecithin compared to normal HDL.[419] As a result of these lipid abnormalities, a number of tissue and organ abnormalities result such as renal failure.[33,85,422,423–425]

Patients who are homozygous for LCAT deficiency have a radioimmunoassay mass of 0–20% of normal amounts compared to heterozygotes which have 50% of the normal values suggesting that the defect is secondary to a structural mutation in the LCAT gene that inactivates the enzymatic function.[426–430] A number of molecular defects underlying some cases of LCAT deficiency have been identified. These include substitutions, frameshifts and insertions.[431–438] Clinical presentations resembling LCAT deficiency may also present in some patients with mutations involving apo A-I as the latter is an activator of LCAT.[93]

Cholesterol Ester Transfer Protein (CETP)

Cholesterol ester transfer protein (CETP), also known as lipid transfer protein I, functions in the transfer of cholesterol esters among lipoproteins thereby playing a very important role in cholesterol homeostasis.[439,440] It is a glycoprotein with a Mr of 74 kD and with one minor and one major isoform with apparent isoelectric points 5.6 and 5.3 respectively.[441,442]

The major sites of CETP synthesis are the liver, small intestine, spleen, adipose tissue, muscle, and the adrenal as well as the kidney.[440,443] Human cDNA has been isolated and characterized.[440] The gene for CETP is 25 kb long, contains 16 exons and 15 introns, and is located in the q12-q21 region of human chromosome 16.[440,444,445]

Deficiencies of CETP activity have been reported and are associated with unusually large HDL particles, hyperalphalipoproteinemia, increased apo A-IV, apo E levels, and decreased LDL cholesterol and plasma apo B levels.[86,446–448] A patient reported by Brown et al. had zero cholesteryl ester and triglyceride transfer activity while the phospholipid transfer activity was 47% of the control values.[86,449] A G-to-A transition at the splice site of intron 14 has been found on molecular analysis of the gene of homozygous CETP patients as well as in heterozygous relatives.[449] An inhibitor of CETP is also present in the serum of these patients.[450]

Lipid Transfer Protein II (LTP-II)

Alberts et al. characterized two distinct lipid transfer proteins, LTP-I and LTP-II.[439] LTP-I is identical to CETP (see previous section). LTP-II is a separate and distinct protein which has been purified and has a molecular mass of 69 kD and apparent isoelectric point of 5.[451] This enzyme facilitates the exchange and transfer of phospholipids from VLDL to HDL; however, it lacks cholesteryl ester and triglyceride transfer activities.[451]

HMG-CoA Reductase

HMG-CoA reductase is the rate-limiting enzyme in cholesterol synthesis. Cellular sterol concentrations regulate its activity at various levels including transcription and turnover.[452–454] The HMG-CoA reductase gene is located on the long arm of chromosome 5 in humans.[257,455] Familial combined hyperlipidemia appears to be associated with increased cholesterol production[456,457] as do a high fraction of individuals with HLP-IV.

Other Candidate Genes

A number of other genes that play a central role in plasma lipid metabolism have been biochemically characterized and may be associated with genetic variations. Among these are: hepatic receptors involved in the apo E-mediated uptake of lipoprotein remnants[458,459]; an HDL receptor which may facilitate uptake or transfer of cholesterol from cellular membranes to HDL[460]; the scavenger receptor which may be involved in foam cell formation[461]; and apo D, a protein complexed with LCAT and comprising about 5% of HDL protein.[462]

Conclusion

It is clear at present that although a great deal is known regarding the molecular and genetic mechanisms of lipoprotein control, there is yet much that is not known. The challenge for the future remains in obtaining an understanding of the complex interactions between environmental influences and genetic predisposition. With such knowledge it may be possible to prevent the development of atherosclerotic injury to the vessels by genetic and environmental modification at an early age.

References

1. American Heart Association. Heart and Stroke Facts, 1994. AHA Statistical Supplement, 1993.
2. Wolf P, Belanger A, D'Agosstino R. Management of risk factors. Neurol Clin 1992;10: 177–191.
3. Tell G, Crouse J, Furberg C. Relation between blood lipids, lipoproteins, and cerebrovascular atherosclerosis: A review. Stroke 1988;19:423–430.
4. Caltelli W, Garrison R, Wilson P, et al. Incidence of coronary heart disease and lipoprotein cholesterol levels. JAMA 1986;256: 2835–2838.
5. Miller G, Miller N. Plasma high-density lipoprotein concentration and development of ischemic heart disease. Lancet 1975;1: 16–19.
6. Gordon T, Castelli W, Hjorland M, et al. High density lipoproteins as a protective factor against coronary heart disease. Am J Med 1977;61:707–716.
7. Krauss R. Relationship of intermediate and low-density lipoprotein subspecies to risk of coronary artery disease. Am Heart J 1987; 113:578–582.
8. Steiner G, Schwartz L, Shumak S, Poapst M. The association of increased levels of intermediate-density lipoproteins with smoking and with coronary artery disease. Circulation 1987;75:124–130.
9. Austin M, King M, Vranizan K, Krauss R. Atherogenic lipoprotein phenotype: A proposed genetic marker for coronary heart disease risk. Circulation 1990;82:495–506.
10. Martin M, Hulley S, Browner W, et al. Serum cholesterol, blood pressure, and mortality: Implications from a cohort of 361,662 men. Lancet 1986;2:933–936.
11. Keys A. Seven Countries: A Multivariate Analysis of Death and Coronary Heart Disease. Cambridge, MA: Harvard University Press, 1980.
12. Gordon T, Kannel W, Castelli W, Dawber T.

Lipoproteins, cardiovascular disease and death: The Framingham study. Arch Int Med 1981;141:1128–1131.

13. Senti M, Pedro-Botet J, Nogues X, et al. Lipoprotein profile in men with peripheral vascular disease: Role of intermediate density lipoproteins and apoprotein E phenotypes. Circulation 1992;85:30–36.

14. Juergens J, Bernatz P. Atherosclerosis of the extremities. In Juergens J, Spittell J Jr, Fairbairn I, (eds). Peripheral Vascular Diseases. New York: WB Saunders Co, 1980, 253–293.

15. Iso H, Jacobs D, Wentworth D, et al. Serum cholesterol levels and 6-year mortality from stroke in 350,977 men screened for the Multiple Risk Intervention Trial. N Engl J Med 1989;320:904–910.

16. Kagan A, Popper J, Rhoads G, Yano K. Dietary and other risk factors for stroke in Hawaiian Japanese men. Stroke 1985;16:34–38.

17. Mendez I, Hachinski V, Wolfe B. Serum lipids after stroke. Neurology 1987;37:507–511.

18. Hollanders F, Shafar J, Burton P. Serum lipid changes following the completed stroke syndrome. Postgrad Med J 1975;51:386–389.

19. Rhoads G, Feinleib M. Serum triglyceride and risk of coronary heart disease, stroke and total mortality in Japanese-American men. Atherosclerosis 1983;3:316–322.

20. Noma A, Matsushita S, Komori T, et al. High and low density lipoprotein cholesterol in myocardial infarction and cerebral infarction. Atherosclerosis 1979;32:327–331.

21. Tilvis R, Erkinjuntii T, Sulkava R, et al. Serum lipids and fatty acids in ischemic strokes. Am Heart J 1987;113:615–619.

22. Hachinski V, Graffagnino C, Beaudry M, et al. Hyperlipidemia is a significant risk factor for stroke and transient ischemic attacks. Arch Neurol 1996;53:303–308.

23. Pedro-Botet J, Senti M, Nogues X, et al. Lipoprotein and apolipoprotein profile in men with ischemic stroke: Role of lipoprotein(a), triglyceride-rich lipoproteins, and apolipoprotein E polymorphism. Stroke 1992;23:1556–1562.

24. Fredrickson D, Levy R, Lees R. Fat transport and lipoproteins: An integrated approach to mechanisms and disorders. N Engl J Med 1967;276:32,94,148,215,273.

25. Goldstein J, Schrott H, Hazzard E, et al. Hyperlipidemia in coronary artery disease. II. Genetic analysis of lipid levels in 176 families and delineating of a new inherited disease, combined hyperlipidemia. J Clin Invest 1973;52:1544–1586.

26. Scanu A, Spector A. Biochemistry and Biology of Plasma Lipoproteins. New York: Dekker, 1986, 514.

27. Goldstein J, Kita T, Brown M. Defective lipoprotein receptors and atherosclerosis: Lessons from an animal counterpart of familial hypercholesterolemia. N Engl J Med 1983;309:288–296.

28. Elovson J, Chatterton J, Bell G, et al. Plasma very low density lipoproteins contain a single molecule of apolipoprotein B. J Lipid Res 1988;29:1461–1473.

29. Bisgair C, Glickman R. Intestinal synthesis, secretion, and transport of lipoproteins. Annu Rev Physiol 1983;45:625–636.

30. Ginsberg H. Lipoprotein physiology and its relationship to atherogenesis. Endocrinol Metab Clin North Am 1990;19:211.

31. Imaizumi K, Fainaru M, Havel R. Composition of proteins of mesenteric lymph chylomicrons in the rat and alterations produced upon exposure of chylomicrons to blood serum and serum proteins. J Lipid Res 1978;19:712–722.

32. Goldberg I, Scheraldi C, Yacoub L, et al. Lipoprotein apoC-II activation of lipoprotein lipase: Modulation by apolipoprotein A-IV. J Biol Chem 1990;265:4266–4272.

33. Norum K, Gjone E, Glomset J. Familial lecithin:Cholesterol acyltransferase deficiency, including fish eye disease. In Scriver C, et al (eds). Metabolic Basis of Inherited Disease. New York: McGraw-Hill, 1989, 1181–1194.

34. Hesler C, Tall A, Swenson T, et al. Monoclonal antibodies to the Mr 74,000 cholesteryl ester transfer protein neutralize all of the cholesteryl ester and triglyceride transfer activities of human plasma. J Biol Chem 1988;263:5020–5023.

35. Mahley R, Hui D, Innerarity T, Beisiegel U. Chylomicron remnant metabolism: Role of hepatic lipoprotein receptors in mediating uptake. Arteriosclerosis 1989;9 (suppl I):I-14–I18.

36. Hui D, Innerarity T, Milne R, et al. Binding of chylomicron remnants and beta-very low density lipoproteins to hepatic and extrahepatic lipoprotein receptors: A process independent of apolipoprotein B48. J Biol Chem 1984;259:15060–15068.

37. Hertz J, Clouthier D, Hammer R. LDL receptor-related protein internalizes and degrades uPA-PAI-1 complexes and is essential for embryo implantation. Cell 1992;71:411–421.

38. Havel R, Kane J. Structure and metabolism of plasma lipoproteins. In Scriver C, et al (eds). Metabolic Basis of Inherited Disease. New York: McGraw-Hill, 1989, 1129–1138.

39. Schumaker V, Lembertas A. Lipoprotein metabolism: Chylomicrons, very-low-density lipoproteins and low-density lipoproteins. In Lusis A, et al (eds). Molecular Genetics of

Coronary Artery Disease. Basel, Switzerland: Karger, 1992.

40. Yamamoto T, Takahashi S, Sakai J, Kawarabayasi Y. The very low density lipoprotein receptor. TCM 1993;(vol 3):144–148.

41. Eisenberg S, Olivecrona T. Very low density lipoprotein: Fate of phospholipids, cholesterol and apolipoprotein during lipolysis in vitro. J Lipid Res 1979;20:614–623.

42. Breckenridge W, Little J, Steiner G, et al. Hypertriglyceridemia associated with deficiency of apolipoprotein C-II. N Engl J Med 1978;298:1265–1273.

43. Patsch J, Gotto AJ, Olivecrona T, Eisenberg S. Formation of high density lipoprotein 2-like particles during lipolysis of very low density lipoproteins in vitro. Proc Natl Acad Sci USA 1978;75:4519–4523.

44. Havel R. The formation of LDL: Mechanisms and regulation. J Lipid Res 1984;25:1570–1576.

45. Demant T, Carlson L, Holmquist L, et al. Lipoprotein metabolism in hepatic lipase deficiency: The effect of hepatic lipase on high density lipoprotein. J Lipid Res 1988;29:1603–1611.

46. Auwerx J, Marzetta C, Hokanson J, Brunzell J. Large buoyant LDL-like particles in hepatic lipase deficiency. Arteriosclerosis 1989;9:319–325.

47. Brown M, Goldstein J. A receptor-mediated pathway for cholesterol homeostasis. Science 1986;232:34–47.

48. Spady D, Bilheimer D, Dietschy J. Rates of receptor dependent and independent low density lipoprotein uptake in the hamster. Proc Natl Acad Sci USA 1983;80:3449–3503.

49. Pittman R, Carew T, Attie A, et al. Receptor-dependent and receptor-independent degradation of low density lipoprotein in normal rabbits and in receptor-deficient mutant rabbits. J Biol Chem 1982;23:7994–8000.

50. Brown M, Dana S, Goldstein J. Regulation of 3-hydroxy-3methylglutaryl coenzyme A reductase activity in cultured human fibroblasts: Comparison of cells from a normal subject and from a patient with homozygous familial hypercholesterolemia. J Biol Chem 1974;249:789–796.

51. Goldstein J, Dana S, Brown M. Esterification of low density lipoprotein human fibroblasts and its absence in homozygous familial hypercholesterolemia. Proc Natl Acad Sci USA 1974;71:4288–4294.

52. Goldstein J, Brown M. The low density lipoprotein pathway and its relation to atherosclerosis. Annu Rev Biochem 1977;46:897–930.

53. Brown M, Goldstein J. Regulation of the activity of the low density lipoprotein receptor in human fibroblasts. Cell 1975;6:307–316.

54. Tall A. Plasma high density lipoproteins: Metabolism and relationship to atherosclerosis. J Clin Invest 1990;86:379.

55. Nishida H, Kata H, Nishida T. Affinity of lipid transfer protein for lipid and lipoprotein particles as influenced by lecithin-cholesterol acyltransferase. J Biol Chem 1990;265:4876–4883.

56. Fielding C, Fielding P. Regulation of human plasma lecithin:cholesterol acyltransferase activity by lipoprotein acceptor cholesteryl ester content. J Biol Chem 1981;256:2102–2104.

57. Karathanasis S, McPherson J, Zannis V, Breslow J. Linkage of human apolipoproteins A-I and C-III genes. Nature 1983;304:371–373.

58. Karathanasis S. Apolipoprotein multigene family: Tandem organization of human apolipoprotein AI, CIII, and AIV genes. Proc Natl Acad Sci USA 1985;82:6374–6378.

59. Cheung P, Kao F, Law M, et al. Localization of the structural gene for human apolipoprotein A-I on the long arm of human chromosome 11. Proc Natl Acad Sci USA 1984;81:508–511.

60. Lusis A. Genetic factors affecting blood lipoproteins: The candidate gene approach. J Lipid Res 1988;29:397–429.

61. Windler E, Chao Y, Havel R. Regulation of the hepatic uptake of triglyceride-rich lipoproteins in the rat: Opposing effects of apolipoprotein E and individual apoproteins. J Biol Chem 1980;255:8303–8307.

62. Gotto A, Pownall Jr H, Havel R. Introduction to the plasma lipoproteins. Methods in Enzymol 1986;128:1–48.

63. Baker H, Gotto JA, Jackson R. The primary structure of human plasma high density apolipoprotein glutamine I (apoA-I). II. The amino acid sequence and alignment of cyanogen bromide fragments IV, III, and I. J Biol Chem 1975;250:2725–2738.

64. Brewer JH, Fairwell T, Larue A, et al. The amino acid sequence of human apoA-I, an apoprotein isolated from high density lipoproteins. Biochem Biophys Res Commun 1978;80:623–630.

65. McLachlin A. Repeated helical pattern in apolipoprotein A-I. Nature 1977;267:465–466.

66. Li W, Tanimura M, Luo C, et al. The apolipoprotein multigene family: Biosynthesis, structure, structure-function relationships, and evolution. J Lipid Res 1988;29:245–271.

67. Segrest J, Jackson R, Morrisett J, Gotto JA. A molecular theory of lipid-protein interactions in the plasma lipoproteins. FEBS Lett 1974;38:247–253.

68. Fukushima D, Kupferberg J, Kokoyama S, et al. A synthetic amphiphilic helical docosapeptide with the surface properties of plasma apolipoprotein A-I. J Am Chem Soc 1979;101: 3703–3704.

69. Kaiser E, Kezdy F. Secondary structures of proteins and peptides in amphiphilic environments. Proc Natl Acad Sci USA 1983; 80:1137–1143.

70. Haddad I, Ordovas J, Fitzpatrick T, Karathanasis S. Linkage, evolution, and expression of the rat apolipoprotein A-I, and A-IV genes. J Biol Chem 1986;261:13268–13277.

71. Tam S, Archer T, Deeley R. Biphasic effects of estrogen on apolipoprotein synthesis in human hepatoma cells: Mechanism of antagonism by testosterone. Proc Natl Acad Sci USA 1986;83:3111–3115.

72. Masumoto A, Koga S, Uchida E, Ibayashi H. Effects of insulin, dexamethasone and glucagon on the production of apolipoprotein A-I in cultured rat hepatocytes. Atherosclerosis 1988;70:217–223.

73. Lin R. Effects of hormones on apolipoprotein secreted in cultured rat hepatocytes. Metab Clin Exper 1988;37:745–751.

74. Apostolopoulos J, LaScala M, Howlett G. The effect of triiodothyronine on rat apolipoprotein A-I and A-IV gene transcription. Biochem Biophys Res Commun 1988; 154:997–1102.

75. Sorci-Thomas M, Prack M, Sashti M, et al. Differential effects of dietary fat on the tissue-specific expression of the apolipoprotein A-I gene: Relationship to plasma concentration of high density lipoproteins. J Lipid Res 1989;30:1397–1403.

76. Zannis V, Karathanasis S, Deutmann H, et al. Intracellular and extracellular processing of human apoA-I. Secreted apoA-I isoprotein 2 is a propeptide. Proc Natl Acad Sci USA 1983;80:2574–2578.

77. Gordon J, Sims H, Lentz S, et al. Proteolytic processing of human preproapolipoprotein A-I: A proposed defect in the conversion of pro A-I to A-I in Tangier's disease. J Biol Chem 1983;258:4037–4044.

78. Bojanovski D, Gregg R, Brewer JH. Tangier disease. In vitro conversion of proapo-A-I Tangier to mature apo-A-I Tangier. J Biol Chem 1984;259:6049–6051.

79. Skipski V. Lipid composition of lipoproteins in normal and diseased states. In Nelson G (ed). Lipids and Lipoproteins: Quantitation, Composition, and Metabolism. New York: Wiley-Interscience, 1972, 471–583.

80. Scanu A, Elelstein C, Keim P. Serum lipoproteins. In Putman F (ed). The Plasma Proteins. New York: Academic Press, 1975, 317–334.

81. Fielding C, Shore V, Fielding P. A protein cofactor of lecithin:cholesterol acyltransferase. Biochem Biophys Res Commun 1972;46: 1943–1949.

82. Fielding C. Lecithin:cholesterol acyltransferase. In Esfahani M, Swaney J (eds). Advances in Cholesterol Research. New Jersey: Telford Press, 1990, 271–314.

83. Fielding C, Fielding P. Evidence for a lipoprotein carrier in human plasma catalyzing sterol efflux from cultured fibroblasts and its relationship to lecithin:cholesterol acyltransferase. Proc Natl Acad Sci USA 1981;78:3911–3914.

84. Glomset J. The plasma lecithin:cholesterol acyltransferase reaction. J Lipid Res 1968;9: 155–167.

85. Glomset J, Norum K. The metabolic role of lecithin:cholesterol acyltransferase: Perspectives from pathology. Adv Lipid Res 1973;2: 1–16.

86. Brown M, Hesler C, Tall A. Plasma enzymes and transfer proteins in cholesterol metabolism. Curr Opin Lipidol 1990;1:122–127.

87. Kovanen P, Schneider W, Hillman G, et al. Separate mechanisms for the uptake of high and low density lipoprotein by mouse adrenal gland in vivo. J Biol Chem 1979; 5498–5505.

88. Makrides S, Ruiz-Opazo N, Hayden M, et al. Sequence and expression of Tangier apoA-I gene. Eur J Biochem 1988;173:465–471.

89. Weisgraber K, Rall S, Bersot T, et al. Apolipoprotein A-I Milano: Detection of normal A-I in affected subjects and evidence for a cysteine for arginine substitution in the variant A-I. J Biol Chem 1983;258: 2508–2513.

90. Franceschini G, Sirtori C, Capurso A, et al. Decreased high density lipoprotein cholesterol levels with significant lipoprotein modification and without clinical atherosclerosis in an Italian family. J Clin Invest 1980;66: 892–900.

91. Franceschini G, Baio M, Calabresi L, et al. Apolipoprotein A-I Milano: Partial lecithin: cholesterol acyltransferase deficiency due to low levels of functional enzyme. Biochim Biophys Acta 1990;1043:1–6.

92. Utermann G, Haas J, Steinmetz A, et al. Apolipoprotein A-IGiessen (Pro143-Arg). A mutant that is defective in activating lecithin:cholesterol acyltransferase. Eur J Biochem 1984;144:325–331.

93. Rall S Jr, Weisgraber K, Mahley R, et al. Abnormal lecithin:cholesterol acyltransferase activation by a human apolipoprotein A-I variant in which a single lysine residue is deleted. J Biol Chem 1984;259:10063–10070.

94. Utermann G, Feussner G, Franceschini G, et al. Genetic variants of group A apolipoproteins: Rapid methods for screening and char-

acterization without ultracentrifugation. J Biol Chem 1982;257:501–507.

95. Funke H, von Eckardstein A, Pritchard P, et al. A frameshift mutation in the human apolipoprotein A-I gene causes high density lipoprotein deficiency, partial lecithin:cholesterol-acyltransferase deficiency, and corneal opacities. J Clin Invest 1991;87:371–376.

96. Nichols W, Dwulet F, Liepnieks J, Benson M. Variant apolipoprotein AI as a major constituent of a human hereditary amyloid. Biochem Biophys Res Commun 1988;156: 762–768.

97. Nichols W, Gregg R, Brewer JH, Benson M. A mutation in apolipoprotein A-I in the Iowa type of familial amyloidotic polyneuropathy. Genomics 1990;8:318–323.

98. Schonfeld G, Krul E. Genetic defects in lipoprotein metabolism. In Goldbourt U, de Faire U, Berg K (eds). Genetic Factors in Coronary Heart Disease. Dordrecht: Kluwer Academic Publishers, 1994, 239–266.

99. Schonfeld G. The genetic dyslipoproteinemias: Nosology update 1990. Atherosclerosis 1990;81:81–93.

100. Zannis V, Breslow J, Ordovas J, Karanthanasis S. Isolation and sequence of Tangier apoA-I gene. Arteriosclerosis 1984;4:562A.

101. Law S, Lackner K, Fojo S, et al. The molecular biology of human ApoA-I, apoA-II, apoC-II and apoB. Adv Exp Med Biol 1986;201:151–162.

102. Schmitz G, Assmann G, Brennhausen B, Schaefer H-J. Interaction of Tangier lipoproteins with cholesteryl ester-laden mouse peritoneal macrophages. J Lipid Res 1987;28:87–99.

103. Hamsten A, Iselius L, Dahlen G, de Faire U. Genetic and cultural inheritance of serum lipids, low and high density lipoprotein cholesterol and serum apolipoproteins A-I, A-II and B. Atherosclerosis 1986;60:199–208.

104. Goldbourt U, Neufeld H. Genetic aspects of arteriosclerosis. Arteriosclerosis 1986;6: 357–377.

105. Moll P, Sing C, Williams R, et al. The genetic determination of plasma apolipoprotein AI levels measured by radioimmunoassay: A study of high-risk pedigrees. Am J Hum Genet 1986;38:361–372.

106. Shulman R, Herbert P, Fredrickson D, et al. Isolation and alignment of the tryptic peptides of alanine apolipoproteins, an apolipoprotein from human plasma very low density lipoproteins. J Biol Chem 1974;249:4969–4974.

107. Wu A, Windmueller H. Relative contribution of liver and intestine to individual plasma apolipoproteins in the rat. J Biol Chem 1979;254:7316–7322.

108. Zannis V, Cole F, Jackson C, et al. Distribution of apoA-I, apoCII, apoCIII, and apoE mRNA in human tissues, time dependent induction of apoE mRNA by cultures of human monocyte-macrophages. Biochemistry 1985;24:4450–4455.

109. Lenich C, Brecher P, Makrides S, et al. Apolipoprotein gene expression in the rabbit: Abundance, size and distribution of apolipoprotein mRNA species in different tissues. J Lipid Res 1988;29:755–764.

110. Hussain M, Zannis V. Intracellular modification of human apolipoprotein AII (apoAII) and sites of apoAII mRNA synthesis: Comparison of apoAII with apoCII and apoCIII isoproteins. Biochemistry 1990;29:209–217.

111. Zannis V, Breslow J. Two dimensional maps of human apolipoproteins in normal and diseased states. In Radola B (ed). Electrophoresis. Berlin: Walter de Gruyter, 1980, 437–473.

112. Kashyap M, Srivastava L, Hynd B, et al. Quantitation of human apolipoprotein C-III and its subspecies by radioimmunoassay and analytical isoelectric focusing: Abnormal plasma triglyceride-rich lipoprotein apolipoprotein C-III subspecies concentrations in hypertriglyceridemia. J Lipid Res 1981;22:800–810.

113. Sparrow J, Gotto JA. Apolipoprotein/lipid interactions: Studies with synthetic polypeptides. CRC Crit Rev Biochem 1982;13:87–107.

114. Sparrow J, Pownall H, Hsu F, et al. Lipid binding by fragments of apolipoprotein C-III obtained by thrombin cleavage. Biochemistry 1977;16:5427–5431.

115. Chung J, Scanu A. Isolation, molecular properties, and kinetic characterization of lipoprotein lipase from rat heart. J Biol Chem 1977;4202–4209.

116. Brown W, Baginsky M. Inhibition of lipoprotein lipase by an apoprotein of human very low density lipoprotein. Biochem Biophys Res Commun 1972;46:375–382.

117. Krauss R, Herbert P, Levy R, Fredrickson D. Further observations on the activation and inhibition of lipoprotein lipase by apolipoproteins. Circ Res 1973;33:403–411.

118. Kowal R, Hertz J, Weisgraber K, et al. Opposing effects of apolipoprotein E and C on lipoprotein binding to low density lipoprotein receptor-related protein. J Biol Chem 1990;265:10771–10779.

119. Weisgraber K, Mahley R, Kowal R, et al. Apolipoprotein C-I modulates the interaction of apolipoprotein E with beta-migrating very low density lipoprotein (beta-VLDL) and inhibits binding of beta-VLDL to low density lipoprotein receptor-related protein. J Biol Chem 1990;265:22453–22459.

120. Windler E, Chao Y, Havel R. Determinants of hepatic uptake of triglyceride-rich lipoproteins and their remnants in the rat. J Biol Chem 1980;255:5475–5480.

121. Quarfordt S, Michalpoulos G, Schirmer B. The effect of human C apolipoproteins on the in vitro hepatic metabolism of triglyceride-rich emulsions in the rat. J Biol Chem 1982;257:14642–14647.

122. Karathanasis S, Oettgen P, Haddad I, Antonarakis S. Structure, evolution and polymorphisms of the human apolipoprotein A4 gene (apoA4). Proc Natl Acad Sci USA 1986;83:8456–8461.

123. Ordovas J, Cassidy D, Civiera F, et al. Familial apolipoprotein A-I, C-III, and A-IV deficiency and premature atherosclerosis due to deletion of a gene complex on chromosome 11. J Biol Chem 1989;264:16339–16342.

124. Karathanasis S, Ferris E, Haddad I. DNA inversion within the apolipoproteins AI/ CIII/ AIV-encoding gene cluster of certain patients with premature atherosclerosis. Proc Natl Acad Sci USA 1987;84:7198–7202.

125. Schonfeld G, George P, Miller J, et al. Apolipoprotein C-II and C-III levels in hyperlipoproteinemia. Metab Clin Exp 1979; 28:1001–1010.

126. Weisgraber K, Bersot T, Mahley R. Isolation and characterization of an apoprotein from the d<1.006 lipoproteins of human and canine lymph homologous with the rat A-IV apoprotein. Biochem Biophys Res Commun 1978;85:287–292.

127. Beisiegel U, Utermann G. An apolipoprotein homolog of rat apolipoprotein A-IV in human plasma. Eur J Biochem 1979;93:601–608.

128. Utermann G, Beisiegel U. Apolipoprotein A-IV: A protein occuring in human mesenteric lymph chylomicrons and free in the plasma: Isolation and quantification. Eur J Biochem 1979;99:333–343.

129. Green P, Glickman R, Riley J, Quinet E. Human apolipoprotein A-IV. Intestinal origin and distribution in plasma. J Clin Invest 1980;65:911–919.

130. Bisgaier C, Sachdev O, Megna L, Glickman R. Distribution of apolipoprotein A-IV in human plasma. J Lipid Res 1985;26:11–25.

131. Green P, Lefkowitch J, Glickman R, et al. Human intestinal lipoproteins. Studies in chyluric subjects. J Clin Invest 1979;64: 233–242.

132. Go M, Schonfeld G, Pfleger B, et al. Regulation of intestinal and hepatic apoprotein synthesis after chronic fat and cholesterol feeding. J Clin Invest 1988;81:1615–1620.

133. Weinberg R, Jordan M. Effects of phospholipid on the structure of human apolipoprotein A-IV. J Biol Chem 1990;265:8081–8086.

134. Weinberg R, Dantzker C, Patton C. Sensitivity of serum apolipoprotein A-IV levels to changes in dietary fat content. Gastroenterology 1990;98:17–24.

135. Steinmetz A, Utermann G. Activation of lecithin:cholesterol acyltransferase by human apolipoprotein A-IV. J Biol Chem 1985; 260:2258–2264.

136. Stein O, Stein U, Lefevre M, Roheim P. The role of apolipoprotein A-IV in reverse cholesterol transport studied with cultured cells and liposomes derived from an ether analog of phosphatidylcholine. Biochim Biophys Acta 1986;878:7–13.

137. Elshourbagy N, Walker D, Boguski M, et al. The nucleotide and derived amino acid sequence of human apolipoprotein A-IV mRNA and the close linkage of its gene to the genes of apolipoproteins A-I and CIII. J Biol Chem 1986;261:1998–2002.

138. Karathanasis S, Yunic I, Zannis V. Structure, evolution and tissue-specific synthesis of human apolipoprotein A-IV. Biochemistry 1986;25:3962–3970.

139. Menzel H, Kovary P, Assmann G. Apolipoprotein A-IV polymorphism in man. Hum Genet 1982;62:349–352.

140. Menzel H, Boerwinkle E, Schrangl-Will S, Utermann G. Human apolipoprotein A-IV polymorphism: Frequency and effect on lipid and lipoprotein levels. Hum Genet 1988;79:368–372.

141. DeKnijff P, Rosseneu M, Beisiegel U, et al. Apolipoprotein A-IV polymorphism and its effect on plasma lipid and apolipoprotein concentrations. J Lipid Res 1988;29: 1621–1627.

142. Lukka M, Metson J, Ehnholm C. Apolipoprotein A-IV polymorphism in the Finnish populations: Gene frequencies and description of a rare allele. Hum Hered 1988;38:359–362.

143. Lohse P, Brewer JH. Genetic polymorphism of apolipoprotein A-IV. Curr Opin Lipidol 1991;2:90–95.

144. Lohse P, Kindt M, Rader D, Brewer JH. Genetic polymorphism of human plasma apolipoprotein A-IV is due to nucleotide substitutions in the apolipoprotein A-IV gene. J Biol Chem 1990;265:10061–10064.

145. Lohse P, Kindt M, Rader D, Brewer JH. Human plasma apolipoproteins A-IV-0 and A-IV-3. Molecular basis for two rare variants of apolipoprotein A-IV-1. J Biol Chem 1990;265:12734–12739.

146. Menzel H, Sigurdsson G, Boerwinkle E, et al. Frequency and effect of human apolipoprotein A-IV polymorphism on lipid and lipoprotein levels in an Icelandic population. Hum Genet 1990;344–346.

147. Eichner J, Kuller L, Ferrell R, Kamboh M. Phenotypic effects of apolipoprotein structural variation on lipid profiles: II Apolipoprotein A-IV and quantitative lipid measures in the healthy women study. Genet Epidemiol 1989;6:493–499.

148. Weinberg R, Jordan M, Steinmetz A. Distinctive structure and function of human apolipoprotein variant apoA-IV-2. J Biol Chem 1990;265:18372–18378.

149. Schaefer E, Ordovas J, Law S, et al. Familial apolipoprotein A-I and C-III deficiency, variant II. J Lipid Res 1985;26:1089–1101.

150. Matsunaga T, Hiasa Y, Yanagi H, et al. Apolipoprotein A-I deficiency due to a codon 84 nonsense mutation of the apolipoprotein A-I gene. Proc Natl Acad Sci USA 1991;88:2793–2797.

151. Kessling A, Hursthemke B, Humphries S. A study of DNA polymorphisms around the human apolipoprotein AI gene in hyperlipidemic and normal individuals. Clin Genet 1985;28:296–306.

152. Antonarakis S, Oettgen J, Shakravati A, et al. DNA polymorphism of the human apolipoprotein AI-CIII-AIV gene cluster. Am J Hum Genet 1986;39:A185.

153. Stocks J, Paul H, Galton D. Haplotypes identified by DNA restriction-fragment-length polymorphism of the human apolipoprotein AI-CIII-AIV gene region and hypertriglyceridemia. Am J Hum Genet 1987;41:106–118.

154. Rees A, Shoulders C, Stocks J, et al. DNA polymorphism adjacent to human apolipoprotein A-I gene: Relation to hypertriglyceridemia. Lancet 1983;1:444–446.

155. Ferns G, Stocks J, Ritchie C, Galton D. Genetic polymorphisms of apolipoprotein C-III and insulin in survivors of myocardial infarction. Lancet 1985;2:300–303.

156. Rees A, Stocks J, Sharpe C, et al. Deoxyribonucleic acid polymorphism in the apolipoprotein A-I-C-III gene cluster. Associations with hypertriglyceridemia. J Clin Invest 1985;76:1090–1095.

157. Rees A, Stocks J, Paul H, et al. Haplotypes identified by DNA polymorphisms at the apolipoprotein A-I and C-III loci and hypertriglyceridemia: A study in a Japanese population. Human Genet 1986;72:168–171.

158. Henderson H, Landon S, Michie J, Berger G. Association of a DNA polymorphism in the apolipoprotein C-III gene with diverse hyperlipidaemic phenotypes. Hum Genet 1987;75:62–65.

159. Hayden M, Kirk H, Clark C, et al. DNA polymorphisms in and around the apo-AI-CIII genes and genetic hyperlipidemias. Am J Hum Genet 1987;40:421–430.

160. Ordovas J, Schaefer E, Salem D, et al. Apolipoprotein A-I gene polymorphism associated with premature coronary artery disease and familial hypoalphalipoproteinemia. N Engl J Med 1986;314:671–677.

161. Anderson R, Benda T, Wallace R, et al. Prevalence and associations of apolipoprotein A-I-linked DNA polymorphisms: Results from a population study. Genet Epidemiol 1986;3:385–397.

162. Brewer JH, Lux S, Ronan R, John K. Amino acid sequence of human apoLp-Gln-II (apoA-II), an apolipoprotein isolated from the high-density lipoprotein complex. Proc Natl Acad Sci USA 1972;69:1304–1308.

163. Jahn C, Osborne Jr J, Schaeffer E, Brewer Jr H. Activation of the enzymatic activity of hepatic lipase by apoAII. Eur J Biochem 1983;131:25–29.

164. Knott T, Priestley L, Urdea M, Scott J. Isolation and characterization of a cDNA encoding the precursor for human apolipoprotein AII. Biochem Biophys Res Commun 1984;120:734–740.

165. Lackner K, Law S, Brewer JH. Human apolipoprotein A-II: Complete nucleic acid sequence of preproapoA-II. FEBS Lett 1984;175:159–164.

166. Moore M, Kao F, Tsao L, Chan L. Human apolipoprotein A-II: Nucleotide sequence of a cloned cDNA, and localization of its structural gene on human chromosome 1. Biochem Biophys Res Commun 1984;123:1–7.

167. Gordon J, Budelier K, Sims H, et al. Biosynthesis of human preapolipoprotein A-II. J Biol Chem 1983;258:14054–14059.

168. Gordon J, Bisgair C, Sims H, et al. Biosynthesis of human preapolipoprotein A-IV. J Biol Chem 1984;259:468–474.

169. Cheung M, Albers J. Characterization of lipoprotein particles isolated by immunoaffinity chromatography. J Biol Chem 1984;259:12201–12209.

170. Ohta T, Hattori S, Nishiyama S, Matsuda I. Studies on the lipid and apolipoprotein compositions of two species of Apo-A-I containing lipoproteins in normolipidemic males and females. J Lipid Res 1988;29:721–728.

171. Fruchart J, Bard J. Lipoprotein particle measurement: An alternative approach to classification of lipid disorders. Curr Opin Lipidol 1991;2:362–366.

172. Bekaert E, Alaupovic P, Knight-Gibson C, et al. Composition of plasma Apo-A-I containing particles in children and adults. Pediatr Res 1991;29:315–321.

173. Albers J, Aladjem F. Precipitation of 125I-labeled lipoproteins with specific polypeptide antisera: Evidence for two populations with differing polypeptide compositions in hu-

man high density lipoproteins. Biochemistry 1971;10:3436–3442.

174. Jackson R, Gotto JA, Lux S, et al. Human high density lipoprotein: Interaction of the cyanogen bromide fragments from apolipoprotein glutamine II (A-II) with phosphatidylcholine. J Biol Chem 1973;248:8449.

175. Mao S, Jackson R, Gotto JA, Sparrow J. Mechanism of lipid-protein interaction in the plasma lipoproteins: Identification of a lipid-binding site in apolipoprotein A-II. Biochemistry 1981;20:1676–1680.

176. Lagocki P, Scanu A. In vitro modulation of the apolipoprotein composition of high density lipoprotein. J Biol Chem 1980;255:3701–3706.

177. Lusis A, Taylor R, Wangenstein R, LeBoeuf R. Genetic control of lipid transport in mice. II. Genes controlling structure of high density lipoproteins. J Biol Chem 1983;258:5071–5078.

178. Scott T, Knott T, Priestley L, et al. High-density lipoprotein composition is altered by a common DNA polymorphism adjacent to apolipoprotein AII gene in man. Lancet 1985;1:771–773.

179. Ferns G, Shelley C, Stocks J, et al. A DNA polymorphism of the apolipoprotein AII gene in hypertriglyceridemia. Hum Genet 1986;74:302–306.

180. Goldstein J, Hobbs H, Brown M. Familial hypercholesterolemia. In Stanbury J, et al (eds). The Metabolic Basis of Inherited Disease. New York: McGraw-Hill, 1982, 1981–2030.

181. Davis R, Boogaerts J. Intrahepatic assembly of very low density lipoproteins: Effect of fatty acids on triacylglycerol and apolipoprotein synthesis. J Biol Chem 1982;257:10908–10913.

182. Suita-Mangan P, Janero D, Lane M. Association and assembly of triglyceride and phospholipid with glycosylated and unglycosylated apoproteins of very low density lipoprotein in the intact liver cell. J Biol Chem 1982;257:11463–11467.

183. Knott T, Rall SJ, Innerarity T, et al. Human apolipoprotein B: Structure of carboxyl-terminal domains, sites of gene expression, and chromosomal localization. Science 1985;230:37–43.

184. Mehrabian M, Sparkes R, Mohandas T, et al. Human apolipoprotein B: Chromosomal mapping and DNA polymorphisms of hepatic and intestinal species. Somatic Cell Mol Genet 1986;12:245–254.

185. Blackhart B, Ludwig E, Pierotti V, et al. Structure of the human apolipoprotein B gene. J Biol Chem 1986;261:15364–15367.

186. Ludwig E, Blackhart B, Pierotti V, et al. DNA sequence of the human apolipoprotein B gene. DNA 1987;6:363–372.

187. Kane J, Chen G, Hamilton R, et al. Remnants of lipoproteins of intestinal and hepatic origin in familial dysbetalipoproteinemia. Arteriosclerosis 1983;3:47–56.

188. Kane J, Hardman D, Paulus H. Heterogeneity of apolipoprotein B: Isolation of a new species from chylomicrons. Proc Natl Acad Sci USA 1980;77:2465–2469.

189. Powell L, Wallis S, Pease R, et al. A novel form of tissue-specific RNA processing produces apolipoprotein-B48 in intestine. Cell 1987;50:831–840.

190. Chen S-H, Habib G, Yang C, et al. Apolipoprotein B-48 is the product of a messenger RNA with an organ-specific in-frame translational stop codon. Science 1987;283:363–366.

191. Hospattankar A, Higuchi K, Law S, et al. Identification of a novel in-frame translational stop codon in human intestine apoB mRNA. Biochem Biophys Res Commun 1987;148:279–285.

192. Higuchi K, Hospattankar A, Law S, et al. Human apolipoprotein B (apoB) mRNA: Identification of two distinct apoB mRNAs, an mRNA with the apoB-100 sequence and an apoB mRNA containing a premature in-frame translational stop codon, in both the liver and intestine. Proc Natl Acad Sci USA 1988;85:1772–1776.

193. Bostrom K, Lauer S, Poksay K, et al. Apolipoprotein B48 RNA editing in chimeric apolipoprotein EB mRNA. J Biol Chem 1989;264:15701–15708.

194. Scott J. The molecular and cell biology of apolipoprotein-B. Mol Biol Med 1989;6:65–80.

195. Scott J. Thrombogenesis linked to atherosclerosis at last? Nature 1989;341:22–23.

196. Chen G, Zhu S, Hardman D, et al. Structural domains of human apolipoprotein B-100. J Biol Chem 1989;264:14396–14375.

197. Chen G, Hardman D, Hamilton R, et al. Distribution of lipid-binding regions in human apolipoprotein B-100. Biochemistry 1989;28:2477–2484.

198. Milne R, Theolis RJ, Maurice R, et al. The use of monoclonal antibodies to localize the low density lipoprotein receptor binding of apolipoprotein B. J Biol Chem 1989;264:19754–19760.

199. Soria L, Ludwig E, Clarke H, et al. Association between a specific apolipoprotein B-100 mutation and familial defective apolipoprotein B-100. Proc Natl Acad Sci USA 1989;86:587–591.

200. Innerarity T, Mahley R, Weisgraber K, et al. Familial defective apolipoprotein B100: A

mutation of apolipoprotein B that causes hypercholesterolemia. J Lipid Res 1990;31: 1337–1349.

201. Lund-Katz S, Innerarity T, Curtiss L, et al. C13-NMR evidence that substitution of Gln for Arg 3500 in familial defective apo-B100 disrupts the conformation of the receptor binding domain. Arteriosclerosis 1989;9: 715A.

202. Lund-Katz S, Innerarity T, Arnold K, et al. C13-NMR evidence that substitution of glutamine for arginine 3500 in familial defective apolipoprotein B-100 disrupts the conformation of the receptor binding domain. J Biol Chem 1991;266:2701–2704.

203. Yang C, Gu Z, Weng S, et al. Structure of apolipoprotein B-100 of human low density lipoproteins. Arteriosclerosis 1989;9:96–108.

204. Yang C, Chen S, Gianturco S, et al. Sequence, structure, receptor-binding domains and internal repeats of human apolipoprotein B-100. Nature 1986;323: 738–742.

205. Robinson M, Schumaker V, Butler R, et al. Ag(c): Recognition by a monoclonal antibody. Arteriosclerosis 1986;6:341–344.

206. Tikkanen M. Ag system re-explored: Studies with monoclonal anti-apolipoprotein antibodies. Adv Exp Med Biol 1987;210:55–62.

207. Butler R, Brunner E. Ag(i): Detection of antithetical factor to Ag(h). Vox Sang 1974;27: 550–555.

208. Butler R, Brunner E, Morganti G. Contribution to the inheritance of the Ag groups: A population genetic study. Vox Sang 1974;26: 485–496.

209. Berg K, Hames G, Dahlem G, et al. Genetic variations in serum low density lipoproteins and lipid levels in man. Proc Natl Acad Sci USA 1976;73:937–940.

210. Law A, Wallis S, Powell L, et al. Common DNA polymorphisms with coding sequence of apolipoprotein B gene associated with altered lipid levels. Lancet 1986;1:1301–1303.

211. Berg K. DNA polymorphism at the apolipoprotein B locus is associated with lipoprotein level. Clin Genet 1986;30:515–520.

212. Hegele R, Huang L-S, Herbert P, et al. Apolipoprotein B-gene DNA polymorphisms associated with myocardial infarction. N Engl J Med 1986;315:515–520.

213. Kane J, Havel R. Disorders of the biogenesis and secretion of lipoproteins containing the B apolipoproteins. In Scriver C, et al (eds). The Metabolic Basis of Inherited Disease. New York: McGraw-Hill, 1989, 1887–1912.

214. Cooper R, Gulbrandsen C. The relationship between serum lipoproteins and red cell membranes in abetalipoproteins: Deficiency of lecithin:cholesterol acyltransferase. J Lab Clin Med 1971;78:323–335.

215. Herbert P, Assmann G, Gotto JA, Fredrickson D. Familial high density lipoprotein deficiencies: Tangier disease. In Stanbury J, et al (eds). The Metabolic Basis of Inherited Disease. New York: McGraw-Hill, 1982, 2053–2072.

216. Forsyth C, Lloyd J, Fosbrooke A. A-beta-lipoproteinemia. Arch Dis Child 1965;40: 47–52.

217. Kostner G, Holasek A, Bohlmann G, Thiede H. Investigation of serum lipoproteins and apoproteins in abetalipoproteinaemia. Clin Sci Mol Med Suppl 1974;46:457–468.

218. Lackner K, Monge J, Gregg R, et al. Analysis of the apolipoprotein B gene and messenger ribonucleic acid in abetalipoproteinemia. J Clin Invest 1986;78:1707–1712.

219. Dullaart R, Speelberg B, Schuurman H-J, et al. Epitopes of apolipoprotein B-100 and B-48 in both liver and intestine: Expression and evidence for local synthesis in recessive abetalipoproteinemia. J Clin Invest 1986;79: 1831–1841.

220. Wetterau J, Aggerbeck L, Bouma M-E, et al. Absence of microsomal triglyceride transfer protein in individuals with abetalipoproteinemia. Science 1992;258:999–1001.

221. Utermann G, Menzi H, Kraft H, et al. Lp(a) glycoprotein phenotypes. J Clin Invest 1987;80:458–465.

222. Fosbrooke A, Choksey S, Wharton B. Familial hypo-B-lipoproteinemia with hypoalphalipoproteinemia and fasting chylomicronemia. Arch Dis Child 1978;48:729–732.

223. Zannis V, Kardassis D, Zanni E. Genetic mutations affecting human lipoproteins, their receptors, and their enzymes. Adv Hum Genetics 1993;21:145–319.

224. Young S, Bertics S, Curtiss L, Witztum J. Characterization of an abnormal species of apolipoprotein B, apolipoprotein B-37, associated with familial hypobetalipoproteinemia. J Clin Invest 1987;79:1831–1841.

225. Young S, Bertics S, Curtiss L, et al. Genetic analysis of a kindred with familial hypobetalipoproteinemia: Evidence for two separate gene defects: One associated with an abnormal apolipoprotein B species, apolipoprotein B-37, and a second associated with low plasma concentrations of apolipoprotein B-100. J Clin Invest 1987;79: 1842–1851.

226. Vega G, von Bergmann K, Grundy S, et al. Increased metabolism of VLDL-apolipoprotein B and synthesis of bile acids in a case of hypobetalipoproteinemia. Metabolism 1987;36:262–269.

227. Malloy M, Kane J, Hardman D, et al. Normotriglyceridemic abetalipoproteinemia. Absence of the B-100 apolipoprotein. J Clin Invest 1981;78:398–410.

228. Kesaniemi Y, Farkkila M, Kervinen K, et al.

Regulation of low density lipoprotein apolipoprotein B levels. Am Heart Journal 1987;113:508–513.

229. Grundy S, Vega G, Kesaniemi Y. Abnormalities in metabolism of low density lipoproteins associated with coronary heart disease. Acta Med Scand 1985;701:23–37.

230. Rauh G, Keller C, Schuster G, et al. Familial defective apolipoprotein B-100: A common cause of primary hypercholesterolemia. Clin Invest 1992;70:77–84.

231. Innerarity T, Balestra M, Arnold K, et al. Isolation of defective receptor-binding low density lipoproteins from subjects with familial defective apoB100. Arteriosclerosis 1988;8:551A.

232. Mahley R. Apolipoprotein E: Cholesterol transport protein with an expanding role in cell biology. Science 1988;240:622–630.

233. Innerarity T, Mahley R. Enhanced binding by cultured human fibroblasts of apo-E-containing lipoproteins as compared with low density lipoproteins. Biochemistry 1987;17:1440–1447.

234. Lusis A, Heinzmann C, Sparkes R, et al. Regional mapping of human chromosome 19: Organization of genes for plasma lipid transport (APOC1, -C2, and -E and LDLR) and the genes C3, PEPD and GPI. Proc Natl Acad Sci USA 1986;83:3929–3933.

235. Blue M, Williams S, Zucker S, et al. Apolipoprotein E synthesis in human kidney, adrenal gland and liver. Proc Natl Acad Sci USA 1983;80:283–287.

236. Reue K, Quon D, O'Donnell K, et al. Cloning and regulation of messenger RNA for mouse apolipoprotein E. J Biol Chem 1984;259:2100–2107.

237. Elshourbagy N, Liao W, Mahley R, Taylor J. Apolipoprotein E mRNA is abundant in the brain and adrenals as well as in the liver, and is present in other peripheral tissues of rats and marmosets. Proc Natl Acad Sci USA 1985;82:203–207.

238. Wernette-Hammond M, Lauer S, Corsini A, et al. Human apolipoprotein E3 in aqueous solution. Evidence for two structural domains. J Biol Chem 1988;263:6240–6248.

239. Weisgraber K, Innerarity T, Harder K, et al. The receptor-binding domain of human apolipoprotein E: Monoclonal antibody inhibition of binding. J Biol Chem 1983;258:12348–12354.

240. Innerarity T, Friedlander E, Rall SJ, et al. The receptor binding domain of human apolipoprotein E: Binding of apolipoprotein E fragments. J Biol Chem 1983;258: 12341–12347.

241. Utermann G, Hees M, Steinmetz A. Polymorphism of apolipoprotein E and occurrence of dysbetalipoproteinaemia in man. Nature 1977;269:604–607.

242. Zannis V, Breslow J, Utermann G, et al. Proposed nomenclature of apoE isoproteins, apoE genotypes, and phenotypes. J Lipid Res 1982;23:911-X.

243. Weisgraber K, Rall SJ, Mahley RW. Human E apoprotein heterogeneity. Cysteine-arginine interchanges in the amino acid sequence of the apo-E isoforms. J Biol Chem 1981;256:9077–9083.

244. Weisgraber K, Innerarity T, Mahley R. Abnormal lipoprotein receptor binding activity of the human E apoprotein due to cysteine-arginine interchange at a single site. J Biol Chem 1982;257:2518.

245. Schneider W, Kovanen P, Brown M, et al. Abnormal binding of mutant apoprotein E to the low density lipoprotein receptors of human fibroblasts and membranes from liver and adrenal of rats, rabbits and cow. J Clin Invest 1981;68:1075–1085.

246. Utermann G. Apolipoprotein E polymorphism in health and disease. Am Heart J 1987;113:433–440.

247. Klasen E, Talmud P, Havekes L, et al. A common restriction fragment length polymorphism of the human apolipoprotein E gene and its relationship to type III hyperlipidemia. Hum Genet 1987;75:244–247.

248. Assmann G, Schmitz G, Menzel H, Schulte H. Apolipoprotein E polymorphism and hyperlipidemia. Clin Chem 1984;30:641–643.

249. Utermann G, Kindermann I, Kaffarnick H, Steinmetz A. Apolipoprotein E phenotypes and hyperlipidemia. Hum Genet 1984;65: 232–236.

250. Ghiselli G, Schaefer E, Zech L, et al. Increased prevalence of apolipoprotien E4 in type V hyperlipoproteinemia. J Clin Invest 1982;70:474–477.

251. Yamamura T, Yamamoto A, Hiramori K, Namb S. A new isoform of apoE-apoE-5 associated with hyperlipidemia and atherosclerosis. Atherosclerosis 1984;50:159–172.

252. Jackson R, Sparrow J, Baker H, et al. The primary structure of apolipoprotein-serine. J Biol Chem 1974;249:5308–5313.

253. Shulman R, Herbert P, Wehrly K, Fredrickson D. The complete amino acid sequence of CI (apo Lp=Ser), an apolipoprotein from human very low density lipoprotein. J Biol Chem 1975;250:182–190.

254. Lauer S, Walker D, Elshourbagy N, et al. Two copies of the human apolipoprotein C-I gene are linked closely to the apolipoprotein E gene. J Biol Chem 1988;263:7277–7286.

255. Smit M, van der Kooij-Meijs E, Frants R, et al. Apolipoprotein gene cluster on chromosome 19. Hum Genet 1988;78:90–93.

256. Francke U, Brown M, Goldstein J. Assignment of the human gene for the low density lipoprotein receptor to chromosome 19: Synteny of a receptor, a ligand and a genetic disease. Proc Natl Acad Sci USA 1984;81: 2826–2830.

257. Lindgren V, Luskey R, Russell D, Francke U. Human genes involved in cholesterol metabolism: Chromosomal mapping of the loci for the low density lipoprotein receptor and 3-hydroxy-3-methylglutaryl-coenzyme A reductase with cDNA probes. Proc Natl Acad Sci USA 1985;82:8567–8571.

258. Knott J, Robertson M, Priestly L, et al. Characterization of mRNAs encoding the precursor for human apolipoprotein CI. Nucleic Acids Res 1984;12:3909–3915.

259. Wei C, Tsao Y, Robbertson D, et al. The structure of the human apolipoprotein C-II gene. J Biol Chem 1985;260:15211–15221.

260. Jackson R, Baker H, Gilliam E, Gotto A. Primary structure of very low density apolipoprotein C-II of human plasma. Proc Natl Acad Sci USA 1977;74:1942.

261. Baggion G, Manzato E, Gabelli C, et al. Apolipoprotein CII deficiency syndrome: Clinical features, lipoprotein characterization, lipase activity and correction of hypertriglyceridemia after apolipoprotein C-II administration in two affected patients. J Clin Invest 1986;77:520–527.

262. Connelly P, Maguire G, Hofmann T, Little J. Structure of apolipoprotein C-II Toronto, a nonfunctional human apolipoprotein. Proc Natl Acad Sci USA 1987;84:270–273.

263. Maguire G, Little J, Kakis G, Breckenridge W. Apolipoprotein C-II deficiency associated with nonfunctional mutant forms of apolipoprotein C-II. Can J Biochem 1984;62: 847–852.

264. Hayden M, Vergani C, Humphries S, et al. The genetics and molecular biology of apolipoprotein CII. Adv Exp Med Biol 1986;201:241–251.

265. Curry M, McConathy W, Alaupovic P. Quantitative determination of human apolipoprotein D by electroimmunoassay and radial immunodiffusion. Biochim Biophys Acta 1977;491:232–241.

266. Kostner G. Studies of the composition and structure of human serum lipoproteins: Isolation and partial characterization of apolipoprotein A-III. Biochim Biophys Acta 1974;336:383–395.

267. McConathy W, Alaupovic P. Studies on the isolation and partial characterization of apolipoprotein D and lipoprotein D of human plasma. Biochemistry 1976;515–520.

268. Drayna D, McLean J, Wion K, et al. Human apolipoprotein D gene: Gene sequence, chromosomal localization, and homology to the alpha 2-μ globulin superfamily. DNA 1987;6:194–204.

269. Drayna D, Fielding C, McLean J, et al. Cloning and expression of human apolipoprotein D cDNA. J Biol Chem 1986; 261: 16535–16539.

270. Peitsch M, Boguski M. Is apolipoprotein D a mammalian bilin-binding protein? New Biol 1990;2:197–206.

271. Boyles J, Notterpek L, Wardell M, Rall JS. Identification, characterization and tissue distribution of apolipoprotein D in the rat. J Lipid Res 1990;31:2243–2256.

272. Morton R, Zilversmit D. The separation of apolipoprotein D from cholesteryl ester transfer protein. Biochim Biophys Acta 1981;663:350–355.

273. Francone O, Gurakar A, Fielding C. Distribution and functions of lecithin:cholesterol acyltransfer protein in plasma lipoproteins: Evidence for a functional unit containing these activities together with apolipoproteins A-I and D that catalyzes the esterification and transfer of cell-derived cholesterol. J Biol Chem 1989;264:7066–7072.

274. Albers J, Lin J, Roberts G. Effect of human plasma apolipoproteins on the activity of purified lecithin:cholesterol acyltransferase. Artery 1979;5:61–75.

275. Berg K. A new serum type system in man - the Lp system. Acta Pathol Microbiol Scand 1963;59:369–382.

276. Albers J, Adolphson J, Hazzard W. Radioimmunoassay of human plasma Lp(a) lipoprotein. J Lipid Res 1977;18:331–338.

277. Sing C, Schultz J, Shreffler D. The genetics of the Lp antigen. II. A family study and proposed models of genetic control. Ann Hum Genet 1974;38:47–56.

278. Armstrong V, Walli A, Seidel D. Isolation, characterization, and uptake in human fibroblasts of an apo(a)-free lipoprotein obtained on reduction of lipoprotein(a). J Lipid Res 1985;26:1314–1323.

279. Utermann G, Weber W. Protein composition of Lp(a) lipoprotein from human plasma. FEBS Lett 1983;154:357.

280. Gaubatz J, Heideman C, Gotto A Jr, et al. Isolation and characterization of the two major apoproteins in human lipoprotein(a). J Biol Chem 1983;258:4582–4589.

281. Karadi I, Kostner G, Gries A, et al. Lipoprotein(a) and plasminogen are immunologically related. Biochim Biophys Acta 1988; 960:91–97.

282. Eaton D, Fless G, Kohr W, et al. Partial amino acid sequence of apolipoprotein(a) shows that it is homologous to plasminogen. Proc Natl Acad Sci USA 1987;84: 3224–3228.

283. McLean J, Tomlinson J, Kuang W, et al. cDNA sequence of human apolipoprotein (a) is homologous to plasminogen. Nature 1987;330:132–137.

284. Utermann G. The mysteries of lipoprotein (a). Science 1989;246:904–910.

285. Drayna D, Hegele R, Hass P, et al. Genetic linkage between lipoprotein(a) phenotype and a DNA polymorphism in the plasminogen gene. Genomics 1988;3:230–236.

286. Weinkamp L, Guttormsen S, Schultz J. Linkage between the loci for the Lp(a) lipoprotein (LP) and plasminogen (PLG). Hum Genet 1988;79:80–82.

287. Berg K. Lp(a) lipoprotein: An overview. In Scanu A (ed). Lipoprotein(a). New York: Academic Press, 1990, 1–23.

288. Utermann G, Kraft H, Menzel H, et al. Genetics of the quantitative Lp(a) lipoprotein trait. I. Relation of Lp(a) glycoprotein phenotypes to Lp(a) lipoprotein concentrations in plasma. Hum Genet 1988;78:41–46.

289. Utermann G, Duba H, Menzel H. Genetics of the quantitative Lp(a) lipoprotein trait. II. Inheritance of Lp(a) glycoprotein phenotypes. Hum Genet 1988;78:47–50.

290. Boerwinkle E, Menzel H, Kraft H, Utermann G. Genetics of the quantitative Lp(a) lipoprotein trait. III. Contribution of Lp(a) glycoprotein phenotypes to normal lipid variation. Hum Genet 1989;82:73–78.

291. Utermann G, Hoppichler F, Dieplinger H, et al. Defects in the low density lipoprotein receptor gene affect lipoprotein (a) levels: Multiplicative interaction of two gene loci associated with premature atherosclerosis. Proc Natl Acad Sci USA 1989;86:4171–4174.

292. Brown M, Goldstein J. Teaching old dogmas new tricks. Nature 1987;330:113–114.

293. Scanu A. Lipoprotein(a): Heterogeneity and biological relevance. J Clin Invest 1990;85:1709–1715.

294. Scanu A. Lp(a) as a marker of coronary heart disease. Clin Cardiol 1991;14:35–39.

295. Miles L, Hoover-Plow J, Levin E, et al. Interaction of Lp(a) with plasminogen receptors. Circulation 1991;84(suppl II):566. Abstract.

296. Rhoads G, Dahlen G, Berg K, et al. Lp(a) lipoprotein as a risk factor for myocardial infarction. JAMA 1986;256:2540–2544.

297. Durrington P, Ishola M, Hunt L, et al. Apolipoproteins (a), AI, and B and parental history in men with early onset ischaemic heart disease. Lancet 1988;1:1070–1073.

298. Rosengren A, Wilhelmsen L, Eriksson E, et al. Lipoprotein(a) and coronary heart disease: A prospective case-control study in a general population sample of middle aged men. Br Med J 1990;301:1248–1251.

299. Murai A, Miyahara T, Fujimoto N, et al. Lp(a) lipoprotein as a risk factor for coronary heart disease and cerebral infarction. Atherosclerosis 1986;59:199–204.

300. Woo J, Lau E, Lam C, et al. Hypertension, lipoprotein(a), and apolipoprotein A-I as a risk factor for stroke in the Chinese. Stroke 1991;22:203–208.

301. Zenker G, Koltringer P, Bone G, et al. Lipoprotein(a) as a strong indicator for cerebrovascular disease. Stroke 1986;17:942–945.

302. Dahlen G. Incidence of Lp(a) lipoprotein among populations. In Scanu A (ed). Lipoprotein(a). New York: Academic Press, 1990, 151–173.

303. Anderson R, Goldstein J, Brown M. Localization of low density lipoprotein receptors on plasma membrane of normal human fibroblasts and their absence in cells from a familial hypercholesterolemia homozygote. Proc Natl Acad Sci USA 1976;73:2434–2438.

304. Anderson R, Brown M, Goldstein J. Role of the coated endocytic vesicle in the uptake of receptor-bound low density lipoprotein in human fibroblasts. Cell 1977;10:351–364.

305. Orci L, Carpenter J, Perrelet K, et al. Occurence of low density lipoprotein receptors within large pits on the surface of human fibroblasts as demonstrated by freeze-etching. Exp Cell Res 1978;113:1–13.

306. Anderson R, Brown M, Beisiegel U, Goldstein J. Surface distribution and recycling of the low density lipoprotein receptor as visualized with anti-receptor antibodies. J Cell Biol 1982;93:523–531.

307. Bersot T, Mahley R, Brown M, Goldstein J. Interactions of swine lipoproteins with the low density lipoprotein receptor in human fibroblasts. J Biol Chem 1976;251:2395–2398.

308. Mahley R, Innerarity T, Weisgraber K, Fry D. Accumulation of lipid by aortomedial cells in vivo and in vitro. Am J Path 1977;87:205–226.

309. Kovanen P, Brown M, Basu S, et al. Saturation and suppression of hepatic lipoprotein receptors: A mechanism for the hypercholesterolemia of cholesterol-fed rabbits. Proc Natl Acad Sci USA 1981;78:1396–1400.

310. Goldstein J, Anderson R, Brown M. Coated pits, coated vesicles, and receptor-mediated endocytosis. Nature 1979;279:679–685.

311. Brown M, Anderson R, Goldstein J. Recycling receptors: The round-trip itinerary and migrant membrane proteins. Cell 1983;32:663–667.

312. Yamamoto T, Davis C, Brown M, et al. The human LDL receptor: A cysteine-rich protein with multiple Alu sequences in its mRNA. Cell 1984;39:27–38.

313. Sudhof T, Goldstein J, Brown M, Russell D. The LDL gene: A mosaic of exons shared

with different proteins. Science 1985;228:815–822.

314. Russell D, Yamamoto T, Schneider W, et al. cDNA cloning of the bovine low density lipoprotein receptor: Feedback regulation of a receptor mRNA. Proc Natl Acad Sci USA 1983;80:7501–7505.

315. Esser V, Limbird L, Brown M, et al. Mutational analysis of the ligand binding domain of the low density receptor. J Biol Chem 1988;263:13282–13290.

316. Russell D, Brown M, Goldstein J. Different combinations of cysteine-rich repeats mediate binding of low density lipoprotein receptor to two different proteins. J Biol Chem 1989;264:21682–21688.

317. Hobbs H, Russell D, Brown M, Goldstein J. The LDL receptor locus in familial hypercholesterolemia: Mutational analysis of a membrane protein. Annu Rev Genet 1990;24:133–170.

318. Hobbs H, Lehrman M, Yamamoto T, Russell D. Polymorphism and evolution of Alu sequences in the human low density lipoprotein receptor gene. Proc Natl Acad Sci USA 1985;82:7651–7655.

319. Lehrman M, Russell D, Goldstein J, Brown M. Exon-Alu recombination deletes 5 kilobases from the low density lipoprotein receptor gene, producing a null phenotype in familial hypercholesterolemia. Proc Natl Acad Sci USA 1986;83:3679–3683.

320. Horsthemke B, Beisiegel U, Dunning A, et al. Unequal crossing-over between two Alu-repetitive DNA sequences in the low-density-lipoprotein-receptor gene. A possible mechanism for the defect in a patient with familial hypercholesterolemia. Eur J Biochem 1987;164:77–81.

321. Hobbs H, Brown M, Russell D, et al. Deletion in the gene for the low-density-lipoprotein receptor in a majority of French Canadians with familial hypercholesterolemia. N Engl J Med 1987;317:734–737.

322. Hobbs H, Leitersdorf E, Goldstein J, et al. Multiple mutations that prevent synthesis of LDL receptors in FH: Evidence for 13 alleles, including 5 deletions. J Clin Invest 1988;81:909–917.

323. Rudiger N, Heinsvig E, Hansen F, et al. DNA deletions in the low density lipoprotein (LDL) receptor gene in Danish families with familial hypercholesterolmia. Clin Genet 1991;39:451–462.

324. Lehrman M, Schneider W, Brown M, et al. The Lebanese allele of the low density lipoprotein receptor locus: Nonsense mutation produces truncated receptor that is retained in endoplasmic reticulum. J Biol Chem 1987;262:401–410.

325. Esser V, Russell D. Transport-deficient mutations in the low density lipoprotein receptor. Alterations in the cysteine-rich and cysteine-poor regions of the protein block intracellular transport. J Biol Chem 1988;263:13276–13281.

326. Leitersdorf E, Hobbs H, Fourie A, et al. Deletion in the first cysteine-rich repeat of low density lipoprotein receptor impairs its transport but not lipoprotein binding in fibroblasts from a subject with familial hypercholesterolemia. Proc Natl Acad Sci USA 1988;85:7912–7916.

327. Hobbs H, Leitersdorf E, Leffert C, et al. Evidence for a dominant gene that suppresses hypercholesterolemia in a family with defective low density lipoprotein receptors. J Clin Invest 1989;84:656–664.

328. Leitersdorf E, van der Westhuyzen D, Coetzee G, Hobbs H. Two common low density lipoprotein receptor gene mutations cause familial hypercholesterolemia in Afrikaners. J Clin Invest 1989;84:954–961.

329. Knight B, Gavigan S, Soutar A, Patel D. Defective processing and binding of low-density lipoprotein receptors in fibroblasts from a familial hypercholesterolemic subject. Eur J Biochem 1989;179:693–698.

330. Soutar A, Knight B, Patel D. Identification of a point mutation in growth factor repeat C of the low density lipoprotein-receptor gene in a patient with homozygous familial hypercholesterolemia that affects ligand binding and intracellular movement of receptors. Proc Natl Acad Sci USA 1989;86:4166–4170.

331. Leitersdorf E, Tobin E, Davignon J, Hobbs H. Common low-density lipoprotein receptor mutations in the French Canadian population. J Clin Invest 1990;85:1014–1023.

332. Tolleshaug H, Hobgood K, Brown M, Goldstein J. The LDL receptor locus in familial hypercholesterolemia: Multiple mutations disrupt transport and processing of a membrane receptor. Cell 1983;32:941–951.

333. Hobbs H, Brown M, Goldstein J, Russell D. Deletion of exon encoding cysteine-rich repeat of low density lipoprotein alters its binding specificity in a subject with familial hypercholesterolemia. J Biol Chem 1986;261:13114–13120.

334. Horsthemke B, Dunning A, Humphries S. Identification of deletions in the human low density lipoprotein gene. J Med Gen 1987;24:144–147.

335. Lehrman M, Goldstein J, Brown M, et al. Internalization-defective LDL receptor produced by genes with nonsense and frameshift mutations that truncate the cytoplasmic domain. Cell 1985;41:735–743.

336. Davis C, van Driel I, Russell D, et al. The low density lipoprotein receptor: Identification

of amino acids in cytoplasmic domain required for rapid endocytosis. J Biol Chem 1987;262:4075–4082.

337. Carrela M, Cooper A. High affinity binding of chylomicron remnants to rat liver plasma membranes. Proc Natl Acad Sci USA 1979; 76:338–342.

338. Hui D, Innerarity T, Mahley R. Lipoprotein binding to canine hepatic membranes. Metabolically distinct apo-E and apoB,E receptors. J Biol Chem 1981;256:5646–5654.

339. Herz J, Hamann U, Rogne S, et al. Surface location and high affinity for calcium of a 500 Kd liver membrane protein closely related to the LDL-receptor suggest a physiological role as lipoprotein receptor. EMBO J 1988;7:4119–4127.

340. Brown M, Herz J, Kowal R, Goldstein J. The low-density lipoprotein receptor-related protein: Double agent or decoy. Curr Opin Lipidol 1991;2:65–72.

341. Herz J, Kowal R, Ho Y, et al. Low density lipoprotein receptor-related protein mediates endocytosis of monoclonal antibodies in cultured cells and rabbit liver. J Biol Chem 1990;265:21355–21362.

342. Sottrup-Jensen L. Alpha2-macroglobulins: Structure, shape, and mechanism of proteinase complex formation. J Biol Chem 1989;264:11539–11542.

343. Willnow T, Goldstein J, Orth K, et al. Low density lipoprotein receptor-related protein and gp330 bind similar ligands, including plasminogen activator-inhibitor complexes and lactoferrin, an inhibitor of chylomicron remnant clearance. J Biol Chem 1992;267: 26172–26180.

344. Chappell D, Fry G, Waknitz M, et al. The low density lipoprotein receptor-related protein/a2-macroglobulin receptor binds and mediates catabolism of bovine milk lipoprotein lipase. J Biol Chem 1992;267: 25764–25767.

345. Karathanasis S. Lipoprotein metabolism: High density lipoproteins. In Lusis A, et al (ed). Monographs in Human Genetics. Basel: Karger, 1992.

346. Aviram M, Bierman E, Oram J. High density lipoprotein stimulates sterol translocation between intracellular and plasma membrane pools in human pools in human monocyte-derived macrophages. J Lipid Res 1989;30:65–76.

347. Oram J, Johnson C, Brown T. Interaction of high density lipoprotein with its receptor on cultured fibroblasts and macrophages: Evidence for reversible binding at the cell surface without internalization. J Biol Chem 1987;262:2405–2410.

348. Oram J, Mendez A, Stotte P, Johnson T. HDL apolipoproteins mediate removal of sterol from intracellular pools but not from plasma membranes of cholesterol-loaded fibroblasts. Arteriosclerosis 1991;11: 403– 414.

349. Slotte J, Oram J, Bierman E. Binding of high density lipoprotein to cell receptors promotes translocation of cholesterol from intracellular membranes to the cell surface. J Biol Chem 1987;262:12904–12907.

350. Graham D, Oram J. Identification and characterization of a high density lipoprotein-binding in cell membrane by ligand blotting. J Biol Chem 1987;262:7439–7442.

351. McKnight G, Reasoner J, Gilbert T, et al. Cloning and expression of a cellular high density lipoprotein-binding protein that is up-regulated by cholesterol loading of cells. J Biol Chem 1992;267:12131–12141.

352. Xia Y-R, Klisak I, Sparkes R, et al. Localization of the gene for high-density lipoprotein binding protein (DHLBP) to human chromosome 2q37. Genomics 1993;524–525.

353. Fielding C, Havel R. Lipoprotein lipase. Arch Pathol 1977;101:225–229.

354. Nilsson-Ehle P, Garfinkel A, Scholtz M. Lipolytic enzymes and plasma lipoprotein metabolism. Annu Rev Biochem 1980;49: 667–693.

355. Chait A, Iverius P-H, Brunzell J. Lipoprotein lipase secretion by human monocyte-derived macrophages. J Clin Invest 1982;36: 490–493.

356. Shimada K, Gill P, Silbert J, et al. Involvement of cell surface heparin sulfate in the binding of lipoprotein lipase to cultured bovine endothelial cells. J Clin Invest 1981; 68:995–1002.

357. Saxena U, Klein M, Goldberg I. Metabolism of endothelial cell-bound lipoprotein lipase: Evidence for heparan sulphate proteoglycan-mediated internalization and recycling. J Biol Chem 1990;265:12880–12886.

358. Havel R, Shore V, Shore B, Bier D. Role of specific glycopeptides of human serum lipoproteins in the activation of lipoprotein lipase. Circ Res 1970;27:595–600.

359. Miller A, Smith L. Activation of lipoprotein lipase by apolipoprotein glutamic acid. J Biol Chem 1973;248:3359–3362.

360. Wion K, Kirchgessner T, Lusis A, et al. Human lipoprotein lipase complementary DNA sequence. Science 1987;235:1638–1641.

361. Kirchgessner T, Leboef R, Langner C, et al. Genetic and developmental regulation of the lipoprotein lipase gene: Loci both distal and proximal to the lipoprotein lipase structural gene control enzyme expression. J Biol Chem 1989;264:1473–1482.

362. Semenkovich C, Chen S, Wims M, et al. Lipoprotein lipase and hepatic lipase mRNA

tissue specific expression, developmental regulation, and evolution. J Lipid Res 1989; 30:423–431.

363. Goldberg I, Soprano D, Wyatt M, et al. Localization of lipoprotein lipase mRNA in selected rat tissues. J Lipid Res 1989;30: 1569–1577.

364. Zinder O, Hamosho M, Fleck T, Scow R. Effect of prolactin on lipoprotein lipase in mammary gland and adipose tissue of rats. Am J Physiol 1974;226:744–748.

365. Eckel R, Fujimoto W, Brunzell J. Gastric inhibitory polypeptide enhanced lipoprotein lipase activity in cultured preadipocytes. Diabetes 1979;28:1141–1142.

366. Wasada T, McCorkle K, Harris V, et al. Effect of gastric inhibitory polypeptide on plasma levels of chylomicron triglycerides in dogs. J Clin Invest 1981;68:1106–1112.

367. Raynolds M, Awald P, Gordon D, et al. Lipoprotein lipase gene expression in rat adipocytes is regulated by isoproterenol and insulin through different mechanisms. Mol Endocrinol 1990;4:1416–1422.

368. Querfeld U, Ong J, Prehn J, et al. Effects of cytokines on the production of lipoprotein lipase in cultured human macrophages. J Lipid Res 1990;31:1379–1386.

369. Jonasson I, Hansson G, Bondjers G, et al. Interferon-gamma inhibits lipoprotein lipase in human monocyte-derived macrophages. Biochim Biophys Acta 1990;1053:43–48.

370. Mackay A, Oliver J, Rogers M. Regulation of lipoprotein lipase activity and mRNA content in rat epididymal adipose tissue in vitro by recombinant tumor necrosis factor. Biochem J 1990;269:123–126.

371. Kern P, Ong J, Saffari B, Carty J. The effects of weight loss on the activity of expression of adipose-tissue lipoprotein in very obese humans. N Engl J Med 1990;322:1053–1059.

372. Savard R, Bouchard C. Genetic effects in the response of adipose tissue lipoprotein lipase activity to prolonged exercise: A twin study. Int J Obesity 1990;14:771–777.

373. Peterson J, Bihain B, Bengstsson-Olivercrona G, et al. Fatty acid control of lipoprotein lipase: A link between energy metabolism and lipid transport. Proc Natl Acad Sci USA 1990;87:909–913.

374. Pradinesfigures A, Barcellinicouget S, Dani C, et al. Transcriptional control of the expression of lipoprotein lipase gene by growth hormone in preadipocyte Ob1771 cells. J Lipid Res 1990;31:1283–1291.

375. Nilsson-Ehle P, Carlström S, Belfrage P. Rapid effect on lipoprotein lipase activity in adipose tissue of humans after carbohydrate and lipid intake. Scand J Clin Lab Invest 1975;35:373–378.

376. Pykalisto O, Smith P, Brunzell J. Determinations of human adipose tissue lipoprotein lipase: Effect of diabetes and obesity on basal and diet-induced activity. J Clin Lab Invest 1975;56:1108–1117.

377. Lithell H, Boberg J, Hellsing K, et al. Lipoprotein-lipase activity in human skeletal muscle and adipose tissue in the fasting and the fed states. Atherosclerosis 1978;30:89–94.

378. Dootlittle M, Ben-Zeev O, Elovson J, et al. The response of lipoprotein lipase to feeding and fasting. Evidence for posttranslational regulation. J Biol Chem 1990;265:4570–4577.

379. Deeb S, Peng R. Structure of the human lipoprotein lipase gene. Biochemistry 1989; 28:4131–4135. Erratum. Biochemistry 1989; 28(16):6786.

380. Sparkes R, Zollman S, Klisak I, et al. Human genes involved in lipolysis of plasma lipoproteins: Mapping of loci for lipoprotein lipase to 8p22 and hepatic lipase to 15q21. Genomics 1987;1:138–144.

381. Kirchgessner T, Chuat J, Heinzmann C, et al. Organization of the human lipoprotein lipase gene evolution of the lipase gene family. Proc Natl Acad Sci USA 1989;86:9647–9651.

382. Brunzell J. Familial lipoprotein lipase deficiency and other causes of chylomicron syndrome. In Stanbury J et al (eds). The Metabolic Basis of Inherited Disease. New York: McGraw-Hill, 1983, 1913–1932.

383. Ferrans V, Buja L, Roberts W, Fredrickson D. The spleen in type I hyperlipoproteinemia. Histochemical, biochemical, microfluorometric and electron microscopic observations. Am J Pathol 1971;64:67–96.

384. Brunzell J, Chait A, Nikkila E, et al. Heterogeneity of primary lipoprotein lipase deficiency. Metab Clin Exper 1980;29:624–629.

385. Brunzell J. Familial lipoprotein lipase deficiency and other causes of the chylomicronemia syndrome. In Scriver C, et al (eds). The Metabolic Basis of Inherited Disease. New York: McGraw-Hill, 1989, 2053–2072.

386. Kern P. Lipoprotein lipase and hepatic lipase. Curr Opin Lipidol 1991;2:162–169.

387. Babirak S, Iverius P, Fujimoto W, Brunzell J. The detection of the heterozygous state for lipoprotein lipase deficiency. Arteriosclerosis 1989;9:326–334.

388. Wilson D, Emi M, Iverius P, et al. Phenotypic expression of heterozygous lipoprotein lipase deficiency in the extended pedigree of a proband homozygous for a missense mutation. J Clin Invest 1990;86:735–750.

389. Brunzell J, Iverius PH, Scheibel M, et al. Primary lipoprotein lipase deficiency. Adv Exp Med Biol 1986;201:227–239.

390. Winkler F, D'Arcy A, Hunzikas W. Struc-

ture of human pancreatic lipase. Nature 1990;343:771–774.

391. Emi M, Wilson D, Iverius P, et al. Missense mutation (Gly-Glu188) of human lipoprotein lipase imparting functional deficiency. J Biol Chem 1990;265:5910–5916.

392. Hayden M, Ma Y, Brunzell J, Henderson H. Genetic variants affecting human lipoprotein and hepatic lipases. Curr Opin Lipidol 1991;2:104–109.

393. Kuusi T, Kinnunen P, Nikkila E. Hepatic endothelial lipase anti-serum influences rat plasma low and high density lipoproteins in vivi. FEBS Lett 1979;104:384–388.

394. Twu J, Garfinkel A, Schotz M. Hepatic lipase purification and characterization. Biochim Biophys Acta 1984;792:330–337.

395. Ben-Zeev O, Ben-Avram C, Wong H, et al. Hepatic lipase: A member of a family of structurally related lipases. Biochim Biophys Acta 1987;919:13–20.

396. Kinnunen P, Virtanen J, Vainio P. Lipoprotein lipase and hepatic lipase: Their role in plasma lipoprotein metabolism. Atheroscler Rev 1983;7:65–105.

397. Dootlittle M, Wong H, Davis R, Schotz M. Synthesis of hepatic lipase in liver and extrahepatic tissues. J Lipid Res 1987;28:1326–1334.

398. Verhoeven A, Jansen H. Secretion of hepatic lipase is blocked by inhibition of oligosaccharide processing at the stage of glucosidase. Int J Lipid Res 1990;31:1883–1893.

399. Kuusi T, Nikkila E, Virtanen I, Kinnunen P. Localization of the heparin-releasable lipase in situ in the rat liver. Biochem J 1979;181:245–246.

400. Jansen H, VanBerkel T, Hulsmann W. Binding of liver lipase to parenchymal and nonparenchymal rat liver cells. Biochem Biophys Res Commun 1978;85:148–152.

401. Komaromy M, Schotz M. Cloning of rat hepatic lipase cDNA: Evidence for a lipase gene family. Proc Natl Acad Sci USA 1987;84:1526–1530.

402. Kirchgessner T, Svenson K, Lusis A, Schotz M. The sequence of cDNA encoding lipoprotein lipase: A member of a lipase gene family. J Biol Chem 1987;262:8463–8466.

403. Datta S, Luo C, Li W, et al. Human hepatic lipase. Cloned cDNA sequence, restriction fragment length polymorphisms, chromosomal localization, and evolutionary relationships with lipoprotein lipase and pancreatic lipase. J Biol Chem 1988;263:1107–1110.

404. Ameis D, Stahnke G, Kobayashi J, et al. Isolation and characterization of the human hepatic lipase gene. J Biol Chem 1990;265:6552–6555.

405. Little J, Connelly P. Familial hcpatic lipase

deficiency. Adv Exp Med Biol 1986;201:253–260.

406. Carlson L, Holmquist L, Nilsson-Ehle P. Deficiency of hepatic lipase activity in post-heparin plasma in familial hyper-α-triglyceridemia. Acta Med Scand 1986;219:435–447.

407. Connelly F, Maguire G, Lee M, Little J. Plasma lipoprotein in familial hepatic lipase deficiency. Arteriosclerosis 1990;10:40–48.

408. Auwerx J, Babirak S, Hokanson J, et al. Coexistence of abnormalities of hepatic lipase and lipoprotein lipase in a large family. Am J Hum Genet 1990;46:470–477.

409. Breckenridge W, Little J, Alaupovic P, et al. Lipoprotein abnormalities associated with a familial deficiency of hepatic lipase. Atherosclerosis 1982;45:161–179.

410. Hegele R, Vezina C, Moorjani S, et al. A hepatic lipase gene mutation associated with heritable lipolytic deficiency. J Clin Endocrinol Metab 1992;72:730–732.

411. Fielding P, Fielding C. A cholesteryl ester transfer complex in human plasma. Proc Natl Acad Sci USA 1980;77:3327–3330.

412. Soutar A, Garner C, Baker H, et al. Effect of the human plasma apolipoproteins and phosphatidylcholine acyl donor on the activity of lecithin:cholesterol acyltransferase. Biochemistry 1975;14:3057–3064.

413. Fukushima D, Yokoyama S, Kroon D, et al. Chain length-function correlation of amphiphilic peptides. Synthesis and surface properties of a tetratetracontapeptide segment of apolipoprotein A-I. J Biol Chem 1980;255:10651–10657.

414. Simon J, Boyer J. Production of lecithin:cholesterol acyltransferase by the isolated perfused rat liver. Biochim Biophys Acta 1971;218:549–551.

415. Warden C, Langner C, Gordon J, et al. Tissue-specific expression, developmental regulation, and chromosomal mapping of the lecithin:cholesterol acyltransferase gene: Evidence for expression in brain and testes as well as liver. J Biol Chem 1989;264:21573–21581.

416. McLean J, Fielding C, Drayna D, et al. Cloning and expression of human lecithin:cholesterol acyltransferase cDNA. Proc Natl Acad Sci USA 1986;83:2335–2339.

417. McLean J, Wion K, Drayna D, et al. Human lecithin:cholesterol acyltransferase gene: Complete gene sequence and sites fo expression. Nucl Acids Res 1986;14:9387–9406.

418. Glomset J, Assmann G, Gjone E. Lecithin:cholesterol acytyltransferase deficiency and fish eye disease. In Stanbury J, et al (eds). The Metabolic Basis of Inherited Disease. New York, McGraw-Hill, 1983, 1933–1952.

419. Glomset J, Norum K, Gjone E. Familial

lecithin:cholesterol acyltransferase deficiency. In Stanbury J et al (eds). The Metabolic Basis of Inherited Disease. New York: McGraw-HIll, 1982, 643–654.

420. Glomset J, Nichols A, Norum K, et al. Plasma lipoproteins in familial lecithin:cholesterol acyltransferase deficiency: Further studies of very low density lipoprotein abnormalities. J Clin Invest 1973;52:1078–1092.

421. Mitchell C, King W, Applegate K, et al. Characterization of apolipoprotein E-rich high density lipoproteins in familial lecithin:cholesterol acyltransferase deficiency. J Lipid Res 1980;21:625–634.

422. Norum K, Gjone E. The influence of plasma from patients with familial plasma lecithin:cholesterol acyltransferase deficiency on the lipid pattern of erythrocytes. Scand J Clin Lab Invest 1968;22:94–98.

423. Hovig T, Gjone E. Familial lecithin:cholesterol acyltransferase deficiency: Ultrastructural aspects of a new syndrome with particular reference to lesions in the kidneys and the spleen. Acta Pathol Microbiol Scand 1973;81:681–697.

424. Alaupovic P, McConathy W, Curry M, et al. Apolipoproteins and lipoprotein families in familial lecithin:cholesterol acyltransferase deficiency. J Clin Lab Invest 1974;33(suppl 137):83–87.

425. Stokke K, Bjerve K, Blomhoff J, et al. Familial lecithin:cholesterol acyltransferase deficiency: Studies on lipid composition and morphology of tissues. Scand J Clin Lab Invest 1974;33(suppl 137):93–100.

426. Albers J, Adolphson J, Chen C. Radioimmunoassay of human plasma lecithin-cholesterol acyltransferase. J Clin Invest 1981;67:141–148.

427. Albers J, Utermann G. Genetic control of lecithin-cholesterol acyltransferase (LCAT): Measurement of LCAT mass in a large kindred with LCAT deficiency. Am J Hum Genet 1981;33:702–708.

428. Frohlich J, Hon K, McLeod R. Detection of heterozygotes for familial lecithin:cholesterol acyltransferase (LCAT) deficiency. Am J Hum Genet 1982;34:65–72.

429. Frohlich J, McLeod R, Pritchard P, et al. Plasma lipoprotein abnormalies in heterozygotes for familiar lecithin:cholesterol acyltransferase deficiency. Metab Clin Exper 1988;37:3–8.

430. Albers J, Chen C-H, Adolphson J. Familial lecithin:cholesterol acyltransferase: Identification of heterozygotes with half-normal enzyme activity and mass. Clin Genet 1981;58:306–309.

431. Taramelli R, Pontoglio M, Candini G, et al. Lecithin:cholesterol acyltransferase deficiency: Molecular analysis of mutated allele. Hum Genet 1990;85:195–199.

432. Assmann G, von Eckardstein A, Funke H. Lecithin:cholesterol acyltransferase deficiency and fish-eye disease. Curr Opin Lipidol 1991;2:110–117.

433. Gotoda T, Yamada N, Murase T, et al. Differential phenotypic expression by three mutant alleles in familial lecithin:cholesterol acyltransferase deficiency. Lancet 1991;778–781.

434. Klein H-G, Lohse P, Duverger N, et al. Two different allelic mutations in the lecithin:cholesterol acyltransferase (LCAT) gene resulting in classic LCAT deficiency: LCAT (Tyr[83]-Stop) and LCAT(Tyr[156]-Asn). J Lipid Res 1993;34:49–58.

435. Maeda E, Naka Y, Matozaki T, et al. Lecithin:cholesterol acyltransferase (LCAT) deficiency with a missense mutation in exon 6 of the LCAT gene. Biochem Biophys Res Commun 1991;178:460–466.

436. Bujo H, Kusonoki J, Ogasawara M, et al. Molecular defects in familial lecithin:cholesterol acyltransferase (LCAT) deficiency: A single nucleotide insertion in LCAT gene causes a complete deficient type of the disease. Biochem Biophys Res Commun 1991;181:933–940.

437. Funke H, vonEckardstein A, Pritchard P, et al. A molecular defect causing fish eye disease: An amino acid exchange in lecithin:cholesterol acyltransferase (LCAT) leads to the selective loss of alpha-LCAT activity. Proc Natl Acad Sci USA 1991;88: 4855–4858.

438. Klein H-G, Lohse P, Pritchard P, et al. Two different allelic mutations in the lecithin-cholesterol acyltransferase gene associated with the fish eye syndrome. J Clin Invest 1992;89:499–506.

439. Albers J, Tollefson J, Chen C, Steinmetz A. Isolation and characterization of human plasma lipid transfer proteins. Arteriosclerosis 1984;4:49–58.

440. Drayna D, Jarnagin A, McLean J, et al. Cloning and sequencing of human cholesteryl ester transfer protein cDNA. Nature 1987;327:632–634.

441. Jarnagin A, Kohr W, Fielding C. Isolation and specificity of a Mr 74,000 cholesteryl ester transfer protein from human plasma. Proc Natl Acad Sci USA 1987;84:1854–1857.

442. Hesler C, Swenson T, Tall A. Purification and characterization of a human plasma cholesteryl ester transfer protein. J Biol Chem 1987;262:2275–2282.

443. Jiang X, Moulin P, Quinet E, et al. Mammalian adipose tissue and muscle are major sources of lipid transfer protein mRNA. J Biol Chem 1991;266:4631–4639.

444. Agellon L, Quinet E, Gillette T, et al. Organization of the human cholesteryl ester transfer protein gene. Biochemistry 1990;29:1373–1376.

445. Lusis A, Zollman S, Sparkes R, et al. Assignment of the human gene for cholesteryl ester-transfer protein to chromosome 16q12–21. Genomics 1987;1:232–235.

446. Koizumi J, Mabuchi H, Yoshimura A, et al. Deficiency of serum cholesterol-ester transfer activity in patients with familial hyperalphalipoproteinemia. Atherosclerosis 1985;58:175–186.

447. Inazu A, Brown M, Hesler C, et al. Increased high-density lipoprotein levels caused by a common cholesteryl-ester transfer protein gene mutation. N Engl J Med 1990;323:1234–1238.

448. Yamashita S, Sprecher D, Sakai N, et al. Accumulation of apolipoprotein E-rich high density lipoproteins in hyperalphalipoproteinemic human subjects with plasma cholesteryl ester transfer protein deficiency. J Clin Invest 1990;86:688–695.

449. Brown M, Inazu A, Hesler C, et al. Molecular basis of lipid transfer protein deficiency in a family with increased high-density lipoproteins. Nature 1989;342:448–451.

450. Groener J, daCol P, Kostner G. A hyperalphalipoproteinemic family with normal cholesterol ester transfer/exchange activity. Biochem J 1987;242:27–32.

451. Tollefson J, Ravnik S, Albers J. Isolation and characterization of a phospholipid transfer protein (LTP-II) from human plasma. J Lipid Res 1988;29:1593–1602.

452. Clarke C, Fogelman A, Edwards P. Transcriptional regulation of the 3-hydroxy-3-methyl-glutaryl coenzyme A reductase gene in rat liver. J Biol Chem 1985;260:14363–14367.

453. Chin D, Gil G, Faust J, et al. Sterols accelerate degradation of hamster 3-hydroxy-3-methylglutaryl coenzyme A reductase encoded by a constitutively expressed cDNA. Mol Cell Biol 1985;5:634–641.

454. Osborne T, Goldstein J, Brown M. 5' end of HMG-CoA reductase gene contains sequences responsible for cholesterol-mediated inhibition of transcription. Cell 1985;42:203–212.

455. Mohandas T, Heinzmann C, Sparkes R, et al. Assignment of human 3-hydroxy-3-methylglutaryl coenzyme A reductase gene to q13-q23 region of chromosome 5. Somatic Cell Mol Genet 1986;12:89–94.

456. Vega G, Beltz W, Grundy S. Low density lipoprotein metabolism in hypertriglyceridemic and normolipidemic patients with coronary heart disease. J Lipid Res 1985;26:115–126.

457. Brunzell J. Obesity and coronary heart disease: A targeted approach. Arteriosclerosis 1984;4:180–182.

458. Hui D, Brecht W, Hall E, et al. Isolation and characterization of the apoE receptor from canine and human liver. J Biol Chem 1986;261:4256–4267.

459. Beisiegel U, Heber W, Ihrke G. Two apoE-binding proteins in human liver. Arteriosclerosis 1987;7:494A.

460. Brinton E, Oram J, Chen C-H, et al. Binding of high density lipoprotein to cultured fibroblasts after chemical alteration of apoprotein amino acid residues. J Biol Chem 1986;261:495–503.

461. Haberland M, Fogelman A. The role of altered lipoproteins in the pathogenesis of atherosclerosis. Am Heart J 1987;113:573–577.

462. Drayna D, Fielding C, McLean J, et al. Cloning and expression of human apolipoprotein C cDNA. J Biol Chem 1986;261:16535–16539.

463. Havel R, Goldstein J, Brown M. Lipoproteins and lipid transport. In Bondy P, Rosenberg L (eds). Metabolic Control and Disease. Philadelphia: WB Saunders, 1980, 393–494.

Chapter 6

Genetics of Atherosclerosis

Thomas J. DeGraba, MD

Recent investigations into the role of heritable disorders in cerebrovascular disease have uncovered a multitude of risk factors and clinical conditions which are influenced not only by gene expression, but also by environmental interactions that weave a complex web resulting in increased stroke risk.[1,2] Traditionally, the greatest clinical efforts in stroke prevention have been directed towards modification of the major risk factors for athero-sclerotic disease, including symptomatic treatment of hypertension, diabetes, hypercholesterolemia and smoking.[3–7] Current advances in molecular biology have opened a vast wealth of information regarding the complex genetic control of the factors that regulate the initiation, progression, maturation, and symptomatic activation of atherosclerotic plaque. New techniques in molecular genetics will afford future opportunities for diagnostic testing and specific gene therapy targeted at cells involved in the progression and activation of atherosclerotic plaque.[8,9,10] This chapter will review existing observations regarding genetic profile and potential molecular therapies in this newly emerging field and relate them to the current understanding of atherosclerotic pathophysiology.

Introduction

The mechanisms involved in the initiation, progression, and maturation of atherosclerotic plaques have previously been well described,[11,12] characterized by an abnormal accumulation of vascular smooth muscle cells, inflammatory cells, and extracellular matrix proteins in the intimal region between the endothelial lining and the media. Processes within the plaque lead to the further disruption of the endothelial cell lining, which normally maintains the homeostatic balance of the vessel through interactions at the blood/endothelial interface and production of factors that regulate vessel tone, leukocyte adhesion and migration, coagulation, and cellular growth factors. Subsequent perturbation of the endothelial integrity and alteration of its phenotype predisposes to intraluminal thrombogenesis and/or vessel occlusion resulting in myocardial infarction, stroke, and peripheral vascular disease.[11,12] Characterization of the pathophysiologic mechanisms of vascular perturbation will ultimately facilitate early prophylactic intervention.

From Alberts MJ (ed). *Genetics of Cerebrovascular Disease.* Armonk, NY: Futura Publishing Company, Inc., © 1999.

Mechanism: Endothelial Injury

Atherosclerosis is initiated by an injury to the endothelium and smooth muscle of the arterial wall. Endothelial injury can be mediated by a number of processes, including blood flow stress in hypertension (commonly seen at the branch points of the arterial tree), hyperlipidemia, cigarette smoke, homocysteine, radiation,[13] oxidized low-density lipoprotein (LDL), oxygen-derived free radicals, immune complexes, and infections.[14] Response to the insult by the endothelium is highlighted by an increased expression of the proinflammatory cytokines, tumor necrosis factor-alpha (TNF-α) and interleukin-1 (IL-1),[15] adhesion molecules ICAM-1, VCAM-1, and E-selectin which promote monocyte recruitment and migration into the subendothelial space. This aggregate of macrophages and T-lymphocytes is involved in the oxidation of LDL via free radical formation and release, leading to accelerated lipid uptake, and forming a foam cell base which constitutes the so-called "fatty streak," the hallmark of early atherosclerotic lesions. Vascular endothelial changes also increase the potential of vasoconstriction with endothelin-1 (ET-1) release and a reduction in prostacyclins.

Vascular Smooth Muscle Proliferation

Activated macrophages and altered endothelial cells release chemoattractants, inducing platelet-derived growth factors, basic fibroblast growth factor (bFBGF), and mitogens such as TGF-β. Release of these factors results in vascular smooth muscle proliferation, continued leukocyte recruitment, lipid oxidation, and further cytokine release, which leads to the progression of the fibrous plaque.

Plaque Maturation and Destabilization

As the plaque progresses, production of a dense connective tissue cap with embedded smooth muscle cells forms overlaying a core of lipid and necrotic debris. Continued release of growth-regulatory molecules and cytokines enhances endothelial changes and macrophage activation. Protease, elastase, and collagenase released from macrophages weaken the extracellular matrix and increase the potential for plaque rupture. Van der Wal et al.[16] found that rupture of coronary artery plaques resulting in fatal myocardial infarction in twenty patients demonstrated marked increase expression of HLA-DR (class II) antigens on both the leukocytes and the smooth muscle cells, strongly implicating an active inflammatory process in association with symptomatic disease. Uninhibited positive feedback of the inflammatory and proliferative process, regulated by activated cells within the lesion, potentiates the further development of the atherosclerotic plaque.

It is now widely accepted that each stage of the lesion formation is potentially reversible and, therefore, halting progression and even plaque regression may be possible if the injurious agents are removed or if inhibitors of plaque progression are utilized.

Evidence for Genetic Influence

Family History

Prospective epidemiological studies have identified a number of familial associated risk factors that increase the occurrence of stroke and myocardial infarction.[17] Kiely et al.,[18] in the Framingham Study, found: (1) that a parental history of atherosclerotic coronary artery disease was associated with stroke in the offspring, and a trend between stroke in the Offspring Study Group and (2) a history of sibling atherothrombotic brain infarction. Direct correlation from these studies, implicating genetic influence of atherosclerotic disease, is not uniformly accepted. Graffagnino et al.[19] found that a family history of vascular disease was not an independent risk factor for stroke and ischemic heart disease in multivariate analysis, as demonstrated by others,[20,21] but a family history for vascular disease had a positive predictive value for the risk factors hypertension, ischemic heart disease, and dyslipidemias. Consideration for explaining stroke aggregation must be given to the possibility that family culture and life style factors, including dietary and eating habits, may account for familial clustering. However, studies showing an increased concordance rate of stroke among monozygotic versus dizygotic twins suggest an important genetic component to stroke risk.[22]

Early epidemiological studies, drawing attention to the role of potential genetic factors regulating vascular pathology, give rise to a rapidly expanding body of literature that focuses on the mechanism of atherogenesis. Those mechanisms which are found to be under the influence of genetic control will be the focus of the remainder of this chapter.

Lipids and Atherosclerosis

The contribution of abnormalities of lipid metabolism and their effects on endothelial cell injury and progression of atherosclerotic plaques have been addressed in the previous chapter. In addition to the heritable disorders leading to elevated LDL and triglycerides and lowered HDL, lipoprotein (a) [Lp(a)], inherited as an autosomal codominant trait has been found to be an independent risk factor for atherothrombotic stroke.[23,24] This lipoprotein, known to be associated with aberrations in thrombogenesis and atherogenesis, is controlled by multiple alleles resulting in at least six reported phenotypes, each carrying different degrees of risk for vascular disease.[25] Utermann et al.[25] and Zhuang et al.[26] showed a highly significant association between the

S1 & S2 phenotypes with the presence of high Lp(a) levels and cerebral and cardiovascular disease. Other variations in lipid expression, such as apolipoprotein E, directed by three codominant alleles, are believed to play a role in stroke risk[27] and should be considered for modification in future gene therapy.

Racial Variance

Numerous studies have investigated the relationship between race and the distribution of atherosclerosis.[28-31] Evidence suggests that intracranial atherosclerosis occurs predominantly in the black population, and extracranial disease occurs more commonly in the white population.[29] Intracranial occlusive lesions are also more common in Asians than in Caucasians.[29-31] One explanation for the difference in disease manifestation is the difference in risk factor occurrence between subpopulations. For example, the higher incidence of hypertension in black persons may predispose them to intracranial disease. However, Inzitari et al.[31] showed that race was an independent predictor of the location of cerebrovascular lesions despite risk factor variation. Therefore, it can be deduced that a demonstrable genetic profile, either alone or in combination with other risk factors, would lead to differential topography of cerebrovascular occlusive disease.

Yatsu et al.[28] investigated restriction fragment length polymorphisms of the apolipoprotein A-I gene in racial groups and found that *Sac I* polymorphism was associated with increased carotid stenosis in blacks but not whites. Since total cholesterol and LDL levels are not associated with an increase of carotid artery stenosis in blacks, *Sac I* polymorphism may identify black subjects at increased risk for atherothrombotic brain infarction. Ichinose and Kuriyama[32] demonstrated that by in vitro amplification, using gene-specific primers, the presence of polymorphisms in the 5' flanking region of the apolipoprotein (a) can be determined. They identified allelic subtype differences between Caucasians and Japanese and in patients with myocardial infarctions. Since the 5'-alleles play a role in Lp(a) levels, the authors hypothesize that, in the future, characterization of an individual's gene subtype will make it possible to predict their apo (a) levels, and thus their stroke risk.

Mediators of Inflammation in Atherosclerosis

The adherence of circulating monocytes and lymphocytes to activated arterial endothelium is one of the earliest detectable events in the pathogenesis of atherosclerosis.[11,12] Recent information regarding the genetic control of the inflammatory process initiating atherosclerosis presents exciting possibilities in the identification of individual patient risk and potential therapies through local gene transfection of affected arterial walls. The remainder of this chapter will focus on the evolving understanding of these mediators of inflammation.

Inflammatory Cytokine Expression

The proinflammatory cytokines TNF-α and IL-1β are key mediators in the initiation of a local vascular inflammatory response. Their action is characterized by the stimulation of adhesion molecule production, thrombogenesis, smooth muscle proliferation, platelet activation, hemorrhagic necrosis, and release of vasoactive agents. Expression of these cytokines helps initiate and promote the maturation of atheromatous plaques. Barath et al. demonstrated a localization of TNF-α in atherosclerotic vessels versus normal, ie. nonatherosclerotic, vessels in humans.[33] Expression of TNF-α was noted in over 88% of tissue samples studied with atherosclerosis and in none of the sections that were histologically identified as normal vessel. The highest concentration of TNF-α was found in the smooth muscle cells but was also expressed by endothelial cells and macrophages. The pattern of expression revealed a trend towards increased levels ap-

proaching the areas of atheromatous ulceration. Lei and Buja,[34] using quantitative reverse transcription-polymerase chain reaction for sensitive measurements of TNF-α levels, found significant elevations in total tissue levels of TNF-α mRNA in advanced atherosclerotic lesions, which suggests an association between TNF-α expression and plaque progression. Studies in our laboratory revealed a strong association between expression of TNF-α mRNA and ischemic symptoms in patients with high grade atherosclerotic stenosis.[35] In 20 patients undergoing carotid endarterectomy for stenosis ≥ 80% at our institution, in situ hybridization was performed on the atherosclerotic plaques for TNF-α mRNA. Plaques from patients (n=12) who experienced a TIA or stroke prior to the endarterectomy had a twofold to threefold increase in mRNA (by mean optical density) versus plaques from asymptomatic patients (n=8). Since the degree of maximal stenosis and risk factor exposure was similar in the symptomatic and asymptomatic groups, the increase in TNF-α mRNA in symptomatic patients suggests a genetic variation in expression response to the classic risk factor profile, thus resulting in a different clinical presentation.

Interleukin-1 (IL-1) gene expression has, along with TNF-α, been associated with infectious and inflammatory processes.[36] Integrated into a network of cytokines, the three major members of the IL-1 family (IL-1α, IL-1β, and IL-1 receptor antagonist), are the focus of great interest in the attempt to understand the key processes influencing cerebrovascular disease. IL-1's potential role in the formation and progression of atherosclerosis has been mentioned above, and work in our laboratories has demonstrated that incubation of rat microvascular endothelial cell cultures with IL-1β shows an up-regulation of ICAM-1 expression.[37] In addition, the studies also reveal that genetically predisposed animals, spontaneously hypertensive rats (SHRs), bind monocytes more avidly than endothelium from normotensive rats when exposed to provocative stimuli.[37,38] Clinton et al.[39] demonstrated that rabbits fed an atherogenic diet showed an enhanced ability to accumulate IL-1α and IL-1β mRNA in aortic tissue in response to intravenous administration of lipopolysaccharides. These studies support the theory that inducible IL-1 gene expression plays an important role in the pathophysiologic mechanism of increased EC monocyte adhesion and migration leading to progression of atherosclerosis. Polymorphisms for IL-1α,[40] IL-1β,[41] and IL-1ra[42–44] have been identified, and growing evidence reveals that specific cytokine alleles are associated with different diseases.[43,44] This understanding has far-reaching implications given the advancing ability to modify gene expression in disease states[45,46] (see Chapter 17).

Transforming growth factor-β (TGF-β) is a multifaceted growth modulatory agent that is also believed to play a pivotal role in atherogenesis and vessel injury repair. In atherogenesis, TGF-β is felt to have an inhibitory effect on smooth muscle proliferation and migration as well as reducing leukocyte adhesion to the endothelium.[47] Inhibition of TGF-β, as can be seen with elevated Apo(a) and PAI-1 levels,[47,49] is linked to plaque maturation involving leukocyte binding, endothelial cell injury, and lipid deposition. TGF-β is believed to be important in maintaining normal vessel wall structure through regulatory interaction with cells supporting the extracellular matrix,[48] which is critical in establishing the hemodynamic behavior of the vessel. In general, TGF-β is believed to inhibit atherosclerotic development. Of interest, Grainger et al.[49] found that women activate > 90% of the TGF-β present while males activate only 38%, leading the authors to ponder the possibility that elevated activated TGF-β in women supports their lower incidence of coronary artery disease. To gather information on the function of TGF-β in an *in vivo* model, Nabel et al.[50] transferred recombinant human gene (TGFB1), encoding for TGF-β1, into porcine arteries and noted substantial extracellular matrix production, including procollagen, and a significantly less robust cellular proliferation when compared to the response to

platelet derived growth factor (PDGF). Nikol et al.[51] found that mRNA for TGF-β was found at 2.5 times the level in postangioplasty coronary arteries versus primary atherosclerotic arteries in patients with coronary artery disease. Restenotic vessels had a fivefold increase of TGF-β mRNA over normal vessels, supporting the view that TGF-β is intimately involved in repair of vessel injury.

Adhesion Molecule Expression

Adherence of circulating monocytes and lymphocytes to activated endothelial cell wall is one of the earliest events in the formation of atherosclerotic plaques.[11,12] Studies have described expression of ICAM-1 and E-selectin on endothelial cells of human vessels resulting in the adhesion and migration of circulating leukocytes.[50,51] Cybulsky et al.[57] also report VCAM-1 to be expressed early in atherosclerotic lesions. Given the rising prominence of cellular activation and interaction in the formation of atherosclerosis, an understanding of the regulation of the inflammatory pathways has become more essential. Recent studies have searched for DNA polymorphisms in patients with known atherosclerotic disease in an attempt to identify genetic profiles that are associated with a higher risk for stroke.[52,53] In examining polymorphisms in genes encoding leukocyte adhesion molecules on vessel endothelium, Wenzel et al.[52] found a significant difference in allele frequency at two sites on the E-selectin gene in young patients with angiographically established severe coronary or peripheral vascular atherosclerosis compared to unselected controls. The S128R mutation in the epidermal growth factor (EGF) domain may play a role in ligand binding. The second mutation, L554F, was found in the membrane domain and affects membrane anchoring and shearing of the soluble protein. Studies are ongoing to determine the effects of these mutations on the ability of E-selectin to recognize the neutrophil and monocyte cell surface carbohydrate, sialyl Lewis-x, essential for endothelial/leukocyte interaction.

In our laboratory, we are interested in identifying a profile of cytokine and adhesion molecule expression on atherosclerotic plaques that is associated with the conversion of the endothelium from an asymptomatic to a proinflammatory and prothrombotic symptomatic state.[35,54] Carotid plaques from symptomatic (n=25) and asymptomatic (n=17) patients undergoing carotid endarterectomy with lesions involving >60% stenosis were snap frozen and immunofluorescent studies were performed to measure endothelial expression of the adhesion molecule ICAM-1. ICAM-1 expression was measured as a percent of luminal endothelial surface of plaque sections. A greater than 2.5-fold increased expression of ICAM-1 was found in the high-grade regions of symptomatic plaque as compared to the low-grade regions (p<0.0001). ICAM-1 expression in high-grade symptomatic plaque was 50% greater than in the high grade asymptomatic plaque (p<0.05). In addition, in situ hybridization revealed an increase in ICAM-1 mRNA through the symptomatic plaque versus plaques from patients who were asymptomatic (Figure 1). Elevation in adhesion molecule expression in symptomatic versus asymptomatic plaque strongly suggests that mediators of inflammation are associated with, if not in part the trigger for, the conversion of carotid plaques to a symptomatic state. Furthermore, percent carotid stenosis and risk factor profile with respect to hypertension, diabetes, and hypercholes-

terolemia were similar in both the asymptomatic and symptomatic groups. This would suggest that a genetic predisposition rather than exposure to risk factor burden alone was involved with the patient's risk of TIA and stroke.

In addition to variations in genotype which lead to an increased risk of developing atherosclerosis, changes in hemodynamic forces and oxidative stress may alter endothelial gene expression.[68–71] Nagel et al.[69] demonstrated that endothelial cell cultures exposed to increased shear stress forces upregulate the expression of ICAM-1 via a transcription regulatory element, called the shear stress response element (SSRE). They hypothesized that this element, present in the promoter region of several genes including ICAM-1, may be a final common pathway by which biomechanical forces influence gene expression and trigger key elements of the atherogenic pathway. The authors also noted that E-selectin and VCAM-1, which lack the SSRE, were not upregulated with the increase of shear stress forces. Marui et al.,[72] however, found that gene regulation of VCAM-1 is affected by a reduction oxidation (redox)-sensitive transcriptional regulatory protein. This regulatory protein is one of a family of transcription regulatory proteins best characterized by NF-κB[56,74] which is activated by oxidative stress. This activated protein binds to the promoter region on the adhesion molecule gene and initiates mRNA production. The process is consistent with

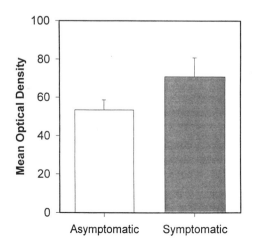

Figure 1. ICAM-1 expression in human carotid plaque—high-grade stenosis. Expression of ICAM-1 mRNA was measured semiquantitatively using a "mean optical density" of silver grains/area plaque section following in situ hybridization. Image acquisition and analysis was performed on *NIH Image*. An increase in ICAM-1 mRNA was noted in high-grade regions of symptomatic plaque versus the high-grade regions of asymptomatic plaque. A similar pattern was seen with TNF-α mRNA (data not shown). ICAM-1, intercellular adhesion molecule-1; mRNA, messenger ribonucleic acid; NIH, National Insitutes of Health; TNF-α, tumor necrosis factor-alpha.

cytokine, IL-1 and TNF-α, induction of cellular synthesis of reactive oxygen species by inflammatory and endothelial cells in the early phase of atherogenesis. Since the proinflammatory cytokines are released in response to risk factor exposure, the data suggest that hypertension, as well as hypercholesterolemia, contributes to the oxidative stress of the arterial wall which in turn leads to adhesion molecule expression and mononuclear leukocyte recruitment.[58,70] Tropea et al.[73] showed that in an in vivo model, areas of vascular constriction and elevated pressure gradient lead to increased adhesion molecule expression and monocyte binding.

The above data support not only the genetic control of the atherogenic pathways, but the influence of modifiable stroke risk factors on gene expression.

Other Related Factors

Potential genetic factors leading to a significant contribution in atherosclerotic plaque formation are numerous. Identification of the key pieces to the puzzle are undoubtedly varied from patient to patient, though common pathogenic mechanisms are likely to be identified within large segments of the population, as proposed above. Included in the mix are genes regulating angiotensin-converting enzyme (ACE), homocysteine, and thrombin receptors. Polymorphisms of the ACE gene have been reported to be a risk factor for coronary and carotid disease,[60,61] believed to be due to a proatherogenic vascular proliferation of vessel wall components as well as to increased vascular tone. Contradictory studies exist which do not find a correlation between the deletion polymorphism and atherosclerosis.[62,63] Castellano et al.[61] reported that the ACE genotype DD is linked with the risk of carotid intima-media thickening, whereas Markus et al.[62] found no correlation of the DD genotype with carotid atheroma. They did, however, find the DD genotype to be a risk factor for lacunar stroke.

Patients homozygous for the autosomal recessive trait homocystinuria will experience arterial and venous thromboembolic events and accelerated vascular disease at a very young age. Although this is a relatively uncommon disorder, it is now apparent that early atherosclerotic disease is also seen in patients heterozygous for homocysteinemia (which is seen in approximately 1% of the population).[64,65] Even mild to moderate elevations (>16 μmol/L) of homocysteine may lead to arterial endothelial cell injury[64,66] with evidence that treatment of elevated levels with folic acid and pyridoxine can lead to normalization. Although not all studies have been able to show direct correlation of homocystinuria to carotid atherosclerosis,[67] the consensus in the literature supports screening for this heritable abnormality particularly in patients with early atherosclerotic disease found to have a paucity of risk factors.

Given the multifactorial nature of plaque development, genes that may play an important role could remain uncovered. By using differential screening from a cDNA library of human normal and atherosclerotic vessels, genes expressed in the diseased state can be identified. The effectiveness of genetic screening is evidenced by work done by Pang et al.[75] who found genes for L-ferritin and H-ferritin (iron-binding proteins), expressed in human aortic atherosclerotic plaques versus normal aortas. To further study the significance of these genes, they compared aortic plaques of rabbits on high cholesterol diets to normal rabbits. Time course studies of plaque development demonstrated a close association between lesion formation and increased ferritin gene expression. Although mRNA for the binding proteins was noted in early lesions, no iron deposition was seen, as opposed to the advanced stages where large iron deposits were observed. Combined with other data, the authors postulated that early expression of the gene may be a protective mechanism directed against toxic oxidation by iron, but that in advanced lesions large amounts of iron-bound ferritin result in superoxide formation and further progression of vessel injury.

The duel nature of a number of genes

and proteins highlights the complexity of the developing atheromatous plaque. Potential therapeutic intervention will undoubtedly need to involve consideration of the stage of development of the atherosclerotic plaque with respect to time of delivery.

Advances in Gene Therapy

The advances in gene transfer, gene modification, and gene suppression combined with an ever growing understanding of the key components in genetic control of atherosclerotic disease make the likelihood of effective therapy for slowing, reversing, and preventing atherosclerosis a realistic possibility in the future. Among the various ideas in approaching gene modification, antisense oligonucleotide therapy and vector delivery gene transfer have been recently reported in the atherosclerotic literature with optimistic results. (see also *Chapter 17.*)

Antisense Oligonucleotide Therapy

A promising tool for the inhibition of gene expression is the use of oligonucleotides which can alter mRNA and DNA synthesis. Antisense nucleotides bind to complementary sequences in mRNA and are effective in probing gene function. However, antisense oligonucleotide technology has recently shown its therapeutic potential to inhibit vascular smooth muscle cell (VSMC) accumulation in both in vitro and in vivo settings.[76,77] The ability to reduce VSMC proliferation could be of potential benefit particularly in the setting of postangioplasty restenosis. Despite limitations, including inconsistent reliability in cellular incorporation of oligonucleotides, Pickering et al.[10] synthesized an antisense oligonucleotide to human proliferating cell nuclear antigen (PCNA) mRNA and demonstrated incorporation into human VSMC culture and human fibrous atherosclerotic plaque fragments with incubation. They found that the S-chimeric oligonucleotide can be successfully delivered into human tissue and significantly inhibit gene expression of VSMCs. These results are en-

couraging although the limitations of specific tissue delivery, rapid turnover of antisense nucleotides, and variable reproducibility require further attention.

Gene Transfer

The challenge in developing a successful gene transfer technique is finding a vector that can facilitate entry and expression of the therapeutic DNA material into the specific target cells. The use of liposome-mediated gene transfer and retroviral vectors has demonstrated poor gene transfer efficiency. Adenoviruses have become the vector of choice due to their versatility and ability to successfully infect target cells. Several techniques have been used to deliver genetic material using the adenovirus vector. Betz et al.[45] demonstrated that enhancement of the anti-inflammatory cytokine system could be used in brain tissue to attenuate the volume of stroke in a rat middle cerebral artery occlusion (MCAO) model. Recombinant adenovirus carrying human IL-1ra cDNA was injected into the lateral ventricle of rats and MCAO was performed 5 days after therapy. Brain concentrations of IL-1ra were 5-fold higher and CSF concentration was 50-fold higher in the IL-1ra cDNA treated rats versus rats injected with an adenoviral vector carrying *E. coli* β-galactosidase gene. This demonstrated that transient gene transfer of components of the cytokine system could be successfully delivered and orchestrate a change in clinical manifestation. Gene transfer to an intra-arterial site requires a slightly different approach. In vivo delivery of genetic material uses a double balloon catheter to temporarily stop flow, then present the vector carrying the inserted gene to the intimal lining of the vessel.[78] This method has met with varying degrees of success with endothelium transfection following intraluminal delivery of 2% to 30% and transfection of the media < 1%. Rios et al.[9] demonstrated effective gene transfer to the adventitial layer using topical administration of an adenovirus vector onto normal and atherosclerotic arteries of cynomolus monkeys. Although the transfer of genetic material was efficient and done without occlusion of the vessel, no genetic transfection occurred beyond the adventitial layer into the media or intima.

Another approach is to employ an *ex vivo* gene transfer method. After the deployment of stents following coronary angioplasty, restenosis and thrombosis are significant complications. To address the problem, Flugelman[79] seeded stents with genetically engineered endothelial cells transduced with a retroviral vector carrying human t-PA gene and implanted them in donor sheep (autologous EC culture). The premise is that production of t-PA on the seeded transfected endothelium would locally maintain blood flow without the need for excessive systemic anticoagulation. Flugelman found that both in vivo and ex vivo methods were effective in gene transfer and recovery.

Numerous obstacles still exist in the eventual use of gene therapy. Cell-specific delivery, the use of short-term or short-lived therapy in a process of a chronic nature, and establishment of reliable criteria for evaluating efficacy in the clinical setting all must be overcome. It is possible, for example, that gene therapy that reduces adhesion molecule expression on activated plaque will have no effect on radiographic morphology. Reliable in vivo markers will be required to check the efficacy of the therapy in the clinical setting. However, the experiments described above, as well as other work, raise intriguing possibilities for human therapy in the future.

Summary

The expanding field of molecular biology has given us extraordinary new ways in which to approach vascular disease. Screening of cDNA libraries to identify pivotal genes involved in atherogenesis will lead to a better understanding of progressive atherosclerotic disease process. Use of transgenic animal models will allow us to study individual gene modifications in the in vivo setting. Finally, gene transfer holds great promise for eventually treating and even "curing" atherosclerosis in humans.

References

1. DeGraba TJ, Penix L. Genetics of ischemic stroke. Curr Opin Neurol 1995;8:24–29.
2. Brass LM, Alberts MJ. The genetics of cerebrovascular disease. Ballieres Clin Neurol 1995;4(2):221–245.
3. MacMahon S, Peto R, Cutler J, et al. Blood pressure, stroke and coronary heart disease. Part I. Effects of prolonged differences in blood pressure: Evidence from nine prospective observational studies corrected for the regression dilution bias. Lancet 1990;335: 765–774.
4. The Diabetes Control and Complication Trial (DCCT) Research Group. Effect of intensive diabetes management on macrovascular events and risk factors in the diabetes control and complication trial. Am J Cardiol 1995;75:894–903.
5. Crouse JR, Byington RP, Bond MG, et al. Provastatin, lipids, and atherosclerosis in the carotid arteries (PLAC-II). Am J Cardiol 1995;75:455–459.
6. Cashin-Hemphill L, Mack WJ, Pogoda JM, et al. Beneficial effects of colestipol-niacin on coronary atherosclerosis: A 4-year follow-up. JAMA 1990;264:3013–3017.
7. Colditz GA, Bonita R, Stampfer MJ, et al. Cigarette smoking and risk of stroke in middle-aged women. N Engl J Med 1988;318:937–941.
8. Gibbons GH, Dzau VJ. Molecular therapies for vascular diseases. Science 1996;272: 689–693.
9. Rios CD, Ooboshi H, Piegors D, et al. Adenovirus-mediated gene transfer to normal and atherosclerotic arteries. Arterioscler Thromb Vasc Biol 1995;15:2241–2245.
10. Pickering JG, Isner JM, Ford CM, et al. Processing of chimeric antisense oligonucleotides by human vascular smooth muscle cells and human atherosclerotic plaque: Implications for antisense therapy of restenosis after angioplasty. Circulation 1996;93:772–780.
11. Ross R. The pathogenesis of atherosclerosis: A perspective for the 1990's. Nature 1993; 362:801–809.
12. DeGraba TJ, Fisher M, Yatsu FM. Atherogenesis and stroke. In Barnett HJM, Mohr JP, Stein BM, Yatsu FM (eds). Stroke: Pathophysiology, Diagnosis, and Management. 2nd ed. New York: Churchill Livingstone, Inc, 1992, 29–48.
13. Ross R. The pathogenesis of atherosclerosis:

Atherogenesis and inflammation. Lab Invest 1988;58:249.

14. Fuster V, Badimon L, Badimon JJ, et al. The pathogenesis of coronary artery disease and the acute coronary syndromes. N Engl J Med 1992;326:242–250.

15. Liu Y, Liu T, McCarron RM, et al. Evidence for activation of endothelium and monocytes in hypertensive rats. Am J Physiol 1996;270: H2125–2131.

16. van der Wal AC, Becker AE, van der Loos CM, et al. Site of intimal rupture or erosion of thrombosed coronary atherosclerotic plaques is characterized by an inflammatory process irrespective of the dominant plaque morphology. Circulation 1994;89:36–44.

17. Wolf PA, Cobb JL, D'Agostino RB: Epidemiology of stroke. In Barnett HJM, Mohr JP, Stein BM, Yatsu FM (eds). Stroke: Pathophysiology, Diagnosis, and Management. 2nd ed. New York: Churchill Livingstone, Inc, 1992, 3–27.

18. Kiely DK, Wolf PA, Cupples LA, et al. Familial aggregation of stroke: The Framingham Study. Stroke 1993;24:1367–1371.

19. Graffagnino C, Gasecki AP, Doig GS, Hachinski VC. The importance of family history in cerebrovascular disease. Stroke 1994;25:1599–1604.

20. Khaw KT, Barrett-Connor CE. Family history of stroke as an independent predictor of ischemic heart disease in men and stroke in women. Am J Epidemiol 1986;123:59–66.

21. Howard G, Evans G, Toole JF, et al. Characteristics of stroke victims associated with early cardiovascular mortality in their children. J Clin Epidemiol 1990;43:49–54.

22. Brass LM, Isaacsohn JL, Merikangas KR, et al. A study of twins and stroke. Stroke 1992;23:221–223.

23. Nagayama M, Shinohara Y, Nagayama T. Lipoprotein (a) and ischemic cerebrovascular disease in young adults. Stroke 1994;25: 74–78.

24. Zenker G, Koltringer P, Bone G, et al. Lipoprotein (a) as a strong indicator for cerebrovascular disease. Stroke 1986;17:942–945.

25. Utermann G, Kraft HG, Menzel HJ, et al. Genetics of the quantitative Lp(a) lipoprotein trait. Hum Genet 1988;78:41–46.

26. Zhuang Y, Li J, Wang J. Apolipoprotein (a) phenotype in cardio-cerebrovascular disease. Clin Med J 1994;107:133–136.

27. Pedro-Botet J, Senti M, Nogues X, et al. Lipoprotein and apolipoprotein profile in men with ischemic stroke. Stroke 1992;23: 1556–1562.

28. Yatsu FM, Kasturi R, Alam R. Gene polymorphism of apolipoprotein AI, the major protein of high density lipoprotein in pre-

dicting stroke risk among white and black subjects. Stroke 1993;24(suppl 1):1–26–1–30.

29. Gorelick PB. Distribution of atherosclerotic cerebrovascular lesions: Effects of age, race, and sex. Stroke 1993;24(suppl):I–16–I–19.

30. Leung SY, Ng THK, Yuen ST, Lauder IJ, Ho FCS. Pattern of cerebral atherosclerosis in Hong Kong Chinese: Severity in intracranial and extracranial vessels. Stroke 1993;24: 779–786.

31. Inzitari D, Hachinski VC, Taylor DW, Barnett HJM. Racial differences in the anterior circulation in cerebrovascular disease: How much can be explained by risk factors? Arch Neurol 1990;47:1080–1084.

32. Ichinose A, Kuriyama M. Detection of polymorphisms in the 5′-flanking region of the gene for apolipoprotein(a). Biochem Biophys Research Comm 1995;209;372–378.

33. Barath P, Fishbein MC, Cao J, et al. Detection and localization of tumor necrosis factor in human atheroma. Am J Cardiol 1990;65: 297–302.

34. Lei X, Buja M. Measurement by quantitative reverse transcription-polymerase chain reaction of the levels of tumor necrosis factor alpha mRNA in atherosclerotic arteries in Watanabe heritable hyperlipidemic rabbits. Lab Invest 1996;74:136–145.

35. Penix L, Siren A-L, Hallenbeck JM, et al. Human carotid endarterectomy plaques show expression of intracellular adhesion molecule-1 and tumor necrosis factor-α. Neurology 1996;46(2) suppl:A194.

36. Lennard AC. Interleukin-1 receptor antagonist. Crit Rev Immunol 1995;15:77–105.

37. McCarron RM, Wang L, Siren A-L, et al. Adhesion molecule expression on normotensive and hypertensive rat brain endothelial cells. Proc Soc Exp Biol Med 1994;205:257–262.

38. McCarron RM, Wang L, Siren A-L, et al. Monocyte adhesion to stimulated cerebromicrovascular EC derived from hypertensive and normotensive rats. Am J Physiol 1994;36:H2491–2497.

39. Clinton SK, Fleet JC, Loppnow H, et al. Interleukin-1 gene expression in rabbit tissue in vivo. 1991;138:1005–1014.

40. Bailly S, di Giovine FS, Blakemore AIF, et al. Genetic polymorphism of human interleukin-1α. Eur J Immunol 1993;23:1240–1245.

41. Di Giovine FS, Takhsh E, Blakemore AIF, et al. Hum Mol Genet 1992;1:450.

42. Blakemore AIF, Tarlow JK, Cork MJ, et al. Interleukin-1 receptor antagonist gene polymorphism as a disease severity factor in systemic lupus erythematosus. Arthritis Rheum 1994;37:1380–1385.

43. Mansfield JC, Holden H, Tarlow, et al. Novel genetic association between ulcerative colitis

and the anti-inflammatory cytokine inter-leukin-1 receptor antagonist. Gastroenterology 1994;106:637–642.

44. Polymorphism in human IL-1 receptor antagonist gene intron 2 is caused by variable numbers of an 86-bp tandem repeat. Hum Genet 1993;91:403–404.

45. Betz AL, Yang GY, Davidson BL. Attenuation of stroke size in rats using an adenoviral vector to induce over expression of interleukin-1 receptor antagonist in brain. J Cereb Blood Flow Metab 1995;15:547–551.

46. Bandara G, Mueller GM, Galea-Lauri J, et al. Intraarticular expression of biologically active interleukin 1-receptor-antagonist protein by ex vivo gene transfer. Proc Natl Acad Sci 1993;90:10764–10768.

47. Grainger DJ, Metcalfe JC. A pivotal role for TGF-β in atherogenesis? Biol Rev 1995;70:571–596.

48. Streuli CH, Schnidhause C, Korbin M, et al. Extracellular matrix regulates expression of the TGF-β gene. J Cell Biol 1993;120:253–260.

49. Grainger DJ, Kemp PR, Metcalfe JC, et al. Active TGF-β is depressed five-fold in triple vessel disease patients compared with syndrome X patients. J Cell Biochem 1994;18A:267.

50. Postron RN, Haskard DO, Coucher JR, et al. Expression of intercellular adhesion molecule-1 in atherosclerotic plaques. Am J Pathol 1992;140:665–673.

51. van der Wal AC, Das PK, Tigges AJ, et al. Adhesion molecules on the endothelium and mononuclear cells in human atherosclerotic lesions. Am J Pathol 1992;141:1427–1433.

52. Wenzel K, Ernst M, Rohde K, et al. DNA polymorphisms in adhesion molecule genes: A new risk factor for early atherosclerosis. Hum Genet 1996;97:15–20.

53. Miettinen HE, Hamalainen L, Kontula K. Polymorphisms of the apolipoprotein and angiotensin converting enzyme genes in young North Karelian patients with coronary heart disease. Hum Genet 1994;94:189–192.

54. DeGraba TJ, Sirén AL, Penix LaRoy, et al. Increased endothelial expression of intracellular adhesion molecule-I (ICAM-I) in symptomatic vs asymptomatic human atherosclerotic plaque. Stroke 1998;29:1405–1410.

55. Howard G, Evans GW, Toole JF, et al. Characteristics of stroke victims associated with early cardiovascular mortality in their children. J Clin Epidemiol 1990;43(1):49–54.

56. Baeuerle PA, Baltimore D. Activation of DNA-binding activity in an apparently cytoplasmic precursor of the NF-kappa B transcription factor. Cell 1988;53:211–217

57. Cybulsky MI, Gimbrone MAJ. Endothelial expression of a mononuclear leukocyte ad-hesion molecule during atherogenesis. Science 1991;251:788–791.

58. Griendling K, Alexander W. Endothelial control of the cardiovascular system: Recent advances. FASEB J 1996;10:283–292.

59. Grainger DJ, Kemp PR, Metcalf JC, et al. Active TGF-B is depressed five-fold in triple vessel disease patients compared with syndrome X patients. J Cell Biochem 1994;18A:267.

60. Cambien F, Poirer O, Lecerf L, et al. Deletion polymorphism in the gene for angiotensin-converting enzyme is a potent risk factor for myocardial infarction. Nature 1992;359: 641–644.

61. Castellano M, Muiesan ML, Rizzoni D, et al. Angiotensin-converting ezyme I/D polymorphism and arterial wall thickness in a general population. Circulation 1995;91:2721–2724.

62. Markus HS, Barley J, Lunt R, et al. Angiotensin-converting enzyme gene deletion polymorphism. Stroke 1995;26:1329–1333.

63. Catto A, Carter AM, Barrett JH, et al. Angiotensin-converting enzyme insertion/deletion polymorphism and cerebrovascular disease. Stroke 1996;27:435–440.

64. Glueck CJ, Shaw P, Lang JE, et al. Evidence that homocysteine is an independent risk factor for atherosclerosis in hyperlipidemic patients. Am J Cardiol 1995;75:132–136.

65. Fryer RH, Wilson BD, Gubler DB, et al. Homocysteine, a risk factor for premature vascular disease and thrombosis, induces tissue factor activity in endothelial cells. Arterioscler Thromb 1993;13:1327–1333.

66. Harker LA, Roos R, Slichter SJ, et al. Homocysteine-induced arteriosclerosis: The role of endothelial cell injury and platelet response in its genesis. J Clin Invest 1976;58:731–741.

67. Rubba P, Mercuri M, Faccenda F, et al. Premature carotid atherosclerosis: Does it occur in both familial hypercholesterolemia and homocysteinuria? Stroke 1994;25:943–950.

68. Ando J, Kamiya A. Flow-dependent regulation of gene expression in vascular endothelial cells. Jpn Heart J 1996;37:19–32.

69. Nagel T, Resnick N, Atkinson WJ, et al. Shear stress selectivity upregulates intercellular adhesion molecule-1 expression in cultured human vascular endothelial cells. J Clin Invest 1994;94:885–891.

70. Alexander RW. Hypertension and the pathogenesis of atherosclerosis. Hypertension 1995;25:155–161.

71. Morigi M, Zoja C, Figliuzzi M, et al. Fluid shear stress modulates surface expression of adhesion molecules by endothelial cells. Blood 1995;85:1696–1703.

72. Marui N, Offerman M, Swerlick R, et al. Vascular cell adhesion molecule-1 (VCAM-1) gene transcription and expression are regu-

lated through an antioxidant sensitive mechanism in human vascular endothelial cells. J Clin Invest 1993;92:1866–1874.

73. Tropea BI, Huie P, Cooke JP, et al. Hypertension-enhanced monocyte adhesion in experimental atherosclerosis. J Vasc Surg 1996;23:596–605.

74. Schreck R, Baeuerle P. A role for oxygen radicals as second messengers. Trends Cell Biol 1991;1:39–42.

75. Pang JHS, Jiang MJ, Chen YL, et al. Increased ferritin gene expression in atherosclerotic lesions. J Clin Invest 1996;97:2204–2212.

76. Speir E, Epstein SE. Inhibition of smooth muscle cell proliferation by antisense oligodeoxynucleotide targeting by mRNA encoding proliferating cell nuclear antigen. Circulation 1992;86:538–547.

77. Morishita R, Gibbons GH, Ellison KE, et al. Single intraluminal delivery of antisense cdc2 kinase and proliferating-cell nuclear antigen oligonucleotides result in chronic inhibition of neointimal hyperplasia. Proc Natl Acad Sci USA 1993;90:8474–8478.

78. Flugelman MY. Inhibition of intravascular thrombosis and vascular smooth muscle cell proliferation by gene therapy. Thromb Haemost 1995;74:406–410.

Chapter 7

Genetics of Coagulation Disorders

Thomas L. Ortel, MD, PhD

Hemostasis is a physiologic process that functions to prevent the loss of blood following vascular injury.[1,2] Platelets form the initial component, adhering to the site of injury and forming a primary platelet plug. This is followed by a series of enzymatic reactions that leads to the formation of an insoluble fibrin clot, strengthening the initial platelet plug.[2] In contrast, thrombosis is a pathologic process that occurs when the hemostatic process is poorly regulated. To prevent this from occurring, several "natural anticoagulant" mechanisms exist to prevent a developing clot from disseminating,[3] and the fibrinolytic pathway functions to remove already formed thrombi.[4]

As would be expected, these mechanisms are maintained in a delicate balance. Inherited deficiency states of the procoagulant components usually result in a hemorrhagic diathesis, whereas deficiencies of the anticoagulant or fibrinolytic components often produce a hypercoagulable state. Conversely, specific point mutations in the procoagulant components result in an increased thrombotic risk due to increased concentrations or resistance to inactivation. In this chapter, we will discuss the individual components of normal hemostasis, and review the clinical and laboratory manifestations of the inherited abnormalities of blood coagulation.

The Hemostatic Mechanism: Primary Hemostasis

Primary hemostasis refers to the initial events that occur following vascular injury. Damage to the endothelial surface exposes subendothelial adhesive glycoproteins such as von Willebrand factor, fibronectin, and collagen. Platelets adhere to this surface via the glycoprotein Ib/IX complex and subsequently become activated by exposure to collagen. This results in the surface exposure of glycoprotein IIb/IIIa, which binds circulating fibrinogen and results in platelet aggregation. Simultaneously, platelet activation results in the release of several substances stored in the platelet, including serotonin, calcium, fibrinogen, factor V, and ADP. This release stimulates the recruitment of additional platelets, and procoagulant surfaces are exposed to support the process of secondary hemostasis. Disorders of primary hemostasis consist of quantitative and qualitative defects of platelets and von Willebrand factor.

Platelets

Quantitative Defects

Congenital thrombocytopenias may be due to either a decrease in the number of

From Alberts MJ (ed). *Genetics of Cerebrovascular Disease*. Armonk, NY: Futura Publishing Company, Inc., © 1999.

129

megakaryocytes or to ineffective thrombocytopoiesis. Congenital megakaryocyte hypoplasia is an autosomal recessive disorder frequently associated with skeletal abnormalities (for example, the thrombocytopenia-absent radius, or TAR, syndrome), cardiac abnormalities, and other congenital defects.[5] The Wiskott-Aldrich syndrome is a sex-linked disorder that is characterized by thrombocytopenia, eczema, and immune defects.[5] The May-Hegglin anomaly and Alport's syndrome, on the other hand, are autosomal dominant syndromes characterized by abnormal megakaryocytopoiesis and thrombocytopenia.[5] Although each of these syndromes differs in its underlying pathophysiology, mode of inheritance, and associated congenital anomalies, all share the common hematologic manifestation of thrombocytopenia. Severity of bleeding is variable with the different disorders, however, and may range from mild to life-threatening. Intracranial hemorrhage is a frequent cause of morbidity and mortality in patients with congenital megakaryocyte hypoplasia and the Wiskott-Aldrich syndrome.[5] In contrast, bleeding complications are much less frequent with the autosomal dominant thrombocytopenias. Laboratory diagnosis depends on visual inspection of the peripheral smear and the bone marrow; coagulation studies are normal, except for a prolonged bleeding time secondary to the decreased platelet count. Treatment consists of platelet transfusions as needed for bleeding episodes.

Qualitative Defects

Abnormalities of the Surface Glycoproteins

Glycoprotein Ib/IX (Bernard-Soulier Syndrome)   Bernard Soulier syndrome is a rare congenital disorder consisting of a prolonged bleeding time, large platelets, and thrombocytopenia. It is due to a deficiency of the platelet glycoprotein Ib/IX complex. This complex is necessary for normal adhesion of a platelet to von Willebrand factor in exposed subendothelium.[6] The clinical deficiency state is inherited as an autosomal recessive trait. Clinical manifestations include ecchymoses, epistaxis, gingival bleeding, menorrhagia, and gastrointestinal bleeding. The severity of hemorrhagic symptoms is variable, although in many patients it is severe. Laboratory studies reveal a prolonged bleeding time, thrombocytopenia, and absent platelet agglutination in the presence of ristocetin. Platelets aggregate normally to other agonists, such as ADP and epinephrine. Therapy consists of transfusion with normal platelets for bleeding episodes. To decrease the risk of alloimmunization with associated platelet transfusion refractoriness, prophylactic platelet transfusions should not be used.[7]

Glycoprotein IIb/IIIa (Glanzmann's Thrombasthenia)   Glanzmann's thrombasthenia is a rare hemorrhagic diathesis that results from a deficiency or abnormality of the platelet glycoprotein IIb/IIIa complex. This surface glycoprotein is a member of the integrins, a family of adhesive protein receptors composed of α/β-heterodimers.[8] Following platelet adhesion to the subendothelium, glycoprotein IIb/IIIa serves as a receptor for fibrinogen and is essential for platelet aggregation.[8] The clinical deficiency state is inherited as an autosomal recessive trait. Clinical manifestations include mucocutaneous bleeding, purpura, gastrointestinal bleeding, and hematuria.[7] Laboratory evaluation includes a normal platelet count and morphology, but a prolonged bleeding time. Platelet aggregation studies reveal absent aggregation to weak agonists such as ADP or epinephrine but normal agglutination to ristocetin. Antifibrinolytic therapy may be useful for gingival bleeding.[7] Therapy for severe bleeding episodes consists of transfusion with normal platelets. As for patients with Bernard-Soulier syndrome, prophylactic platelet transfusions should be avoided.

Secretion Defects   Inherited disorders of platelet secretion result from either deficiencies of one or more types of platelet granules, or from defects in the secretory mechanism itself. In general, these disorders result in a mild to moderate bleeding diathesis, characterized by easy bruising, epistaxis, menorrhagia, and excessive postoperative and postpartum bleeding. The gray platelet syndrome is a rare inherited disorder characterized by an absence of platelet α-granules.[7] These granules contain a variety of proteins that are released during the activation process, including platelet factor 4, platelet derived growth factor, coagulation factor V, von Willebrand factor, and fibrinogen. The mode of inheritance is uncertain. A deficiency of dense granules, which contain intracellular stores of ADP, ATP, calcium, pyrophosphate, and serotonin, results in δ-storage pool disease.[7] Two inherited disorders in which deficiencies of platelet dense granules have been observed include the Hermansky-Pudlak syndrome and the Chediak-Higashi syndrome.[7] Both are inherited in an autosomal recessive manner. Defects in platelet secretion represent a heterogenous collection of abnormalities of stimulus-response coupling.[7] In this case, even though the various platelet granules may be present, appropriate secretion does not occur. The inheritance pattern, when studied, is generally autosomal dominant.

Clinical laboratory analysis generally reveals a prolonged bleeding time, platelet counts that vary from normal to slightly decreased, and abnormal platelet aggregation studies. In general, qualitative platelet defects result in aggregation studies characterized by the absence of the second wave of aggregation when stimulated by ADP or epinephrine, and decreased aggregation following stimulation with collagen. Bleeding symptoms are usually mild and can be easily treated with platelet transfusions.

von Willebrand Factor

von Willebrand factor (vWF) is an adhesive glycoprotein composed of multiple subunits held together by disulfide bonds.[9]

In the plasma, vWF circulates as a complex with factor VIII, protecting the latter from degradation. Besides the plasma, vWF is also found in platelets, endothelial cells, and in the basement membrane of blood vessels. In these locations, the protein functions to promote attachment of platelets to the subendothelium in the event of vessel injury. vWF is critical for the glycoprotein Ib/IX-mediated adherence of platelets to a subendothelial surface. vWF also binds to glycoprotein IIb/IIIa to contribute to platelet spreading and aggregation.

The gene for vWF has been characterized, and localized to chromosome 12. A deficiency or abnormality of vWF results in von Willebrand's disease (vWD), a genetically and clinically heterogenous disorder with an estimated prevalence of 1 in 100.[10] Based on clinical and laboratory parameters, the disorder can be classified into three types.[11] Type 1 vWD is the most common, accounting for ~70–80% of cases, and is due to a quantitative decrease in the level of functional vWF. This type is generally inherited as an autosomal dominant trait, with heterozygotes demonstrating a bleeding tendency. Type 2 vWD differs in that it is a qualitative defect in vWF, with abnormal multimer patterns observed on agarose electrophoresis. This type accounts for ~15–30% of cases and is also generally inherited as an autosomal dominant trait. Type 3 vWD represents a severe quantitative defect with undetectable levels of vWF activity and antigen, and factor VIII levels that are frequently less than 10%. This type is relatively rare and is inherited as either an autosomal recessive or as a compound heterozygote.

One variant of von Willebrand's disease deserves special mention at this point because the clinical phenotype resembles classic hemophilia. The Normandy variant of von Willebrand factor has a mutation in the factor VIII binding domain of von Willebrand factor, thereby resulting in decreased factor VIII and the clinical manifestations of hemophilia A.[12] This differs from true hemophilia A, however, because of its autosomal inheritance, which is the genetic pattern of von Willebrand's disease.

Hemorrhagic symptoms of vWD include epistaxis, easy bruising, gingival bleeding, gastrointestinal hemorrhage, and menorrhagia. Bleeding is also common following surgical procedures or dental extractions. The bleeding tendency seems to vary in any individual patient, however, and may vary between afflicted family members. In general, type 1 vWD patients have a milder disease. Type 3 vWD patients, on the other hand, have the most severe hemorrhagic tendency, and, because of the associated decrease in factor VIII, these patients may also develop hemarthroses and deep hematomas. In general, however, death from bleeding is a rare event in vWD.

The laboratory diagnosis of vWD can be difficult due to natural variation in the course of the disease.[13] Classically, there would be reduced plasma levels of vWF antigen, vWF activity, and factor VIII, and a prolonged bleeding time. More characteristically, however, an individual patient will not manifest all of the laboratory abnormalities, and repeat testing will need to be performed. Once the diagnosis has been confirmed, the type of vWD is determined by multimer analysis on agarose electrophoresis. Additional laboratory abnormalities that may be observed include a prolonged aPTT when the factor VIII level is reduced below ~30%, and thrombocytopenia may be observed with type 2b vWD.

The aim of therapy is to correct the prolonged bleeding time and any associated coagulation factor abnormalities.[14] For type 1 vWD patients, DDAVP is the initial treatment of choice. DDAVP induces the release of intracellular vWF from endothelial cells, resulting in a transient increase in circulating vWF levels. DDAVP is relatively contraindicated in patients with type 2b vWD, due to the potential risk of exacerbating the thrombocytopenia and possible thrombosis, and is generally of little use in type 3 vWD, due to the lack of intracellular stores. For the latter patients, and for type 1 patients who do not respond to DDAVP, intermediate purity factor VIII preparations contain the full range of vWF multimers and should be used for therapy or as prophylaxis preceding operative procedures.[14]

The Hemostatic Mechanism: Secondary Hemostasis

Following formation of a platelet plug at the site of injury, the stage is set for initiation of the coagulation cascade (Figure 1).[1,2] This consists of a series of reactions that ultimately result in an insoluble fibrin clot.[2] Initiation of the procoagulant response occurs through the extrinsic pathway, involving tissue factor and factor VIIa. This pathway is quickly downregulated by the tissue factor pathway inhibitor, however, and continued growth of the fibrin clot requires recruitment of the intrinsic pathway.[15] As fibrin is formed, it is cross-linked by factor XIIIa to form a stable, covalently bonded clot which is more resistant to lysis. Except for deficiencies of the initiation step of contact phase activation (factor XII, high molecular weight kininogen, and prekallikrein), deficiencies of any of the plasma components of the procoagulant response result in a bleeding diathesis. Inheritance is either sex-linked or autosomal recessive, and hemorrhagic strokes have been described with most of the deficiency states (Table 1).

Initiation of Secondary Hemostasis

Tissue Factor

Tissue factor is an integral membrane protein located on the plasma membrane of most vascular cells.[15] Upon tissue injury, tissue factor binds to factor VII or factor VIIa to initiate the extrinsic pathway of blood coagulation (Figure 1). The tissue factor/factor VIIa complex activates factor X to factor Xa and factor IX to factor IXa. The gene for tissue factor has been localized to chromosome 1. No patient with a deficiency of this protein has been described.

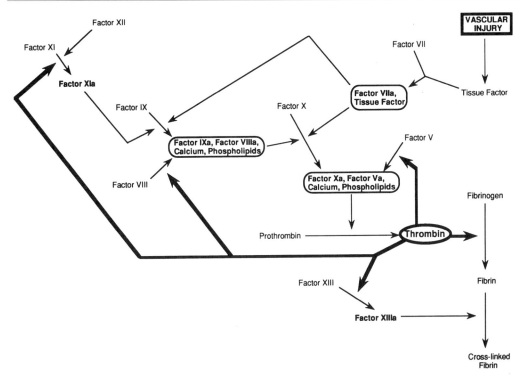

Figure 1. Schematic representation of the coagulation cascade. Hemostasis is initiated following the exposure of tissue factor by vascular injury, which results in initiation of the extrinsic arm of the pathway. The tissue factor/factor VIIa complex activates factor X to factor Xa and factor IX to factor IXa. Factor Xa is the enzymatic component of the prothrombinase complex (factor Xa, factor Va, calcium ions, and a phospholipid membrane surface), and factor IXa is the enzymatic component of the factor X-ase complex (factor IXa, factor VIIIa, calcium ions, and a phospholipid membrane surface). The prothrombinase complex activates prothrombin to thrombin, a critical step in the procoagulant response. Thrombin has numerous functions, including the conversion of fibrinogen to fibrin, activation of factor V, factor VIII, and factor XIII, and the feedback activation of factor XI (bold lines). The latter activity results in initiation of the intrinsic arm of the procoagulant response. The final result is the formation of an insoluble fibrin clot and the cessation of bleeding.

Factor VII

Factor VII, a vitamin K-dependent protein, is the initial *plasma* component of the extrinsic pathway of blood coagulation. As described above, factor VII forms a one-to-one complex with exposed tissue factor following vascular injury. This complex activates factor X and factor IX, which initiates the common pathway as well as the mechanism to maintain and amplify the procoagulant response.

The gene for factor VII has been characterized, and localized to chromosome 13, adjacent to the gene for factor X. The deficiency state is inherited as an autosomal recessive trait with an estimated incidence of 1 in 500,000.[16] Clinical manifestations include easy bruising, epistaxis, soft tissue hemorrhage, and menometrorrhagia in women.[16,17] Intracranial hemorrhagic events are not infrequent and include subarachnoid hemorrhage, intraventricular hemorrhage, and intracerebral hemorrhage (Table 1).[17] These events can be associated with significant morbidity and mortality, and may occur without antecedent trauma or underlying central nervous system abnormality. Characteristically, factor VII deficiency will present with a prolonged prothrombin time (PT), a

Table 1

Inherited coagulation factor abnormalities that have been associated with intracranial hemorrhage.

Factor	Function	Inheritance	Chromosomal Location	Neurologic Manifestations Associated with Deficiency State	Abnormal Screening Test
Procogulant Factors					
Factor VII	Proteinase zymogen	AR	13	Intracranial hemorrhage common	PT
Factor X	Proteinase zymogen	AR	13	Cerebral hemorrhage reported	PT, aPTT
Factor V	Procofactor	AR	1	Cerebral hemorrhage reported	PT, aPTT
Fibrinogen	Structural protein		4		
Afibrinogenemia		AR		Intracranial hemorrhage common	PT, aPTT, TCT
Dysfibrinogenemia		AR or AD		Subarachnoid hemorrhage reported	PT, aPTT, TCT
Factor IX	Proteinase zymogen	X-linked	X	Intracranial hemorrhage common in severe deficiency states (Intracerebral, subarachnoid, and subdural)	aPTT
Factor VIII	Procofactor	X-linked	X	Intracranial hemorrhage common in severe deficiency states (Intracerebral, subarachnoid, and subdural)	aPTT
Factor XI	Proteinase zymogen	AR	4	Cerebral hemorrhage reported	aPTT
Factor XIII	Transglutaminase	AR	6 (α subunit)	Intracranial hemorrhage common	—
Fibrinolytic Pathway					
α_2-antiplasmin	Proteinase inhibitor	AR	17	Subarachnoid hemorrhage; epidural hemorrhage	—
Tissue plasminogen activator	Proteinase	?	8	Single report of congenital excess resulting in intracerebral hemorrhage	—

AR, autosomal recessive; AD, autosomal dominant; PT, prothrombin time; aPTT, activated partial thromboplastin time; TCT, thrombin clot time.

normal activated partial thromboplastin time (aPTT), a normal thrombin clotting time (TCT), and a normal bleeding time. Therapy for hemorrhagic episodes can be difficult because of the short biological half-life of factor VII, although only 10–20% of factor VII activity may be necessary for adequate hemostasis. Recombinant factor VII may eventually be a therapeutic alternative in these patients.

Factor X

Factor X (Stuart factor) is a vitamin K-dependent procoagulant protein that represents the convergence of the intrinsic and extrinsic pathways to the common pathway (Figure 1).[1,2] Factor X circulates as an inactive zymogen that can be activated by either the factor VIIa-tissue factor complex (extrinsic pathway) or by factor IXa with factor VIIIa, calcium ions, and an anionic phospholipid surface (intrinsic pathway). The activated factor is a serine proteinase that converts prothrombin to thrombin in the presence of factor Va, calcium ions, and anionic phospholipids (the "prothrombinase complex"; cf. Figure 1).

The gene for factor X has been characterized, and localized to chromosome 13. Deficiency of factor X is a rare disorder with an estimated incidence of less than 1 in 500,000.[16] The deficiency state is inherited as an autosomal recessive trait. Similar to hemophilia A, the severity of the bleeding diathesis is related to the amount of factor X clotting activity present. Patients with less than 1% factor X clotting activity have severe hemorrhagic problems, including bleeding from the umbilical stump, mucosal membrane bleeding, hematomas, hemarthroses, and spontaneous intracranial and intraspinal hemorrhage.[16,18,19] On the other hand, patients with 10% factor X activity have significantly fewer bleeding complications. In the clinical laboratory, factor X deficiency will usually present with a prolonged PT and aPTT, but a normal TCT, consistent with its presence in the common arm of the pathway. Variant deficiencies with other patterns of laboratory abnormalities have been reported, however. Management of bleeding episodes consists of fresh frozen plasma (FFP) or prothrombin complex concentrates (PCCs).

Factor V

Factor V is an essential cofactor that interacts with the serine proteinase factor Xa in the presence of calcium ions and a procoagulant surface to accelerate the conversion of prothrombin to thrombin.[1,2,20] Factor V circulates as a single-chain procofactor that possesses little, if any, procoagulant activity. Full procoagulant activity is expressed following activation by thrombin or factor Xa. Factor V is similar structurally and functionally to factor VIII, which is a procoagulant cofactor in the "factor X-ase complex."[2]

The gene for factor V has been characterized, and localized to chromosome 1. Parahemophilia, the hemorrhagic disorder resulting from factor V deficiency, is inherited as an autosomal recessive trait, with an estimated incidence of the homozygous state of less than one in one million.[16,20] Clinical manifestations of factor V deficiency consist of posttraumatic bruising, menorrhagia, epistaxis, and bleeding from mucosal surfaces.[16] Unlike factor VIII deficiency, spontaneous hemarthroses, dissecting hematomas, and intracranial hemorrhages have been reported but are uncommon.[16,21] Hemorrhagic complications are variable in severity from kindred to kindred, which may correlate better with the platelet level of factor V than with the plasma level. Platelets contain approximately 20% of the total circulating amount of factor V and may serve as a reservoir for this essential cofactor. The laboratory manifestations of factor V deficiency consist of a prolonged PT and aPTT, since this factor is present in the common part of the pathway. Treatment consists of fresh frozen plasma for bleeding episodes and surgical prophylaxis.

Prothrombin (Factor II)

Prothrombin is a vitamin K-dependent zymogen that is activated by factor Xa to produce the serine proteinase thrombin.[1,2] Thrombin has numerous roles in blood coagulation, including activation of platelets, feedback activation of factors V and VIII, feedback activation of factor XI in the presence of a negatively charged surface, activation of factor XIII, and conversion of fibrinogen to fibrin (Figure 1).[1,2] Thrombin can also exert an anticoagulant effect when bound to the surface protein thrombomodulin, which changes the substrate specificity of the enzyme.[3] Thrombin bound to thrombomodulin activates protein C but is no longer capable of activating platelets nor cleaving factor V or fibrinogen.[3]

The gene for prothrombin has been localized to chromosome 11 and the clinical deficiency state is inherited as an autosomal recessive trait.[16] As with many of the coagulation factors, the deficiency state may be due to either absence of the specific protein or expression of a dysfunctional molecule.[16] Unlike most other coagulation factor deficiencies, however, the severity of bleeding does not always correlate well with the prothrombin level. Hemorrhagic symptoms include easy bruising, epistaxis, menorrhagia, and postoperative bleeding. Hemarthroses are rare. As with factor X deficiency, these patients will present with a prolonged PT and aPTT. Therapy is dependent on plasma infusions (FFP) or PCCs for bleeding complications. Circulating levels of 10–20% appear to be sufficient for hemostasis in most cases.

Fibrinogen

Fibrinogen is a complex molecule composed of a dimer of two identical halves, with each half composed of three separate nonidentical polypeptide chains, termed Aα, Bβ, and γ chains.[1,2] Plasma fibrinogen is converted to insoluble fibrin by a process that can be separated into three distinct phases. First, thrombin releases two small fibrinopeptides from the Aα and Bβ chains. Second, the resulting fibrin monomers form polymers. Lastly, the fibrin polymers are stabilized by covalent cross-linking by factor XIIIa. The resulting fibrin clot is more compact and resistant to mechanical disruption and/or lysis by plasmin.

The genes for the three fibrinogen chains have been characterized, and localized to chromosome 4. Congenital abnormalities of fibrinogen can be separated into quantitative and qualitative defects, although there is some overlap. Congenital afibrinogenemia is inherited as an autosomal recessive trait; congenital hypofibrinogenemia is also generally recessive, although some families appear to have a dominant inheritance pattern.[22] The afibrinogenemic patients have a very high incidence of hemorrhagic complications, and intracranial bleeding is a leading cause of death.[22,23] Other hemorrhagic manifestations include gingival bleeding, epistaxis, menorrhagia, gastrointestinal hemorrhage, and hemarthrosis. Bleeding usually occurs when fibrinogen concentrations are below 50 mg/dL, and is often associated with trauma or surgical procedures. In the clinical laboratory, these patients will have a prolongation of all tests dependent on the formation of a fibrin clot, including the PT, aPTT, TCT, and the reptilase time. Replacement therapy with cryoprecipitate is used to control bleeding episodes and in preparation for surgical procedures.

The congenital dysfibrinogenemias represent an inherited *qualitative* defect of the fibrinogen molecule.[22] These defects result in abnormal molecules with defective fibrinopeptide release, abnormal polymerization, or abnormal cross-linking, and may also be associated with a quantitative decrease in antigen (referred to as a hypodysfibrinogenemia). Most of these represent heterozygous conditions and thus the coagulopathy is inherited in an autosomal dominant manner.[22] Slightly more than half of these patients are entirely asymptomatic, with only laboratory abnormalities documenting a coagulopathy.[22] Almost a third present with hemorrhagic conditions, as with the hypofibrinogenemias.[22] Interestingly, about 20% of patients with dysfibrinogenemias have presented with thromboembolic complications.[22,24] As with the hypofibrinogenemias, laboratory diagnosis reveals prolongation of the PT, aPTT, TCT, and reptilase times. An important distinguishing feature is that these patients will have a discrepancy between functional and immunologic fibrinogen levels, due to the circulating abnormal protein. Therapy for hemorrhagic complications consists of replacement therapy, as for hypofibrinogenemia. Occasionally, dysfibrinogenemias may be associated with other coagulation factor deficiencies (e.g., fibrinogen St. Louis and factor VIII deficiency), and the possibility of concurrent coagulopathies should be determined before initiation of therapy.[25]

Amplification and Propagation of the Procoagulant Response

As mentioned above, the extrinsic pathway is rapidly and effectively shut down by tissue factor pathway inhibitor binding to the factor VIIa/tissue factor/factor Xa complex.[15] To maintain a procoagulant response, therefore, the components of the intrinsic pathway must be incorporated into the clotting process (Figure 1). Deficiencies of two of the components involved at this stage (factor VIII and factor IX) result in the bleeding diatheses known as hemophilia A and hemophilia B.

Factor VIII

Factor VIII is an essential cofactor that interacts with factor IXa in the presence of calcium ions and a procoagulant surface to accelerate the conversion of factor X to factor Xa (the "factor X-ase" complex; Figure 1). Factor VIII circulates in the plasma as a complex, calcium-dependent heterodimer that is stabilized by von Willebrand factor.[26] The full procoagulant activity of factor VIIIa is expressed following limited proteolysis by thrombin or factor Xa. The activated cofactor is subsequently inactivated by further proteolysis by thrombin or factor Xa, or by the natural anticcagulant protein C. Structurally and functionally, coagulation factor VIII is very similar to coagulation factor V, which is a component of the "prothrombinase complex."[1,2]

The gene for factor VIII has been characterized, and localized to the X chromosome. Deficiency of factor VIII results in hemophilia A, which represents approximately 85% of the cases of "true hemophilia", with an estimated incidence of 1/6,000 to 1/10,000.[16] The genetics of the disorder were first described in the early 1800s, and large pedigrees dating back to as early as the sixteenth century have been described.[27] Inheritance is sex-linked; most affected individuals are males while females are carriers of the defect. Females with congenital hemophilia A have been reported, which may result from the union of an affected male with a female carrier, Turner's syndrome with a single X chromosome, or in carrier females with extreme lyonization and expression of the defective gene, but these cases are rare.

The clinical manifestations of hemophilia A are directly related to the circulating level of factor VIII in the affected individual.[16] Severe hemophiliacs have less than 1% of the normal level of factor VIII and manifest severe recurrent bleeding episodes. Spontaneous hemorrhagic events include hemarthroses, intramuscular hematomas, pseudotumors, hematuria, gingival bleeding, epistaxis, and intracranial bleeding. The latter accounts for approximately 25% of the hemorrhagic deaths in severe hemophilia A, and includes subarachnoid, subdural, and intracerebral hemorrhage.[16,28,29] Mild hemophiliacs, on the other hand, have greater than 5% of the normal level of factor VIII. These patients seldom develop spontaneous bleeding complications but have an increased bleeding tendency following trauma or surgical procedures. Moderate hemophiliacs have 1% to 5% of the normal level of factor VIII and display hemorrhagic tendencies between the two extremes.

The laboratory diagnosis of hemophilia A includes a normal PT, TCT, and bleeding time, but a prolonged aPTT that corrects when patient plasma is mixed with normal pooled plasma. Confirmation requires determination of a factor VIII level, which is also useful for predicting the clinical manifestations of the disease. Carrier detection and prenatal diagnosis can be performed by restriction fragment length polymorphism analysis of the affected family.[16] Treatment of bleeding in hemophilia A consists of replacement therapy to raise and maintain the level of circulating factor VIII at a level where hemostasis can be achieved. Currently, this is best achieved using either monoclonal antibody-purified or recombinant factor VIII preparations. In patients with severe hemophilia A and recurrent bleeding, programs of self-therapy and/or prophylaxis are useful. A complication that can develop in hemophilic patients is the development of alloantibodies (or factor inhibitors) to infused factor VIII, which can greatly complicate their management.

Factor IX

Factor IX is a vitamin K-dependent zymogen of a serine proteinase that activates factor X to factor Xa. Factor IX can be activated by either factor XIa or the factor VIIa/tissue factor complex (Figure 1). When activated, factor IXa interacts with factor VIIIa in the presence of calcium ions and an anionic phospholipid surface to form the "factor X-ase" complex.

The gene for factor IX has been characterized and, like the gene for factor VIII, has been localized to the X chromosome.[16] Inheritance of factor IX deficiency, or hemophilia B, is a sex-linked trait with an estimated incidence of 1 in 30,000.[16] Clinically, the bleeding manifestations of hemophilia B are indistinguishable from those of hemophilia A. As with hemophilia A, the severity of clinical manifestations is related to the level of factor IX activity. Severe hemophilia B patients have less than 1% of the normal level of factor IX and manifest recurrent hemarthroses, soft tissue hemorrhages, pseudotumors, hema-

turia, and intracranial hemorrhage.[16,28,29] As with hemophilia A, intracranial hemorrhage in a hemophilia B patient represents a medical emergency that requires rapid treatment with replacement therapy. Mild hemophilia B patients, on the other hand, have greater than 5% of the normal level of factor IX and rarely develop spontaneous hemorrhagic events but do have increased bleeding in association with trauma or surgical procedures.

Hemophilia B is detected in the clinical laboratory by a prolonged aPTT with a normal PT, TCT, and bleeding time. Confirmation of the diagnosis requires determination of a factor IX level. Treatment of hemorrhagic episodes requires replacement therapy with either monoclonal antibody-purified or recombinant factor IX preparations. Duration of therapy is determined by the severity of the bleeding complication. As with factor VIII deficiency, patients may develop factor IX inhibitors which can complicate therapy.

Factor XI

Until recently, the role of factor XI in normal hemostasis was unclear. However, it was known that patients deficient in this protein did manifest a true bleeding state. Recently, it has been shown that factor XI can be activated by thrombin in the presence of a negatively charged surface, thereby promoting the continued procoagulant response of the intrinsic pathway after the extrinsic arm of the pathway has been stopped by tissue factor pathway inhibitor.[15,30] Factor XIa is also capable of activating factor XI in an autocatalytic manner in the presence of dextran sulfate, thereby maintaining the procoagulant response.[15]

The gene for factor XI has been localized to chromosome 4.[31] Factor XI deficiency is inherited as an autosomal recessive trait, although it has been referred to as an "incomplete" autosomal recessive since heterozygotes may manifest minor hemorrhagic complications.[30] Many factor XI deficient patients are of Ashkenazi Jewish descent, with an estimated incidence of the homozygous deficiency state in Israel of 1 in every 190 individuals.[30] Factor XI deficiency has also been described in multiple other ethnic groups, although the incidence is considerably lower. Spontaneous hemorrhagic events are relatively uncommon, with most bleeding events being associated with surgical and/or dental procedures.[16,32] The risk of hemorrhage does not correlate with the factor XI level, and there is also variability in the bleeding experienced by an individual patient in response to various surgical procedures. Therapy is generally prophylactic, attempting to obtain levels above 30% before surgical procedures, and consists of fresh frozen plasma.

Fibrin Stabilization

Factor XIII

Following the initial formation of fibrin polymers at a site of injury, the nascent clot is strengthened by factor XIIIa, which cross-links adjacent polymers to form a mesh that is resistant to mechanical disruption or proteolytic lysis.[33] Factor XIII circulates as an inactive tetramer ($\alpha_2\beta_2$) that is activated by thrombin to form factor XIIIa (α subunit). Factor XIIIa is a transamidase that catalyzes the cross-linking between γ-glutamyl and ϵ-lysinyl groups from different fibrin strands.[34] Factor XIIIa also cross-links α_2-antiplasmin to the fibrin strands, thereby enhancing the resistance to lysis.

The gene for the α subunit of factor XIII has been localized to chromosome 6, and the deficiency state is inherited as an autosomal recessive trait.[34] The gene for the β subunit has been localized to chromosome 1.[34] Clinically, these patients develop bleeding complications early, with prolonged bleeding from the umbilical stump.[33] Other bleeding complications include intramuscular hematomas, pseudotumors, recurrent abortions in females, delayed wound healing, and intracranial hemorrhage with little or no antecedent trauma.[33] Laboratory evaluation reveals normal screening tests (PT, aPTT, TCT, and bleeding time) but an abnormal urea solubility test. Non-cross-linked fibrin clots are soluble in 5 M urea within minutes, whereas cross-linked clots remain insoluble for at least 24 hours. Because of the long biological half-life of factor XIII (6–10 days) and the fact that very little factor XIII activity is necessary for normal hemostasis, prophylactic cryoprecipitate infusions can be used to reduce the incidence of hemorrhage, especially intracranial bleeding.

Contact Phase Pathway

The initial components of the contact phase include factor XII, prekallikrein, and high molecular weight kininogen.[1,2] When plasma comes into contact with a foreign surface, such as glass, factor XII changes from an inert form to an enzymatically active state. Factor XI, circulating in a noncovalent complex with prekallikrein and high molecular weight kininogen, is activated by factor XIIa, initiating the intrinsic pathway. Factor XIIa also activates prekallikrein to kallikrein, which can then activate additional factor XII to factor XIIa. In addition, kallikrein releases the biologically active bradykinin from high molecular weight kininogen, thereby intertwining the kinin and coagulation mechanisms. The contact phase pathway can also convert single-chain urokinase plasminogen activator to two-chain plasminogen activator, intertwining the coagulation and fibrinolytic mechanisms.

A deficiency state of any of the contact phase pathway components will result in a markedly prolonged aPTT. In contrast to the other components of the intrinsic pathway, however, there is no associated hemorrhagic risk. Therefore, the role of these proteins in normal hemostasis is unknown, although it is felt that they may be important in the initiation of the fibrinolytic pathway. The deficiency states of all three are inherited as autosomal recessive traits. Interestingly, these patients do experience thromboembolic events, including pulmonary emboli, myocardial infarcts, and arterial embolic strokes, indicating that the prolonged in vitro clotting times are not protective from thromboembolism.[35,36] At present, there are no conclusive data indicating that these patients are at an increased risk for thromboembolism compared to the general population. However, laboratory confirmation of a contact phase pathway deficiency is important, predominantly to rule out other deficiency states that do carry an increased hemorrhagic risk.

Combined Deficiencies

Kindreds including individuals with deficiencies of two or more coagulation factors have been described but are rare.[37] Combined factor V and factor VIII deficiency is the most common of these multiple factor deficiency states.[37] A positional cloning approach has linked combined factor V and factor VIII deficiency to a 2.5 cM span on the long arm of chromosome 18.[38] As with the individual deficiencies, the more severe the deficiency, the more frequent the hemorrhagic complications. These patients exemplify why it is important to completely evaluate the hemostatic mechanism in a patient with a newly diagnosed coagulation factor deficiency. Determination of the appropriate therapy is dependent on accurate knowledge of the specific deficiency state.

Regulation of the Procoagulant Response: The "Natural Anticoagulant" Mechanisms

The generation of an explosive procoagulant response is essential to prevent excessive hemorrhage from a damaged blood vessel. Conversely, it is also essential that the procoagulant response be contained and not allowed to disseminate throughout the vasculature. To provide this critical regulatory function, a series of natural anticoagulants is operative at essentially every step in hemostasis.[3] A deficiency state, therefore, would be expected to result in an increased prothrombotic risk, although this has not been observed for all of the anticoagulant proteins (Table 2).

Inhibition of the Extrinsic Pathway

Tissue Factor Pathway Inhibitor

Tissue factor pathway inhibitor is a Kunitz-type protease inhibitor that regulates the initiation phase of the procoagulant response.[15] This inhibitor forms a quaternary complex with factor Xa and the factor VIIa/tissue factor complex to effectively block the extrinsic pathway-mediated clotting mechanism. A continued procoagulant response requires initiation of the intrinsic pathway via feedback activation of factor XI by thrombin. The gene for tissue factor pathway inhibitor has been localized to chromosome 2. Although a deficiency of tissue factor pathway inhibitor would be predicted to result in a hypercoagulable state, no patient with tissue factor pathway inhibitor deficiency has been identified.

Inhibition of Procoagulant Proteases

Antithrombin III

The primary physiologic role of antithrombin III is the regulation of the procoagulant proteinases thrombin, factor Xa, and factor IXa.[39] Antithrombin III is a member of the serine proteinase inhibitor (serpin) family, which also includes α_1-antitrypsin, plasminogen activator inhibitor, and heparin cofactor II. Inhibition occurs by the formation of an equimolar complex between the inhibitor and an individual proteinase, a process that is markedly accelerated by the presence of heparin.[39]

The gene for antithrombin III has been characterized, and localized to chromosome 1.[39] Inheritance of the deficiency state is an

Table 2

Inherited coagulation factor abnormalities that have been associated with intracranial thromboembolic events.

Factor	Function	Inheritance	Chromosomal Location	Neurologic Manifestations	Prevalence of the Hypercoagulable State
Natural Anticoagulants					
Antithrombin III	Serine proteinase inhibitor	AD	1	Cerebral venous thrombosis; thromboembolic events	1/2–5,000
Heparin cofactor II	Serine proteinase inhibitor	AD	22	Thromboembolic events	Case reports
Protein C	Protease zymogen	AD	2	Cerebral venous thrombosis; thromboembolic events	1/16–36,000
				Neonatal purpura fulminans	1/200*
Protein S	Cofactor	AR / AD	3	Cerebral venous thrombosis; thromboembolic events	1/33,000
Fibrinolytic Pathway					
Plasminogen	Protease zymogen	AD	6	Cerebral venous thrombosis	Rare
Procoagulant Factors					
Elevated von Willebrand factor	Primary hemostasis protein	?	12	Thromboembolic events	?
Factor V Leiden	Procoagulant procofactor	AD	1	Cerebral venous thrombosis; thromboembolic events uncommon but have been reported in certain families	~1/15–30 (Caucasian)
Prothrombin G20210A 3'-untranslated mutation	Proteinase zymogen	AD	11	Cerebrovascular thromboembolic events in patients < 50 years of age	~1/25–50 (Caucasian)
Fibrinogen polymorphisms	Structural protein	?	4	Thromboembolic stroke	~1/50–1/100 (Bβ448)
Elevated factor VIII	Procoagulant procofactor	?	X	Thromboembolic stroke	~1/10 (single study)

AD, autosomal dominant; AR, autosomal recessive.
*Prevalence of the heterozygous state.

autosomal dominant trait, with an estimated incidence of 1/2,000 to 1/5,000.[39,40] The overall incidence of thrombosis in heterozygous individuals is estimated at about 50%, although this is variable depending on the specific type of abnormal protein. Most patients present with venous thrombosis involving the lower extremities, but other frequent symptomatic sites include the inferior vena cava, the mesenteric veins, the hepatic and portal veins, the axillary veins, and the cerebral veins[41,42] Arterial thrombosis has been reported, including cerebral thromboembolic events,[43–45] but this is uncommon. In many of these patients the first thrombotic event is triggered by a known risk factor, such as pregnancy or childbirth, prolonged immobilization, major trauma, or estrogen therapy.[41]

Laboratory diagnosis of antithrombin III deficiency requires evaluation of antigenic levels as well as a functional analysis of the molecule. Functional studies should optimally look at both the ability of antithrombin III to inhibit procoagulant proteinases independent of heparin and in the presence of heparin. Patients with a specific defect in the heparin binding domain of antithrombin III (as opposed to the reactive site) have a more benign clinical course.[40] Therapy consists of anticoagulation for thromboembolic events and prophylactic anticoagulation and/or antithrombin III replacement therapy in asymptomatic patients during periods of high risk (e.g., pregnancy, surgery).[39] Heparin can be used, although higher doses may be necessary to achieve a therapeutic effect.

Heparin Cofactor II

Heparin cofactor II is a second anticoagulant protein that is capable of inhibiting thrombin.[46] Similar to antithrombin III, the inhibitory effect of heparin cofactor II is accelerated by heparin. The affinity of heparin cofactor II for heparin is lower than that for antithrombin III, however, and nonheparin glycosaminoglycans, such as dermatan sulfate, which has no effect on antithrombin III, have a more pronounced effect. Heparin co-

factor II also has little effect on coagulation proteinases other than thrombin.

The gene for heparin cofactor II has been characterized, and localized to chromosome 22.[46] Although the precise role of heparin cofactor II in normal hemostasis is unclear, two kindreds have been reported with heterozygous heparin cofactor II deficiency and an associated increased risk for thrombosis.[47,48] Inheritance was autosomal dominant. Laboratory diagnosis requires specific testing to document a deficiency of heparin cofactor II. Therapy consists of anticoagulation for thromboembolic events.

Inactivation of Procoagulant Cofactors

Thrombomodulin

Thrombomodulin is an integral endothelial cell membrane protein that binds thrombin and converts it from a procoagulant protein to an anticoagulant one.[49] When bound to thrombomodulin, the ability of thrombin to activate platelets, activate factor V, or convert fibrinogen to fibrin is diminished. Instead, the thrombin-thrombomodulin complex activates protein C, which can then inactivate the procoagulant cofactors Va and VIIIa. Thrombin bound to thrombomodulin is also more rapidly inactivated by antithrombin III, providing a mechanism to prevent continued protein C activation. The gene for thrombomodulin has been characterized, and localized to chromosome 20. Several point mutations have been identified in patients with venous thrombosis.[50] Recently, a 42-year-old woman was described with sagittal sinus thrombosis who was found to have a point mutation in the thrombomodulin gene, changing the alanine at position 25 to a threonine.[51] The same mutation was also found in the patient's daughter, who had not sustained any thrombotic events.[51]

Protein C

Protein C is a vitamin K-dependent protein that functions as an antithrombotic agent by inactivating factor Va and factor VIIIa.[3] It circulates as a zymogen that is activated by thrombin bound to thrombomodulin. In the presence of protein S, anionic phospholipids, and calcium ions, activated protein C proteolytically cleaves factors Va and VIIIa, thereby effectively stopping the generation of the procoagulant proteins thrombin and factor Xa, respectively.

Patients with protein C deficiency manifest two distinct clinical phenotypes. A clinically "overt" form of protein C deficiency is inherited as an autosomal dominant disorder, with a prevalence in the normal population of 1/16,000 to 1/36,000.[52] In patients with this form of protein C deficiency, as many as 50% of the heterozygous patients will develop a venous thrombosis before the age of 45 years.[52] The most common sites of thrombosis are in the deep veins of the legs, the iliofemoral veins, and the mesenteric veins. Cerebral vein thrombosis and thromboembolic strokes have also been described.[53,54]

In contrast, a much more frequent "covert" form of protein C deficiency has been described which may have a frequency in the general population as high as 1 in 200 individuals.[55] This appears to be an autosomal recessive disorder, since venous thrombosis is rarely seen in heterozygotes.[55,56] The homozygous state, on the other hand, is associated with a severe decrease in the level of protein C and neonatal purpura fulminans.[56] Interestingly, DNA sequence analysis of patients with protein C deficiency has determined that certain mutations may be found in symptomatic (overt) as well as asymptomatic (covert) forms of protein C deficiency.[52,57] These observations suggest that in some kindreds, the phenotypic expression of heterozygous protein C deficiency may be dependent on additional risk factors, either inherited or acquired.

Laboratory diagnosis of protein C deficiency requires demonstration of a decreased level of functional protein C in a patient who is not being treated with warfarin. Immunologic levels can be performed in patients with a functional deficiency to determine the type of protein C deficiency. Since mutations are located throughout the protein C gene, molecular diagnostic approaches are generally not useful. Prophylactic anticoagulation is not recommended for asymptomatic patients. Treatment of a thromboembolic event is the same as for a patient who is not protein C deficient, but caution is required when initiating warfarin because of the risk for coumarin-induced skin necrosis.[3] Therefore, the patient should be fully anticoagulated with heparin before initiating warfarin, and large doses of warfarin should be avoided.

Protein S

Protein S is a vitamin K-dependent protein that interacts with activated protein C to accelerate the inactivation of factor Va and factor VIIIa.[3] Unlike the other vitamin K-dependent coagulation proteins, protein S is a not a zymogen of a serine proteinase but a cofactor. In the circulation, ~60% of the total protein S may be complexed to the complement component C4b-binding protein. Only the free protein S is functionally active as a cofactor for activated protein C.

The gene for protein S has been localized to chromosome 3. As with antithrombin III and protein C, the heterozygous state is associated with an increased risk for thromboembolic events and thus the disorder is inherited in an autosomal dominant fashion.[3,40,58] Predominant thrombotic events include deep venous thrombosis, superficial thrombophlebitis, and pulmonary embolism, but thrombosis of axillary, mesenteric, and cerebral veins has also been described.[59] Protein S deficiency has also been associated with an increased frequency of arterial thromboembolic strokes, particularly in younger adults.[60–63] About half the thromboembolic events are associated with an identifiable risk factor, with the remainder being "spontaneous". The risk of developing a thromboembolic complication increases with age, and in one study, more than 67% of the protein S deficient patients (136 members from 12 families) had sustained a thromboembolic event by age 35 years.[59] As with protein C deficiency, the homozygous state may be associated with neonatal purpura fulminans.

Laboratory diagnosis requires demonstration of a reduced free protein S at a time when the patient is clinically stable. Because C4b-binding protein is an acute phase reactant, it will be elevated with acute inflammation, thereby resulting in a decrease in the amount of free protein S. The gene for protein S is large, and mutations in patients with protein S deficiency have been identified throughout the gene. Prophylactic anticoagulation is not recommended for these patients, but, once they have sustained a thrombotic complication, they are at an increased risk for recurrence and may require lifetime anticoagulation.[59] Warfarin-induced skin necrosis has been reported in several protein S deficient patients, but it is rare.

Protein C Inhibitor (Plasminogen Activator Inhibitor 3)

Protein C inhibitor is a plasma glycoprotein that inhibits activated protein C.[64] Similar to antithrombin III, protein C inhibitor is a member of the serpin superfamily. In addition to inhibiting activated protein C, protein C inhibitor has also been reported to inhibit several components of the procoagulant pathway, including thrombin and factor Xa. Furthermore, the inhibitory effect of protein C inhibitor is enhanced by heparin.[64] Consequently, the specific physiologic role of protein C inhibitor in normal hemostasis is unclear. The gene has been characterized, and localized to chromosome 14. A deficiency state of this protein has not been described.

The Hemostatic Mechanism: Inherited Abnormalities Associated with an Increased Thrombotic Risk

All of the inherited disorders of hemostasis discussed above have been related to functional deficiencies of specific procoagulant or anticoagulant proteins. As would be expected, loss of a procoagulant protein results in a hemorrhagic disorder, whereas loss of an anticoagulant protein results in a hypercoagulable state. In contrast, many inherited disorders of hemostasis may result in an apparent *increase* in the functional activity of a specific protein. Consequently, specific polymorphisms in certain procoagulant proteins will be associated with an increased tendency for thrombosis (Table 2).

Glycoprotein IIIa PL^{A1}/PL^{A2} Polymorphism

A polymorphism located at position 33 of the mature platelet surface glycoprotein IIIa is a common cause of alloimmune thrombocytopenias. In addition, the PL^{A2} polymorphism has been associated with an increased risk for coronary thrombosis in certain studies,[65,66] although not in others.[67] An association with stroke has not been reported.

von Willebrand Factor

Elevated levels of vWF have been demonstrated to be risk factors for transient ischemic attacks and stroke.[68,69] In addition, elevated vWF levels were also associated with an increased risk for mortality during the first 6 months after a stroke.[68] No specific mutations have been identified in the genes for vWF from these patients, and the prevalence of persistently elevated vWF levels in the general population is unknown.

Factor VII

Elevated factor VII levels have been associated with an increased risk for myocardial infarction.[70] Certain polymorphisms in the factor VII gene are associated with increased levels of circulating factor VII. These polymorphisms have also been associated with an increased risk for myocardial infarction,[71] but an association has not been demonstrated for stroke.[72]

Factor V Leiden

As described above, factor Va is inactivated by limited proteolysis by activated protein C in the presence of protein S, calcium ions, and a phospholipid membrane surface.[73] In 1993, Dahlbäck et al.[74] described several families with an inherited thrombophilia who were found to manifest resistance to the anticoagulant activity of activated protein C in in vitro assays. This was subsequently found to be due to a single-point mutation in the factor V gene, converting the arginine at amino acid position 506 to a glutamine,[75] thereby blocking cleavage by activated protein C at this site.[76] It is important to recognize that this point mutation has no effect on the procoagulant activity of factor Va and there is no detectable change in the level of factor V in the circulation. The frequency of this mutation is very high in the Caucasion population (ranging from 3% to 8%), but is very uncommon in the Japanese and other Eastern populations, as well as in the African population.[73] In patients with deep venous thrombosis, however, the frequency of activated protein C resistance has been reported to range from 21% to 64%, depending on the methodology of the study.[77,78]

Patients who are heterozygous for factor V Leiden manifest a three- to sixfold increase in the incidence of a venous thromboembolic event compared to normal individuals.[79] In addition, the risk for thrombosis increases with age.[79] Patients who are homozyous for the mutant factor V may manifest up to a seventyfold increase in thromboembolic risk.[79] Environmental risk factors can contribute to the risk in these patients, with the combined effect of oral contraceptives and heterozygosity for factor V Leiden increasing the thromboembolic risk to a level similar to that of the homozygous patient.[80] In general, factor V Leiden is associated with an increased risk of venous thromboembolism and not arterial thromboembolism.[81] However, one study found an association between activated protein C resistance and stroke, although there was no correlation between factor V Leiden

and stroke.[82] In addition, three kindreds have been described with inherited activated protein C resistance and stroke occurring at a young age.[83] Several recent reports have also described an association between the presence of activated protein C resistance and cerebral venous thrombosis, although several of these individuals may also have an additional hypercoagulable state, either inherited or acquired.[84-87]

Diagnosis of activated protein C resistance is based on the ability of exogenously added activated protein C to prolong a clotting assay.[74] Specificity of the assay can be increased by diluting the patient's sample into factor V deficient plasma.[88] Factor V Leiden analysis is performed by using the polymerase chain reaction (PCR) and restriction analysis with the enzyme Mnl I on genomic patient DNA.[75]

Prothrombin G20210A 3'-Untranslated Polymorphism

Recently, Poort and colleagues[89] described a common genetic variation in the 3'-untranslated region of the prothrombin gene that was associated with elevated levels of prothrombin and an increased risk of venous thrombosis. The overall prevalence of this polymorphism in one study was 2.3%, and, as with factor V Leiden, initial studies suggest it is less common in non-Caucasian populations.[90] Several studies have confirmed the association of this polymorphism with an increased risk for venous thrombosis.[89,91] Although an association with arterial thrombosis has generally not been observed, a recent study reported that the prothrombin G20210A genotype was associated with an increased risk for cerebrovascular ischemic disease in patients younger than 50 years of age.[92] Diagnosis of this polymorphism is performed by PCR and restriction endonuclease analysis.[89]

Fibrinogen

Elevated fibrinogen levels are associated with an increased risk for ischemic heart disease, peripheral vascular disease, and stroke.[93–95] Certain polymorphisms in the gene encoding the β-fibrinogen chain have been associated with increased fibrinogen concentrations and stroke.[96,97] For patients with one of these fibrinogen polymorphisms, all screening coagulation assays (PT, aPTT, TCT) will most likely be normal. In contrast, patients with a dysfibrinogenemia and an associated thrombotic tendency are relatively uncommon and more likely to have venous thrombotic complications.[24] These patients will have abnormal screening coagulation assays (Table 1).

Factor VIII

Elevated levels of factor VIII have been associated with an increased risk for venous as well as arterial thromboembolic disease, including stroke.[68,98] The molecular basis of the elevated levels of factor VIII is unknown, although familial clustering of elevated factor VIII levels suggests an inherited disorder.[99]

Therapy for patients with one of these disorders is the same as for patients with an abnormality in one of the natural anticoagulants who have sustained a thromboembolic event.

The Fibrinolytic Pathway

In contrast to the natural anticoagulant system, which functions to downregulate and prevent dissemination of an ongoing procoagulant response, the fibrinolytic pathway provides the mechanism whereby already-formed fibrin clots can be actively removed.[4] In this way, a fibrin clot that initially formed to stop bleeding from a torn blood vessel is remodeled and gradually removed to restore normal blood flow. As with every phase of blood coagulation, the fibrinolytic process is carefully regulated to prevent widespread dissemination and systemic fibrin(ogen)olysis.[4]

Plasminogen

Plasminogen circulates in the plasma as an inactive zymogen that is activated by limited proteolysis to the serine proteinase plasmin.[4] Under normal physiologic conditions, plasmin action is restricted to the site of fibrin deposition by plasma inhibitors that prevent plasmin-mediated proteolysis of other plasma proteins. When plasmin degrades fibrin that has been cross-linked by factor XIIIa, a unique fragment referred to as the D-dimer is released. This fragment results from the cross-linking of fragment D moieties from adjacent fibrin monomers, and is observed in clinical conditions characterized by ongoing clotting and lysis, such as disseminated intravascular coagulation. Plasmin is also capable of proteolytically cleaving fibrinogen, factor V, and factor VIII, processes more likely to occur following the systemic infusion of plasminogen activators for the treatment of thrombotic disorders.[4]

The gene for plasminogen has been localized to chromosome 6. Inheritance of the deficiency state is, in most cases, autosomal dominant.[100] However, clinical presentation is variable in the heterozygous state with no clear association between functional plasminogen levels and prothrombotic risk.[100] Symptomatic patients exhibit recurrent venous thrombotic events, including deep venous thrombosis, pulmonary embolism, mesenteric venous thrombosis, and intracranial thrombosis.[100] Diagnosis is dependent on documentation of decreased plasmin activity, since many of the patients may have normal antigenic levels of a dysfunctional protein. Therapy consists of anticoagulation for thrombotic events.

Tissue Plasminogen Activator

Tissue type-plasminogen activator (t-PA) is a serine proteinase that activates plasminogen to plasmin.[4] It is synthesized in endothelial cells as a single-chain protein that can be cleaved to a two-chain form by plasmin. Both forms are capable of activating plasminogen, but this occurs at a slow rate in the absence of fibrin. Fibrin provides a surface to which both plasminogen and t-PA bind, with a resultant marked increase in the enzymatic activity of t-PA.

The gene for t-PA has been characterized, and localized to chromosome 8. There have been no documented reports of a patient with t-PA deficiency, although several kindreds have been described with an inherited defect in t-PA release.[100–102] This disorder, inherited as an autosomal dominant trait, is associated with a hypercoagulable state characterized by venous thromboembolic events. Subsequent studies in at least one of these families, however, revealed that the true molecular defect was an imbalance between t-PA and a t-PA inhibitor rather than a defect in t-PA release.[100,103] Therapy consists of anticoagulation for thromboembolic events.

In contrast to the procoagulant defect seen with decreased t-PA activity, there have also been several reports of patients with hemorrhagic disorders due to *excess* t-PA.[100,104] The inheritance pattern is unknown. Hemorrhagic symptoms in one patient included recurrent bleeding following trauma or surgery and a spontaneous intracerebral hemorrhage.[104] Fibrinolytic inhibitors (e.g., ϵ-aminocaproic acid) may be useful for hemorrhagic episodes.

Urokinase

Urokinase is also a serine proteinase that can activate plasminogen to plasmin.[4] It circulates as a single-chain form which can be converted to a two-chain form by plasmin or kallikrein.[4] As with t-PA, single-chain urokinase does not significantly activate plasminogen in the absence of fibrin. Two-chain urokinase, on the other hand, has no specific affinity for fibrin and can activate plasminogen in the presence or absence of fibrin. Two-chain urokinase can be rapidly inhibited by plasminogen activator inhibitor 1 or plasminogen activator inhibitor 2, whereas single-chain urokinase is

not significantly inhibited by these inhibitors. The urokinase gene is located on chromosome 10. Abnormal fibrinolytic activity resulting from an inherited defect of urokinase has not been described.

α_2-Antiplasmin

α_2-antiplasmin is a serine proteinase inhibitor that forms a 1:1 complex with plasmin and thereby inhibits its fibrinolytic activity.[4] It is cross-linked to the fibrin α-chain by factor XIIIa, thereby increasing the resistance of the nascent clot to proteolysis by plasmin. Hereditary deficiency is a rare disorder that is inherited as an autosomal recessive trait, although some heterozygotes manifest symptoms of mild bleeding. Clinical symptoms include easy bruisability, epistaxis, prolonged bleeding following surgical and/or dental procedures, muscle hematomas, hemarthroses, and umbilical stump bleeding.[100,105,106] Intracranial hemorrhage has been reported (Table 1).[100] Bleeding may start several hours after initial hemostasis has been achieved, consistent with an increased susceptibility of nascent clots to spontaneously occurring fibrinolysis.[100] Laboratory diagnosis is dependent on a specific assay for α_2-antiplasmin activity; the PT, aPTT, and bleeding time will all be normal. Whole blood clot lysis time will be short, due to the enhanced fibrinolytic activity. Patients with this deficiency have been successfully treated with antifibrinolytic agents.[100]

Plasminogen Activator Inhibitor I

Plasminogen activator inhibitor 1 (PAI-1) is a specific inhibitor of the serine proteinases tissue plasminogen activator and urokinase.[4] Similar to many of the coagulation factor inhibitors, PAI-1 is a member of the serpin family. It is present both in plasma, although in low amounts, and the extracellular matrix, where it may influence cell migration. The gene has been characterized, and localized to chromosome 7. A deficiency state of PAI-1 has been described in several patients, characterized by postoperative and posttraumatic bleeding.[107,108] One large kindred with PAI-1 deficiency has been described, with an autosomal recessive pattern of inheritance.[109] Antifibrinolytic agents were effective in treating and preventing bleeding episodes.

Several studies have investigated whether elevated levels of PAI-1 are associated with venous and/or arterial thromboembolic events. Elevated levels of PAI-1 are associated with a specific sequence length dimorphism located in the promoter region of the PAI-1 gene (the 4G/5G dimorphism). In one study, elevated PAI-1 levels, but not the 4G/5G promoter polymorphism, were associated with cerebrovascular disease.[110] Ridker and colleagues[111] found no association of the 4G/5G polymorphism with either venous or arterial thromboembolic disease, whereas Margaglione and coworkers[112] found an association of the polymorphism with a family history of coronary artery disease. In some studies, the PAI-1 genotype appears to enhance the phenotypic expression of thrombophilia in patients with a separate hypercoagulable state, such as protein S deficiency, but does not produce a hypercoagulable state when present as an isolated abnormality.[113]

Plasminogen Activator Inhibitor 2

Plasminogen activator inhibitor 2 (PAI-2) is a second specific inhibitor of tissue plasminogen activator and urokinase.[4] PAI-2 is also a member of the serpin superfamily. This inhibitor was initially identified in the placenta and in the plasma of pregnant women. A deficiency state of PAI-2 has not been described.

Hyperhomocysteinemia

Homocysteine is a non-protein-forming, sulfur-containing amino acid whose metabolism is at the intersection of remethylation and transsulfuration.[114] Deficiencies in the enzymes involved in homocysteine metabolism can result in elevated homocysteine levels, which are associated with an increased risk for vascular disease.[114] Recently, several studies have shown that mild hyperhomocysteinemia is also a risk factor for venous thromboembolism.[115,116] Furthermore, mild hyperhomocysteinemia is frequently identified in association with a thermolabile mutation in the methylenetetrahydrofolate reductase gene that results in diminished activity of the enzyme.[117] In certain populations, the prevalence of the homozygous state for this thermolabile mutation in normal individuals is ~10–12%.[118] Treatment of these patients with folic acid and pyridoxine can rapidly decrease the homocysteine levels and thereby potentially decrease the risk for thrombosis.[114] Although hyperhomocysteinemia is not a primary defect in the hemostatic mechanism per se, the high frequency of the thermolabile mutation in the general population and its association with venous thrombosis confirms that it should be considered during the diagnostic work-up of these patients.

Lupus Anticoagulants and the Primary Antiphospholipid Antibody Syndrome

Lupus anticoagulants are acquired antibodies that prolong phospholipid-dependent coagulation tests (e.g., aPTT). Antiphospholipid antibodies are detected in solid-phase enzyme-linked immunosorbent assays (ELISA) by their ability to bind to a wide variety of anionic phospholipids (e.g., phosphatidylserine, phosphatidic acid, and cardiolipin). In general, lupus anticoagulants and antiphospholipid antibodies represent overlapping subgroups of a heterogenous polyclonal population of antibodies.[119] These antibodies are directed against an endogenous phospholipid and/or a phospholipid-protein complex containing β_2-glycoprotein I and/or prothrombin as the protein component.[120] Although these antibodies result in a prolongation of clotting studies in vitro, they are usually not associated with an increased hemorrhagic risk. In fact, they are often associated with recurrent arterial and/or venous thromboembolism, recurrent fetal loss, and thrombocytopenia.[121] The mechanism for this prothrombotic risk remains unclear, although proposed mechanisms include direct endothelial cell damage, antibody-mediated platelet activation, and inhibition of the natural anticoagulants thrombomodulin, protein C, protein S, antithrombin III, and prostacyclin.[121]

Several studies have demonstrated that family members of patients with systemic lupus erythematosus or the primary antiphospholipid antibody syndrome frequently have an increased incidence of elevated antiphospholipid antibody levels.[122,123] In addition, several kindreds have been described with a familial form of the antiphospholipid antibody syndrome,[124,125] as well as a familial form of Sneddon's syndrome, a disorder characterized by livedo reticularis and ischemic cerebrovascular lesions frequently in association with antiphospholipid antibodies.[126,127] Associations with the major histocompatibility complex have been investigated in multiple studies, but these studies are difficult to interpret due to small sample size, disease heterogeneity, and questions regarding appropriate ethnically matched control populations. Nevertheless, these results suggest that some kindreds possess an inherited predisposition to develop antiphospholipid antibodies and the associated clinical manifestations. Additional inherited disorders as well as associated environmental factor(s) may be necessary for full expression of the clinical phenotype, however.

Patients with lupus anticoagulants and/or antiphospholipid antibodies have a significantly increased risk of venous and arterial thromboembolic events.[128] Venous thromboembolic events include deep venous thrombosis, axillary vein thrombosis, pulmonary embolism, and retinal vein thrombosis. Arterial thrombosis involves the peripheral arteries, most commonly in the upper and lower extremities. Neurologic syndromes associated with antiphospho-

lipid antibodies include thromboembolic strokes, transient ischemic attacks, retinal artery occlusion, cerebral venous thrombosis, and amaurosis fugax.[129,130] Associated clinical manifestations include livedo reticularis, recurrent fetal loss, and thrombocytopenia.[121]

The laboratory diagnosis of a lupus anticoagulant is determined by the following criteria, as recommended by the Subcommittee on Lupus Anticoagulants/Antiphospholipid Antibodies of the International Society of Thrombosis and Haemostasis[131]: (1) prolongation of a phospholipid-dependent screening assay, such as the aPTT or dilute Russell's viper venom time (DRVVT) ratio; (2) lack of correction of a prolonged screening assay after a 1:1 mix with pooled normal plasma; and (3) correction of a prolonged screening assay by the addition of excess phospholipid, such as the platelet neutralization procedure (PNP). The presence of anticardiolipin antibodies is determined by a solid-phase ELISA. Additional laboratory studies that should be considered in these patients include a complete blood count, a VDRL, and, if indicated, serologies for lupus. Treatment consists of therapeutic anticoagulation for thromboembolic episodes. These patients are at an increased risk for recurrent thromboembolic events and may require lifelong anticoagulation.[132] In addition, the PT is not infrequently abnormal in these patients, which can lead to difficulties with monitoring warfarin therapy with the international normalized ratio (INR).[133]

Combined Abnormalities and Prothrombotic Risk

Extensive characterization of families with inherited thrombophilia has led to the realization that in many of these patients, thrombosis is a multigenic disorder.[134] Families have been described with factor V Leiden and inherited protein C deficiency,[135] protein S deficiency,[136] and antithrombin III deficiency.[137] For example, a study of families with a combination of protein C deficiency and factor V Leiden found that 73% of the carriers of both risk factors sustained a thrombotic episode, compared to 31% of isolated protein C deficient individuals and 13% of isolated factor V Leiden individuals.[135] The association of factor V Leiden and antiphospholipid antibodies may also result in a more severe clinical phenotype, although this is not uniformly observed. Interestingly, it has also been reported that the presence of factor V Leiden may modulate the hemorrhagic manifestations of patients with hemophilia A who have the same genetic mutation but a variable clinical phenotype.[138] The effects of acquired risk factors must also be considered, in particular the combined effect of oral contraceptives and factor V Leiden.[80] All of these parameters need to be taken into consideration when evaluating the individual patient.

Diagnostic Evaluation of the Patient with a Suspected Disorder of Hemostasis

It cannot be emphasized enough that the initial evaluation of any patient with a suspected disorder of hemostasis must include a detailed history and physical examination. Prior hemorrhagic episodes should be documented in regard to the source and extent of bleeding, associations with surgical and dental procedures, pregnancy, or trauma, concurrent medications, and the need for blood transfusions. Prior thromboembolic episodes should be documented in regard to arterial and/or venous thrombosis, frequency of events, and prior and current therapeutic interventions. A detailed family history should also be obtained, especially emphasizing hemorrhagic and/or thromboembolic events in younger individuals, recurrent miscarriages, bleeding during surgical or dental procedures. Physical examination includes a careful assessment for petechiae, ecchymoses, and other evidence for hemorrhage.

Patients presenting with an initial thromboembolic event should be evaluated for hepatosplenomegaly, lymphadenopathy, or other signs suggesting malignancy.

The initial laboratory examination of a patient with a hemorrhagic tendency should include a complete blood count with a blood film to assess the number and appearance of the platelets. A bleeding time may also be useful in the diagnostic evaluation of primary hemostasis (e.g., platelets and von Willebrand factor) and to direct further testing. The PT, aPTT, and TCT will assess the components of the intrinsic, extrinsic, and common pathways. If a clotting study is prolonged, a mixing study (1:1 with normal pooled plasma) should be obtained to differentiate between a deficiency state and a possible inhibitor (e.g., a specific coagulation factor inhibitor or a lupus anticoagulant). If a deficiency is identified, specific factor assays should be obtained to accurately identify the deficient factor(s). If the PT, aPTT, and TCT are all normal in a patient with recurrent hemorrhagic events, factor XIII and α_2-antiplasmin levels should be determined.

The laboratory evaluation of a patient with a suspected hypercoagulable state should start with the same screening tests as for the hemorrhagic diatheses (complete blood count and smear, PT, aPTT, and TCT). If the aPTT is prolonged and does not correct with a 1:1 mix with pooled normal plasma, a lupus anticoagulant should be suspected. Additional tests that can be done to identify and confirm a lupus anticoagulant include the DRVVT and the PNP. The anticardiolipin antibody is identified by an ELISA. Specific functional and/or antigenic assays are available for protein C, protein S, antithrombin III, and plasminogen, and the physician should know how the test is performed at his or her institution. As mentioned above, free protein S antigen levels may be decreased during an acute inflammatory state, but this is an acquired deficiency state and should be repeated when the patient is clinically stable. Activated protein C resistance is determined by a functional assay, and, if positive, factor V genotyping is performed to determine if the patient has factor V Leiden. The G20210A polymorphism in the 3'-untranslated region of the prothrombin gene is performed by molecular diagnostics.

Acknowledgements: This review was completed during the tenure of a Clinician Scientist Award (#91–419) from the American Heart Association and Genentech and a Pew Scholar Award. I wish to thank Dr. Frank G. Keller and Dr. B. Gail Macik for critically reading an early version of the manuscript and providing useful suggestions and recommendations.

References

1. Furie B, Furie BC. Molecular basis of blood coagulation. In Hoffman R, Benz EJ, Shattil SJ, et al (eds). Hematology: Basic Principles and Practice. 2nd ed. New York: Churchill Livingstone, 1995, 1566–1587.
2. Davie EW, Fujikawa K, Kisiel W. The coagulation cascade: Initiation, maintenance, and regulation. Biochemistry 1991;30:10363-10370.
3. Esmon CT. Regulatory mechanisms in hemostasis: Natural anticoagulants. In Hoffman R, Benz EJ, Shattil SJ, et al (eds). Hematology: Basic Principles and Practice. 2nd ed. New York: Churchill Livingstone, 1995, 1597–1605.
4. Lijnen HR, Collen D. Molecular and cellular basis of fibrinolysis. In Hoffman R, Benz EJ, Shattil SJ, et al (eds). Hematology: Basic Principles and Practice. New York: Churchill Livingstone, 1995:1588–1596.
5. George JN. Thrombocytopenia due to diminished or defective platelet production. In Beutler E, Lichtman MA, Coller BS, Kipps TJ (eds). Williams Hematology. 5th ed. New York: McGraw-Hill, 1995, 1281–1290.
6. Lopez JA. The platelet glycoprotein Ib-IX complex. Blood Coagul Fibrinolysis 1994;5:97–119.
7. Coller BS. Hereditary qualitative platelet disorders. In Beutler E, Lichtman MA, Coller BS, Kipps TJ (eds). Williams Hematology. 5th ed. New York: McGraw-Hill, 1995, 1364–1385.
8. Phillips DR, Charo IF, Parise LV, Fitzgerald LA. The platelet membrane glycoprotein IIb-IIIa complex. Blood 1988;71:831–843.
9. Ginsburg D, Bowie EJW. Molecular genetics of von Willebrand disease. Blood 1992;79:2507–2519.
10. Rodeghiero F, Castaman G, Dini E. Epidemiological investigation of the prevalence of von Willebrand's disease. Blood 1987;69:454–459.
11. Sadler JE. A revised classification of von Willebrand disease. Thromb Haemost 1994;71:520–525.
12. Jorieux S, Tuley EA, Gaucher C, et al. The mutation Arg (53)→Trp causes von Willebrand disease Normandy by abolishing binding to factor VIII: Studies with recombinant von Willebrand factor. Blood 1992;79:563–567.
13. Triplett DA. Laboratory diagnosis of von Willebrand's disease. Mayo Clin Proc 1991;66:832–840.
14. White GC, Montgomery RR. Clinical aspects of and therapy for von Willebrand disease. In Hoffman R, Benz EJ, Shattil SJ, et al (eds). Hematology: Basic Principles and Practice. 2nd ed. New York: Churchill Livingstone, 1995, 1725–1736.
15. Broze GJ Jr. The role of tissue factor pathway inhibitor in a revised coagulation cascade. Semin Hematol 1992;29:159–169.
16. Hoffman M, Roberts HR. Hemophilia and related conditions: Inherited deficiencies of prothrombin (factor II), factor V, and factors VII to XII. In Beutler E, Lichtman MA, Coller BS, Kipps TJ (eds). Williams Hematology. 5th ed. New York: McGraw-Hill, 1995, 1413–1438.
17. Ragni MV, Lewis JH, Spero JA. Factor VII deficiency. Am J Hematol 1981;10:79–88.
18. Sumer T, Ahmad M, Sumer NK. Severe congenital factor X deficiency with intracranial hemorrhage. Eur J Pediatr 1986;145:119–120.
19. Bachmann F. Mode of inheritance of factor X deficiency. Thromb Diath Haemorrh 1965;17 (suppl):191–199.
20. Ortel TL, Keller FG, Kane WH. Factor V. In High KA, Roberts HR (eds). Molecular Biology of Thrombosis and Hemostasis. New York: Marcel Dekker, 1995, 119–146.
21. Wadia RS, Sangle SA, Kripalaney S, et al. Familial intracranial haemorrhage due to factor V deficiency. J Neurol Neurosurg Psychiatry 1992;55:227–228.
22. Gralnick HR, Connaghan DG. Hereditary abnormalities of fibrinogen. In Beutler E, Lichtman MA, Coller BS, Kipps TJ (eds). Hematology: Basic Principles and Practice. 5th ed. New York: McGraw-Hill, 1995, 1439–1454.
23. Montgomery R, Natelson SE. Afibrinogenemia with intracerebral hemorrhage: Report of a successfully treated case. Am J Dis Child 1977;131:555–556.
24. Haverkate F, Samama M. Familial dysfib-

rinogenemia and thrombophilia. Thromb Haemost 1995;73:151–161.

25. Sherman LA, Gaston LW, Kaplan ME. Fibrinogen St. Louis: A new inherited fibrinogen variant, coincidentally associated with hemophilia A. J Clin Invest 1972;51:590–597.

26. Tuddenham EGD. Factor VIII. In High KA, Roberts HR (eds). Molecular Basis of Thrombosis and Hemostasis. New York: Marcel Dekker, 1995, 167–196.

27. Brinkhous KM. A short history of hemophilia, with some comments on the word "hemophilia". In Brinkhous KM, Hemker HC (eds). Handbook of Hemophilia. New York: American Elsevier Publishing Co, 1975, 3–20.

28. Larsson SA, Wiechel B. Deaths in swedish hemophiliacs, 1957–1980. Acta Med Scand 1983;214:199–206.

29. Eyster ME, Gill FM, Blatt PM. Central nervous system bleeding in hemophiliacs. Blood 1978;51:1179–1188.

30. Walsh PN. Factor XI: A renaissance. Semin Hematol 1992;29:189–201.

31. Fujikawa K, Chung DW. Factor XI. In High KA, Roberts HR (eds). Molecular Basis of Thrombosis and Hemostasis. New York: Marcel Dekker, 1995, 257–268.

32. Henry EI, Rosenthal RL, Hoffman I. Spontaneous hemorrhages caused by plasma thromboplastin antecedent deficiency: Report of a case. J Am Med Assoc 1956;162:727–729.

33. Kitchens CS, Newcomb TF. Factor XIII. Medicine 1979;58:413–429.

34. Lai T-S, Greenberg CS. Factor XIII. In High KA, Roberts HR (eds). Molecular Basis of Thrombosis and Hemostasis. New York: Marcel Dekker, 1995, 287–308.

35. Goodnough LT, Saito H, Ratnoff OD. Thrombosis or myocardial infarction in congenital clotting factor abnormalities and chronic thrombocytopenias: A report of 21 patients and a review of 50 previously reported cases. Medicine (Baltimore) 1983; 62:248–255.

36. Harris MG, Exner T, Rickard KA. Multiple cerebral thrombosis in Fletcher factor (prekallikrein) deficiency: A case report. Am J Hematol 1985;19:387–393.

37. Soff GA, Levin J. Familial multiple coagulation factor deficiencies. 1. Review of the literature: Differentiation of single hereditary disorders associated with multiple factor deficiencies from coincidental concurrence of single factor deficiency states. Semin Thromb Hemost 1981;7:112–148.

38. Nichols WC, Seligsohn U, Zivelin A, et al. Linkage of combined factors V and VIII deficiency to chromosome 18q by homozygosity mapping. J Clin Invest 1996;99:596–601.

39. Lane DA, Olds RR, Thein S-L. Antithrombin and its deficiency states. Blood Coagul Fibrinolysis 1992;3:315–341.

40. De Stefano V, Finazzi G, Mannucci PM. Inherited thrombophilia: Pathogenesis, clinical syndromes, and management. Blood 1996; 87:3531–3544.

41. Demers C, Ginsberg JS, Hirsh J, et al. Thrombosis in antithrombin-III-deficient persons: Report of a large kindred and literature review. Ann Intern Med 1992;116:754–761.

42. Hirsh J, Piovella F, Pini M. Congenital antithrombin III deficiency: Incidence and clinical features. Am J Med 1989;87(suppl 3B): 34S–38S.

43. Johnson EJ, Prentice CRM, Parapia LA. Premature arterial disease associated with familial antithrombin III deficiency. Thromb Haemostas 1990;63:13–15.

44. Ueyama H, Hashimoto Y, Uchino M. Progressing ischemic stroke in a homozygote with variant antithrombin III. Stroke 1989; 20:815–818.

45. Vomberg PP, Breederveld C, Fleury P. Cerebral thromboembolism due to antithrombin III deficiency in two children. Neuropediatrics 1987;18:42–44.

46. Church FC, Shirk RA, Phillips JE. Heparin cofactor II. In High KA, Roberts HR (eds). Molecular Basis of Thrombosis and Hemostasis. New York: Marcel Dekker, 1995, 379–392.

47. Tran TH, Marbet GA, Duckert F. Association of hereditary heparin cofactor II deficiency with thrombosis. Lancet 1985;413–414.

48. Sie P, Dupouy D, Pichon J, Boneu B. Constitutional heparin co-factor II deficiency associated with recurrent thrombosis. Lancet 1985;2:414–416.

49. Clouse LH, Comp PC. The regulation of hemostasis: The protein C system. N Engl J Med 1986;314:1298–1304.

50. Ohlin A-K, Norlund L, Marlar RA. Thrombomodulin gene variations and thromboembolic disease. Thromb Haemost 1997;78:396–400.

51. Norlund L, Zoller B, Ohlin A-K. A novel thrombomodulin gene mutation in a patient suffering from sagittal sinus thrombosis. Thromb Haemost 1997;78:1164–1166.

52. Reitsma PH, Bernardi F, Doig RG, et al. Protein C deficiency: A database of mutations, 1995 update. Thromb Haemost 1995;73: 876–889.

53. Camerlingo M, Finazzi G, Casto L, et al. Inherited protein C deficiency and nonhemorrhagic arterial stroke in young adults. Neurology 1991;41:1371–1373.

54. Wintzen AR, Broekmans AW, Bertina RM, et al. Cerebral haemorrhagic infarction in young patients with hereditary protein C de-

ficiency: Evidence for "spontaneous" cerebral venous thrombosis. Br Med J [Clin Res] 1985;290:350–352.

55. Miletich J, Sherman L, Broze G Jr. Absence of thrombosis in subjects with heterozygous protein C deficiency. N Engl J Med 1987; 317:991–996.

56. Seligsohn U, Berger A, Abend M, ct al. Homozygous protein C deficiency manifested by massive venous thrombosis in the newborn. N Engl J Med 1984;310:559–562.

57. Tait RC, Walker ID, Reitsma PH, et al. Prevalence of protein C deficiency in the healthy population. Thromb Haemost 1995;73:87–93.

58. Simmonds RE, Ireland H, Lane DA, et al. Clarification of the risk for venous thrombosis associated with hereditary protein S deficiency by investigation of a large kindred with a characterized gene defect. Ann Intern Med 1998;128:8–14.

59. Engesser L, Broekmans AW, Briet E, et al. Hereditary protein S deficiency: Clinical manifestations. Ann Intern Med 1987;106: 677–682.

60. Green D, Otoya J, Oriba H, Rovner R. Protein S deficiency in middle-aged women with stroke. Neurology 1992;42:1029–1033.

61. Israels SJ, Seshia SS. Childhood stroke associated with protein C or S deficiency. J Pediatr 1987;111:562–564.

62. Sacco RL, Owen J, Mohr JP, et al. Free protein S deficiency: A possible association with cerebrovascular occlusion. Stroke 1989;20: 1657–1661.

63. Girolami A, Simioni P, Lazzaro AR, et al. Severe arterial cerebral thrombosis in a patient with protein S deficiency (moderately reduced total and markedly reduced free protein S): A family study. Thromb Haemostas 1989;61:144–147.

64. Shirk RA, Phillips JE, Church FC. Protein C inhibitor. In High KA, Roberts HR (eds). Molecular Basis of Thrombosis and Hemostasis. New York: Marcel Dekker, 1995, 447–458.

65. Weiss EJ, Bray PF, Tayback M, et al. A polymorphism of a platelet glycoprotein receptor as an inherited risk factor for coronary thrombosis. N Engl J Med 1996;334:1090–1094.

66. Zotz RB, Winkelmann BR, Nauck M, et al. Polymorphism of platelet membrane glycoprotein IIIa: Human platelet antigen 1b (HPA-1b/PIA2) is an inherited risk factor for premature myocardial infarction in coronary artery disease. Thromb Haemost 1998;79: 731–735.

67. Herrmann S-M, Poirier O, Marques-Vidal P, et al. The Leu33/Pro polymorphism (PIA1/PIA2) of the glycoprotein IIIa (GPIIIa) receptor is not related to myocardial infarc-

tion in the ECTIM study. Thromb Haemost 1997;77:1179–1181.

68. Catto AJ, Carter AM, Barrett JH, et al. von Willebrand factor and factor VIII:C in acute cerebrovascular disease: Relationship to stroke subtype and mortality. Thromb Haemost 1997;77:1104–1108.

69. Qizilbash N, Duffy S, Prentice CR, et al. von Willebrand factor and risk of ischemic stroke. Neurology 1997;49:1552–1556.

70. Meade TW, Mellows S, Brozovic M, et al. Hemostatic function and ischaemic heart disease: Principal results of the Northwick Park Heart Study. Lancet 1986;2:533–537.

71. Iacoviello L, Di Castelnuovo A, De Knijff P, et al. Polymorphisms in the coagulation factor VII gene and the risk of myocardial infarction. N Engl J Med 1998;338:79–85.

72. Heywood DM, Carter AM, Catto AJ, et al. Polymorphisms of the factor VII gene and circulating FVII:C levels in relation to acute cerebrovascular disease and poststroke mortality. Stroke 1997;28:816–821.

73. Hillarp A, Zoller B, Dahlbäck B. Activated protein C resistance as a basis for venous thrombosis. Am J Med 1996;101:534–540.

74. Dahlbäck B, Carlsson M, Svensson PJ. Familial thrombophilia due to a previously unrecognized mechanism characterized by poor anticoagulant response to activated protein C: Prediction of a cofactor to activated protein C. Proc Natl Acad Sci USA 1993;90: 1004–1008.

75. Bertina RM, Koeleman BPC, Koster T, et al. Mutation in blood coagulation factor V associated with resistance to activated protein C. Nature 1994;369:64–67.

76. Kalafatis M, Bertina RM, Rand MD, Mann KG. Characterization of the molecular defect in factor VR506Q. J Biol Chem 1995;270: 4053–4057.

77. Koster T, Rosendaal FR, De Ronde H, et al. Venous thrombosis due to poor anticoagulant response to activated protein C: Leiden Thrombophilia Study. Lancet 1993;342: 1503–1506.

78. Griffin JH, Evatt B, Wideman C, Fernández JA. Anticoagulant protein C pathway defective in majority of thrombophilic patients. Blood 1993;82:1989–1993.

79. Price DT, Ridker PM. Factor V Leiden mutation and the risks for thromboembolic disease: A clinical perspective. Ann Intern Med 1997;127:895–903.

80. Vandenbroucke JP, Koster T, Briet E, et al. Increased risk of venous thrombosis in oral-contraceptive users who are carriers of factor V Leiden mutation. Lancet 1994;344:1453–1457.

81. Ridker PM, Hennekens CH, Lindpaintner K,

et al. Mutation in the gene coding for coagulation factor V and the risk of myocardial infarction, stroke, and venous thrombosis in apparently healthy men. N Engl J Med 1995;332:912–917.

82. van der Bom JG, Bots ML, Haverkate F, et al. Reduced response to activated protein C is associated with increased risk for cerebrovascular disease. Ann Intern Med 1996; 125:265–269.

83. Simioni P, De Ronde H, Prandoni P, et al. Ischemic stroke in young patients with activated protein C resistance: A report of three cases belonging to three different kindreds. Stroke 1995;26:885–890.

84. Zuber M, Toulon P, Marnet L, Mas J-L. Factor V Leiden mutation in cerebral venous thrombosis. Stroke 1996;27:1721–1723.

85. Dulli DA, Luzzio CC, Williams EC, Schutta HS. Cerebral venous thrombosis and activated protein C resistance. Stroke 1996;27: 1731–1733.

86. Deschiens M-A, Conard J, Horellou MH, et al. Coagulation studies, factor V Leiden, and anticardiolipin antibodies in 40 cases of cerebral venous thrombosis. Stroke 1996;27: 1724–1730.

87. Provenzale JM, Barboriak DP, Ortel TL. Dural sinus thrombosis associated with activated protein C resistance: MR imaging findings. Am J Roentgenol 1998;170:499–502.

88. Le DT, Griffin JH, Greengard JS, et al. Use of a generally applicable tissue factor-dependent factor V assay to detect activated protein C resistant factor Va in patients receiving warfarin and in patients with a lupus anticoagulant. Blood 1995;85:1704–1711.

89. Poort SR, Rosendaal FR, Reitsma PH, Bertina RM. A common genetic variation in the 3'-untranslated region of the prothrombin gene is associated with elevated plasma prothrombin levels and an increase in venous thrombosis. Blood 1996;88:3698–3703.

90. Rosendaal FR, Doggen CJM, Zivelin A, et al. Geographic distribution of the 20210G to A prothrombin variant. Thromb Haemost 1998;79:706–708.

91. Hillarp A, Zoller B, Svensson PJ, Dahlbäck B. The 20210A allele of the prothrombin gene is a common risk factor among Swedish outpatients with verified deep venous thrombosis. Thromb Haemost 1997;78:990–992.

92. De Stefano V, Chiusolo P, Paciaroni K, et al. Prothrombin G20210A mutant genotype is a risk factor for cerebrovascular disease in young adults. Blood 1998;91:3562–3565.

93. Thompson SG, Kienast J, Pyke SD, et al. Hemostatic factors and the risk of myocardial infarction or sudden death in patients with angina pectoris: European Concerted Action on Thrombosis and Disabilities Angina Pectoris Study Group. N Engl J Med 1995; 332: 635–641.

94. Fowkes FGR, Connor JM, Smith FB, et al. Fibrinogen genotype and risk of peripheral atherosclerosis. Lancet 1992;339:693–696.

95. Wilhelmsen L, Svardsudd K, Korsan-Bengtsen K, et al. Fibrinogen as a risk factor for stroke and myocardial infarction. N Engl J Med 1984;311:501–505.

96. Kessler C, Spitzer C, Stauske D, et al. The apolipoprotein E and β-fibrinogen G/A-455 gene polymorphisms are associated with ischemic stroke involving large-vessel disease. Arterioscler Thromb Vasc Biol 1997;17: 2880–2884.

97. Carter AM, Catto AJ, Bamford JM, Grant PJ. Gender-specific associations of the fibrinogen Bβ 448 polymorphism, fibrinogen levels, and acute cerebrovascular disease. Arterioscler Thromb Vasc Biol 1997;17:589–594.

98. Koster T, Blann AD, Briet E, et al. Role of clotting factor VIII in effect of von Willebrand factor on occurrence of deep-vein thrombosis. Lancet 1995;345:152–155.

99. Kamphuisen PW, Houwing-Duistermaat JJ, van Houwelingen HC, et al. Familial clustering of factor VIII and von Willebrand factor levels. Thromb Haemost 1998;79: 323–327.

100. Aoki N. Hereditary abnormalities of the fibrinolytic system. In Francis JL (ed). Fibrinogen, Fibrin Stabilization, and Fibrinolysis: Clinical, Biochemical and Laboratory Aspects. Chicester: Ellis Horwood Ltd, 1988, 264–276.

101. Johansson L, Hedner U, Nilsson IM. A family with thromboembolic disease associated with deficient fibrinolytic activity in vessel wall. Acta Med Scand 1978;203:477–480.

102. Stead NW, Bauer KA, Kinney TR, et al. Venous thrombosis in a family with defective release of vascular plasminogen activator and elevated plasma factor VIII/von Willebrand's factor. Am J Med 1983;74:33–39.

103. Pizzo SV, Fuchs HE, Doman KA, et al. Release of tissue plasminogen activator and its fast-acting inhibitor in defective fibrinolysis. Arch Intern Med 1986;146:188–191.

104. Booth NA, Bennett B, Wijngaards G, et al. A new life-long hemorrhagic disorder due to excess plasminogen activator. Blood 1983; 61:267–275.

105. Aoki N, Saito H, Kamiya T, et al. Congenital deficiency of α2-plasmin inhibitor associated with severe hemorrhagic tendency. Blood 1982;59:1246–1251.

106. Miles LA, Plow EF, Donnelly KJ, et al. A

bleeding disorder due to deficiency of α2-antiplasmin. Blood 1982;59:1246–1251.

107. Lee MH, Vosburgh E, Anderson K, McDonagh J. Deficiency of plasma plasminogen activator inhibitor 1 results in hyperfibrinolytic bleeding. Blood 1993;81:2357–2362.

108. Fay WP, Shapiro AD, Shih JL, et al. Brief report: Complete deficiency of plasminogen activator inhibitor type 1 due to a frame shift mutation. N Engl J Med 1992;327:1729–1733.

109. Fay WP, Parker AC, Condrey LR, Shapiro AD. Human plasminogen activator inhibitor-1 (PAI-1) deficiency: Characterization of a large kindred with a null mutation in the PAI-1 gene. Blood 1997;90:204–208.

110. Catto AJ, Carter AM, Stickland M, et al. Plasminogen activator inhibitor-1 (PAI-1) 4G/5G promoter polymorphism and levels in subjects with cerebrovascular disease. Thromb Haemost 1997;77:730–734.

111. Ridker PM, Hennekens CH, Lindpaintner K, et al. Arterial and venous thrombosis is not associated with the 4G/5G polymorphism in the promoter of the plasminogen activator inhibitor gene in a large cohort of US men. Circulation 1997;95:59–62.

112. Margaglione M, Cappucci G, Colaizzo D, et al. The PAI-1 gene locus 4G/5G polymorphism is associated with a family history of coronary artery disease. Arterioscler Thromb Vasc Biol 1998;18:152–156.

113. Zoller B, de Frutos PG, Dahlback B. A common 4G allele in the promoter of the plasminogen activator inhibitor-1 (PAI-1) gene as a risk factor for pulmonary embolism and arterial thrombosis in hereditary protein S deficiency. Thromb Haemost 1998;79:802–807.

114. Welch GN, Loscalzo J. Homocysteine and atherothrombosis. N Engl J Med 1998;338:1042–1050.

115. den Heijer M, Koster T, Blom HJ, et al. Hyperhomocysteinemia as a risk factor for deep-vein thrombosis. N Engl J Med 1996;334:759–762.

116. den Heijer M, Blom HJ, Gerrits WMJ, et al. Is hyperhomocysteinemia a risk factor for recurrent venous thrombosis? Lancet 1995;345:882–885.

117. Engberson AMT, Franken DG, Boers GHJ, et al. Thermolabile 5,10-methylenetetrahydrofolate reductase as a cause of mild hyperhomocysteinemia. Am J Hum Genet 1995;56:142–150.

118. Frosst P, Blom HJ, Milos R, et al. A candidate genetic risk factor for vascular disease: A common mutation in methylenetetrahydrofolate reductase. Nature Genet 1995;10:111–113.

119. Petri M. Diagnosis of antiphospholipid antibodies. Rheum Dis Clin North Am 1994;20:443–469.

120. Roubey RAS. Autoantibodies to phospholipid-binding plasma proteins: A new view of lupus anticoagulants and other "antiphospholipid" autoantibodies. Blood 1994;84:2854–2867.

121. Shapiro SS. The lupus anticoagulant/antiphospholipid syndrome. Annu Rev Med 1996;47:533–553.

122. Goldberg SN, Conti-Kelly AM, Greco TP. A family study of anticardiolipin antibodies and associated clinical conditions. Am J Med 1995;98:473–479.

123. Mackworth-Young C, Chan J, Harris N, et al. High incidence of anticardiolipin antibodies in relatives of patients with systemic lupus erythematosus. J Rheumatol 1987;14:723–726.

124. Bansal AS, Hogan PG, Gibbs H, Frazer IH. Familial primary antiphospholipid antibody syndrome. Arthritis Rheum 1996;39:705–706.

125. Ford PM, Brunet D, Lillicrap DP, Ford SE. Premature stroke in a family with lupus anticoagulant and antiphospholipid antibodies. Stroke 1990;21:66–71.

126. Lousa M, Sastre JL, Cancelas JA, et al. Study of antiphospholipid antibodies in a patient with Sneddon's syndrome and her family. Stroke 1994;25:1071–1074.

127. Pettee AD, Wasserman BA, Adams NL, et al. Familial Sneddon's syndrome: Clinical, hematologic, and radiographic findings in two brothers. Neurology 1994;44:399–405.

128. Sammaritano LR, Gharavi AE, Lockshin MD. Antiphospholipid antibody syndrome: Immunologic and clinical aspects. Semin Arthritis Rheum 1990;20:81–96.

129. Asherson RA, Khamashta MA, Gil A, et al. Cerebrovascular disease and antiphospholipid antibodies in systemic lupus erythematosus, lupus-like disease, and the primary antiphospholipid syndrome. Am J Med 1989;86:391–399.

130. Trimble M, Bell DA, Brien W, et al. The antiphospholipid syndrome: Prevalence among patients with stroke and transient ischemic attacks. Am J Med 1990;88:593–597.

131. Brandt JT, Triplett DA, Alving B, Scharrer I. Criteria for the diagnosis of lupus anticoagulants: An update. Thromb Haemostas 1995;74:1185–1190.

132. Khamashta MA, Cuadrado MJ, Mujic F, et al. The management of thrombosis in the antiphospholipid-antibody syndrome. N Engl J Med 1995;332:993–997.

133. Moll S, Ortel TL. Monitoring warfarin ther-

apy in patients with lupus anticoagulants. Ann Intern Med 1997;127:177–185.

134. Lane DA, Mannucci PM, Bauer KA, et al. Inherited thrombophilia: Part 1. Thromb Haemost 1996;76:651–662.

135. Koeleman BPC, Reitsma PH, Allaart CF, Bertina RM. Activated protein C resistance as an additional risk factor for thrombosis in protein C-deficient families. Blood 1994;84: 1031–1035.

136. Zoller B, Berntsdotter A, de Frutos PG, Dahlback B. Resistance to activated protein C as an additional genetic risk factor in hereditary deficiency of protein S. Blood 1995;85:3518–3523.

137. van Boven HH, Reitsma PH, Rosendaal FR, et al. Factor V Leiden (FV R506Q) in families with inherited antithrombin deficiency. Thromb Haemostas 1996;75: 417–421.

138. Nichols WC, Amano K, Cacheris PM, et al. Moderation of hemophilia A phenotype by the factor V R506Q mutation. Blood 1996;88:1183–1187.

Part III

Genetics of Specific Stroke Etiologies and Syndromes

Chapter 8

Genetic Epidemiology and Family Studies of Stroke

Lawrence M. Brass, MD, Mark J. Alberts, MD

Introduction

The epidemiology and common risk factors for cerebrovascular disease have been well documented in cross-sectional and longitudinal population surveys. The results of these studies have helped to advance strategies for the prevention of vascular disease. In spite of improvement in the stratification of risk of stroke for individuals,[1,2] significant variability still exists and many risk factors are as yet unknown or poorly understood.

Most studies of stroke have generally involved unrelated individuals with little information on family history. Although there are reported associations in small numbers of cases with uncommon Mendelian diseases,[3] few studies have investigated the role of genetics in the occurrence of stroke. Despite a significant role in cardiovascular risk, relatively few studies have investigated familial or genetic contributions to stroke risk. Most of the evidence has been circumstantial, the results have often been conflicting,[4] and the exact importance of genetic factors in the pathogenesis of stroke in the population is unknown.

Most cases of ischemic vascular disease are not likely to be associated with a single genetic abnormality. The risk for stroke appears to be due to complex interactions between genetically determined risk factors (i.e., hypertension, diabetes, lipid levels) and various environmental factors (i.e., diet, smoking, exercise).

The field of genetic epidemiology is in a stage of rapid growth. Our goal is to update our previous work and provide an overview of selected topics related to familial stroke.[5] Since the subject matter is very broad, not all aspects could be covered. We have tried, however, to review each of the major areas with appropriate references for those interested in exploring a topic in greater detail.

Genetic Epidemiology

Genetic epidemiology, a new field that has been fueled by advances in molecular biology, is defined as[6,7]: a "science that deals with the etiology, distribution, and control of disease in groups of relatives and with inherited causes of disease in populations".[8] It is the study of interactions of single genes and environmental factors, through the use of conventional epidemiological designs such as case-control methods (described below).

From Alberts MJ (ed). *Genetics of Cerebrovascular Disease.* Armonk, NY: Futura Publishing Company, Inc., © 1999.

Common diseases like stroke, have been a special challenge for those investigating genetic contributions to disease. Genetic analysis of complex traits is complicated by many factors: incomplete penetrance, genetic heterogeneity, the number of loci contributing to disease (and their interactions), mode of inheritance, presence of phenocopies (nongenetic causes), and others.[9] Classic segregation analysis to separate out these factors is unlikely to succeed because there may not be sufficient family resources to overcome the large number of unknown parameters. New strategies to study genetically complex traits have been proposed, but they are beyond the scope of this review. For example, Risch has developed a ratio of the risk of relatives of a proband versus population prevalence of a quantitative trait as a direct indicator of the mode of inheritance.[10]

Types of Studies

The types of epidemiological studies most commonly used for the study of familial and genetic contributions to stroke are similar to those for other vascular risk factors. Most are based on family history and all are observational. Only a few types of observational studies have been applied to stroke. The most common are ecological, cohort, and case-control studies; cluster, migration, and twin studies are used less frequently.[11]

Ecological Studies

In ecological studies, the rate of occurence of a disease is compared among specific geographic, racial, or ethnic groups. Ecological studies have suggested a genetic contribution to stroke largely through racial studies which have indicated higher rates of ischemic and hemorrhagic stroke among African-Americans,[12,13] higher rates of intracranial hemorrhage among Japanese persons,[14] and racial effects on the location of anterior circulation vascular disease.[15] Although useful in providing initial clues for familial or genetic associations, these stud-

ies are rarely definitive. Local habits such as diet may be difficult to control for in the analysis. Migration studies can help separate genetics from local (environmental) factors. For example, when Japanese persons migrate to the United States, over time their risk of stroke changes to reflect their new geographic location.[16]

In ecological studies, it is important that the data be community-based, with proper and standardized definitions. Investigators should ensure that the environmental contributions to risk are as homogeneous as possible between the two groups to ascertain that the observed differences are due to racial and cultural factors.[17]

Cohort Studies

In cohort studies, a group of individuals (cohort) are selected who have not yet had the disease of interest (stroke). The cohort is then followed over time for the development of the disease. The two major advantages of this method are: (1) potential confounding factors can be well characterized and accounted for, and (2) screening for endpoints can be done in a systemic and well-controlled fashion. This can be especially important for stroke where the estimates of the reliability of self-reporting vary widely.[18,19] Examples of cohort studies that have reported an increased risk of stroke associated with a positive family history include the Framingham Study[20] and a study from Sweden.[21] The major disadvantages of cohort studies are time (many years of follow-up are often required) and the associated cost (because of the large number of people who need to be followed to ensure there are enough new strokes occur to adequately address a study's objectives). The less common the disease, the larger the cohort that is needed and the longer the duration of follow-up that is required.

More recently, the Family Heart Study,[22] an NIH-supported multicenter study of familial, genetic, and nongenetic determinants of cardiovascular disease, showed a robust familial association between parental history and prevalent stroke. Several cohorts contributed patients to this study including: the Forsyth and Minneapolis cohorts of the Atherosclerotic Risk in the Community (ARIC) study, the Framingham Offspring Study, and the Utah Family Tree Study.[23] In this study, Liao and colleagues found an adjusted (age, gender, race) odds ratio for prevalent stroke of 2.00 (95% CI: 1.13–3.54) for those with a paternal history of stroke and 1.41 (85% CI: 0.80–2.50) for those with a maternal history of stroke. This association was not changed after further adjustment for common vascular risk factors (cholesterol, cigarette smoking, coronary artery disease, hypertension, or diabetes). The familial effect was similar for African-Americans and those of European descent.

Another recent cohort study, by Jousilahti and colleagues, demonstrated that a positive parental history of stroke predicted the risk of stroke independently from other vascular risk factors.[24] They also found a positive parental history of coronary artery disease was associated with the risk of ischemic stroke among women, but not men.

Case-Control Studies

In this method, a group of individuals with the disease trait or suspected risk factor (cases) is compared to those without the trait or risk factor (controls). For studies using stroke patients as cases, there are important concerns about potential biases. Certain patients who have had a stroke are more likely to have vascular risk factors; however, caution must be used in estimating the risk associated with an individual factor. Particular attention must be paid to ensure that the selected cases are representative, the controls are comparable to the cases, and potential biases are minimized. For example, in stroke studies relying on hospitalized patients, cases may reflect biases of stroke severity, socioeconomic status, race, or other factors known to influence the use of hospital care with acute stroke. Cross-sectional or retrospective studies of vascular risk factors are often a good starting point, but are very different from the determination of the risk of an individual carrying a particular risk factor (including suspected genotypes).

Although cohort studies are often feasible for stroke studies in older populations, case-control studies may be the only practical method of study for stroke among younger patients or for less common types of strokes.

Case-control studies may also be the only practical alternative when collecting samples for genetic testing because the costs associated with obtaining samples across a large study cohort may be prohibitive. In this situation, a nested case-control study may be used; cases are matched for controls within the larger cohort study. The number of samples, and associated costs, can thus be significantly reduced.

Although a genotype is present since conception, issues of directions of causality are not a significant problem for genetic cross-sectional studies; however, they are not free from bias. Estimates of risk may be confounded by contributions of a particular genotype to early (or later) stroke-related deaths or recovery (which would allow them to participate in a study). To overcome this type of potential confounding, investigators should ideally look at the genotype of healthy individuals in cohorts followed for many years. It is interesting to note that the genetic contributions to familial stroke studies, molecular biology, and epidemiology are highly dependent on each other.[6]

Twin Studies

The use of the twin method was first proposed by Galton, who is considered the father of behavioral genetics. With increased understanding of the mechanisms of inheritance, twin studies have become popular, especially in the study of behavioral medicine and cardiovascular diseases.[25] The twin study paradigm holds considerable promise for the detection of genetic influences on stroke occurrence, subtype, severity, and recovery. Mono- and dizygotic twins are a unique resource for studies of the origin and natural history of various diseases because the genetic similarities are well defined and easily understood.[26]

Twin studies, comparing the concordance rate between monozygotic and dizygotic twins, have long been used to estimate the heritability of a trait or disorder.[26,27] Assumptions need to be made about the environmental similarity between types of twins (dizygotic twins tend to be associated with older mothers, prior siblings, and the use of fertility drugs). These differences do not appear to significantly impact on studies of cardiovascular risk.[28] Twin studies have been useful in understanding the heritability of complex traits such as cardiovascular risk factors,[29,36] diabetes,[30] hypertension,[31] lipid (cholesterol and lipoprotein) disorders,[31] obesity,[32] personality traits,[33] cognitive functions,[34] and cardiac mortality.[35] Results from these studies have paved the way for many of the current investigations into the molecular biology of vascular risk factors. Twin registries continue to provide important information on the role of genetics in vascular disease.

Twins studies of coronary heart disease have reported concordance rates of 0.25 to 0.28 among dizygotic twins and 0.48 to 0.90 among monozygotic twins.[25,37,38] This excess among the monozygotic twins indicates a strong genetic component for cardiovascular disease. Death from coronary heart disease is also influenced by genetic factors; however, the genetic effect decreases in older age groups.[35]

Shared environmental risk for cardiovascular disease has been addressed via twin studies. Since twins strongly tend to have a similar environment (including culture, socioeconomic class), differences in the concordance rates of vascular disease between monozygotic twins and dizygotic twins are thus likely due to their different genotypes.

The familial occurrence of stroke could indicate either genetic or common environmental factors. For stroke mechanisms where there is little evidence directly relating to environmental factors (e.g., cerebral aneurysms), a family history is more suggestive for a genetic etiology. For stroke types where environmental risk factors have been well described, determining a genetic contribution is more difficult (e.g., carotid disease where cigarette smoking and diet have been more strongly associated with the presence of disease). Twins studies can also help in this type of situation.

One major twin study looked at the impact of genetics on cerebrovascular disease. Using data from the National Academy of Sciences Veteran Twin Registry and a mailed health survey, Brass et al.[39] found the prevalence of stroke in the entire twin cohort was 3.1% (292/9475). The rates of stroke were similar for monozygotic twins (131/4200) and dizygotic twins (143/4585) and approximated expected population-based rates. (The age range was 58–68 years.) The proband concordance rates was 17.7% for monozygotic pairs and 3.6% for dizygotic pairs ($X^2 = 4.94$, 1 d.f.; relative risk 4.3, $p<0.05$). This approximately fivefold increase in the prevalence of stroke among the monozygotic compared to the dizygotic twin pairs suggests that genetic factors are involved in the etiology of stroke; however, the number of strokes was too large to make a reliable estimate of the overall heritability.

New data is expected from this same group. Recently, more than 14,000 of the surviving twins responded to a cooperative telephone survey. Questions were included on stroke, myocardial infarction, angina, carotid disease, atrial fibrillation, and peripheral vascular disease. Longitudinal information dates back to 1971 and the prevalence of stroke is a little more than 8%. The results of this survey should help provide an estimate of the genetic contribution to stroke and the major vascular risk factors.[27]

Twin studies also provide an opportunity to control for the genetic contributions to a disease and examine environmental factors in isolation. This can be done by comparing the rates of disease among monozygotic twins discordant for a candidate environmental risk factor. This done by comparing differences in environmental risk factors among monozygotic twins. Because twins tend to share environmental factors also, potential biases for factors not specifically included in the planned analysis are also reduced. Using this technique, Haapanen and colleagues compared the degree of carotid atherosclerosis between smoking and nonsmoking members of monozygotic twin pairs.[40] They found that among the brothers who smoked (compared to their monozygotic twin), the prevalence of carotid stenosis was higher and the area of stenosis was also increased. This association remained after adjusting for common risk factors (age, cholesterol, blood pressure, and body mass index) in a multiple, logistic, regression analysis.

Review of Genetic Epidemiology Studies

Cardiovascular and cerebrovascular diseases share many of the same risk factors but with differing importance. There is also a strong correlation between transient ischemic attacks (TIA), stroke, and coronary artery disease. Genetic influences are estimated to account for over half of the risk for cardiac disease.[25,36] It seems likely that there will also be a similar contribution for cerebrovascular diseases.

The genetic role in cardiovascular disease is apparent from several lines of evidence. Many familial aggregation studies have been reported for cardiovascular disease.[4] This was, in part, due to clustering a risk factors. When focusing on younger patients (<55 years old), Nora et al. found a tenfold increased risk of coronary artery disease among first-degree relatives.[41] This type of clustering could still be accounted for by shared environmental factors.

As mentioned above, for younger patients, genetic factors appear to play a larger role in cardiovascular disease. It may be that these factors serve to accelerate atherosclerosis. Much less data exist for ischemic cerebrovascular disease. In a hospital-based survey of patients with TIA, however, preliminary data suggest an opposite age effect

(i.e., a less strong history among younger patients).[42] Of note, this was present with the expected association of family history of myocardial infarction among younger patients. These results could represent a bias in survival or recall, but the opposite finding of MI and stroke, within a single study, emphasized the need to consider these two diseases separately.

By comparison, significantly less work has been done on the genetics of cerebrovascular disease, and the role of family history in stroke risk remains unclear. Only a limited number of epidemiological studies have addressed familial contributions to stroke (Table 1). In some respects, this is due to the complex nature of stroke. Stroke has numerous subtypes (i.e., ischemic, hemorrhagic; large vessel, small vessel; atherothrombotic and embolic) which can produce similar clinical phenotypes from very diverse mechanisms and underlying genotypes. Most studies have had serious flaws, such as a lack of differentiation between stroke types and mechanisms, family histories that were obtained retrospectively from unvalidated questionnaires or chart reviews, and no correction for age and other risk factors as part of the analyses.[4] Just as one would not lump together all types of cancer for genetic studies, stroke subtypes must be viewed as distinct diseases and processes. This makes the study of the genetic contribution to stroke especially challenging. In the following section, we examine broad categories of stroke and investigations related to familial risk.

Stroke

If genetic factors do, in fact, influence the risk for stroke, epidemiological studies would be expected to demonstrate an independent effect for family history. Recently, two of the larger, better designed epidemiological studies reported similar results regarding family history and stroke. The Framingham Offspring Study[20] found that a parental history of stroke, TIA, or myocardial infarction was associated with an increased risk of atherothrombotic brain infarction (pa-

Table 1
Family Studies of Stroke

Study	Prospective	No. of Patients	Controls	Age Correction	Stroke Subtypes	Multivariate Analysis	Genetic/Familial Factors for Stroke	Other
Gifford 1966[50]	Yes	255	Yes	No	No	No	Yes	
Alter 1967[176]	Yes	239	Yes	No	No	No	Yes	
Heyden 1969[61]	Yes	120	Yes	No	No	No	Yes (risk factors)	
Marshall 1971[173]	Yes	1,307	Yes	?	Yes	No	Yes	Angiogram
Alter 1972[174]	Yes	544	Yes	No	No	No	No	
Marshall 1973[49]	No	180	No	No	Yes	No	No	
Herman 1983[161]	Yes	371	Yes	Yes	No	Yes	No	
Diaz 1986[54]	Yes	131	Yes	No	No	No	Yes (risk factors)	
Khaw 1986[51]	Yes	3,415	No	Yes	No	Yes	Yes (women)	Cohort, 9-yr follow-up CT used for DX
Welin 1987[21]	Yes	792	No	?	Yes	Yes	Yes	Cohort, 8.5-yr follow-up; CT used for Dx
Boysen et al. 1988[162]	Yes	295	Yes	Yes	Yes	Yes	No	Low rate of CT
Thompson et al. 1989[56]	Yes	603	Yes	No	No	Yes	No (yes for MI)	
Howard et al 1990[45]	Yes	55	No	?	Yes	Yes	Yes	Higher risk with parental DM and early age of stroke

Study								
Brass 1991[42]	No	117	No	Yes	Yes	No	No (yes for ≥ 70)	Subjects enrolled after TIA, incomplete ascertainment
Brass 1992[39]	No	292	Yes	No	No	No	Yes	Twins study
Kiely 1993[20]	Yes	34	Yes	Yes	Yes	Yes	Yes	Cohort study with neurologist exam
Carrieri 1994[175]	Yes	164	Yes	Yes	Yes	Yes	Yes (young pts)	
Wang 1995[76]	Yes	149	Yes	Yes	Yes	Yes	Yes	
Vitullo 1996[58]	Yes	237	Yes	Yes	Yes	Yes	Yes	
Bromberg 1996[75]	Yes	163	No	Yes	Yes	Yes	Yes	
Bogousslvsky 1996[59]	Yes	822	No	Yes	Yes	Yes	Yes	Hypertensive, noncardioembolic strokes
Brass 1996[47]	Yes	545	Yes	No	No	No	Yes (not older pts)	Twins study
Kubota 1997[55]	Yes	502	Yes	Yes	Yes	Yes	Yes-SAH ICH (risk factors) No-ischemic stroke	Neurologist/ neurosurgeon evaluations
Liao 1997[22]	Yes	105	3039	Yes	No	Yes	Yes	Parental
Jousilahti 1997[24]	Yes	14,371	Yes	Yes	Offspring, but not in parents	Yes	Yes	Higher risk of stroke for those with paternal h/o stroke Parental h/o CAD also assocaited with stroke (among women)

ternal history of stroke relative risk = 2.4, for maternal history the relative = 1.4). This association was not statistically significant when adjusted for age and the presence of other common risk factors. This suggests that genetic contributions to stroke etiology are acting through stroke risk factors[43,44] in the individual or family members.[45]

In a second study from London, Ontario, investigators prospectively assessed the prevalence of a family history of stroke and ischemic heart disease among patients with well-characterized atherothrombotic stroke or transient ischemic attacks. Those with stroke were more likely to have a positive family history for cerebrovascular disease in a first-degree relative (odds ratio = 2.33) or ischemic heart disease in a first-degree relative (odds ratio = 2.14).[43] The authors did not note a difference between maternal or paternal family history, as had previously been reported.[21] They also looked at vascular risk factors including hypertension, ischemic heart disease, hyperlipidemia, and smoking. A positive family history correlated with the presence of risk factors. When all of these factors where included in a multivariate analysis, family history was not an independent risk factor, but merely a marker for the presence of other vascular risk factors. The identification of sufficiently large family pedigrees could also provide the clinical material needed for the application of molecular biological techniques to identify candidate genes.[4]

This apparent lack of an independent association between family history and stroke does not mean a lack of genetic contribution to stroke, as many of the most common and potent vascular risk factors have a significant genetic contribution. Other limitations to these types of studies are described later in this chapter. Even genetic variations in the metabolism of cardiovascular drugs may impact on the risk of primary or recurrent stroke.[46]

The previously mentioned results from the National Academy of Sciences Veteran Twin Registry[39] strongly suggest an important genetic contribution to stroke, although there was insufficient power to make a reli-

able estimate of heritability. In a more recent analysis based on a follow-up survey of the same cohort, now a decade older, the impact of genetic contributions was significantly less.[47] This is consistent with cardiovascular disease where genetic influences appear to have the greatest impact among younger patients.

Stroke Mortality

Parental history of coronary artery disease is a risk factor for both coronary artery disease and cardiac-related death, especially among those less than age 60 years.[48] Early studies failed to demonstrate an overall increased risk of death related to cerebral hemorrhage among siblings of patients with a proven cerebral hemorrhage.[49] There was one difference (increased risk) noted, among the brothers of women patients, that did achieve statistical significance.

Gifford found excessive death from cerebrovascular disease among parents and relatives of stroke patients. The diagnosis of stroke in this study included thrombosis, embolism, and hemorrhage.[50] More recently, a family history of stroke among first-degree relatives was associated with an increased risk of stroke-related death among women, even after adjusting for other major risk factors.[51] A maternal history of stroke was associated with an increased risk of stroke among men in a Swedish study.[21] Maternal and parental history of stroke have been associated with an increased risk of stroke among the offspring in the Framingham Study.[20] Conversely, specific genotypes may be associated with improved survival and better clinical outcome.[52,53]

Ischemic Stroke

Cerebral infarction has been noted to cluster in families, but few studies have examined independent familial risks. This is especially important because, as for other forms of stroke, there may be a genetic influence on vascular risk factors.[44] In a study by Diaz et al. on the siblings of stroke patients, there was a nonsignificant association between stroke patients and stroke in siblings. There was, however, a strong sibling association between the combined factors of hypertension, heart disease, or diabetes.[54]

Although an independent contribution of family history to the risk of both subarachnoid hemorrhage and intracerebral hemorrhage was demonstrated in a case-control study of stroke by Kubota and colleagues, the authors were not able to support this association for either ICH or infarction in multivariate analyses.[55]

Among a group of patients with transient ischemic attacks admitted to a single hospital, Brass and Shaker demonstrated a positive association between a personal and family association of stroke only among patients over age 70. In this same study, the association between personal and family history of myocardial infarction was present, as previous reported[42] among younger patients. The reason for these different age-related associations remain unclear. In addition, there was an association between family history of myocardial infarction and personal history of stroke. This report also suggests that, for investigators of the genetics of stroke (or myocardial infarction), a combined endpoint ischemic vascular endpoint may be more appropriate.[56] This idea is also supported by other reported associations between a family history of stroke and an increased risk for myocardial infarction[20,57,58] and family history of myocardial infarction and an increased risk for stroke,[56,59] suggesting that for some studies myocardial infarction and stroke should be considered within the same phenotype.[60]

Heyden compared the cause of death among parents of patients with carotid occlusive disease undergoing endarterectomy to controls with and without vascular disease. The parents of those with carotid disease had higher rates of vascular death than either of the control groups; however, the rate of hypertension was also higher among those with carotid disease.[61] It is possible that familial effects of hypertension (or other vascular risk factors) accounted for part of this association.

Within ischemic stroke, there may be a differential effect of family history among different stroke subtypes. Bogousslavksy and colleagues reported that family history was a more important risk factor for ischemic stroke (compared with hemorrhagic stroke) and, within ischemic stroke, for large artery (carotid) stroke (compared with small-vessel disease).[59]

Intracerebral Hemorrhage

Hypertension is the most potent and most common risk factor for intracerebral hemorrhage (ICH). However, only about half of the cases have a clear history of hypertension.[62] This suggests that other factors, possibly genetic, are likely to play a role. Few studies have examined family history of ICH. In the case-control study by Kubota and colleagues, the odds ratio for a family history of ICH was 2.4, but this did not remain significant after adjusting for common vascular risk factors.[55] The family history of ICH, however, did correlate with two risk factors for intracerebral hemorrhage: hypertension and alcohol consumption. As with cerebral aneurysms, many types of vascular malformations also have a genetic contribution.

Hypertension is the most common association with ICH, and there is a genetic contribution to the development of hypertension. About 30% of blood pressure variance can be attributed to genetic factors.[63] The second most common cause in adults is probably cerebral amyloid angiopathy (CAA). CAA is characterized by the deposition of amyloidgenic proteins within the walls of cerebral vessel. There are two common inherited forms of CAA. The first is the Icelandic form of CAA, which is inherited as an autosomal dominant trait, but with the onset of ICHs around the age of 20–30 years.[64] Icelandic CAA is due to deposition of cystatin C, a protease inhibitor, in the vessel walls.[65] A single base mutation in exon 2 of cystatin C causes Icelandic CAA.[66] The second form of CAA, the Dutch form, is also an autosomal dominant disease due to several different mutations in the amyloid precursor protein (APP) gene.[67,68] These disorders are reviewed in more detail in *Chapter 11*. Even in the more commonly seen "sporadic" CAA, there appears to be an association with specific alleles such as apo E e4[69] or the endoglin gene.[70]

These specific etiologies appear to be uncommon in sporadic cases of ICH. In a study of sporadic cases of presumed hypertensive ICH, Graffagnino and colleagues failed to identify any mutations in exons 16 and 17 of APP or exon 2 of cystatin C. However, in their series of 48 consecutive patients admitted with ICH, a family history of stroke was found in 63% of patients; a family history of ICH was found in 15%.[71] This suggests that additional genetic factors contribute to the familial occurrence of intracerebral hemorrhage.

Subarachnoid Hemorrhage

Aneurysmal subarachnoid hemorrhage (SAH) is a common stroke mechanism among in the 40–55 age group. Most cases are due to rupture of intracranial aneurysms (IAs). Females are affected more commonly than males, with a 2:1 to 3:1 ratio. A familial aggregation of SAH/IAs has been found in 10% to 15% of all cases.[72] The pattern of inheritance varies among the studies, although some families demonstrate an autosomal dominant pattern.[73,74]

Subarachnoid hemorrhage is most commonly due to rupture of berry aneurysms. One of the most potent risk factors is hypertension. The genetic contribution to these specific factors is discussed in the following sections. In clinical studies, genetic factors appear to play a clear and important part in the pathogenesis of SAH.[55,75,76] In a case-control study, Kubota and colleagues reported an increased risk of SAH among those with a family history of SAH with an odds ratio of 15.88.[55] This association persisted even after adjusting for potential risk factors (hypertension, diabetes, hyperlipidemia, obesity, alcohol use, and smoking), and remained the most potent risk factor for SAH. In a population-based, case-control study reported by Wang and colleagues, only 11.4% of cases had a first-degree relative with a history of SAH compared with 6.4% for controls.[76] The odds ratio of 2.4 did not significantly change after adjusting for smoking, hypertension, and size of family. Similar results were reported by Bromber who also demonstrated that closer (first-degree) relatives had a great risk for SAH and hypertension (compared to second-degree relatives).[75]

Familial SAH has been associated with several diseases, most strongly the autosomal dominant form of adult polycystic kidney disease. The gene for this disease, mapped to chromosome 16, has recently been cloned and identified, although its function remains unknown at present.[77] Other inherited diseases associated with familial SAH include[78]: Ehlers-Danlos type IV, Marfan's syndrome, coarctation of the aorta, and neurofibromatosis. However, most familial SAH cases are not associated with these diseases. Genetic factors related to SAH and these diseases are reviewed in detail in *Chapter 12*.

A detailed study of large sibships with familial SAH not associated with other diseases has shown a pattern of inheritance most consistent with an autosomal dominant trait.[79] A segregation analysis of all familial SAH cases produced conflicting findings and no clear pattern of inheritance that could adequately describe all cases.[78] There is a suggestion that cases with multiple intracranial aneurysms are more likely to be familial than cases with single aneurysms. It has also been suggested that aneurysms in the anterior communicating artery region are less likely to be familial than aneurysms in other intracranial locations.

Risk Factors and Systemic Disorders

Further evidence supporting a genetic contribution to stroke can be found in studies of vascular pathology and stroke risk factors. These studies not only support a genetic contribution to stroke, but also help point toward new therapeutic strategies for stroke prevention.

Risk Factors

Over the past 10 years there have been significant, and in some cases dramatic, advances in our understanding of genetics of risk factors responsible for many types of vascular disease, particularly atherosclerosis.[44] Much progress has been made in identifying the genetic basis for some types of hypertension, diabetes, and lipid disorders. Genetic epidemiological studies have also identified familial clustering for atherosclerotic coronary artery disease.[80]

The example of atherosclerosis is also germane to understanding the genetics of stroke. In many cases, the development of atherosclerosis is a multifactorial process, with input from genetic and environmental factors. The numerous risk factors for atherosclerosis such as hypertension, diabetes, and hyperlipidemia, can also have genetic and environmental etiologies. In addition, some individuals may have significant atherosclerosis but no clinical symptoms for many years. This creates the possibility of labeling an asymptomatic individual "unaffected", even though there be a high-grade atherosclerotic lesion that has not produced any symptoms.

Hypertension is a disease with many causes,[81] including constitutional (age, sex, body mass), genetic, and environmental factors.[82] It has been estimated that genetic factors account for one-third to one-half of the total risk[63,83]; however, this effect decreases with age.[84]

Recently, relatively common specific genetic changes have been associated with hypertension. The angiotensin I-converting enzyme gene is one of the major genes of the renin-angiotensin and kallikrein-kinin systems and is a candidate gene for several cardiovascular diseases.[85] This includes separate associations between the deletion (D) polymorphism of the gene encoding angiotensin-converting enzyme (ACE) and a the angiotensin II AT1 receptor gene with myocardial infarction.[86] There have been several negative studies of the association between the ACE gene deletion polymorphism and ischemic stroke[87]; however, one study was positive when restricted to lacunar stroke.[88] The genetics of hypertension is reviewed in detail in *Chapter 4*.

There are also clear genetic components for other traditional vascular risk factors such as diabetes and obesity[32,89,90] which can contribute to familial stroke.

Cigarette Smoking

Cigarette smoking is a potent environmental risk factor for vascular disease including ischemic stroke[91] and carotid atherosclerosis.[40] Genetic factors may also profoundly influence the effect of an environmental factor for vascular disease. For example, in a family study of the risk of smokers at risk for cardiovascular disease, Khaw et al. estimated that two-thirds of the risk associated with cigarette smoking was due to genetic factors (a susceptibility to the toxic effects of cigarette smoking).[51] In addition, the habit of cigarette smoking and smoking cessation has genetic influences.[92]

Lipids and Lipoproteins

Lipid abnormalities are a well-established risk for coronary artery disease, especially premature atherosclerosis.[93] The abnormalities that have more recently been associated with the risk of stroke include high LDL, low HDL, elevated cholesterol and increased Lp(a).[94,95]

Cholesterol levels appear to be influenced by both genetic and environmental effects. Elevated cholesterol is a potent risk factor for coronary artery disease and contributes to cerebrovascular disease. In a study of twins raised apart, a strong genetic contribution was demonstrated for the total cholesterol level.[96] The heritability of plasma triglycerides is also high and calculated to be 68%[97]; however, there is debate about the overall genetic contribution to lipid abnormalities.[98]

Lipoproteins influence the development of premature coronary atherosclerosis and may play a role in cerebrovascular diseases. Over half of younger patients with coronary artery disease have a familial lipoprotein disorder.[99] Four main types of lipoprotein abnormalities have been described: (1) elevated LDL cholesterol; (2) reduced HDL cholesterol; (3) elevated chylomicron remnants and intermediate- density lipoproteins; and (4) elevated levels of lipoprotein (a). Lipoprotein transport genes have been implicated in each of these abnormal lipoprotein phenotypes.[100]

Recently, low apolipoprotein E (apo E) levels were strongly associated with ischemic stroke[101] as were increased serum lipoprotein (a) [Lp(a)] levels and intermediate-density lipoproteins and decreased high-density lipoprotein levels.[102] The apo E E3/E3 phenotypes appeared to protect against early vascular morbidity.[101]

Although not reported in all studies, increased levels of Lp(a) have, in case-control studies, been found to be an independent risk factor for coronary artery disease and stroke.[103,102,104] The levels of Lp(a), an LDL-like particle vary in the population; however, 90% of this variability is attributable to the Lp(a) gene.[105] The genetics of lipid metabolism is reviewed in *Chapter 5*.

The exact contribution of lipids to the risk for stroke remains unclear. It may, as with cardiac disease, be most potent for those with premature atherosclerosis. In addition, it may only contribute to the risk for ischemia due to large- and medium-sized vessel atherosclerosis. Much work remains to be done, but given its strong contribution to premature atherosclerosis and the recent association of specific abnormalities and stroke, lipoprotein abnormalities may emerge as a potent pathway for the expression of the genetic contributions to stroke.

Coagulation Deficits

Numerous coagulation defects can be inherited and may produce strokes.[106] Many genetic disorders of coagulation are not rare in the population.[80] For example, the heterozygous state for antithrombin III occurs in approximately 1 in 2000[107] and the G1691A mutation in the factor V gene is present in up to 6% of the U.S. population.[108] Both have been associated with an increased risk for thrombotic disease. Deficiencies in antithrombin III, protein S and protein C are inherited as autosomal domi-

nant traits and have been reported to produce arterial thrombosis (including strokes) and venous thrombosis, particularly in children and young adults (although they can affect older individuals also). However, these inherited deficiencies are quite rare causes of stroke, even among children.

Coagulation disorders, however, directly account for a small minority of strokes.[3,109] For example, resistance to activated protein C (APC-R) was thought initially to be an important and common cause of thrombotic events.[110] The vast majority of APC-R cases are due to an inherited mutation in factor V (the factor V Leiden mutation). Studies have found that heterozygosity for the factor V Leiden mutation is associated with a 3.5- to 10-fold increased risk of thrombosis; the homozygous form has a 50- to 100-fold increased risk.[108]

Despite these findings, recent large population-based studies have failed to show a strong association between APC-R due to factor V Leiden and arterial thrombotic events such as myocardial infarction and stroke.[108,111,112] However, APC-R and factor V Leiden do appear to be associated with venous thrombosis, including deep venous thrombosis and cerebral venous thrombosis. A series of studies have found APC-R/factor V Leiden in about 20% of cases of cerebral venous thrombosis compared to 2% of controls.[112] It is found in 20% to 60% of patients with deep venous thrombosis depending on the selection criteria, the assay system, and the population studied.[113]

Another group of disorders that has received increasing attention over the past 5 years is the antiphospholipid antibody (APA) syndromes.[114] These syndromes include the lupus anticoagulants (idiopathic, drug-induced, or associated with lupus), the anticardiolipin antibodies, and Sneddon's syndrome (stroke and livedo reticularis).[115] These syndromes have been associated with arterial and venous thrombotic events, as well as migraine headaches, Raynaud's syndrome, spontaneous abortions, and connective tissue symptoms. Some patients with a lupus anticoagulant have an elevated PTT on presentation. This abnormality is an artifact of the *in vitro* assay used for PTT determinations; in reality, the disease produces a hypercoagulable state. Formal genetic linkage studies have not been performed; however, it appears as if antiphospholipid antibodies occur with increased frequency in relatives, but not spouses of aCL-positive probands. This data suggests that anticardiolipin illnesses may be familial.[116] Several recent studies have found a strong familial association of these disorders, with a pattern generally consistent with autosomal inheritance (with incomplete penetrance).[117,118] Coagulation disorders are discussed also in *Chapter 7*.

Although under genetic control, coagulation proteins are also strongly influenced by environmental and systemic factors. Acquired deficiencies of coagulation proteins are common.[107] Interestingly, those individuals with the heterozygote states of some conditions could be at increased susceptibility for developing an acquired deficiency and possibly an increased risk for stroke.[107] This, however, is still speculative.

Similarly, platelets have a central role in thromboembolic disease. Preliminary reports in patients with coronary ischemia suggested an association with the Pl[A2] allele (a twofold increase in risk) and generated great interest in the role of inherited platelet risk factors for vascular disease.[119] An increased risk was also found for coronary stent thrombosis.[120] However, in a much larger cohort, the Physicians' Health Study, healthy men with the GPIIIa Pl[A2] allele were not associated with any increase in the risk of myocardial infarction, stroke, or venous thrombosis.[121] Additional work on the Pl[A2] allele and other polymorphisms is still needed.[122]

Homocysteine

Elevated serum homocysteine levels are recognized as an important factor in producing premature atherosclerotic lesions and thrombotic events. Both environmental and genetic factors influence the levels of levels of homocysteine.[123] Two of the most common disorders are homocystinuria and hyperhomocysteinemia. The classic form of homocystinuria is an autosomal recessive disorder due to mutations in the cystathionine β-synthase (CBS) gene. More than a dozen different mutations in this gene have been described, with a mutation in exon 8 responsible for approximately 50% of all cases.[124] Disorders of homocysteine metabolism are discussed in greater detail in *Chapter 14*.

In clinical studies, untreated patients with homocystinuria have a 33% to 50% risk of thrombotic events by age 25(125). Arterial or venous thrombotic events causing strokes account for about one-third of all thrombotic complications. Treatment for patients with disorders of homocysteine metabolism depends on the underlying disorder. Some patients are responsive to high doses of vitamin B6 supplementation while other patients will benefit from folate and B12 supplementation.[123]

Hemoglobinopathy

Hemoglobinopathies are a very common group of disorders with widespread systemic effects, including stroke. Sickle cell anemia occurs when patients are homozygous for hemoglobin S. This mutant hemoglobin is produced by a single nucleotide change (adenine to thymine). Once systemic symptoms are produced, the illness is referred to as sickle cell disease. Cerebrovascular events occur in 10% to 17% of patients, and up to 10% of patients may have asymptomatic strokes detected by brain MRI.[126,127] Although early hypotheses were that the ischemic events were due to sludging of blood cells leading to thrombosis and occlusion, recent studies have shown this to be a less common mechanism of stroke. A more frequent process appears to be a vasculopathy produced by intimal hyperplasia, which leads to occlusion of the large intracranial vessels.[128,129] These occlusions can produce ischemic strokes. However, stenosis of these large vessels can lead to formation of small collateral vessels, termed "moyamoya" disease. These vessels tend to rupture, producing an ICH or an SAH in some cases. Sickle cell disease can also produce intracranial aneurysms which can rupture, producing a subarachnoid hemorrhage.[130] Thrombosis of intracranial veins can also occur, producing venous infarctions. (See *Chapter 14* for more detail.)

Individuals with sickle cell trait (heterozygous for hemoglobin S) do not have a significantly increased risk of stroke, although a few documented cases have been reported related to dehydration or hypoxia.[131] Persons with hemoglobin SC (double heterozygotes) have an increased rate of stroke, approximately 2% to 5%.[126]

Mitochondrial Disorders

Genetic disorders affecting the mitochondria are characterized with increasing frequency. Many of these disorders have systemic as well as neurologic effects. One type in particular, the MELAS syndrome (mitochondrial encephalomyopathy, lactic acidosis, and stroke-like episodes) can produce stroke-like events.[132] Like most mitochondrial disorders, MELAS shows a maternal pattern of inheritance with incomplete penetrance. About 75% of cases are due to a point mutation in the gene for the transfer RNA for leucine.[133] Patients with MELAS tend to have strokes affecting the parietal-occipital region. Most of these strokes do not respect typical vascular territories. Other symptoms may include migraine headaches, seizures, myoclonus, ataxia, and dementia.[132] The changes are seen most readily on T2-weighted MRI scans, and tend to effect deep cortical and superficial white matter layers. The underlying pathophysiology is unclear, although large-vessel occlusions are not present. Recent studies have shown evidence of a small-vessel angiopathy in some patients.[134]

Other Vascular Risk Factors

Almost all vascular risk factors have a genetic contribution.[135] These factors include diabetes,[136] obesity,[32] personality traits,[33] susceptibility to cigarette smoking,[51] familial Sneddon's syndrome,[137] and antiphospholipid antibody syndromes.[138] Many other risk factors are under investigation for their genetic contribution.

Vascular Pathologies and Cardiac Disorders Associated with Stroke

The following section is a sampling of pathological conditions associated with vascular diseases. Not all have been directly associated with an increased risk of symptomatic cerebrovascular disease. The field is still young, however, and holds tremendous potential for understanding the genetic mechanisms of familial stroke, and may ultimately lead to more effective therapies for the prevention of stroke.

Atherosclerosis

Atherosclerotic disease of the carotid arteries has been strongly associated with coronary artery disease and the risk for stroke. It is not surprising that similar risk factors have been demonstrated for both carotid and coronary artery disease.

In epidemiological studies, higher HDL cholesterol (and lower LDL cholesterol) levels have been associated with lower rates of cardiovascular disease.[139] Restriction fragment length polymorphism techniques[140] have been used to study polymorphisms of the apolipoprotein A-I gene which encodes for the most prominent apoproteins of high-density lipoprotein (HDL), and it appears that there are strong genetic associations with an abnormal lipid profile. This suggests that genetic factors influence the development of symptomatic coronary atherosclerosis.[141] A similar influence is likely for the large-and medium-size vessels of the head and neck.

Genetic factors have been shown to influence carotid wall thickness.[142] Yatsu and colleagues also found that the *Sac* I polymorphism S2 allele was significantly elevated in blacks with carotid stenosis, but not in whites with carotid stenosis. Furthermore, this association in blacks was not associated with an abnormal lipid profile. These findings and techniques may begin to explain the observed racial differences in the prevalence of carotid atherosclerosis. It may also help in the formulation of new strategies for the prevention of carotid atherosclerosis.[143]

Dissections

Individuals with a family history of cervical artery dissection are at risk for recurrent arterial dissection.[144,145] Syndromes associated with an increased risk for dissection have been described both as parts of larger clinical syndromes,[146,147] and with specific defects which may result in a weaker adventia such as type III collagen deficiency and fibromuscular dysplasia.[148]

Vascular Malformations

There is convincing evidence that one type of vascular malformation, cavernous hemangioma, is frequently inherited with an autosomal dominant pattern. Several large kindreds, some of Mexican-American origin, have been described in the literature. Recent genetic linkage studies have mapped a gene causing cerebral cavernous malformations to chromosome 7q11.2-q21. There is evidence of genetic heterogeneity on the same chromosome and a probable "founder effect" among Hispanic patients (i.e., a common ancestor among reported families).[149]

There have been several cases of familial arteriovenous malformations (AVMs) described.[149a] Usually, these families are too small and isolated to allow any firm conclusions about the mode of inheritance. In rare cases, AVMs have been noted to occur with a familial clustering. An autosomal dominant pattern of inheritance has been observed in large series. In general, familial cases tend to have an earlier age of onset, 10–19 years, compared to 21–40 years in sporadic cases.

Rendu-Osler-Weber, also known as hereditary hemorrhagic telangiectasia (HHT), is characterized by telangiectasia affecting mucous membranes and skin throughout the body, producing epistaxis, GI hemorrhages, and other bleeding events.[136] Patients with HHT are prone to develop pulmonary AVMs which can act as a source of emboli that can produce strokes. In addition to telangiectasias, cerebral AVMs, and angiomas, aneurysms and carotid-cavernous sinus fistulae have been reported in HHT patients. Two genetic loci have been identified for HHT. One is the endoglin gene on chromosome 9, the other is the activin receptor-like kinase gene on chromosome 12.[125,137] A variety of mutations in these genes that alter the protein size or function have been identified in cases of HHT. Both endoglin and ALK1 are involved in binding transforming growth factor-β. This protein system is important for vascular development and stability. These disorders are also reviewed in *Chapter 11.*

Cardiomyopathy

Heart failure and cardiomyopathy increase the risk for stroke.[150] Emerging evidence suggests that cardiomyopathies in children and young adults are often due to genetic causes.[151] In older patients, vascular diseases that injure the myocardium play a greater role; however, these etiologic mechanisms may also have a significant genetic contribution. The most common etiology for cardiomyopathy in adults is ischemic and often associated with vascular risk factors such as hypertension, diabetes, hyperlipidemia which have a genetic contribution. The situation may be even more complex in that genetic factors may influence the heart's response to hypertension. For example, heritable difference in the contractile proteins may result in an individual more prone to develop hypertensive changes.

Familial hypertrophic cardiomyopathy (HCM) is inherited as an autosomal dominant disorder. It shows genetic heterogeneity with loci on chromosomes 1, 7, 11, 14, and 15.[101,102] Those cases linked to chromosome 14 are due to a variety of mutations in the β-myosin heavy chain gene.[103] Cases that map to chromosomes 15 and 1 are due to mutations in the alpha-tropomyosin and cardiac troponin genes, respectively.[104] Cases linked to chromosome 7 appear to be due to mutations in the cardiac myosin binding protein-C.[105]

There are specific forms of familial cardiomyopathy such as familial dilated cardiomyopathy (FDCM) which appear to be genetically heterogeneous in terms of inheritance mode and localization. Linkage has been reported to loci on chromosomes 1, 3, 9, and 10.[106,107] Some forms of X-linked FDCM are due to mutations in the dystrophin gene.[108] Other cases may be due to mitochondrial mutations. However, these cases account for a small minority of all of the familial cases. (Further information can be found in *Chapter 9.*)

Congenital Heart Disease

Structural defects of the heart and disease of the great vessels are increasingly recognized as risk factors for cerebrovascular disease in adults. Examples include atrial septal defects and atherosclerotic disease of the aortic arch. Congenital heart malformations are among the most common causes of birth defects, with an estimated prevalence of 5 to 10 per 1000 live births.[152] The rate is likely to be even higher as many of the occult lesions, such as patent foramen ovale, are diagnosed with greater frequency.

The congenital heart diseases were thought to be predominantly multifactorial, with less than 10% due to single gene defects. This may be a gross underestimation with the recognition that a common genetic defect may cause several apparently different forms of congenital heart diseases.[153] For example, in Keeshond hounds, an inbred dog, there is a high frequency of tetralogy of Fallot. Relatives have an increased incidence of subarterial ventricular septal defects. Additional detailed studies demonstrated a variety of conotruncal defects including truncus arteriosus.[154]

CADASIL

Although there is evidence of a genetic contribution to ischemic stroke, there are few genetic defects that specifically affect the cerebral circulation resulting in cerebral infarction in adults. In 1994, a cerebral autosomal dominant arteriopathy with subcortical infarcts and leukoencephalopathy (CADASIL) was described.[155] As the name suggests, it has an autosomal dominant mode of inheritance, and more recently, the condition has been localized to chromosome 19q12 in 15 unrelated families.[156,157]

CADASIL usually presents in the third to seventh decade with recurrent strokes and an absence of traditional vascular risk factors. A period of remission or a slow steady progression may also be seen. Vascular headaches and psychiatric symptoms, especially depression, are associated with the disease. In later stages, pseudobulbar palsy and dementia occur.

Magnetic resonance imaging show widespread white matter changes similar to those described for hypertensive subcortical ischemia. Pathologically, there is concentric thickening of the vascular wall and narrowing of the vessel lumen within small arteries and arterioles. Hyaline degeneration of the intima or subendothelial fibrous proliferation may be present, similar to lacunar disease.[158] (See *Chapter 10* for a detailed review of CADASIL.)

The identification of this disease represents an important advance not only for understanding a specific genetic cause of ischemic stroke, but also for molecular biological studies of cerebrovascular disease, especially small-vessel disease. A better understanding of the processes associated with small-vessel disease may have a significant impact on understanding how and why cerebrovascular diseases occur and become symptomatic.

Other Vascular Diseases

Several types of vasculitides may exhibit a familial occurrence in some cases. One of the most common is fibromuscular dysplasia (FMD), which can affect the renal and cerebral vessels. In FMD, the media of the vessels is affected most frequently (90–95% of cases) with a noninflammatory proliferation of the fibrous tissue or smooth muscle. When FMD affects the cerebral vasculature, bilateral involvement of the internal carotid arteries is found in 60% to 70% of cases.[148] The vertebral arteries may also be involved. Vessel dissection is seen commonly in patients with FMD. Females are affected much more frequently than males (5:1 in some series). The overall pattern of inheritance suggests autosomal dominant with incomplete penetrance (particularly in males).

Fabry's disease is a rare metabolic disorder associated with skin lesions (angiokeratomas) and strokes. It is inherited as an X-linked dominant trait. An enzymatic defect in alpha-galactosidase A leads to accumulation of various glycosphingolipids in endothelial and epithelial cells.[159] Cerebrovascular symptoms typically occur between the ages of 30 and 40 years. The vertebrobasilar circulation is involved in about two-thirds of cases, producing brainstem signs and symptoms. Large vessels can be affected and become elongated and tortuous, leading to thrombus formation. Small-vessel strokes can be caused by either stenosis due to deposition of the glycosphingolipid in the vessel wall, or obstruction of the vessel ostia due to the elongation of the parent large vessel.[160]

A number of other inherited connective tissue disorders can also have cerebrovascular manifestations. These disorders include neurofibromatosis (type 1), Marfan's syndrome, Ehlers-Danlos disease, and pseudoxanthoma elasticum.

Interpretation and Limitations of Epidemiological Studies

Not all epidemiological reports have found an association between family history and stroke.[56,161-163] Existing studies are may be limited by several common biases.[4,164,165] Stroke is often self-reported in the patient and through a surrogate family member for relatives. The reliability of self-reported stroke varies widely[55]; however, for surrogate data the results are especially poor.[166] This is true even for dramatic stroke syndromes such as subarachnoid hemorrhage.[167]

Research in genetic epidemiology poses much greater practical problems than standard epidemiological studies of risk factors because there are often significant biases in how families are ascertained. In addition, reports of disease in relatives is known to be subject to error. For stroke, the problem is substantial. Stroke patients often have deceased parents whose strokes predominantly occurred prior to the era of computed tomography (CT). The diagnosis of stroke-related death, before CT, has been shown to be highly unreliable. This is true even when based on death certificate data where there may be many false-positive[168] and false-negative[169] reported diagnoses.

Even when accurate historical information is provided, the original diagnosis may have been incorrect. Among parents and grandparents, the diagnosis of stroke, in most studies, was in the pre-CT era and based solely on clinical criteria. When there was an occurrence of stroke, it was common to classify intracerebral hemorrhage as an ischemic stroke.[170] Incomplete data or low response rates can significantly alter the estimate of relative risk.[171] Incomplete data can also result from the long latency period between the initial onset of the disease process and a symptomatic stroke or "silent" stroke that was not recognized by the patient or physician. Many other types of biases[172] can impact these studies and may account for those reports which did not show a significant association between family history and stroke risk.[5]

Many potential biases exist in the study of complex genetic disease which may have influenced some of the reports of family history and stroke. The more common biases and problems in studying the genetics of cerebrovascular disease include[4]: (1) *Ascertainment bias*, where identification methods can influence the outcome. For example, in a hospital-based study, individuals with less severe syndromes might be excluded. Similarly ethnic, socioeconomic, racial, and cultural factors may have an impact on admissions and confound results. (2) *Genetic heterogeneity*, where mutations at one of several gene loci can produce the same phenotype. (3) *Pathological heterogeneity* due to the many mechanisms that can result in a stroke. Even within a single stroke subtype (e.g., cardioembolism), many different types of cardiac disease can predispose to the formation of a blood clot within the heart and secondary embolism. (4) *Interactions among several genes and environmental factors* may be important for producing the "trait" of stroke. Gene defects may have a cumulative effect which makes the association with an individual genetic defect especially difficult. Atherosclerosis may be this type of polygenic trait. (5) *Phenocopies,* when purely nongenetic (environmental) factors produce a phenotype similar to one produced by genetic factors. (6) *Other risk factors* need to be considered (e.g., hypertension, diabetes, cigarette smoking). The impact of these factors varies with age and may interact with candidate genetic traits under study.

Conclusions

Genetic etiologies are recognized with increasing frequency as a cause of all types of stroke. Epidemiological studies have proven the importance of inherited factors, and molecular genetic studies have identified specific mechanisms and genes responsible for strokes. Identification of specific causative genes may increase our understanding of previously unknown or unrecognized stroke mechanisms, and facilitate the development of treatment strategies that focus on preventative measures.

A key step in determining if a genetic mechanism is the cause of a patient's stroke is to obtain a comprehensive family history. This should begin with construction of a pedigree listing all known living and deceased family members. One or more knowledgeable informants should be questioned in detail about the medical conditions and causes of death for each family member. Supporting documentation from physicians and hospitals should be obtained for individuals presumed to be affected.

There are several other clinical clues that may reflect the presence of a familial disorder. In general, individuals presenting at a young age or having multiple lesions may raise suspicion of an underlying genetic disorder. The evaluation of such patients depends on the type of stroke being investigated. A careful general examination, with special emphasis on the vasculature, skin, and eyes may detect clues that point to an underlying genetic disorder. For ischemic stroke, tests for a hypercoagulable state, lactic acid levels, a hemoglobin electrophoresis, and an angiogram are often indicated.

As more is learned about genetic factors that can produce or contribute to stroke, and additional genetic epidemiological data is collected, presymptomatic screening and perhaps treatment may be possible. This has been the case with other inherited diseases. Although the clinical heterogeneity of stroke presents some challenges, the application of new research techniques and methodologies may be helpful for overcoming these problems.

References

1. Wolf PA, D'Agostino RB, Belanger AJ, Kannel WB. Probability of stroke: A risk profile from the Framingham Study. Stroke 1991;22:312–318.
2. Kernan WN, Horwitz RI, Brass LM, et al. A prognostic system for patients with TIA or minor stroke. Ann Int Med 1991;114:552–557.
3. Natowicz M, Kelly RI. Mendelian etiologies of stroke. Ann Neurol 1987;22:175–192.
4. Alberts MJ. Genetics of cerebrovascular disease. Stroke 1990;21(suppl III):III-127–III-130.
5. Brass LM, Alberts MJ. The genetics of cerebrovascular disease. Baillieres Clin Neurol 1995;4:221–246.
6. Magnus P. Genetic epidemiology: Possibilities and problems. Scand J Soc Med 1992;20:193–195.
7. Morton NE. Genetic epidemiology. Annu Rev Genet 1993;27:523–538.
8. Hopper JL. The epidemiology of genetic epidemiology. Acta Genet Med Gemellol 1992;41:261–273.
9. Ott J. Cutting the Gordian knot in the linkage analysis of complex human traits. Am J Hum Genet 1990;46:219–221.
10. Risch N. Linkage strategies for genetically complex traits. I. Multilocus models. Am J Hum Genet 1990;46:222–228.

11. Mortimer JA, Graves AB. Observational studies in neuroepidemiology: Methodological issues. In Gorelick PB, Alter M (eds). Handbook of Epidemiology. New York: Marcel Dekker, Inc, 1994, 1–26.

12. Kittner SJ, White LR, Losonczy K, et al. Black-white differences in stroke incidence in a national sample: The contribution of hypertension and diabetes mellitus. JAMA 1990;264:1267–1270.

13. Blarajan R. Ethnic differences in mortality from ischemic heart disease and cerebrovascular disease in England Wales. BMJ 1991;302:560–564.

14. Tanaka H, Tanaka Y, Hayashi M, et al. Secular trends for mortality for cerebrovascular disease in Japan, 1960 to 1979. Stroke 1982;13:574–589.

15. Inzitari D, Hachinski VC, Taylor DW, Barnett HJM. Racial differences in the anterior circulation in cerebrovascular disease. Arch Neurol 1990;47:1080–1084.

16. Kagan A, Popper JOS, Rhoads GC. Factors related to stroke incidence in Hawaii Japanese men: The Honolulu heart study. Stroke 1980;11:14–21.

17. Bruno A. Are there racial differences in vascular disease between ethnic and racial groups? Stroke 1998;29:2–3.

18. O'Mahony PG, Dobson R, Rodgers H, et al. Validation of a population screening questionaire to assess prevalence of stroke. Stroke 1995;26:1334–1337.

19. Wyller TB, Ranhoff AH, Bautz-Holter E. Validity of questionaire information from old people on previous cerebral stroke. Cerebrovasc Dis 1994;4:57–58.

20. Kiely DK, Wolf PA, Cupples LA, et al. Familial aggregation of stroke: The Framingham Study. Stroke 1993;24:1366–1371.

21. Welin I, Svardsudd K, Wilhelmsen L, et al. Analysis of risk factors for stroke in a cohort of men born in 1913. N Engl J Med 1987;317:521–526.

22. Liao D, Myers R, Hunt S, et al. Familial history of stroke and stroke risk: The Family Heart Study. Stroke 1997;28:1908–1912.

23. Higgins M, Province M, Heiss G, et al. NHLBI Family Heart Study: Objectives and design. Am J Epidemiol 1996;143:1219–1228.

24. Jousilahti P, Rastenyte D, Tuomilehto J, et al. Parental history of cardiovascular diseases and risk of stroke: A prospective follow-up of 14,371 middle-aged men and women in Finland. Stroke 1997;28:1361–1366.

25. Berg K. Genetics of coronary heart disease and its risk factors. Prog Clin Biol Res 1985;177:351–374.

26. Hrubec Z, Robinette CD. The study of twins in medical research. N Engl J Med 1984;310:435–441.

27. Braun MM, Haupt R, Caporaso NE. The National Academy of Sciences National Research Council Veteran Twin Registry. Acta Genet Med Gemellol 1994;43:89–94.

28. Kendler KS, Holm NV. Differential enrollment in twin registries: Its effect on prevalence and concordance rates and estimates of genetic parameters. Acta Genet Med Gemellol 1985;34:125–140.

29. Feinleib M, et al. The NHLBI twin study of cardiovascular disease risk factors: Methodology and summary of results. Am J Epidemiol 1977;106:284–285.

30. Barnett AH, et al. Diabetes in identical twins: A study of 200 pairs. Diabetologia 1981;20:87–93.

31. Berg K. Twin studies of coronary heart disease and its risk factors. Acta Genet Med Gemellol (Roma) 1984;33:349–361.

32. Stunkard AJ, Harris JR, Pedersen NL, McClearn GE. The body-mass index of twins who have been raised apart. N Engl J Med 1990;322:1483–1487.

33. Pedersen NL, Lichtenstein P, Plomin R, et al. Genetic and environmental influences for type A-like and related traits: A study of twins reared apart and twins reared together. Psychosom Med 1989;51:428–440.

34. Swan GE, Carmelli D, Reed T, et al. Heritability of cognitive performance in aging twins. Arch Neurol 1990;47:259–262.

35. Marenberg ME, Risch M, Berkman LF, et al. Genetic susceptibility to death from coronary heart disease in a study of twins. N Engl J Med 1994;330:1041–1046.

36. Berg K. Twin studies of coronary heart disease and its risk factors. Acta Genet Med Gemellol 1987;36:439–453.

37. de Faire U. Ischaemic heart disease in death discordant twins: A study on 205 male and female pairs. Acta Med Scand Suppl 1974;5 68:1–109.

38. Lilijefors I. Coronary heart disease in male twins: Hereditary and environmental factors in concordant and discordant pairs. Acta Med Scand Suppl 1970;511:9–90.

39. Brass LM, Isaacsohn JL, Merikangas KR, Robinette CD. A study of stroke in twins. Stroke 1992;23:221–223.

40. Haapanen A, Koskenvuo M, Kaprio J, et al. Carotid arteriosclerosis in identical twins discordant for cigarette smoking. Circulation 1989;80:10–16.

41. Nora J, Lortscher R, Spangler R, et al. Genetic epidemiological study of early-onset ischemic heart disease. Circulation 1980;61:503–508.

42. Brass LM, Shaker LA. Family history in patients with transient ischemic attacks. Stroke 1991;22:837–841.

43. Graffagnino C, Gasecki AP, Doig GS, Hachinski VC. The importance of family

history in cerebrovascular disease. Stroke 1994;25:1599–1604.

44. Pullicino P, Greenberg S, Trevisan M. Genetic stroke risk factors. Curr Opin Neurol 1997;10:58–63.

45. Howard G, Evans GW, Toole JF, et al. Characteristics of stroke victims associated with early cardiovascular mortality in their children. J Clin Epidemiol 1990;43:49–54.

46. Arcavia L, Benowitz NL. Clinical significance of genetic influences on cardiovascular drug metabolism. Cardiovasc Drugs Ther 1993;7:311–324.

47. Brass LM, Carrano D, Hartigan PM, et al. Genetic risk for stroke: A follow-up study of the NAS/VA twin registry. Neurology 1996;46: A212. Abstract.

48. Schildkraut JM, Myers RH, Cupples LA, et al. Coronary risk associated with age and sex of parental heart disease in the Framingham Study. Am J Cardiol 1989;64:555–559.

49. Marshall J. Familial incidence of cerebral hemorrhage. Stroke 1973;4:38–41.

50. Gifford AJ. An epidemiological study of cerebrovascular disease. Am J Public Health 1966;56:452–461.

51. Khaw KT, Barrett CE. Family history of heart attack: A modifiable risk factor? Circulation 1986;74:239–244.

52. Alberts MJ, Graffagnino C, Mclenny C, et al. ApoE genotype and survival from intracerebral hemorrhage (letter). Lancet 1995;346:575.

53. Dyker AG, Weir CJ, Lees KR. Influence of cholesterol on survival after stroke: Retrospective study. BMJ 1997;314:1584–1588.

54. Diaz JF, Hachinski VC, Pederson LL, Donald A. Aggregation of multiple risk factors for stroke in siblings of patients with brain infarction and transient ischemic attacks. Stroke 1986;17:1239–1242.

55. Kubota M, Yamaura A, Ono J, et al. Is family history an independent risk factor for stroke? J Neurol Neurosurg Psychiatry 1997;62:66–70.

56. Thompson SG, Greenberg G, Meade T. Risk factors for stroke and myocardial infarction in women in the United Kingdom as assessed in general practice: A case-control study. Br Heart J 1989;61:403–409.

57. Khaw K, Barrett-Connor EB. Family history of stroke as an independent predictor of ischemic heart disease in men and stroke in women. Am J Epidemiol 1986;123:59–66.

58. Vitullo F, Marchioli R, Di Mascio R, et al. Family history and socioeconomic factors as predictors of myocardial infarction, unstable angina and stroke in an Italian population. PROGETTO 3A Investigators. Eur J Epidemiol 1996;12:177–185.

59. Bogousslavsky J, Castillo V, Kumral E, et al. Stroke subtypes and hypertension: Primary hemorrhage vs infarction, large- vs small-artery disease. Arch Neurol 1996;53:265–269.

60. Brass LM, Hartigan PM, Page WF, Concato J. The importance of cerebrovascular disease in studies of myocardial infarction. Stroke 1994;25:247.

61. Heyden S, Heyman A, Camplong L. Mortality patterns among parents of patients with atherosclerotic cerebrovascular disease. J Chonic Dis 1969;22:105–110.

62. Brott T, Thalinger K, Hertzberg V. Hypertension as a risk factor for spontaneous intracerebral hemorrhage. Stroke 1986;17:1078–1083.

63. Corvol P, Jeunemaitre X, Charru A, Soubrier F. Can the genetic factors influence the treatment of systemic hypertension? The case of the renin-angiotensin-aldosterone system. Am J Cardiol 1992;70:14D-20D.

64. Luyendijk W, Bots GT, Vegter van dVM, et al. Hereditary cerebral haemorrhage caused by cortical amyloid angiopathy. J Neurol Sci 1988;85:267–280.

65. Jensson O, Gudmundsson G, Arnason A, et al. Hereditary cystatin C (gamma-trace) amyloid angiopathy of the CNS causing cerebral hemorrhage. Acta Neurol Scand 1987;76:102–114.

66. Levy E, Lopez OC, Ghiso J, et al. Stroke in Icelandic patients with hereditary amyloid angiopathy is related to a mutation in the cystatin C gene, an inhibitor or cysteine proteases. J Exp Med 1989;169:1771–1778.

67. Wei LH, Walker LC, Levy E. Cystatin C: Icelandic-like mutation in an animal model of cerebrovascular beta-amyloidosis. Stroke 1996;27:2080–2085.

68. Fernandez MI, Levy E, Marder K, Frangione B. Codon 618 variant of Alzheimer amyloid gene associated with inherited cerebral hemorrhage. Ann Neurol 1991;30:730–733.

69. Greenberg S, Rebeck W, Vonsattel J, et al. Apolipoprotein E e4 and cerebral hemorrhage associated with amyloid angiopathy. Ann Neurol 1995;38:254–259.

70. Alberts M, Davis J, Graffagnino C, et al. Endoglin gene polymorphism as a risk factor for sporadic intracerebral hemorrhage. Ann Neurol 1997;41:683–686.

71. Graffagnini C, Herbstreith MH, Roses AD, Alberts MJ. A molecular genetic study of intracerebral hemorrhage. Arch Neurol 1994; 51:981–984.

72. Ronkainen A, Hernesniemi J, Ryynanen M. Familial subarachnoid hemorrhage in east Finland, 1977–1990. Neurosurgery 1993;33: 787–797.

73. Alberts MJ, Quinones A, Graffagnino C, et al. Risk of intracranial aneurysms in families with subarachnoid hemorrhage. Can J Neurol Sci 1995;22:121–125.

74. Schievink WI, Schaid DJ, Rogers HM, et al. On

the inheritance of intracranial aneurysms. Stroke 1994;25:2028–2037.

75. Bromberg JE, Rinkel GJ, Algra A, et al. Hypertension, stroke, and coronary heart disease in relatives of patients with SAH. Stroke 1996;27:7–9.

76. Wang PS, Longstreth Jr WT, Koespsell TD. Subarachnoid hemorrhage and family history. Arch Neurol 1995;52:202–204.

77. Ryu SJ. Intracranial hemorrhage in patients with polycystic kidney disease. Stroke 1990;21:291–294.

78. ter Berg HWM, Dippel DWJ, Limburg M, et al. Familial intracranial aneurysms. Stroke 1992;23:1024–1030.

79. Shinton R, Palsingh J, Williams B. Cerebral haemorrhage and berry aneurysm: Evidence from a family for a pattern of autosomal dominant inheritance. J Neurol Neurosurg Psychiatry 1991;54:838–840.

80. Steeds RP, Channer KS. How important is family history in ischaemic heart disease? Q J Med 1997;90:427–430.

81. Harrap SB. Hypertension: Genes versus environment. Lancet 1994;344:169–171.

82. Tishler PV, Lewitter FI, Rosner B, Speizer FE. Genetic and environmental control of blood pressure in twins and their family members. Acta Genet Med Gemelloi 1987;36:455–466.

83. Carmelli D, Robinette D, Fabsitz R. Concordence, discordance, and prevalence of hypertension in World War II male vetran twins. J Hypertens 1994;12:323–328.

84. Sims J, Hewitt JK, Kelly KA, et al. Familial and individual influences on blood pressure. Acta Genet Med Gemellol 1986;35:7–21.

85. Soubrier F, Cambien F. The angiotensin I-converting enzyme gene polymorphism: Implications in hypertension and myocardial infarction. Curr Opin Nephrol Hypertens 1994;3:25–29.

86. Tiret L, Bonnardeaux A, Poirier O, et al. Synergistic effect of angiotensin-converting enzyme and angiotensin-II type 1 receptor gene polymorphisms on risk of myocardial infarction. Lancet 1994;344:910–913.

87. Agerholm-Larsen B, Tybjaerg-Hansen A, Frikke-Schmidt R, et al. ACE gene polymorphism as a risk factor for ischemic cerebrovascular disease. Ann Intern Med 1997; 127:346–355.

88. Markus HS, Barley J, Lunt R, et al. Angiotensin-converting enzyme gene deletion polymorphism: A new risk factor for lacunar stroke but not carotid atheroma. Stroke 1995;26:1329–1333.

89. Slyper A, Schectman G. Coronary artery disease risk factors from a genetic and developmental perspective. Arch Intern Med 1994; 154:633–638.

90. Kyvik KO, Green A, Beck-Nielsen H. Concordance rates of insulin dependent diabetes mellitus: A population based study of young Danish twins. BMJ 1995;311:913–917.

91. Wolf PA, DAgostino RB, Kannel WB, et al. Cigarette smoking as a risk factor for stroke: The Framingham Study. JAMA 1988;259: 1025–1029.

92. Carmelli D, Swan GE, Robinette D, Fabsitz R. Genetic influence on smoking: A study of male twins. N Engl J Med 1992;327:829–833.

93. Genest Jr JJ, Martin-Munley SS, McNamara JR, et al. Familial lipoprotein disorders in patients with premature coronary artery disease. Circulation 1992;85:2025–2033.

94. Gorelick PB, Schneck M, Berglund LF, et al. Status of lipids as a risk factor for stroke. Neuroepidemiology 1997;16:107–115.

95. Jürgens G, Taddei-Peters WC, Költringer P, et al. Lipoprotein(a) serum concentration and apolipoprotein(a) phenotype correlate with severity and presence of ischemic cerebrovascular disease. Stroke 1995;26:1841–1848.

96. Heller DA, de Faire U, Pedersen NL, et al. Genetic and environmental influences on serum lipids in twins. N Engl J Med 1993;328:1150–1156.

97. Christian JC, Feinleib M, Hulley SB, et al. Genetics of plasma cholesterol and triglycerides: A study of adult male twins. Acta Genet Med Gemellol 1976;25:145–149.

98. Dammerman M, Breslow JL. Genetic basis of lipoprotein disorders. Circulation 1995; 91: 505–512.

99. Genest JJ, Martin-Munley SS, McNamara JR, et al. Familial lipoprotein disorders in patients with premature coronary artery disease. Circulation 1992;85:2025–2033.

100. Dammerman M, Breslow JL. Genetic basis of lipoprotein disorders. Circulation 1995; 91:505–512.

101. Couderc R, Mahieux F, Bailleul S, et al. Prevalence of apolipoprotein E phenotypes in ischemic cerebrovascular disease: A case-control study. Stroke 1993;24:661–664.

102. Pedro-Botet J, Senti M, Nogues X, et al. Lipoprotein and apolipoprotein profile in men with ischemic stroke: Role of lipoprotein(a), triglyceride-rich lipoproteins and apolipoprotein E polymorphism. Stroke 1992;23:1556–1562.

103. Ridker PM, Hennekens CH, Stampfer MJ. A prospective study of lipoprotein(a) and the risk of myocardial infarction. JAMA 1993; 270:2195–2199.

104. Rhoads GG, Dahlen G, Berg K, et al. Lp(a) lipoprotein as a risk factor for myocardial infarction. JAMA 1986;256:2540–2544.

105. Boerwinkle E, Leffert CC, Lin J, et al. Apolipoprotein(a) gene accounts for greater

than 90% of the variation in plasma lipopro-tein(a) concentrations. J Clin Invest 1992;90: 52–60.

106. Ravine D, Cooper DN. Adult-onset genetic disease: Mechanisms, analysis, and predic-tion. Q J Med 1997;90:83–103.

107. Lane DA, Olds RR, Thein SL. Antithrombin and its deficiency states. Blood Coag 1992;3:315–341.

108. Ridker PM, Hennekens CH, Lindpainter K, et al. Mutation in the gene coding for coag-ulation factor V and the risk of myocardial infarction, stroke, and venous thrombosis in apparently healthy men. N Engl J Med 1995;332:912–917.

109. Love BB. Rare genetic disorders predispos-ing to stroke. In Biller J, Mathews KD, Love BB (eds). Cerebrovascular Disease in Chil-dren. Newton, MA: Butterworth-Heine-mann, 1994, 147–164.

110. Bertina R, Koelman B, Koster T, et al. Muta-tion in blood coagulation factor V associ-ated with resistance to activated protein C. Nature 1994;369:64–67.

111. Deschiens MA, Conard J, Horellou MH, et al. Coagulation studies, factor V Leiden, and anticardiolipin antibodies in 40 case of cerebral venous thrombosis. Stroke 1996;27: 1724–1730.

112. Zuber M, Toulon P, Marnet L, Mas J-L. Fac-tor V Leiden mutation in cerebral venous thrombosis. Stroke 1996;27:1721–1723.

113. Dahlbäck B. Resistance to activated protein C, the Arg506 to Gln mutation in the factor V gene and venous thrombosis. Thromb Haemost 1995;73:739–742.

114. Feldmann E, Levine SR. Cerebrovascular dis-ease with antiphospholipid antibodies: Im-mune mechanisms, significance, and thera-peutic options. Ann Neurol 1995;37(suppl 1):S114-S1130.

115. Levine SR, Brey RL, Sawaya KL, et al. Recur-rent stroke and thrombo-occlusive events in the antiphospholipid syndrome. Ann Neurol 1995;38:119–124.

116. Goldberg SN, Conti-Kelly AM, Greco TP. A family study of anticardiolipin antibodies and associated clinical conditions. Am J Med 1995;99:473–479.

117. Lossos A, Ben-Hur T, Ben-Nariah Z, et al. Familial Sneddon's syndrome. J Neurol 1995;242:164–168.

118. Rosove M, Brewer P. Antiphospholipid thrombosis: Clinical course after the first thrombotic event in 70 patients. Ann Int Med 1992;117:303–308.

119. Weiss EJ, Bray PF, Tayback M, et al. A poly-morphism of a platelet glycoprotein receptor as an inherited risk factor for coronary throm-bosis. N Engl J Med 1996;334:1090–1094.

120. Walter DH, Schächinger V, Elsner M, Dim-meler S, Zeiher AM. Platelet glycoprotein IIIa polymorphisms and risk of coronary stent thrombosis. Lancet 1997;350:1217–1219.

121. Ridker PM, Hennekens CH, Schmitz C, et al. PIA1/A2 polymorphism of platelet glyco-protein IIIa and risks of myocardial infarc-tion, stroke, and venuos thrombosis. Lancet 1997;349:385–388.

122. Nurden AT. Platelet glycoprotein IIIa poly-morphisms and coronary thrombosis. Lancet 1997;350:1189–1190.

123. Moghadasiam MH, McManus BM. Homo-cyst(e)ine and coronary artery disease. Arch Intern Med 1997;157:2299–2308.

124. Hu F, Gu Z, Kozich V, et al. Molecular basis of cystathionine beta-synthase deficiency in pyrodixine responsive and nonresponsive homocystinuria. Hum Mol Genet 1993;2: 1857–1860.

125. McCully K. Homocysteine and vascular dis-ease. Nat Med 1996;2:386–389.

126. Adams RJ, Kutlar A, McKie V, et al. Alpha thalassemia and stroke risk in sickle cell anemia. Am J Hematol 1994;45:279–282.

127. Zimmerman R, Gill F, Goldberg H, et al. MRI of sickle cell cerebral infarction. Neu-roradiology 1987;29:232–235.

128. Ohene-Frempong K. Stroke in sickle cell disease: Demographic, clinical, and thera-peutic considerations. Semin Hematol 1991; 28:213–219.

129. Francis RB. Large-vessel occlusion in sickle cell disease: Pathogenesis, clinical conse-quences, and therapeutic implications. Med Hypotheses 1991;35:88–95.

130. Wadia RS, Sangle SA, Kripalaney S, et al. Fa-milial intracranial haemorrhage due to fac-tor V deficiency. J Neurol Neurosurg Psy-chiatry 1992;55:227–228.

131. Radhakrishnan K, Thacker A, Maloo J, et al. Sickle cell trait and stroke in the young adult. Postgrad Med J 1990;66:1078–1084.

132. Ciafaloni E, Ricci E, Shanske S, et al. MELAS: Clinical features, biochemistry, and molecu-lar genetics. Ann Neurol 1992;31:391–398.

133. Goto Y, Nonaka I, Horai S. A mutation in the tRNALeu(UUR) gene associated with the MELAS subgroup of mitochondrial en-cephalomyopathies. Nature 1991;346:651–653.

134. Forster C, Hubner G, Muller-Hocker J, et al. Mitochondrial angiopathy in a family with MELAS. Neuropediatrics 1992;23:165–168.

135. Hunt SC, Hasstedt SJ, Kuida H, et al. Ge-netic heritability and common environmen-tal components of resting and stressed blood pressures, lipids, and body mass in-dex in Utah pedigrees and twins. Am J Epi-demiol 1989;129:625–638.

136. Newman B, Selby JV, King MC, et al. Concordance for type 2 (non-insulin dependent) diabetes mellitus in male twins. Diabetologia 1987;30:763–768.

137. Pettee AD, Wasserman BA, Adams NL, et al. Familial Sneddon's syndrome: Clinical, hematologic, and radiographic findings in two brothers. Neurology 1991;443(Part 1): 399–405.

138. Brey RL. Antiphospholipid antibodies and ischemic stroke. Heart Dis Stroke 1992;1: 379–382.

139. Hoeg JM. Familial hypercholesterolemia: What the zebra can tell us about the horse. JAMA 1994;271:543–546.

140. Payne CS, Roses AD. The molecular genetic revolution: Its impact on clinical neurology. Arch Neurol 1988;45:1366–1376.

141. Hoeg JM. Can genes prevent atherosclerosis? JAMA 1996;276:989–992.

142. Duggirala R, Villalpando CG, O'Leary DH, et al. Genetic basis of variation in carotid artery wall thickness. Stroke 1996;27:833–837.

143. Yatsu FM, Kasturi R, Alam R. Gene polymorphisms of apolipoprotein AI, the major protein of high density lipoprotein in predicting stroke risk among white and black subjects. Stroke 1993;24(suppl):I26–I30.

144. Schievink WI, Mokri B, Piepgras DG, Kuiper JD. Recurrent spontaneous arterial dissections: Risk in familial versus nonfamilial disease. Stroke 1996;27:622–624.

145. Schievink WI, Mokri B. Familial aorto-cervicocephalic arterial dissections and congenitally bicuspid aortic valve. Stroke 1995; 26:1935–1940.

146. Schievink WI, Michels W, Mokri B, et al. A familial syndrome of arterial dissections with lentiginosis. N Engl J Med 1995;322: 576–579.

147. Schievink WI, Michels W, Piepgras DG. Neurovascular manifestations of heritable connective tissue disorders: A review. Stroke 1994;25:889–903.

148. Healton E. Cerebrovascular fibromuscular dysplasia. In Barnett H, Mohr J, Stein B, Yatsu F, (eds). Stroke: Pathophysiology, Diagnosis, and Management. New York: Churchill Livingstone, 1992, 749–760.

149. Gunel M, Awad IA, Anson J, Lifton RP. Mapping of a gene causing cerebral cavernous malformations to 7q11.2-q21. PNAS 1995;92:6620–6624.

149a. Yokoyama K, Asano Y, Murakawa T, et al. Familial occurrence of arteriovenous malformation of the brain. J Neurosurg 1991;74: 585–589.

150. Brass LM, Lichtman JH, Chen YT, et al. Risk of stroke among elderly patients with heart failure. Neurology 1998;50(suppl 4):A247.

151. Kelly DP, Strauss AW. Inherited cardiomyopathies. N Engl J Med 1994;330:913–919.

152. Moller JH, Allen HD, Clark EB, et al. Report of the task force on children and youth. Circulation 1992;88:2479–2486.

153. Payne RM, Johnson MC, Grant JW, Strauss AW. Toward a molecular understanding of congenital heart disease. Circulation 1995; 91:494–504.

154. Patterson DF, Pexieder T, Schnarr WR, et al. A single major-gene defect underlying cardiac conotruncal malformations interferes with myocardial growth during embryonic development: Studies in the CTD line of keeshond dogs. Am J Hum Genet 1993;52: 388–397.

155. Bowler JV, Hachinski V. Progress in the genetics of cerebrovascular disease: Inherited subcortical arteriopathies. Stroke 1994;25: 1696–1699.

156. Tournier-Lasserve E, Joutel A, Chabriat H, et al. CADASIL: Clinical phenotypes and genetic data in 15 unrelated families. Neurology 1995;45(suppl 4):A273.

157. Tournier-Lasserve E, Joutel A, Melki J, et al. Cerebral autosomal dominant arteriopathy with subcortical infarcts and leukoencephalopathy maps to chromosome 19q12. Nat Genet 1993;3:256–259.

158. Baudrimont M, Dubas F, Joutel A, et al. Autosomal dominant leukoencephalopathy and subcortical ischemic stroke: A clinicopathological study. Stroke 1993;24: 122–125.

159. Desnick R, Ioannou Y, Eng C, et al. Alpha-galactosidase deficiency: Fabry disease. In Scriver C, Beaudet A, Sly W, et al (eds). The Metabolic and Molecular Basis of Inherited Disease. New York: McGraw Hill, 1995, 2741–2784.

160. Mitsias P, Levine SR. Cerebrovascular complications of Fabry's disease. Ann Neurol 1996;40:8–17.

161. Herman B, Schmitz PIM, Leyten ACM. Multivariate logistric analysis of risk factors for stroke in Tiburg, the Netherlands. Am J Epidemiol 1983;118:514–525.

162. Boysen G, Nyboe J, Appleyard M, et al. Stroke incidence and risk factors for stroke in Copenhagen, Denmark. Stroke 1988;19: 1345–1353.

163. Whitfield JB, Martin NG. Plasma lipids in twins: Environmental and genetic influence. Atherosclerosis 1983;48:265–277.

164. Sackett DL, Gent M. Controversy in counting and attributing events in clinical trails. N Engl J Med 1979;301:1410–1412.

165. Rosenbaum PR. Discussing hidden bias in observational studies. Ann Int Med 1991; 115:901–905.

166. Weiss A, Fletcher AE, Palmer AJ, et al. Use

of surrogate respondents in studies of stroke and dementia. J Clin Epidemiol 1996; 49:1187–1194.

167. Bromberg JE, Rinkel GJ, Algra A, et al. Validation of family history in subarachnoid hemorrhage. Stroke 1996;27:630–632.

168. Iso H, Jacobs DR, Goldman L. Accuracy of death certificate diagnosis of intracranial hemorrhage and nonhemorrhagic stroke: The Minnesota Heart Survey. Am J Epidemiol 1990;132:993–998.

169. Corwin LI, Wolf PA, Kannel WB, McNamara AB. Accuracy of death certification of stroke: The Framingham Study. Stroke 1982;13:818–821.

170. Phillips SJ, Whisnant JP. Hypertension and the brain. Arch Int Med 1992;152: 343– 349.

171. Kreiger N, Nishri ED. The effect of nonresponse on estimation of relative risk in a case-control study. Ann Epidemiol 1997;7: 194–199.

172. Sackett DL. Bias in analytic research. J Chronic Dis 1979;32:51–63.

173. Marshall J. Familial incidence of cerebrovascular disease. J Med Gen 1971;8:84–89.

174. Alter M, Kluznik J. Genetics of cerebrovascular accidents. Stroke 1972;3:41–48.

175. Carrieri PB, Orefice G, Maiorino A, et al. Age-related risk factors for ischemic stroke in Italian men. Neuroepidemiology 1994;13: 28–33.

176. Alter M. Genetic factors in cerebrovascular disease. Trans Am Neurol Assoc 1967;92: 205–208.

Chapter 9

Inherited Cardiac Diseases that Cause Stroke

Mark J. Alberts, MD

Introduction

Heart disease is one of the major underlying causes for some cases of stroke. Two broad categories of cardiac disorders that can lead to a stroke include: (1) processes leading to emboli, and (2) disturbances of cardiac pump function, leading to global or focal cerebral hypoperfusion.

Emboli arising from the heart are one of the leading causes of stroke. Such emboli typically occlude the middle cerebral artery or a branch of this artery. They can, however, occlude almost any cerebral vessel, including those in the vertebrobasilar system, aa well as small penetrating arteries in some cases. There are several broad categories of cardiac disease that can lead to intracardiac thrombus formation with subsequent embolization. These are listed in Table 1.

Abnormalities of cardiac pump function, which can occur secondary to many lesions, can cause cerebral hypoperfusion. Such hypoperfusion may produce global cerebral ischemia or focal strokes, particularly in watershed territories. The three common processes leading to cardiac pump failure are cardiac ischemia (myocardial infarction), cardiac arrest, and severe arrhythmias that lead to markedly reduced cardiac output or functional cardiac arrest.

This chapter will review the common and important inherited diseases that are precursors to the two main processes listed above. In some cases these categories overlap. For example, some arrhythmias may lead to cardiac emboli as well as hypoperfusion.

The focus of this discussion will be on primary cardiac diseases. Systemic diseases that cause cardiac dysfunction as well as dysfunction of many other organ systems (i.e., muscular dystrophy, systemic amyloidosis) will not be reviewed here. This organization is based on clinical utility, as it is usually quite apparent if a patient has a widespread systemic disorder, such as muscular dystrophy, or an isolated primary cardiac disorder. On some occasions patients with a systemic disorder present initially with cardiac dysfunction. A careful and thorough evaluation will usually discern among these possibilities. By focusing first on primary cardiac disorders in this chapter, we will eliminate any confusion with the systemic illnesses discussed in *Chapter 14*.

Cardiomyopathies

The inherited cardiomyopathies can be subdivided into different categories based on either etiology or pathophysiology. In

From Alberts MJ (ed). *Genetics of Cerebrovascular Disease.* Armonk, NY: Futura Publishing Company, Inc., © 1999.

Table 1
Broad Categories of Cardiac Diseases that
Lead to Strokes

• Cardiomyopathies
• Arrhythmias
• Valvular disease
• Other lesions (i.e. tumors)

this section, we will consider three major categories of cardiomyopathy based more on their physiologic abnormalities: ischemic damage, disorders of the structural proteins, and defects in energy metabolism.

Ischemic Cardiomyopathy

Coronary artery atherosclerosis resulting in myocardial infarction or gradual cardiac dysfunction is a leading cause of a cardiomyopathy. Since this disease is more properly considered under the category of atherosclerosis, it is discussed in detail in *Chapter 6*. This is not to minimize the importance of genetic factors in coronary artery disease (CAD); in fact, genetic epidemiological studies have shown that inherited factors play a very strong role in producing coronary atherosclerosis. This is particularly true for early onset CAD. First-degree relatives of a proband who dies from CAD or sudden death before the age of 55 have a tenfold increased risk of suffering a similar fate.[1] In many cases there are no obvious risk factors for coronary atherosclerosis in such families, implying that one or more as yet unidentified genetic factors may be of critical importance in the pathogenesis of this disease.

Hypertrophic Cardiomyopathy

Hypertrophic cardiomyopathy (HCM) in most cases is a familial disease (FHCM) with an autosomal dominant pattern of inheritance and incomplete penetrance.[2–4] In other cases it appears to be sporadic, although, in some of those cases, mutations have been found (see below). This is the first inherited primary cardiac disease for which

pathogenic genetic mutations have been identified (see Table 2).

In 1989, genetic linkage of FHCM was found to the long arm of chromosome 14.[5] Subsequent studies have confirmed linkage of some families to that locus.[6] However, several other families have shown linkage to other chromosomal loci including chromosomes 1q3, 15q2, 7q3, and 11q11.[7–9] Other families have not shown linkage to any of these loci, indicating yet further genetic heterogeneity for FHCM.[9]

Two excellent candidate genes were identified on the long arm of chromosome 14, the alpha and beta myosin heavy chain (MHC) genes.[3] The beta MHC gene is 23kb in size, comprises 40 exons and codes for an mRNA of 6kb.[10–12] Subsequent studies demonstrated a missense mutation in exon 13 of the beta MHC gene, which results in the substitution of an adenine for guanine.[13] The mechanism by which this mutation produces FHCM is the subject of intense study. It has been suggested that impairment of contractile function causes myofibrils to undergo compensatory hypertrophy.[3,14] Use of transgenic animal models will help to examine this hypothesis.

Since the description of this missense mutation, many other mutations in the beta MHC gene have been described (see Table 3 and references 3 and 14 for a partial listing). For the most part, these mutations are located in the first 23 exons of the gene which code for the globular head of the myosin protein. A deletion mutation in the 3′ end of the gene has also been described.[15] In general, mutations in the beta MHC gene occur in approximately 30–50% of FHCM cases.[3,13,16,17]

Table 2
Pathogenic Genes for Familial
Hypertrophic Cardiomyopathy

chromosome 1	cardiac troponin T
chromosome 7	unknown
chromosome 11	cardiac myosin binding protein-C
chromosome 14	beta-myosin heavy chain
chromosome 15	alpha-tropomyosin

Table 3
Mutations in the Beta-Myosin Heavy Chain Gene Leading to Familial Hypertrophic Cardiomyopathy*

Exon	Mutation Type	Number of Mutations
3	single base	3
5	single base	3
7	single base	1
8	unknown	1
9	single base	2
13	single base	3
14	single base	1
15	single base	1
16	single base	5
19	single base	2
20	single base	5
21	single base	1
22	single base	1
23	single base	5
40	deletion	?

*This table was adapted from the table in reference 3.

Several studies have attempted to correlate the clinical heterogeneity of FHCM with the genetic heterogeneity of the beta MHC mutations. One study found that mutations that changed the charge of the amino acid resulted in significantly shorter life expectancy compared to mutations that did not change the charge of the amino acid.[17] Another study found that families with the 908 mutation (leucine for valine) had a disease penetrance of only 61%, a later age of onset, and a lower incidence of cardiac events. However, a kindred with the 403 mutation (arginine for glutamine) had 100% penetrance, and a high rate of cardiac symptoms including sudden death.[18] Other studies have found that kindreds with the missense mutation in exon 13 have a higher incidence of severe disease and sudden death while other mutations may be associated with a more benign form of the disease, although there appears to be significant clinical heterogeneity despite similar mutations among some individuals.[14]

Another study found that affected patients who were homozygous for the mutation had more severe disease than affecteds who were heterozygous for the same mutation.[19] One study found that different kindreds with the same mutation may have dif-

ferent clinical symptoms, possibly due to differences in other aspects of their genetic background, although environmental factors may also be important.[20] These differences are likely to be due to other aspects of an individual's genetic or environmental background.

Studies are underway to determine the pathogenic genes for the forms of FHCM that map to other chromosomes. Recent studies have identified mutations in the alpha-tropomyosin gene for cases that map to CH15, and mutations in the cardiac troponin T gene for cases mapping to CH1. Several different mutations in the cardiac troponin T gene have been described in FHCM families. Most are missense mutations, although a three nucleotide deletion has been found, as well as an intronic mutation that causes a truncated cDNA transcript.[21,22] Some patients with mutations in troponin T had minimal clinical symptomatology, yet carried a poor prognosis for sudden death.[22] Two distinct missense mutations in the alpha-tropomyosin gene have been identified as causes of FHCM.[21,22]

Some forms of FHCM are linked to chromosome 11p11.2. Studies have identified splice donor mutations and duplications in the cardiac myosin binding protein

C (MyBP-C) gene as the cause of FHCM in these families.[22a] The MyBP-C gene is quite large, with more than 21,000 base pairs and 35 exons.[22b] Subsequent studies have reported a variety of mutations in the MyBP-C gene in patients with FHCM. Some of these mutations produce truncated proteins.[22b] Still other cases of FHCM appear to be linked to chromosome 7q3, although the causative gene has not been identifed.[22c]

A compilation of FHCM cases shows that approximately 15% are due to mutations in the cardiac troponin T gene, while less than 3% of cases are caused by mutations in the alpha-tropomyosin gene, and 30% to 50% are due to mutations in the β-MHC gene.[22] It is unclear what percentage of the total cases is due to mutations of MyBP-C or are linked to chromosome 7. Another candidate gene, cardiac actin, is located on the long arm of chromosome 15. This gene was excluded as the causative gene in the families studied.[23]

In some cases where HCM appears to be sporadic, de novo mutations have been identified in the beta MHC gene.[24] In such cases the disease can be transmitted to offspring with an autosomal dominant pattern.[24] These findings support the concept that apparent sporadic forms of inherited diseases may in some cases be due to new mutations. This has implications for diagnosis, since molecular genetic approaches may be very useful in these circumstances.

In addition to the mutations in the gene cited above, there have been reports of deletions in mitochondrial DNA (mtDNA) within the cardiomyocytes of patients with hypertrophic cardiomyopathy.[25] Point mutations have also been identified in transfer RNA genes encoded by mtDNA in association with inherited hypertrophic cardiomyopathy.[26,27] These may be causative for the disease in some cases. A recent report documented a heteroplasmic T to C mutation in the mitochondrial tRNA[gly] in a large family with cardiomyopathy. In this family, the degree of mtDNA heteroplasmy tended to correlate with the severity of symptoms.[28] In addition, this mutation was not found in control or unaffected family members, thereby strongly supporting the fact that it is a true mutation, as opposed to a benign polymorphism.

However, mtDNA has been shown to be quite polymorphic. In addition, mtDNA tends to accumulate various mutations and deletions as aging occurs and in association with cardiac damage from various types of injuries.[29,30] Therefore, the definitive proof that these mtDNA mutations are causative for some cases of sporadic or FHCM will require further study. Based on available data, it seems likely that mutations of mtDNA will be found to account for a very small minority of inherited cardiomyopathies.[3]

Dilated Cardiomyopathy

For many years, dilated cardiomyopathy (DCM) was thought to be entirely a sporadic disorder of unknown etiology.[31] However, several studies have demonstrated the familial occurrence of DCM, which has led to further studies showing that approximately 20% of all DCM cases are familial.[32,33] Familial DCM (FDCM) can be inherited with an autosomal dominant pattern (which is the most common), autosomal recessive, X-linked, or mitochondrial inheritance.[32,34–36] Incomplete penetrance of clinical symptomatology (65% in one study) may further complicate the genetic study of this disease.[37] Preliminary results suggest that FDCM is likely to be a heterogeneous disease (much as FHCM is), and there may be several genes involved in its pathogenesis.[3,38] Penetrance of the disease gene appears to be age-dependent, affecting 10% of individuals less than 20 years of age, and 90% of individuals over the age of 40.[39]

Genetic linkage studies are underway to determine the loci responsible for the familial forms of DCM. Currently, there is evidence of genetic linkage for autosomal dominant FDCM to loci on chromosomes 1 (1p1–1q1), 3 (3p25–3p22), 9 (9q13–9q22), and 10 (10q21–10q23).[38a,39a] As with FHCM, various deletions and point mutations in mtDNA have been found in some patients with FDCM.[25,36] It is unclear if these mutations are causative for FDCM or just polymorphisms in mtDNA-DNA. In either case, such mtDNA mutations do not appear to be a common cause for FDCM.[3]

The X-linked form of FDCM has been studied extensively in recent years. In this disease, males have a much earlier age of onset (early 20s), and the female carriers become symptomatic in the fifth decade.[40] Genetic linkage studies of X-linked FDCM showed linkage to the dystrophin locus on the X chromosome.[35] Mutations in the dystrophin gene are responsible for both Duchenne and Becker muscular dystrophy. Initial genetic analysis of the dystrophin gene failed to detect any significant mutations.[35] However, Western blot analysis of the dystrophin protein showed abnormalities in the N-terminal portion of the protein in only cardiac muscle.[35,39] Subsequent studies in one family found a single point mutation at the boundary between exon 1 and the first intron of dystrophin. [39b] This mutation appears to selectively involve cardiac muscle, perhaps due to isoform switching and compensation in skeletal muscle (which is absent in cardiac muscle).[39,41] The result is almost total absence of dystrophin from cardiac muscle.[39b] Subsequent genetic studies have reported the presence of a missense mutations in exon 9 of dystrophin which changes a threonine to an alanine.[39c] This leads to the cardio-specific phenotype of X-linked dilated cardiomyopathy.

These findings offer one possible explanation for organ-specific involvement caused by specific mutations in a gene that is expressed in many different tissues. Some have suggested that X-linked FDCM is another form of Duchenne and Becker muscular dystrophy that is simply limited to cardiac muscle.[3] Further analysis of the very large dystrophin gene may reveal additional abnormalities. No mutations of the dystrophin gene were found in a group of 27 patients with idiopathic (nonfamilial) DCM.[42]

Abnormal Energy Cardiomyopathy

Various defects in the metabolism of fatty acids are a well-recognized cause of inherited cardiomyopathies. Patients with these disorders may have a range of symptomatology, from totally asymptomatic to sudden death. Episodes of heart failure and arrhythmias may also occur. Such disorders are typically inherited as autosomal recessive traits, and can affect a number of enzymes and proteins involved in fatty-acid metabolism, including the carnitine transporter, carnitine palmitoyltransferase, and various dehydrogenases.[34] Various mutations responsible for several of these disorders have been identified. For example, a single base change in the gene for medium-chain acyl-CoA dehydrogenase causes a guanine for adenine substitution and changes an amino acid in the enzyme.[43]

Another class of energy disorders involves defects in mitochondrial oxidative phosphorylation. These diseases may produce skeletal as well as cardiac myopathies, seizures, and stroke-like episodes.[44] A listing of these mitochondrial syndromes is in Table 4. Two major types of mtDNA mutations have been described: point mutations in transfer RNA (tRNA) genes and deletions of mt-DNA.[45] The mitochondrial genome has separate tRNA genes involved in the translation of mitochondrial mRNA. Either of the mutations described above can lead to a cardiomyopathy.[26,36] As with most mito-

---

Table 4
Inherited Mitochondrial Diseases
that Can Cause a Cardiomyopthy*

1. Myoclonic epilepsy and ragged red fibers (MERRF).
2. Mitochondrial myopathy, encephalopathy, lactic acidosis, stroke-like episodes (MELAS).
3. Kearns-Sayre syndrome.
4. Maternally inherited myopathy and cardiomyopathy.
5. Inherited cardiomyopathy with multiple deletions of mitochondrial DNA.

*This table is adapted from reference 34.

chondrial disorders, inheritance is usually maternal, meaning that affected individuals will have an affected mother.

Arrhythmias

Various cardiac arrhythmias can lead to stroke either through producing stasis of blood within the heart chambers (eventually forming clots that may embolize), or by causing global cerebral hypoperfusion which produces either watershed strokes or diffuse cerebral ischemia.

The most common arrhythmia producing focal stroke is atrial fibrillation. Many cardiomyopathies (as already discussed) can produce secondary disturbances in rhythm, leading to the complications cited above. In addition to the primary cardiomyopathies, systemic diseases such as various forms of muscular dystrophy can produce cardiomyopathies leading to secondary arrhythmias. This section will focus on primary cardiac diseases leading to arrhythmias.

The long QT syndrome (LQTS) is a well-described inherited form of severe arrhythmia,[46] and a rare but serious disorder. It leads to episodes of syncope and severe cardiac arrhythmias (typically ventricular fibrillation) causing sudden death.[47] There are two commonly recognized familial forms of LQTS, the Jervell and Lange-Nielsen syndrome (characterized by autosomal recessive inheritance and congenital neural deafness) and the Romano-Ward syndrome (autosomal dominant inheritance and normal hearing).[48–51] So-called sporadic cases of LQTS in small families may actually represent the result of two heterozygous (but asymptomatic) parents with the recessive form of LQTS producing a homozygous, symptomatic offspring.[47]

Genetic linkage studies of several families have mapped one gene for the autosomal dominant form of LQTS to chromosome 11p15.5.[52] This locus is near the Harvey *rass*-1 gene, which is considered a candidate gene for this condition.[53] Further linkage studies using multipoint analysis

have excluded this region for linkage in several additional families, indicating some genetic heterogeneity for this form of LQTS.[52,54] Subsequent studies have cloned a voltage-gated potassium channel gene, KVLQT1, as the cause of chromosome 11-linked LQTS. A variety of mutations in this gene have been identified, including missense mutations and deletions.[54a]

Recent studies have identified several other loci, including 7q35–36, 3p21–24, and 4q25–27 as other linked loci for this disease in some families.[55,55a] Additional loci are likely to exist for this disorder. A report by Curran et al.[56] found several mutations in a potassium channel gene, HERG, on chromosome 7q35–36, in six long QT families. In one family, a mutation appeared to arise spontaneously. The HERG gene is expressed in cardiac muscle.[56] Thus, mutations in the HERG gene appear to be responsible for one type of inherited long QT syndrome. Abnormalities in potassium conductance are a plausible mechanism to explain the arrhythmias that accompany that syndrome. Studies of the form of LQTS linked to chromosome 3 have shown that mutations in the cardiac sodium channel gene SCN5A are responsible for this disorder.[56a]

There are no reports of genetic linkage for the autosomal recessive form of LQTS, although we anticipate that progress on linking this disease will be made in the near future.[46]

Atrial fibrillation/atrial flutter (AF) are much more common arrhythmias than LQTS, yet there are only very rare cases of the familial occurrence of either arrhythmia.[57–59] In one five generation family there were 22 individuals affected by AF, with an inheritance pattern consistent with an autosomal dominant trait.[59] A genetic linkage study of familial AF has been performed using three families. This study found evidence for linkage on chromosome 10, at 10q22-q24. An 11 centimorgan region has been identified that is likely to contain this familial AF gene.[59a] There is one report of atrial flutter in a family.[60] Familial atrial standstill has also been rarely reported and can lead to embolic strokes.[61]

An autosomal dominant form of ventricular tachycardia without underlying structural disease or a long QT syndrome has been reported.[62] Affected individuals in the four generation kindred ranged in age from 17–77 years. The recognition of the importance of genetic factors in at least some cases of idiopathic ventricular tachycardia is important for two reasons: (1) it may further increase understanding of the underlying pathogenesis of this arrhythmia; and (2) it indicates the importance of screening asymptomatic family members for this potentially fatal arrhythmia.

Valvular Disease

Valvular disease involving the left side of the heart is a well-recognized cause of cardioembolic stroke. Although many types of valvular disease are acquired, there is growing recognition that in some cases valvular disease can be inherited.

One of the most common valvular abnormalities that may be associated with an increased risk of stroke is mitral valve prolapse (MVP). Two varieties of MVP can be identified: syndromic and nonsyndromic. Syndromic MVP is typically a component of another inherited disease; examples include Ehlers-Danlos syndrome, Marfan's syndrome, and osteogenesis imperfecta. MVP can also occur as part of many rare inherited or congenital conditions which will not be reviewed here. Nonsyndromic MVP can occur in either a familial or sporadic manner. An extensive study of the inheritance of MVP by Devereux et al.[63] showed that approximately 2/3 of cases had a familial component. The familial cases were inherited as an autosomal dominant trait with incomplete penetrance and variable expression. There was a significant effect of age and sex on MVP prevalence. MVP was present in 8% of children less than 16 years of age but 36% of adults. Evaluation of MVP and sex effects showed that MVP was present in 19% of males and 41% of females. There was evidence that both symptomatic and asymptomatic MVP were inherited in an autosomal dominant pattern. Other reports have shown the inheritance of MVP among both members of monozygotic twins.[64,65]

Familial stroke has been associated with MVP. A report by Rice et al.[66] described a 27 member family with 8 individuals having MVP. Four of the eight individuals had cerebral ischemic events before the age of 40. Another report by Fisher and Budnitz[64] described the occurrence of MVP and ischemic stroke in monozygotic twins. It is clear that not every individual with MVP has a stroke, suggesting that other environmental, medical, or genetic factors may be important.

Several candidate genes have been proposed for study in familial MVP. Two studies reported an association between MVP and antigens HLA-A3 and BW35.[67,68] However, a more extensive study of these antigens in a large kindred with familial stroke and MVP failed to show any clear association.[66]

The mitral valve is composed of collagen types I, III, and V.[69] Two studies in the late 1980s used genetic linkage techniques to study collagen as a candidate gene in familial MVP.[70,71] These studies used restriction fragment length polymorphisms which are usually not as informative as the more recently described dinucleotide repeats (discussed in more detail in *Chapters 2 and 3*). Nevertheless, these studies failed to find genetic linkage between familial MVP and the collagen genes studied. However, in some cases the pedigrees lacked sufficient power to exclude linkage.[70,71] Using more informative polymorphic markers and larger pedigrees, it will be possible to definitively determine whether familial MVP is linked to one of these collagen genes.

Other Lesions

Atrial myxomas are rare but well-recognized cardiac tumors that can lead to embolization and stroke in some patients. There are a handful of cases in the literature of the familial occurrence of these myxomas (which may not necessarily occur within the atria).[72-74] In cases of familial cardiac myxoma, they tend to occur most commonly in siblings, although other inherited patterns have been observed.

A detailed review of 51 patients seen at the Mayo Clinic, Rochester, Minnesota, with cardiac myxoma highlighted some of the differences between nonfamilial and familial forms.[75] In general, the familial form tended to occur at a younger age (mean age 24 years), affect males more commonly than females (67% males in the familial form; 24% males in the nonfamilial form), and be multicentric in 1/3 of familial cases versus only 6% in the nonfamilial form. Other abnormalities such as peripheral myxomas, various dysplasia and other tumors, and various skin lesions tended to occur more commonly in the familial versus nonfamilial form. However, there were no significant histological differences between the familial and nonfamilial

cases for the cardiac myxomas. The author suggested that in cases of proven familial cardiac myxoma, if the myxoma has features suggestive of a familial case (as described above), then first-degree relatives (especially siblings) should be studied to determine if they also harbor a cardiac myxoma.[75] Transthoracic echocardiography is probably sufficient for this type of work-up.

Conclusion

Cardiac diseases of all types are a major cause of stroke. Progress on unraveling the underlying molecular basis for several inherited cardiac diseases, particularly the hypertrophic cardiomyopathies and long QT syndrome, is proceeding at a rapid pace. Gene therapy for these disorders is on the horizon. Further studies are underway to discern the genetic basis for a host of other common cardiac disorders that lead to stroke.

References

1. Nora J, Lortscher R, Spangler R, et al. Genetic-epidemiological study of early-onset ischemic heart disease. Circulation 1980;61:503–508.
2. Maron B, Nichols PI, Pickle L, et al. Patterns of inheritance in hypertrophic cardiomyopathy: Assessment by M-mode and two-dimensional echocardiography. Am J Cardiol 1984;53:1087–1094.
3. Marian AJ, Roberts R. Molecular basis of hypertrophic and dilated cardiomyopathy. Tex Heart Inst J 1994;21:6–15.
4. Greaves SC, Roche AH, Neutze JM, et al. Inheritance of hypertrophic cardiomyopathy: A cross sectional and M mode echocardiographic study of 50 families. Br Heart J 1987;58:259–266.
5. Jarcho JA, McKenna W, Pare JA, et al. Mapping a gene for familial hypertrophic cardiomyopathy to chromosome 14q1. N Engl J Med 1989;321:1372–1378.
6. Hejtmancik JF, Brink PA, Towbin J, et al. Localization of gene for familial hypertrophic cardiomyopathy to chromosome 14q1 in a diverse US population. Circulation 1991;83:1592–1597.
7. Watkins H, MacRae C, Thierfelder L, et al. A disease locus for familial hypertrophic cardiomyopathy maps to chromosome 1q3. Nat Genet 1993;3:333–337.
8. Thierfelder L, MacRae C, Watkins H, et al. A familial hypertrophic cardiomyopathy locus

maps to chromosome 15q2. Proc Natl Acad Sci USA 1993;90:6270–6274.
9. Carrier L, Hengstenberg C, Beckmann JS, et al. Mapping of a novel gene for familial hypertrophic cardiomyopathy to chromosome 11. Nat Genet 1993;4:311–313.
10. Jaenicke T, Diederich K, Haas W, et al. The complete sequence of the human beta-myosin heavy chain gene and a comparative analysis of its product. Genomics. 1990;8:194–206.
11. Liew C, Sole M, Yamauchi-Takihara K, et al. Complete sequence and organization of the human cardiac beta-myosin heavy chain gene. Nucleic Acids Res 1990;18:3647–3651.
12. Matsuoka R, Yoshida M, Kanda N, et al. Human cardiac myosin heavy chain gene mapped within chromosome region [14q11.2-q13]. Am J Med Genet 1989;32:279–284.
13. Gelsterfer-Lowrance A, Kass S, Tanigawa G, et al. A molecular basis for familial hypertrophic cardiomyopathy: A beta cardiac myosin heavy chain missense mutation. Cell 1990;62:999–1006.
14. Marian AJ, Roberts R. Molecular genetics of cardiomyopathies. Herz 1993;18:230–237.
15. Marian AJ, Yu QT, Mares AJ, et al. Detection of a new mutation in the beta-myosin heavy chain gene in an individual with hypertrophic cardiomyopathy. J Clin Invest 1992;90:2156–2165.
16. Solomon SD, Wolff S, Watkins H, et al. Left

ventricular hypertrophy and morphology in familial hypertrophic cardiomyopathy associated with mutations of the beta-myosin heavy chain gene. J Am Coll Cardiol 1993;22:498–505.

17. Watkins H, Rosenzweig A, Hwang DS, et al. Characteristics and prognostic implications of myosin missense mutations in familial hypertrophic cardiomyopathy. N Engl J Med 1992;326:1108–1114.

18. Epstein ND, Cohn GM, Cyran F, et al. Differences in clinical expression of hypertrophic cardiomyopathy associated with two distinct mutations in the beta-myosin heavy chain gene. A [908]Leuˇ1AVal mutation and a [403]Argˇ1AGln mutation. Circulation 1992;86:345–352.

19. Nishi H, Kimura A, Harada H, et al. Possible gene dose effect of a mutant cardiac beta-myosin heavy chain gene on the clinical expression of familial hypertrophic cardiomyopathy. Biochem Biophys Res Commun 1994;200:549–556.

20. Fananapazir L, Epstein ND. Genotype-phenotype correlations in hypertrophic cardiomyopathy: Insights provided by comparisons of kindreds with distinct and identical beta-myosin heavy chain gene mutations. Circulation 1994;89:22–32.

21. Thierfelder L, Watkins H, MacRae C, et al. Alpha-tropomyosin and cardiac troponin T mutations cause familial hypertrophic cardiomyopathy: A disease of the sarcomere. Cell 1994;77:701–712.

22. Watkins H, McKenna W, Thierfelder L, et al. Mutations in the genes for cardiac troponin T and alpha-tropomyosin in hypertrophic cardiomyopathy. N Engl J Med 1995;332:1058–1064.

22a. Watkins H, Conner D, Thierfelder L, et al. Mutations in the cardiac myosin binding protein-C gene on chromosome 11 causes familial hypertrophic cardiomyopathy. Nat Genet 1995;11:434–437.

22b. Carrier L, Bonne G, Bahrend E, et al. Organization and sequence of human cardiac myosin binding protein C gene (MYBPC3) and identification of mutations predicted to produce truncated proteins in familial hypertrophic cardiomyopathy. Circ Res 1997;80:427–434.

22c. MacRae CA, Ghaisas N, Kass S, et al. Familial hypertrophic cardiomyopathy with Wolff-Parkinson-White syndrome maps to a locus on chromosome 7q3. J Clin Invest;1996:1216–1220.

23. Schwartz K, Beckmann J, Dufour C, et al. Exclusion of cardiac myosin heavy chain and actin gene involvement in hypertrophic cardiomyopathy of several French families. Circ Res 1992;71:3–8.

24. Watkins H, Thierfelder L, Hwang DS, et al. Sporadic hypertrophic cardiomyopathy due to de novo myosin mutations. J Clin Invest 1992;90:1666–1671.

25. Ozawa T, Tanaka M, Sugiyama S, et al. Multiple mitochondrial DNA deletions exist in cardiomyocytes of patients with hypertrophic or dilated cardiomyopathy. Biochem Biophys Res Commun 1990;170:830–836.

26. Taniike M, Fukushima H, Yanagihara I, et al. Mitochondrial tRNA[ILE] mutation in fatal cardiomyopathy. Biochem Biophys Res Comm 1992;186:47–53.

27. Yoon KL, Ernst SG, Rasmussen C, et al. Mitochondrial disorder associated with newborn cardiopulmonary arrest. Pediatr Res 1993;33:433–440.

28. Merante F, Tein I, Benson L, et al. Maternally inherited hypertrophic cardiomyopathy due to a novel T-to-C transition at nucleotide 9997 in the mitochondrial tRNA(glycine) gene. Am J Hum Genet 1994;55:437–446.

29. Hattori K, Tanaka M, Sugiyama S, et al. Age-dependent increase in deleted mitochondrial DNA in the human heart: Possible contributory factor to presbycardia. Am Heart J 1991;121:1735–1742.

30. Corral-Debrinski M, Stepien G, Shoffner J, et al. Hypoxemia is associated with mitochondrial DNA damage and gene induction. JAMA 1991;266:1812–1816.

31. Dec G, Fuster V. Idiopathic dilated cardiomyopathy. N Engl J Med 1994;331:1564–1575.

32. Michels V, Driscoll D, Miller FJ. Familial aggregation of idiopathic dilated cardiomyopathy. Am J Cardiol 1985;55:1232–1233.

33. Michels VV, Moll PP, Miller FA, et al. The frequency of familial dilated cardiomyopathy in a series of patients with idiopathic dilated cardiomyopathy. N Engl J Med 1992;326:77–82.

34. Kelly D, Strauss A. Inherited cardiomyopathies. N Engl J Med 1994;330:913–919.

35. Towbin JA, Hejtmancik JF, Brink P, et al. X-linked dilated cardiomyopathy: Molecular genetic evidence of linkage to the Duchenne muscular dystrophy (dystrophin) gene at the Xp21 locus. Circulation 1993;87:1854–1865.

36. Suomalainen A, Paetau A, Leinonen H, et al. Inherited idiopathic dilated cardiomyopathy with multiple deletions of mitochondrial DNA. Lancet 1992;340:1319–1320.

37. Keeling PJ, McKenna WJ. Clinical genetics of dilated cardiomyopathy [Review]. Herz 1994;19:91–96.

38. Zachara E, Caforio AL, Carboni GP, et al. Familial aggregation of idiopathic dilated cardiomyopathy: Clinical features and pedigree analysis in 14 families. Br Heart J 1993;69:129–135.

38a. Bowles KR, Gajarski R, Porter P, et al. Gene mapping of familial autosomal dominant dilated cardiomyopathy to chromosome 10q21–23. J Clin Invest 1996;98:1355–1360.

39. Mestroni L, Krajinovic M, Severini GM, et al. Molecular genetics of dilated cardiomyopathy. Herz 1994;19:97–104.

39a. Mestroni L, Milasin J, Vatta M, et al. Genetic factors in dilated cardiomyopathy. Arch Mal Coeur Vaiss 1996;89:15–20.

39b. Milasin J, Muntoni F, Severini GM, et al. A point mutation in the 5′ splice site of the dystrophin gene first intron responsible for X-linked dilated cardiomyopathy. Hum Mol Genet 1996;5:73–79.

39c. Ortiz-Lopez R, Li H, Su J, et al. Evidence for a dystrophin missense mutation as a cause of X-linked dilated cardiomyopathy. Circulation 1997;95:2434–2440.

40. Berko BA, Swift M. X-linked dilated cardiomyopathy. N Engl J Med 1987;316: 1186–1191.

41. Muntoni F, Can M, Ganau R, et al. Brief report: Deletion of the dystrophin muscle-promoter region associated with X-linked dilated cardiomyopathy. N Engl J Med 1993; 329:921–925.

42. Michels VV, Pastores GM, Moll PP, et al. Dystrophin analysis in idiopathic dilated cardiomyopathy. J Med Genet 1993;30:955–957.

43. Tanaka K, Yokota I, Coates P, et al. Mutations in the medium chain acyl-CoA dehydrogenase (MCAD) gene. Hum Mutat 1992; 1:271–279.

44. Wallace D. Diseases of the mitochondrial DNA. Annual Rev Biochem 1992;61:1175–1212.

45. Moraes C, Schon E, DiMauro S. Mitochondrial diseases: Toward a rational classification. In Appel S (ed). Current Neurology. St. Louis: Mosby Year Book, 1991, 83–119.

46. Keating M. Genetics of the long QT syndrome. J Cardiovasc Electrophysiol 1994; 5:146–153.

47. Guntheroth W, Motulsky A. Inherited primary disorders of cardiac rhythm and conduction. Prog Med Genet 1983;5:381–402.

48. Fraser G, Froggatt P, James T. Congenital deafness associated with electrocardiographic abnormalities, fainting attacks, and sudden death. Quart J Med 1964;131:361–384.

49. Fraser G, Froggatt P, Murphy T. Genetical aspects of the cardioauditory syndrome of Jervell and Lange-Nielsen (congenital deafness and electrocardiographic abnormalities). Ann Hum Genet 1964;28:133–139.

50. Romano C, Gemme G, Pongiglione R. Aritmie cardiache rare dell'eta pediatrica. Clin Pediatr 1963;45:656–683.

51. Ward O. A new familial cardiac syndrome in children. J Irish Med Assoc 1964;54:103–107.

52. Curran M, Atkinson D, Timothy K, et al. Locus heterogeneity of autosomal dominant long QT syndrome. J Clin Invest 1993;92: 799–803.

53. Keating M, Atkinson D, Dunn C, et al. Linkage of a cardiac arrhythmia, the long QT syndrome, and the Harvey ras-1 gene. Science 1991;252:704–706.

54. Dean JC, Cross S, Jennings K. Evidence of genetic and phenotypic heterogeneity in the Romano-Ward syndrome. J Med Genet 1993;30:947–950.

54a. Wang Q, Curran ME, Splawski I, et al. Positional cloning of a novel potassium channel gene: KVLQT1 mutations cause cardiac arrhythmias. Nat Genet 1996;12:17–23.

55. Jiang C, Atkinson D, Towbin J, et al. Two long QT syndrome loci map to chromosome 3 and 7 with evidence for further heterogeneity. Nat Genet 1994;8:141–147.

55a. Schott J, et al. Mapping of a gene for long QT syndrome to chromosome 4q25–27. Am J Hum Genet 1995;57:1114–1122.

56. Curran M, Splawski I, Timothy K, et al. A molecular basis for cardiac arrhythmias: HERG mutations cause long QT syndrome. Cell 1995;80:795–803.

56a. Wang Q, et al. SCN5A mutations associated with an inherited cardiac arrhythmia, long QT syndrome. Cell 1995;80:805–811.

57. Beyer F, Paul T, Luhmer I, et al. Familial idiopathic atrial fibrillation with bradyarrhythmia. Zeitschrift fur Kardiologie 1993;82:674–677.

58. Phair W. Familial atrial fibrillation. Can Med Assoc J 1963;89:1274–1276.

59. Gould W. Auricular fibrillation: Report on a study of a familial tendency, 1920–1956. Arch Intern Med 1957;100:916–926.

59a. Brugada R, Tapscott T, Czernuszewicz GZ, et al. Identification of a genetic locus for familial atrial fibrillation. N Engl J Med 1997;336:905–911.

60. Gillor A, Korsch E. Familial manifestation of idiopathic atrial flutter. Monatsschrift Kinderheilhunde 1992;140:47–50.

61. Shah MK, Subramanyan R, Tharakan J, et al. Familial total atrial standstill. Am Heart J 1992;123:1379–1382.

62. Rubin D, O'Keeke B, Kay R, et al. Autosomal dominant inherited ventricular tachycardia. Am Heart J 1992;123:1082–1084.

63. Devereux RB, Brown WT, Kramer FR, et al. Inheritance of mitral valve prolapse: Effect of age and sex on gene expression. Ann Intern Med 1982;97:826–832.

64. Fisher M, Budnitz E. Focal cerebral ischemia and mitral valve prolapse in monozygotic twins. Arch Intern Med 1983;143:2180–2181.

65. Jerasty R. Mitral Valve Prolapse. New York: Raven Press, 1979.

66. Rice GP, Boughner DR, Stiller C, et al. Familial stroke syndrome associated with mitral valve prolapse. Ann Neurol 1980;7: 130–134.

67. Braun W, Roman J, Schacter B. HLA antigens in mitral valve prolapse. Transplant Proc 1977;9:1869–1871.

68. Kachro R, Telischi M, Cruz J, et al. HLA and ABO blood groups in blacks with mitral valve prolapse. N Engl J Med 1978;299:1467.

69. Cole W, Chan D, Hickey A, et al. Collagen composition of normal and myomatous human mitral valves. Biochemistry 1984;214: 451–460.

70. Henney AM, Tsipouras P, Schwartz RC, et al. Genetic evidence that mutations in the COL1A1, COL1A2, COL3A1, or COL5A2 collagen genes are not responsible for mitral valve prolapse. Br Heart J 1989;61:292–299.

71. Wordsworth P, Ogilvie D, Akhras F, et al. Genetic segregation analysis of familial mitral valve prolapse shows no linkage to fibrillar collagen genes. Br Heart J 1989;61:300–306.

72. Siltanen P, Tuuteri L, Norio R, et al. Atrial myxoma in a family. Am J Cardiol 1976;38: 252–256.

73. Liebler G, Magovern G, Park S, et al. Familial myxomas in four siblings. J Thorac Cardiovasc Surg 1976;71:605–608.

74. Kleid J, Klugman J, Haas J, et al. Familial atrial myxoma. Am J Cardiol 1973;32:361–364.

75. Carney J. Differences between nonfamilial and familial cardiac myxoma. Am J Surg Path 1985;9:53–55.

Chapter 10

Vasculopathies

Hugues Chabriat, MD, Elisabeth Tournier-Lasserve, MD,
Marie-Germaine Bousser, MD

CADASIL

The acronym CADASIL, for "cerebral autosomal dominant arteriopathy with subcortical infarcts and leukoencephalopathy," has recently been suggested for a hereditary disease of the small arteries of the brain characterized by recurrent strokes and leading progressively to dementia with pseudobulbar palsy.[1,2] CADASIL was possibly first described by Van Bogaert in 1955 as "Binswanger's disease with a rapid course in two sisters".[3] In 1977, the condition was reported by Sourander and Walinder under the term of "hereditary multi-infarct dementia".[4] The same year, Stevens et al. described similar findings in a Scottish family.[5] Up to 1992, seven other families were reported.[6–12] One of these was a very large family from Loire Atlantique in France.[9] In 1993, a genetic analysis by a sequential approach was performed on this large pedigree. The results showed that the CADASIL gene was located to chromosome 19.[1] This was confimed in the pedigree reported by Mas et al.[1,11] Today, however, CADASIL prbobaly remains largely underdiagnosed.

Main Clinical Features

The families affected by CADASIL have been initially identified on the basis of the following criteria: (1) unexplained history of recurrent subcortical ischemic strokes; (2) extensive white-matter signal abnormalities at MRI examination; and (3) autosomal dominant pattern of inheritance. The main clinical parameters of the disease have been established on families with confirmed linkage to the CADASIL locus. CADASIL is a disease of midadulthood. Age at onset of symptoms is between 30 and 50 years.[2,13] Four types of symptoms have been reported: strokes and/or transient ischemic attacks (TIAs), attacks of migraine with aura, psychiatric disturbances and dementia.[13] Eighty percent of CADASIL patients present with ischemic symptoms of various temporal profile: TIAs, reversible ischemic neurologic deficits (RINDs), and completed strokes. Most often, these manifestations occur in the absence of vascular risk factors, particularly in the absence of hypertension. They are always of the subcortical type and mainly consist of lacunar syndromes such as pure motor, pure sensory and sensorimotor strokes, ataxic hemiparesis, and dysarthria. Visual field defects

From Alberts MJ (ed). *Genetics of Cerebrovascular Disease.* Armonk, NY: Futura Publishing Company, Inc., © 1999.

and signs of brainstem or cerebellar invovement are rare. Ischemic manifestations in CADASIL are usually recurrent, leading to subcortical dementia 10 to 20 years later. In contrast to the very high frequency of ischemic stroke, intracerebral hemorrhage is rare, and has been reported in only one case so far.[14] Another frequent symptom of the disease is recurrent headache that mimics attacks of migraine with aura. In a series of five families, 30% of the patients suffered attacks of migraine with aura, satisfying the corresponding diagnostic criteria of the International Headache Society (IHS)[15] except that these patients had signal abnormalities at MRI.[13] Some patients suffered severe attacks with confusion, stupor, and fever. The frequency of migraine symptoms varies from family to family: one third of the symptomatic members of the family reported by Tournier-Lasserve et al.,[9] 90% of those belonging to the family reported by Verin et al.[12] Mood disturbances are the third main manifestation of the disease and can even be inaugural. Their frequency again varies from family to family. Some CADASIL patients suffer severe depression of melancholic type sometimes alternating with manic episodes (DSMIII criteria).[16] The fourth leading manifestation is dementia which occurs progressively at the end stage of the disease, and is present before death in more than 90% of cases. Dementia is observed in 20–30% of the overall symptomatic subjects, and is a "subcortical dementia" with predominating frontal-like symptoms and memory impairment. Pseudobulbar palsy and gait difficulties are frequently associated. Most often, cognitive impairment occurs in a stepwise fashion after recurrent strokes but, in few subjects, it appears progressively in the absence of ischemic events.[2] One important point is that no spinal cord, peripheral nerve, autonomic system or muscle involvement has been clinically observed in CADASIL.

MRI is always abnormal in symptomatic subjects.[1,2,13] It shows small areas of hypointensity on T1- and hyperintensity on T2-weighted images, in the subcortical white-matter and basal ganglia, typical of small infarcts (Figure 1). These abnormalities are associated on T2-weighted images with more or less extensive and confluent areas of T2-hypersignals of symetrical distribution in the subcortical white-matter. This leukoencephalopathy most often spares the U fibers,[2] and the cerebral and cerebellar cortex are usually spared. Infratentorial lesions are rare. Computed tomography of the brain can also show images of small deep infarcts and leukoaraiosis but it is less sensitive than MRI. One crucial finding is that the white-matter MRI signal abnormalities are observed not only in symptomatic subjects, but also in asymptomatic subjects belonging to affected families (Figure 2). In the large pedigree described by Tournier-Lasserve et al., these signal abnormalities were present in 17 out of 45 subjects studied, among whom 9 were symptomatic and 8 were asymptomatic.[9] This was essential to define the status of subjects (affected or nonaffected) for genetic linkage studies. Ultrasound investigations, cerebral angiography, echocardiography, electrocardiography, and extensive coagulation studies failed to disclose any potential cause of cerebral infarcts.[2] In the family reported by Tournier-Lasserve et al., two symptomatic subjects had IgG monoclonal gammapathy.[9] A discrete lipidosis was observed on muscular biopsy in four-fifths of the members of this family, but was not observed in other families.

Pathological data in the only subject so far studied belonging to a family linked to chromosome 19 showed multiple small deep infarcts with underlying vasculopathy[14] very similar to that described by Sourander et al.,[4] and more recently by Davous et al.[6] and Guttierez et al.[17] The leptomeningeal and medullary arteries of the brain that have diameters between 100 to 400 μm are involved. The media was thickened by an eosinophilic, granular, and electron-dense material, the identity of which remains unknown. The same lesions are observed in the small arteries of the skin and muscle[18] and occasionally in other organs.

In summary, CADASIL is a disease of midadulthood most often leading to TIAs

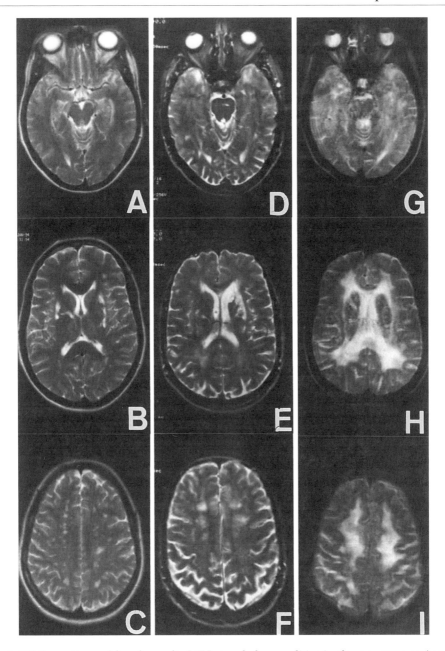

Figure 1. White-matter and basal ganglia MRI signal abnormalities in three symptomatic patients with CADASIL (T2-weighted images). **A, B, and C:** A female patient with a history of TIAs and recurrent episodes mimicking attacks of migraine with aura. **D, E, and F:** A male patient with severe mood disorders and behavior disturbances. **G, H, and I:** A female patient who had a completed stroke and a progressive subcortical dementia with pseudobulbar palsy.

and/or ischemic strokes. Attacks of migraine and mood disorders are other frequent manifestations of the disease. There is a progressive or stepwise cognitive deterioration associated with the recurrence of strokes leading to subcortical dementia and pseudobulbar palsy. Age at death is between 50 and 70 years. All symptomatic patients have MRI

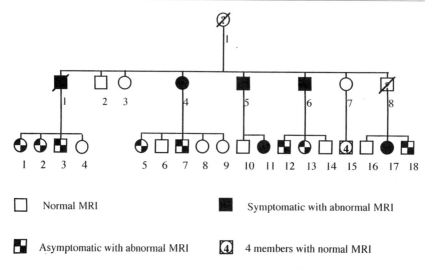

Figure 2. Genealogical tree of one pedigree affected by CADASIL.

signal abnormalities typical of small deep infarcts located in the white-matter and basal ganglia associated with a more or less diffuse leukoencephalopathy. These signal abnormalities are also observed in some asymptomatic subjects born of an affected subject.

Genetics

CADASIL follows an autosomal dominant pattern of inheritance. After 30 years of age, about 50% of the subjects born of an affected parent have typical signal abnormalities at MRI examination. The majority (65%) of these subjects already have clinical signs although some of them remain asymptomatic. After 60 years of age, all subjects with signal abnormalities, but one, are symptomatic. Based on these results, the penetrance of the disease appears complete. The mapping of CADASIL was conducted on the first large pedigree extensively studied.[1,9,14] All consenting adults belonging to this family participated in the study. The disease status was established on the basis of the MRI data. All symptomatic and asymptomatic subjects with prominent and typical MRI abnormalities born of an affected subject were considered as "affected" for the analysis. Asymptomatic offspring of an affected individual

having a normal MRI were considered as "healthy" only when they were older than 35. Below 35 years of age, they were considered as having an unknown status. The affected gene was mapped on chromosome 19 and linkage was confirmed in a second pedigree.[1,19] Genetic analysis of 10 additional families demonstrated the genetic homogeneity of this condition (Tournier-Lasserve, personal communication). These results confirm the value of the clinical and MRI criteria used for diagnosis.

Based on the observation of the high frequency of migraine in CADASIL patients, Joutel et al. raised the hypothesis that another autosomal dominant neurologic condition, familial hemiplegic migraine (FHM), might be due to the alteration of the same gene. They demonstrated by genetic linkage analysis that a gene responsible for FHM was located within the same region on chromosome 19.[20] Clinically, CADASIL and FHM differ widely: FHM starts earlier, is not associated with stroke, depression, dementia, or signal abnormalities at MRI, and neurologic examination is normal between attacks.[21] The identification of the CADASIL gene will provide new insights into the exact relationship between these two disorders. Whatever the results, the locus assignment of the CADASIL gene provides a new tool to investigate familial or sporadic cases of subcortical infarcts of undetermined etiology, as well as cases of migraine with aura[22–25] or bipolar mood disorders[26] associated with white-matter signal abnormalities at MRI.

The natural history of the disease has become more precise after gene mapping. CADASIL starts between 20 and 30 years in one-fifth of the patients with attacks of migraine with aura. Ischemic manifestations observed in four-fifths of patients mainly occur during the fourth and fifth decades and are sometimes associated with severe mood disturbances. Dementia presents between 50 and 60 years and is found almost always before death occurring at a mean age of 65 years.[26a] The genetic analysis of additional families allowed us to reduce the interval on chromosome 19 from 12 to 2 centimorgans between markers D19S226 and D19S199.[26b] Within this interval, Joutel et al. recently found mutations of the Notch 3 gene to be responsible for CADASIL.[26c] At the same time, Ophoff et al. also found mutations in the Ca2+ channel gene CACNL1A4 on chromosome 19 causing familial hemiplegic migraine.[26d] Finally, these major findings support that CADASIL and familial hemiplegic migraine are not allelic disorders.

Familial Arteriosclerotic Leukoencephalopathy with Alopecia and Skeletal Disorders

Fukutake recently summarized the clinical, radiological and pathological features of a rare familial small artery disease of the brain observed in 5 sibships of 12 Japanese patients,[27–29] the parents of whom were nonaffected and always first cousins.

Main Clinical Features

In the 12 subjects (9 males, 3 females), age of onset was between 20 and 38 years. In some cases, duration of the disease was less than 5 years, in others, the follow-up reached 23 years. Fifty percent of these subjects presented with ischemic strokes (most often with one-sided motor deficits) in the absence of vascular risk factors, particularly of hypertension. All patients progressively developed a subcortical dementia with pseudobulbar palsy associated with diffuse pyramidal signs. Half of the patients had brainstem signs such as vestibular symptoms or ophtalmoplegia. Two important associated signs were diffuse alopecia reported in 9 of 12 cases and lumbago with lumbar disc herniation in 7 of 12. Miscellaneous dermatological signs (keratosis, ulcer, xeroderma, pigmentatory macula) or skeletal system disorders (deformity of elbows, megalocephaly and high-arched palates) were observed inconsistently.

CT scans and MRI showed density or signal abnormalities typical of small infarcts in basal ganglia, thalamus and white-matter associated with a diffuse leukoencephalopathy sparing the subcortical U fibers and corpus callosum. Cerebral angiography revealed, in some cases, a narrowing of the small peripheral arteries. Autopsy performed in 2 of 12 cases[28] disclosed arteriosclerosis of the small penetrating arteries (100 and 400μ) in the white-matter, basal ganglia, pons, medulla, and spinal cord. No vessel abnormality was found in other organs. Fibrous intimal proliferation and marked splitting of the internal elastic membrane of these arteries were observed. Electron microscopy confirmed the presence of collagenous fibers in the subintimal areas without endothelial cells, media, and adventitia abnormalities.[28]

Genetics

All of these familial cases had parents who were cousins. Nine other isolated cases with very similar clinical and pathological presentation were reported in Japan; four had parents who were first cousins.[27] The frequency of consanguinity in these families suggests that this very rare Japanese condition is transmitted with an autosomal recessive pattern of inheritance. The responsible gene remains undetermined.

Fibromuscular Dysplasia

Fibromuscular dysplasia is a segmental and systemic angiopathy of unknown cause mainly involving the renal and carotid arteries. In a large consecutive autopsy study, the frequency of FMD of the renal arteries was about 1%.[29a] In patients who underwent cephalic angiograms, the frequency of FMD was estimated to be between 0.25 to 1%.[30–32] It is thus a frequent arterial disease but a very rare cause of stroke. Very few cases of familial FMD have been reported so far.

Main Clinical Features

The definitive diagnosis of fibromuscular dysplasia (FMD) is histologic. It is a nonatherosclerotic, noninflammatory disease of the intermediate-sized arteries characterized by fibrous tissue proliferation or smooth muscle hyperplasia leading to elastic fiber destruction within the arterial wall.[30–34] Three histologic types have been observed depending on the arterial layer where the lesions predominate. Medial FMD is the most frequent (90–95% of cases). Intimal fibroplasia or adventitial fibroplasia is observed in less than 5% of cases. FMD is a multifocal disease mainly affecting the carotid and renal arteries and, less frequently, the vertebral, visceral, iliac, femoral, axillary, subclavian, internal mammary arteries or the aorta.[34] Most often, the diagnosis of FMD is based both on the typical angiographic appearance ("string of beads" or tubular aspect; see Figure 3), and the topography and distribution of the arterial lesions.[30] In a series of 92 patients with FMD, approximately three-fourths had renovascular disease, one fourth, cerebrovascular disease and one fourth, multivessel involvement.[34] Healton estimated that cerebrovascular and renal FMD might coexist in 50% of patients when complete angiography is performed.[33] When the disease is located at the cephalic

level, internal carotid arteries are concerned in 95% of cases and most often bilaterally (60–85%). Vertebral arteries are involved in 12% to 43% of cases and usually in association with the carotid lesions.[33] The carotid and vertebral abnormalities occur at the portion of the arteries opposite to the second cervical vertebra. External carotid artery lesions have been reported. The intracranial portion of the arteries is rarely affected by stenotic lesions. By contrast, intracranial aneurysms of the usual arterial distribution are reported in 20% to 50% of cases of cephalic FMD.[33,35–37]

FMD is often an incidental angiographic finding and usually remains asymptomatic.[32] Clinical manifestations are protean and of various pathophysiologies: stenoses or dissection of the affected arteries, compression of surrounding structures, or rupture of associated aneurysms.[33] FMD of the renal artery can lead to renovascular hypertension and very rarely to renal arteriovenous fistula or to renal infarction. FMD of the mesenteric or coeliac arteries can lead to an occlusive intestinal arterial disease. Ischemic symptoms of one arm can occur when the subclavian and/or the axillary arteries are affected. At the brain level, TIAs and/or srokes have been reported with internal carotid artery and vertebral FMD, but a causal relationship between the ischemic manifestations and the arterial lesions is certain in only a few cases (severe stenosis, occlusion, dissection).[33] A thrombus formation leading to cerebral emboli has also been observed with septal lesions.[37] In the series of cerebrovascular FMD of the literature, 0% to 28% of patients had cerebral infarction and 7% to 68% had TIAs.[33] Other manifestations of the disease are secondary to direct effects of the arterial wall lesions: cervical bruits (0–23%), anterior cervical pain, or Horner's syndrome (0–9%). Dissection of the wall is possible. Recently, Biousse et al. found 21% of cephalic FMD in 65 patients with internal carotid artery dissections.[38] Other manifestations are due to the rupture of the arterial wall, which can lead to fistula or subarachnoid hemorrhage or to compres-

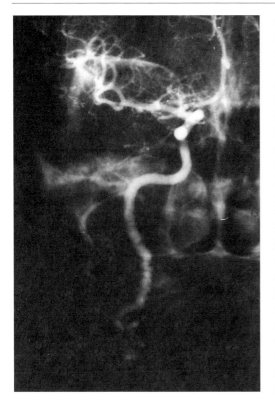

Figure 3. Angiography showing a typical "string of beads" appearance in one patient with FMD of the internal carotid artery.

sion sometimes secondary to aneurysms (cavernous sinus syndrome).[33,34,39]

Genetics

The first familial cases of FMD were reported in 1965 by Halpern et al. and by Hansen et al.[40,41] as three and two hypertensive sisters, respectively, with renal artery abnormalities. Other cases of FMD concerning only the renal arteries were later reported among siblings.[42–44] Based on these findings, Rushton, in 1980, interviewed 21 patients with FMD, 5 of them with cephalic arterial disease diagnosed on the basis of angiography.[45] He found that 12 had relatives of less than 50 years of age with symptoms that suggested an underlying arterial disease, and possibly FMD (hypertension, claudication, myocardial infarction, and/or

stroke[s]). Only one of the 12 patients had cephalic FMD. This patient had a left carotid and right middle cerebral artery stenosis. There was a family history of hypertension associated with claudication in his mother, his mother's sister and brother, and his maternal grandmother. Another interesting finding was the presence of a documented cerebral aneurysm in two subjects (one cousin and her father) belonging to the same branch of the family. Ouchi et al. recently reported cerebral aneurysms in two siblings with renovascular hypertension due to FMD.[46] In 1982, Mettinger and Ericson interviewed 37 patients with cephalic FMD and their relatives and found that one-third of the first-degree relatives (parents, siblings or children) had a history suggestive of an underlying arterial disease[47]: 16 had strokes, 10 were hypertensive, 2 had both stroke and hypertension, and 5 suffered myocardial infarction but none had claudication. Since the mean age at stroke occurence (53 years) and the sex ratio (15 females/3 males) were close to those observed in FMD, the authors suggested that strokes in these relatives could be secondary to FMD. In 1982, Petit et al. described a 32-year-old woman who had a transient episode of left hemiparesis with headache and who presented with a right carotid artery dissection associated with a typical aspect of "string of beads" at the middle third of the left internal carotid artery and a moderate narrowing of the left renal artery at angiography.[48] Her father, at age 33, had had a right sensorimotor deficit with anarthria associated with a left Horner's syndrome secondary to the occlusion of the left carotid artery with a "radish tail" aspect suggestive of arterial dissection and signs of FMD at angiography. These cases are the only documented familial cases of FMD at the cephalic level.

Finally, much remains to be studied about the genetics of FMD. The first step would be to confirm the inheritance pattern of FMD, which Gladstein et al. presumed to be a dominant trait with reduced penetrance in males,[49] and to identify the precise phenotypic expression of the disease by complete clinical, radiological, and ultrasound investigations in the relatives of affected subjects. Such investigations should include the study of renal, limb, cervical, and intracranial arteries. Since FMD can produce protean clinical manifestations, involve different organs, and cause arterial dissections or intracranial aneurysms most often considered as different disorders, it appears crucial to investigate the familial aspects of FMD on the basis of the complete clinical and radiological spectrum of the disease.

Dissections

Cervical arterial dissections result from a bleeding into the vessel wall,[50] secondary either to a neck or head trauma or to the presence of an underlying disease, and possibly to both. They account for 1% of all ischemic strokes and for up to 20% of strokes in the young.[51,52] Cervical arterial dissections affect the anterior circulation more often than the posterior and the extracranial arteries much more frequently than the intracranial arteries. Very few familial cases have been reported so far. These cases are possibly related to the presence of FMD. Dissection of the cervicocerebral arteries is also one of the hallmarks of some very rare hereditary elastic tissue disorders.[51]

Main Clinical Features

The extracranial part of the internal carotid artery is the most frequent site of dissection. Spontaneous cervical carotid dissection often occurs in otherwise healthy subjects. The term spontaneous is used for dissections occuring without a history of obvious trauma and the term traumatic for those occurring after a cervical trauma with or without bony injuries.[51] However, in many cases, it is difficult to differentiate these two varieties; a minor or apparently trivial head or neck trauma is frequently seen in so-called spontaneous dissections. Spontaneous carotid dissection occurs in middle age, between 35 and 50 years of age in 70% of cases. The major clinical presentation includes minor or major strokes or TIAs associated with pain in the ipsilateral neck, face, or head, often preceding the ischemic symptoms by several hours, days, or even weeks.[38] An ispsilateral Horner's syndrome occurs in 40% of cases.[51] Other local symptoms are: pulsatile tinnitus, dysgueusia, facial dysesthesia, and lingual paresis. Although none of these clinical features is specific, their association is very suggestive. The diagnosis is based on arterial investigations: angiography, ultrasounds, and MRI. Internal carotid artery dissection is usually observed from 2 cm distal to the origin of the artery up to the base of the skull. Various patterns have been reported at angiography: irregular narrowing, tapered occlusion, intimal flaps, and distal extraluminal pouch[53] (Figure 4). MRI is now a particularly useful diagnostic tool because of its ability to show both the narrowed lumen and the hematoma within the dissected arterial wall. Vertebral artery dissection is less frequent. It is commonly associated with rotation of the neck and usually involves the V3 segment at the C1C2 level where the artery is most mobile and highly susceptible to mechanical injury. Clinically, the presentation most often includes severe lateral neck pain associated with or followed by ischemic symptoms in the vertebrobasilar or posterior cerebral artery territories. Intracranial carotid or basilar artery dissections are rare and far more severe. Ipsilateral severe headache preceding or coexisting with major neurologic deficits is the usual presentation. Death or major sequelae are the rule. One important point is that half of the cases in the basilar system are associated with subarachnoid hemorrhage.[51] The intravitam diagnosis is very difficult and it is possible that less severe cases are overlooked.

Genetics

Very few cases of familial clustering of dissections have so far been reported. The first familial cases are probably those of Petit et al. in two subjects with FMD[48] (see pre-

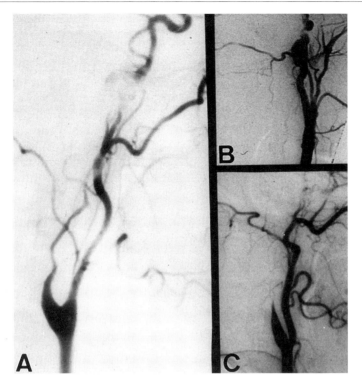

Figure 4. Various angiographic features of dissection of the internal carotid artery. **A:** Long dissection beginning near the carotid sinus. **B:** Pseudoaneurysm of the distal cervical carotid. **C:** Tapering occlusion.

vious paragraph). Mokri et al. later reported a mother and her daughter, and also a father and his son with spontaneous dissection of the internal carotid artery.[54] In these four cases, age at clinical onset was between 19 and 76 years. The clinical presentation included: unilateral pain of the face, head or neck (4/4); partial Horner's syndrome (4/4); diplopia (1/4); unilateral 9th-12th cranial nerves palsy (1/4); and subjective bruits (2/4). The dissection was unilateral in three cases, bilateral in one and associated with subcranial aneurysms of the internal carotid artery in two cases. In one of these cases, spontaneous dissections of the renal arteries occured 8 years later suggesting the presence of a more diffuse arteriopathy. In this case, the authors reported luminal irregularities in the vertebral arteries at the C2 level, supporting the diagnosis of arterial fibromuscular dysplasia.

A familial history of intracranial aneurysms or subarachnoid hemorrhage has re-

cently been reported in 3 of 175 subjects with spontaneous dissections of the cervical arteries.[55] All had another member (one or two) of their family with one or two intracranial aneurysm(s) involving the supraclinoid portion of internal carotid, the middle cerebral artery, or the anterior communicating artery. The occurrence of cervical artery dissections and intracranial aneurysms within the same family suggests the presence of a common underlying vascular defect. The most likely candidate is FMD which can lead to both arterial dissections and intracranial aneurysms.[45,56] Other potential candidates are elastic tissue diseases (Ehlers-Danlos or Marfan's syndromes; see *Chapter 12*) which can also lead to dissection and aneurysm formation. However, no mutation of the type III collagen (Ehlers-Danlos type IV) was found in a series of 18 subjects with cervical artery dissection(s), 5 of whom had one first-degree relative with either dissection or intracranial arterial aneurysms suggesting an underlying hereditary arterial disease.[57] In summary, very little is known about the genetics of cerebral artery dissections, a subject which has so far been widely neglected. Further studies are needed but they are likely to be difficult to perform because of the genetic heterogeneity of this condition which is probably not an entity per se but the mere complication of an underlying arterial injury or disease.

Moyamoya

Moyamoya disease, or the spontaneous occlusion of the circle of Willis, was first described in 1957 by Takeuchi and Shimizu.[58] Since then, various eponyms have been used, such as "cerebral juxtabasilar telangiectasia", "cerebral arterial rete", and "Nishimoto-Kudo disease", "Nishimoto disease".[59] However, the Japanese term "moyamoya," meaning "puff of smoke" for the angiographic appearance of the abnormal vascular network developed in the vicinity of the occluded intracranial arteries, is the most frequently used. Moyamoya was first reported in Japan where the disease is most frequently encountered. The Research Committee on Spontaneous Occulusion of the Circle of Willis, organized in Japan to investigate the disease, estimated the number of Japanese patients to be approximately 2600 to 3300 in 1990.[60] In Japan, the incidence of moyamoya disease was estimated to be $1/10^6$ per year. However, more than one thousand cases have been reported outside Japan, 50% of them in Asiatic countries.[61] Two peaks of age incidence are observed in Japan: children under 10 years of age and young adults in their third decade.[61] Based on the data from literature, moyamoya disease might be familial in approximately 1 of 10 cases.

Main Clinical Features

The diagnosis of moyamoya disease is based on three angiographic criteria according to the Research Committee: (1) stenosis or occlusion at the terminal portion of the intracranial internal carotid artery and/or at the proximal portion of the anterior cerebral artery and/or the anterior cerebral artery; (2) abnormal vascular moyamoya networks in the above-mentioned areas; and 3) bilaterality of these findings. There are no specific symptoms or signs of the disease.[61] The clinical manifestations are secondary to cerebral ischemia, hemorrhage, or epilepsy. In children, ischemic symptoms are most common (69–87% of cases) while in adults, hemorrhage is more prevalent (66% of cases).[61] TIAs, RINDs, and minor or major strokes are observed in various vascular territories. One-sided motor deficits are the most frequent symptoms observed. Intracerebral and/or ventricular hemorrhage seem to be more frequent than subarachnoid hemorrhage. Epilepsy is observed in 5% of cases.[61] Mental retardation is found in 50% of patients. The natural history of the disease is not certain, but it seems that the underlying process is active for a 10-year-period only and stabilizes thereafter. A fatal outcome has been the result in some cases, most frequently because of massive bleeding. At angiography, Suzuki described six phases of evolution of the disease: (1) stenosis of the intracranial bifurcation of the internal carotid artery; (2) first appearance of moyamoya (dilatation of the intracerebral arteries); (3) increasing of moyamoya (disappearance of the middle and anterior cerebral arteries; (4) finer formation of moyamoya (disappearance of the posterior cerebral artery); (5) shrinking of moyamoya (disappearance of the intracerebral arteries); (6) disappearance of moyamoya and presence of a collateral circulation coming only from the external carotid system[62] (Figure 5). Pathological studies show a major fibrous intimal thickening with a frequently stratified and/or tortuous elastic lamina underlying the severe stenosis of the intracranial bifurcation of the internal carotid arteries. The media is normal or thinned. The adventitia is spared. Stenosis and/or dilatation are observed in the small perforating arteries. Multiple microaneurysms are observed in these vessels.[61,63] Leptomeningeal

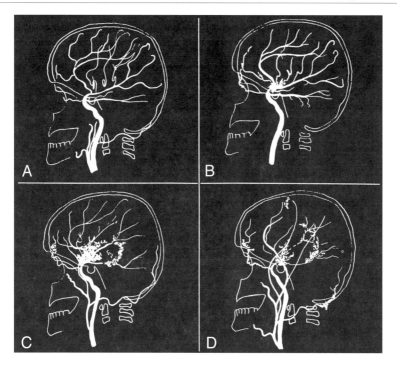

Figure 5. Different phases of evolution of moyamoya at angiography. **A:** Phase 1, stenosis of the intracranial bifurcation of the internal carotid arteries. **B:** Phase 2, first appearance of moyamoya. **C:** Phase 3, increasing of moyamoya with disappearance of the middle and anterior cerebral arteries. **D:** Phase 5, shrinking of moyamoya with disappearance of intracerebral arteries.

vessels mainly consist of dilatated preexisting vessels.[64]

Genetics

Whether moyamoya disease is of acquired or congenital origin is still debated. A major problem is the confusing terminology of moyamoya when based only on the peculiar aspect of the disease at angiography. Today, the term "moyamoya disease" should be distinguished from "moyamoya phenomenon," used for the angiographic aspect of "puff of smoke" whatever the underlying process, and also from "moyamoya syndrome" corresponding to the acquired occlusions of the internal carotid arteries of a known etiology (atherosclerosis, skull base tumor, meningitis, postirradiation) leading to moyamoya phenomenon at angiography. According to the Research Committe on

Spontaneous Occlusion of the Circle of Willis,[65] the diagnosis of "definite" moyamoya disease should be based on the presence of the above-mentionned angiographic criteria and also on the absence of other known etiologies. In Japan, the familial incidence of moyamoya disease is estimated to be between 7% to 12%.[61] Familial cases have also been reported outside of Japan.[66a,67a,68a] In the largest reported study, there was at least one affected member other than the proband in 38 of 508 families (7.54%).[60] Elsewhere, 79 affected parents had 169 children among whom, 6 (3.59%) had the disease.[60] This incidence is 51 times higher than that in the Japanese population. The father-child occurence (5.77%) is more frequent than the mother-child (2.61%) and the sibling (2.22%) occurences.[60] Interestingly, 76 of 79 cases who had children had the adult form of the disease. When only siblings are affected, age of onset in a given family seems very similar;

the difference of age of onset between affected siblings was less than 10 years in 11 of 14 families reviewed by Kitahara et al.[66a] This also supports a hereditary origin of the disease.[67] Another major argument is that 11 of the 13 pairs of monozygotic twins reported in the literature were concordant for the disease.[61,68–71] Clinically, the presentation of the few familial cases detailed in the literature does not differ from that of nonfamilial cases. In a given family, the distribution of the stenosed arteries at angiography is always identical in the affected subjects. In the family reported by Praud et al.,[68] there was a bilateral narrowing of the proximal portion of MCA and ACA sparing the internal carotid arteries in the two affected siblings. There are thus major arguments in support of a hereditary origin of moyamoya disease. Despite the three affected siblings reported by Narumi et al., born of parents who were cousins,[72] the presence of a vertical transmission does not favor an autosomal recessive heredity. However, whether moyamoya disease is an autosomal dominant disease of incomplete penetrance or consists of a heterogeneous group of monogenic diseases of similar presentation, or is a heterogeneous multifactorial disease as previously suggested,[61] remains to be determined.

References

1. Tournier-Lasserve E, Joutel A, Melki J, et al. Cerebral autosomal dominant arteriopathy with subcortical infarcts and leukoencephalopathy maps on chromosome 19q12. Nature Genetics 1993;3:256–259.
2. Bousser MG, Tournier-Lasserve E. Summary of the first International workshop on CADASIL. Stroke 1994;25:704–707.
3. Van Bogaert L. Encephalopathie sous-corticale progressive (Binswanger) à évolution rapide chez deux soeurs. Med Hellen 1955;24:961–972.
4. Sourander P, Walinder J. Hereditary multi-infarct dementia: Morphological and clinical studies of a new disease. Acta Neuropathol (Berl) 1977;39(3):247–254.
5. Stevens DL, Hewlett RH, Brownell B. Chronic familial vascular encephalopathy. Lancet 1977;2:1364–1365.
6. Davous P, Fallet-Bianco C. Démence sous-corticale familiale avec leucoencéphalopathie artériopathique: Observation clinicopathologique. Rev Neurol (Paris) 1991;5:376–384.
7. Colmant HJ. Familiäre zerebrale Gefäberkrankung. Zbl. Allgemein. Pathologie Bd 1980;124(1/2):163.
8. Sonninen V, Savontaus ML. Hereditary multi-infarct dementia. Eur Neurol 1987;27:209–215.
9. Tournier-Lasserve E, Iba-Zizen MT, Romero N, et al. Autosomal dominant syndrome with stroke-like episodes and leukoencephalopathy. Stroke 1991;22:1297–1302.
10. Salvi F, Michelucci R, Plasmati R, et al. Slowly progressive familial dementia with recurrent strokes and white matter hypodensities on CT scan. Ital J Neurol Sci 1992;13:135–140.
11. Mas JL, Dilouya A, De Recondo J. A familial disorder with subcortical ischemic strokes, dementia and leukoencephalopathy. Neurology 1992;42:1015–1019.
12. Verin M, Rolland Y, Landgraf F, et al. New phenotype of CADASIL with migraine as prominent clinical feature. Neurology 1994;44:288.
13. Chabriat H, Tournier-Lasserve E, Nibbio A, et al. Cerebral autosomal dominant arteriopathy with subcortical infarcts and leukoencephalopathy (CADASIL): Main clinical findings in five families. Stroke 1994;25:258.
14. Baudrimont M, Dubas F, Joutel A, et al. Autosomal dominant leukoencephalopathy and subcortical ischemic strokes: A clinicopathological study. Stroke 1993;24:122–125.
15. International Headache Society. Classification and diagnostic criteria for headache disorders, cranial neuralgias and facial pain. Cephalalgia 1988;(suppl)7:8.
16. Diagnostic and Statistical Manual of Mental Disorders, DSMIII. Ed 3 of the American Psychiatric Association, 1987, Washington DC.
17. Gutierrez-Molina M, Caminero-Rodriguez A, Martinez Garcia C, et al. Small arterial granular degeneration in familial Binswanger's syndrome. Acta Neuropath 1994;87:98–105.
18. Ruchoux MM, Chabriat H, Baudrimont M, et al. Presence of ultrastructural arterial lesions in muscle and skin vessels of patients with CADASIL. Stroke 1994;25:2291–2292.
19. Weissenbach J, Gyapay G, Dib C, et al. A second generation linkage map on the human genome. Nature 1992;359:794–801.
20. Joutel A, Bousser M.G, Biousse V, et al. A gene for familial hemiplegic migraine maps to chromosome 19. Nat Gen 1993;5:40–45.
21. Blau JN. Familial hemiplegic migraine. Lancet 1955:1115–1116.

22. Ferbert AN, Busse D, Thron A. Microinfarc-
    tion in classic migraine? A study with mag-
    netic resonance imaging findings. Stroke
    1991;22:1010–1014.
23. Igarashi H, Sakai F, Kan S, et al. Magnetic
    resonance imaging of the brain in patients
    with migraine. Cephalalgia 1991;11:69–74.
24. Osborn RE, Alder DC, Mitchell CS. MR
    Imaging of the brain in patients with mi-
    graine headaches. AJNR 1991;12:521–524.
25. Soges LJ, Cacayorin ED, Petro GR, et al. Mi-
    graine evaluation by MR. AJNR 1988;9:
    425–429.
26. Aylward ED, Roberts-Willie JV, Barta PE, et
    al. Basal ganglia volume and white matter
    hyperintensities in patients with bipolar dis-
    order. Am J Psych 1994;5:687–693.
26a.Chabriat H, Vahedi K, Iba-Zizen MT, et al.
    Clinical spectrum of CADASIL: A study of 7
    families. Lancet 1995;346:934–939.
26b.Ducros A, Nagy T, Alamowitch S, et al.
    CADASIL, genetic homogeneity and map-
    ping of the locus within a 2 cM interval. Am
    J Hum Genet 1996;58:171–181.
26c.Joutel A, Corpechot C, Ducros A, et al.
    Notch3 mutations in CADASIL, a hereditary
    adult-onset condition causing stroke and de-
    mentia. Nature 1996;383:707–710.
26d.Ophoff R A, Terwindt G M, Vergouwe MN,
    et al. Familial hemiplegic migraine and
    episodic ataxia type-2 are caused by muta-
    tions in the Ca2+ channel gene CACNL1A4.
    Cell 1996;87:543–552.
27. Fukutake T, Hirayama K. Familial young-
    adult-onset arteriosclerotic leukoencepha-
    lopathy with alopecia and lumbago without
    arterial hypertension. Eur Neurol 1994;35:-
    69–79.
28. Maeda S, Yokoi S, Nakayama H, et al. Famil-
    ial unusual encephalopathy of Binswanger's
    type without hypertension. Adv Neurol Sci
    (Tokyo) 1976;30:165–177.
29. Fukutake T, Hirayama K. Familial juvenile
    arteriosclerotic leukoencephalopathy with
    diffuse baldness and skeletal disorders. Adv
    Neurol Sci 1992;36:70–80.
29a. Heffelfinger MJ, Holley KE, Harrison EG, et
    al. Arterial fibromuscular dysplasia studied
    at autopsy. Am J Clin Pathol 1970;54:274. Ab-
    stract.
30. Manelfe C, Clarisse J, Fredy D, et al. Dys-
    plasies fibromusculaires des artères cervico-
    céphaliques: A propos de 70 cas. J. Neurora-
    diologie 1974;1:149–231.
31. Houser OW, Baker HL, Sandok BA, et al.
    Cephalic arterial fibromuscular dysplasia.
    Radiology 1971;101:605–611.
32. Corrin LS, Sandok BA, Houser W. Cerebral
    ischemic events in patients with carotid
    artery fibromuscular dysplasia. Arch Neurol
    1981;38:616–618.
33. Healton EB. Cerebrovascular fibromuscular
    dysplasia. In Barnett HJM, Mohr JP, Stein
    BM, Yatsu FM (eds). Stroke: Pathophysiol-
    ogy, Diagnosis, and Management. New
    York: Churchill Livingstone, 1993, 749–760.
34. Lüscher TF, Lie JT, Stanson AW, et al. Arter-
    ial fibromuscular dysplasia. Mayo Clin Proc
    1987;62:931–952.
35. Bolander H, Hassler O, Liliequist B, et al.
    Cerebral aneurysm in an infant with fibro-
    muscular hyperplasia of the renal arteries.
    J Neurosurg 1978;49:756–759.
36. Osborn A, Anderson RE. Angiographic spec-
    trum of cervical and intracranial fibromus-
    cular dysplasia. Stroke 1977;8:617–619.
37. So EL, Toole JF, Moody DM, et al. Cerebral
    embolism from septal fibromuscular dyspla-
    sia of the common carotid. Ann Neurol
    1979;6:75–78.
38. Biousse V, D'Anglejean-Chatillon J, Massiou
    H, et al. Head pain in non-traumatic carotid
    artery dissection: A series of 65 patients.
    Cephalalgia 1994;14:33–36.
39. Havelius U, Hindfelt B, Brismar J, et al.
    Carotid fibromuscular dysplasia and paresis
    of lower cranial nerves (Collet-Sicard syn-
    drome). J Neurosurg 1982;56:850–853.
40. Halpern MM, Sanford HS, Viamonte M Jr.
    Renal artery abnormalities in three hyper-
    tensive sisters: Probable familial fibromuscu-
    lar hyperplasia. JAMA 1965;194:512–513.
41. Hansen J, Holten C, Thorborg JV. Hyperten-
    sion in two sisters caused by so called fibro-
    muscular hyperplasia of the renal arteries.
    Acta Med Scand 1965;178:461–474.
42. Plagnol P, Gille JM, Cambuzat JM, et al. Hy-
    pertension renovasculare familiale. J Radiol
    Electrol Med Nucl 1975;56:173–174.
43. Major P, Genest J, Cartier P, et al. Hereditary
    fibromuscular dysplasia with renovascular
    hypertension. Ann Intern Med 1977;86:583.
44. Morimoto S, Kuroda M, Uchida K et al. Oc-
    currence of renovascular hypertension in
    two sisters. Nephron 1976;17:314–320.
45. Rushton AR. The genetics of fibromuscular
    dysplasia. Arch Intern Med 1980;140;233–236.
46. Ouchi Y, Tagawa H, Yamakado M et al. Clin-
    ical significance of cerebral aneurysm in ren-
    ovascular hypertension due to fibromuscu-
    lar dysplasia: Two cases in siblings.
    Angiology 1989;40:581.
47. Mettinger KL, Ericson K. Fibromuscular dys-
    plasia and the brain: Observations on angio-
    graphic, clinical and genetic chracteristics.
    Stroke 1982;13:46–58.
48. Petit H, Bouchez B, Destee A, et al. Forme fa-
    miliale de dysplasie fibromusculaire des
    artères carotides internes. J Neuroradiology
    1982;9:15–22.
49. Gladstein K, Rushton AR, Kidd KK. Pene-
    trance estimates and recurrence risks for fi-

bromuscular dysplasia. Clin Genet 1980;17: 115–116.

50. Bogousslavsky J. Dissection of the cerebral arteries. Curr Opin Neurol Neurosurg 1988; 1:63–68.

51. Saver JL, Easton JD, Hart RG. Dissections and trauma of cervicocerebral arteries. In Barnett HJM, Mohr JP, Stein BM, Yatsu FM. (eds). Stroke: Pathophysiology, Diagnosis, and Management. New York: Churchill Livingstone, 1993, 671–688.

52. Bogousslavsky J, Regli F. Ischemic stroke in adults younger than 30 years of age: Cause and prognosis. Arch Neurol 1987;44:479–482.

53. Fisher CM, Ojeman RG, Roberson GH. Spontaneous dissection of cervicocerebral arteries. Can J Neurol Sci 1978;5:9.

54. Mokri B, Sundt TM, Houser OW, et al. Familial occurrence of spontaneous dissection of the internal carotid artery. Stroke 1987;18: 246–251.

55. Schievink WI, Mokri B, Michels W, et al. Familial association of intracranial aneurysms and cervical artery dissections. Stroke 1991; 22:1426–1430.

56. Ouchi Y, Tagawa H, Yamakado M, et al. Clinical significance of cerebral aneurysm in renovascular hypertension due to fibromuscular dysplasia: Two cases in siblings. Angiology 1989;40:581–589.

57. Kuivaniemi H, Prockop DJ, Madhatheri SL, et al. Exclusion of mutations in the gene for type III collagen (COL3A1) as a common cause of intracranial aneurysms or cervical artery dissections: Results from sequence analysis of the coding sequences of type III collagen from 55 unrelated patients. Neurology 1993;43:2652–2658.

58. Takeuchi K, Shimizu K. Hypoplasia of the bilateral internal carotid arteries. Brain and Nerve (Tokyo) 1957;9:37.

59. Nishimoto A, Takeuchi S. Abnormal cerebrovascular network related to the internal carotid arteries. J Neurosurg 1968;29:255–260.

60. Fukuyama Y, Sugahara N, Osawa M. A genetic study of idiopathic spontaneous multiple occlusion of the circle of Willis. In Yonekawa Y (ed). Annual Report (1990) by Research Committee on Spontaneous Occlusion of the Circle of Willis. Ministry of Health and Welfare, Japan, 1991, 139.

61. Yokenawa Y, Goto Y, Ogata N. Moyamoya disease: Diagnosis, treatment, and recent achievment. In Barnett HJM, Mohr JP, Stein BM, Yatsu FM (eds). Stroke: Pathophysiology, Diagnosis, and Management. Churchill Livingstone, 1993, 721–747.

62. Suzuki J, Takaku A. Cerebrovascular Moyamoya disease. Arch Neurol 1969;20:288–289.

63. Hanakita J, Kondo A, Ishikawa J, et al. An autopsy case of moyamoya disease. Neurol Surg 1982;10:531.

64. Shinji K, Kazunari O, Katsuo S. Histopathologic and morphometric studies of leptomeningeal vessels in moyamoya disease. Stroke 1990;21:1044–1050.

65. Gotoh F. Research committee on spontaneous occlusion of the circle of Willis. Ministry of Health and Welfare, Japan, 1983.

66. Austin JH, Stears JC. Familial hypoplasia of both internal carotid arteries. Arch Neurol 1971;24:1–10.

66a. Kitahara T, Ariga N, Yamaura A, et al. Familial occurrence of moyamoya disease: Report of three Japanese families. J Neurol Neurosurg Psych 1979;42:208–214.

67. Doolittle TH, Myers RH, Lehrich JR, et al. Multiple sclerosis sibling pairs: Clustered onset and familial presentation. Neurology 1990;40:1516–1552.

67a. Sogaard I, Jorgensen J. Familial occurence of bilateral intracranial occlusion of the internal carotid arteries. Acta Neurochir (Wien) 1975; 31:245–252.

68. Praud E, Labrune B, Lyon G, et al. Une observation familiale d'hypoplasie progressive et bilaterale des branches de l'hexagone de Willis avec examen anatomique. Arch Franç Ped 1972;29:397–409.

68a. Yamada H, Nakamura S, Kageyama N. Moyamoya disease in monovular twins. J Neurosurg 1980;53:109–112.

69. Sonobe M, Takayashi S, Urakawa Y et al. Moyamoya disease found in identical twins. Neurol Surg (Tokyo) 1980;8:1183.

70. Srinivas K, Narayanan M, Gopal S, et al. Carotid arterial occlusion in identical twins with Moyamoya. Arch Neurol 1979;36: 253–254.

71. Asami N, Miyahara S, Veda T, et al. Moyamoya disease in fraternal twins. No-To-Shinkei 1990;42:1093–1096.

72. Narumi S, Nishimura K, Fuchizawa K, et al. Three cases of the moyamoya disease found in an inbred family. Brain and Nerve (Tokyo) 1976;28:1201.

Chapter 11

Intracerebral Hemorrhage and Vascular Malformations

Mark J. Alberts, MD

Introduction

Intracerebral hemorrhage (ICH) is the third most common cause of stroke, accounting for 8–12% of all strokes.[1,2] ICH is also one of the most deadly types of stroke, with a mortality of 50% in some series.[1,3–6] Most cases of ICH are caused by rupture of a small artery within the brain.[7] In the majority of these cases, termed hypertensive or spontaneous ICH, the vessel rupture and subsequent bleed are presumed to be due to the long-standing effects of hypertension (HTN).[8,9] However, recent studies have called this association into question.

Various vasculopathies, including the amyloid angiopathies, are being recognized with increasing frequency as a cause of ICH, particularly in the elderly.[10–12] Inherited forms of amyloid angiopathy have been recognized and well characterized in the past 5 years, and provide a focal point for investigating the role of genetic factors in the pathogenesis of some cases of ICH. Vascular malformations such as arteriovenous malformations (AVM), cavernous hemangiomas (CH), and hereditary telangiectasias can also cause an ICH or subarachnoid hemorrhage.[5] These malformations may be inherited alone or as part of a specific syndrome with other vascular or systemic abnormalities. This chapter will review the genetic vasculopathies and vascular malformation syndromes that can produce an ICH.

Spontaneous/Hypertensive Intracerebral Hemorrhage

This is the most common form of ICH, and typically occurs in the basal ganglia, particularly the putamen and thalamus.[13] In some studies, lobar ICH (i.e., those at the gray-white matter junction) is either the second or third most common location, after putaminal bleeds. Based on epidemiological data, there is a suggestion that lobar ICHs are different from the typical spontaneous/hypertensive ICH. Lobar ICHs tend to occur in older patients (>70 years), while hypertensive ICHs occur between 40 and 69 years.[5] There is evidence that lobar ICHs in some cases are due to amyloid angiopathy (discussed below in more detail), but this is not universally true.[14] One study found that 9 of 29 (31%) lobar ICHs were due to amyloid angiopathy—the remainder were due to various vascular malformations, micro-

From Alberts MJ (ed). *Genetics of Cerebrovascular Disease*. Armonk, NY: Futura Publishing Company, Inc., © 1999.

aneurysms, and other entities.[15] Therefore there may be some overlap between hypertensive basal ganglia ICHs and lobar ICHs. The remainder of this section will focus on spontaneous hypertensive ICHs. Lobar ICHs due to amyloid angiopathy will be discussed in a subsequent section.

Some studies have found that the vast majority of patients with ICH have HTN, implying that HTN plays a major role in the pathogenesis of this type of ICH. These studies found that the frequency of HTN in ICH patients ranged between 72–89%.[8,9] However, several recent studies have suggested a more modest role for HTN in ICH patients, citing a frequency of 45–59%.[5,16] What is clear from these studies is that although HTN is clearly a risk factor for some cases of ICH, not every hypertensive individual develops an ICH, nor can every case of ICH be ascribed to the effects of HTN. The genetics of HTN are discussed in more detail in *Chapter 4*. But these facts suggest that other factors, either environmental or genetic, are important in the pathogenesis of ICH. Ethnic factors may also be important, as several studies have shown an increased incidence of ICH among African-Americans.[3,6,17]

Pathologic examination of ICH cases typically shows no obvious lesion except for the hematoma. A few careful anatomic studies have documented small aneurysmal dilatations of deep penetrating arteries in regions in or near the bleed.[7,18] It is presumed that the long-standing effects of HTN weakened the vessel wall, leading eventually to arterial rupture and an ICH. In some cases, small vascular malformations (AVMs, CHs, etc.) are found.[7,15,19] In other cases, depositions of amyloid are found in the vessel walls.[20] This type of amyloid angiopathy is discussed below in more detail.

A prospective study of ICH patients was carried out at Duke University Medical Center, Durham, NC, to address the potential role of genetic factors in this disease.[21] Among a cohort of 48 prospectively ascertained ICH patients (some hypertensive, deep hemorrhages and others with lobar bleeds), 7 had a family history of ICH in a first-degree relative. In one striking case the proband, his father, and his grandfather all had an ICH (Figure 1A). In another case, 3 of 6 siblings had an ICH (Figure 1B). Among the 7 families with ICH, 4 did not have a history of HTN or evidence of HTN among the affected individuals. The numbers of affecteds in these families was too small to draw any conclusions about the mode of inheritance. These data provide some intriguing, though preliminary, evidence that genetic factors, in addition to HTN, may be important in some cases of "spontaneous" ICH.

Since specific gene mutations have been found to cause hereditary forms of ICH (discussed below in more detail) the Duke group investigated whether patients with spontaneous ICH, with or without a family history of ICH, may harbor these same mutations. Genomic DNA from affected individuals was collected, and by using PCR technology and DNA sequencing it was screened for mutations in exon 2 of cystatin C, and exons 16 and 17 of the amyloid precursor protein (APP). In this cohort of patients, there were no cases where mutations in these exons were detected.[21] This finding does not rule out the existence of mutations in other exons of the same genes, or in other genes.

As part of searching for candidate genes that may be involved in the pathogenesis of sporadic ICH, the Duke group has investigated endoglin. This protein is involved in the development and integrity of vessel walls. Mutations in endoglin are known to cause some cases of hereditary hemorrhagic telangiectasia (described below in more detail).[21a] A six base insertion between exons 7 and 8 of endoglin was found in some ICH patients and controls. When present in the homozygous form, the insertion was found in approximately 9% of ICH patients compared to 2% of controls, which is a significant difference.[21b] When present in the heterozygous form, the insertion was not associated with ICH. The insertion did not appear to alter the endoglin mRNA.[21b] The mechanisms behind its association with ICH is unclear. It may function as a susceptibility locus in a subset

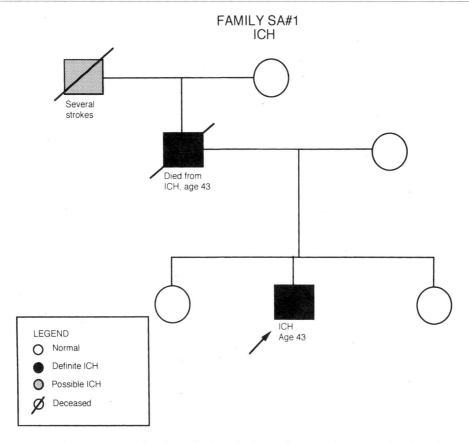

FAMILY SA#1
ICH

Several
strokes

Died from
ICH, age 43

ICH
Age 43

LEGEND
○ Normal
● Definite ICH
◉ Possible ICH
⊘ Deceased

Figure 1A. A multigenerational family with ICHs at identical ages in the proband and his father. The proband's grandfather may have had an ICH also.

of ICH cases, since most ICH patients do not have this mutation.

Another genetic factor is the effect of apo E genotype on the occurrence and outcome of ICH. Much of this work has been done in patients with an ICH due to amyloid angiopathy, which is discussed below in more detail. However, the Duke group did report an overall increase in the apo E e4 allele frequency (20.4%) in 125 patients with ICH (hypertensive, CAA, or due to lytic agents).[21c] The increase in the e4 allele was particularly dramatic in African-Americans with an ICH (28.3%)(Table 1). Whether this association explains in part the increased frequency of ICH in African-Americans is unclear. Perhaps African-Americans with an e4 allele are at higher risk for ICH, particularly if other risk factors are present. The study also found that African-American ICH patients had an age of onset that was approximately 7 years younger than Caucasians.[21c]

A significant association between apo E genotype and recovery/survival following an ICH has also been reported. Patients with an e3/e4 genotype had a mortality of 69%, compared with a mortality of 20% for ICH patients with a 2/3 or 3/3 genotype. Surviving patients with the 3/4 genotype also had a lower functional status compared to those with a 3/3 genotype (Barthel score of 60 vs. 83, respectively).[21d] These effects were independent of the influences of age, sex, and race.[21e] The importance of apo E genotype on the development of lobar ICH and CAA is discussed below in more detail.

At present the extent to which genetics play a role in the pathogenesis of most cases

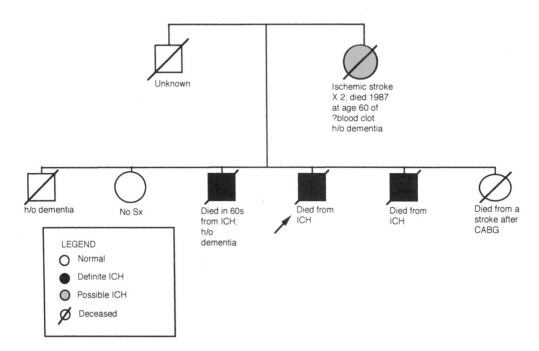

Figure 1B. A six-member sibship where three of the six siblings died of an ICH.

of spontaneous ICH remains unclear. But the absence of HTN in about half of ICH patients, the finding of some mutations described above, and the evidence of a familial aggregation in some cases suggests that further investigation of genetic factors is warranted.

The Amyloid Angiopathies

Cerebral amyloid angiopathy (CAA) is recognized with increasing frequency as a cause of ICH. CAA can be divided into two broad categories: (1) hereditary forms due to specific gene mutations, and (2) sporadic forms that affect mainly elderly patients.

The two most striking varieties of familial CAA are the so called Dutch cerebral hemorrhage (commonly referred to as hereditary cerebral hemorrhage with amyloidosis–Dutch type; HCHWA-D)[22,23] and the Icelandic type of angiopathy (abbreviated hereditary cerebral hemorrhage with amyloidosis-Icelandic type; HCHWA-I).[24] Both of these hereditary diseases cause multiple ICHs, although they have very different epidemiological and molecular genetic features.

It should be remembered that the term "amyloid" does not refer to a specific protein. Several different proteins are amyloidogenic, and deposition of these proteins can lead to formation of an amyloid angiopathy. As discussed below, the different amyloid-producing proteins can produce different clinical syndromes.

Dutch Form

Patients with HCHWA-D appear to be centered around two Dutch villages, Katwijk

Table 1
Apolipoprotein E Allele Frequencies Found in Various Types of Intracerebral Hemorrhages*

Study	Group	Allele Frequency e2 e3 e4	#Patients
Alberts et al.[21c]	ICH (HTN, CAA, lytics)	0.08, 0.71, 0.20	125
Greenberg et al.[93a]	definite/probably CAA*	0.20, 0.49, 0.31	27
Nicoll et al. [93c]	definite CAA**	0.35, 0.58, 0.08	13
Greenberg et al[93b]	deep ICHs	—, —, 0.14	18
Hyman et al. [93e]	control population	0.09, 0.77, 0.14	1899

*CAA, cerebral amyloid angiopathy.
**Data for CAA group does not include patients with AD.

and Scheveningen.[25] The age of onset for ICH in these patients is typically between 45 and 65 years.[25] The ICHs tend to be localized in the temporal-parietal and occipital regions.[22] Two thirds of the patients die because of their first ICH. It is not uncommon for patients to experience a second and even third ICH over the subsequent months and years.[25] The mode of inheritance is clearly autosomal dominant with apparent complete penetrance.[25,26] For reasons that are unclear, affected females had a higher mortality compared to males (relative risk 8.0 and 2.6, respectively).[26a]

Some cases of HCHWA-D can be associated with multiple cerebral infarctions.[22,27] These typically involve the deep subcortical white matter, although cortical infarctions can occur. One study found that 87% of patients with HCHWA-D had ICH, while 13% had ischemic strokes.[22] In some cases it is possible that a hemorrhagic infarction was misdiagnosed as a primary ICH.

The pathologic changes seen with HCHWA-D consist of a patchy distribution of amyloid deposits in somewhat sclerotic arteriolar walls. Approximately 1/4 of the cortical arterioles are involved.[25,28] The angiopathy does not involve other parts of the brain or other visceral organs. In some cases, senile plaques can also be seen, but neurofibrillary tangles (which are typical of Alzheimer's disease [AD]), are not seen.[25] The neuritic plaques seen in HCHWA-D tend to be more diffuse than the plaques seen in classic AD.[29]

These plaques stain with antibodies to Aβ, but do not stain with antibodies to APP (whereas the classic AD plaques do stain for both Aβ and APP [see below for descriptions of Aβ and APP]).[29] Leptomeningeal and small cortical blood vessels from HCHWA-D brains stained strongly for Aβ and APP, while such vessels from AD brains were less positive for Aβ and APP.[29] There is no significant cortical atrophy in cases of HCHWA-D (except that which is due directly to the effects of a chronic ICH), while significant atrophy is common in AD brains. Overall, while these results suggest some pathologic overlap between HCHWA-D and AD, there appears to be substantial areas of dissimilarity between these two entities.

Some patients develop a dementia due to the effects of multiple ICHs.[22] However, there are several documented cases of progressive dementia following a single ICH.[22,30] These clinical findings, combined with some of the molecular studies discussed below, raise the possibility that the mutations causing some cases of HCHWA-D may cause a dementing illness without producing multiple ICHs.

Detailed protein and molecular analyses of the amyloid deposited in the brains of patients with HCHWA-D have shown it to be identical to the amyloid protein deposited in most cases of AD.[31] At this time, it is appropriate to discuss the molecular biology of this type of amyloid, since this is central to the pathogenesis of HCHWA-D.

The amyloid beta protein, commonly referred to as Aβ, is derived from a larger precursor protein, termed the amyloid precursor protein (APP).[32] The gene for APP has been mapped to the proximal long arm of chromosome 21.[33] The gene for APP is quite large, covering at least a million bases, and composed of 18 exons.[34] Through alternative splicing, APP can be produced in one of three forms, designated APP[695], APP[751], and APP[770].[35,36] The two longer forms of APP contain a Kunitz serine protease inhibitor that may be important in its metabolism (or the metabolism of other proteins).[35,36] Exon 1 of APP is the signal peptide, and probably plays a role in its transportation (see below for more detail). The Aβ amyloid that is deposited in plaques and vessel walls is derived from parts of exons 16 and 17 (Figure 2).[33,34,37,38]

APP is synthesized in most organs, including the liver, kidney, muscle, and brain.[39] In the brain, APP is normally synthesized in the neuronal cell body, then axonally transported to the end of the neuron.[40] APP is then inserted into the membrane, with the amino terminus being extracellular and the carboxyl terminus intracellular (Figure 3).[32] The normal function of APP remains somewhat unclear. It has been postulated that APP is important for neuronal growth and perhaps cell-to-cell interactions.[41]

The Aβ fibril that is deposited in HCHWA-D and in some cases of AD is a 39–43 amino acid protein with a weight of approximately 4 kilodaltons.[31,37] These fibrils can be produced by cleavage of APP at two locations, one in the extracellular domain and one in the intramembrane domain (Figure 3). Normally, cleavage of APP takes place within the Aβ moiety at an extracellular site designated glycine 16 (the number 16 refers to the amino acid residue position within the Aβ protein).[42,43] One theory suggests that if this normal cleavage takes place, then the typical Aβ protein will not be deposited. However, if the cleavage of APP occurs before it is inserted into the membrane, then the intramembranous site will be exposed (Figure 3) and Aβ fibrils may be released.[43] The exact steps involved in the intracellular processing of APP and the release of the Aβ fragment is the focus of intense molecular biology research, since it may be important in the pathogenesis of some cases of AD.

As described above, under normal circumstances, the extracellular portion of APP is cleaved, producing a protein analogous to protease nexin-2 (PN-2).[42,43] PN-2 has been found to be an inhibitor of two coagulation proteins, factor XIA and factor IXA.[44,45] In fact, PN-2 is an extremely potent inhibitor of factor IXA, with 71 times more

SCHEMATIC REPRESENTATION OF THE AMYLOID PRECURSOR PROTEIN

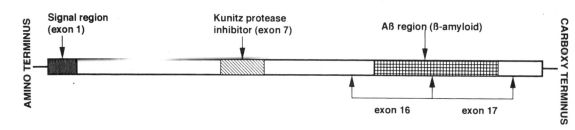

Figure 2. A schematic of the amyloid precursor protein molecule, showing the relationships among some selected exons and the Aβ fragment.

SCHEMATIC REPRESENTATION OF THE AMYLOID PRECURSOR PROTEIN AND ITS INTRAMEMBRANOUS LOCATION

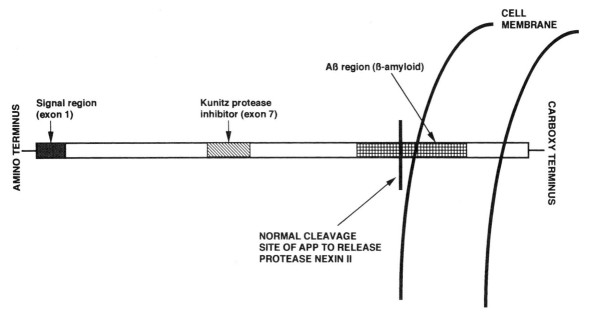

Figure 3. The postulated orientation of the amyloid precursor within the cell membrane, the site of normal cleavage, and the Aβ domain.

inhibition then antithrombin 3.[45] This has obvious implications in the pathophysiology of hemorrhagic diseases, particularly with respect to the amyloidoses.

Two major loci for mutations in the APP gene have been described in patients with some cases of early onset familial AD (Figure 4). One type of point mutation occurs in the three amino acids distal to the carboxyl end of the Aβ fragment. These mutations are termed the "APP[717] mutations". The APP[717] mutations convert valine to either isoleucine, phenylalanine, or glycine.[46–48] Another mutation occurs in the two amino acids directly proximal to the beginning of the Aβ moiety, resulting in the change of a lysine and methionine to an asparagine and leucine, respectively.[49] It is important to note that both classes of mutations do not occur within the Aβ moiety, but instead are immediately proximal or distal to this fragment.

This situation is in contrast to the mutation found to be responsible for HCHWA-D. This mutation is a cytosine to guanine transversion resulting in a single amino acid substitution (glutamic acid changed to glutamine) at position 22 (codon 693) of Aβ (Figure 4).[50] This is close to the middle of the Aβ moiety, as compared to the mutations causing AD which are outside of the Aβ domain.

The extensive study of HCHWA-D has shown that the APP[693] mutation can cause a cerebral amyloid angiopathy resulting in multiple ICHs.[51] In some rare cases it is possible that these patients develop a dementia independent of ICHs, although the data supporting this are inconclusive.[30] An interesting issue is whether the disease manifests itself in a similar manner if the mutation is present in the heterozygous versus homozygous state. This issue has been examined in 5 families with both parents affected. Unfortunately, the numbers are too small to determine conclusively whether the disease has similar effects in homozygotes and heterozygotes, versus whether the homozy-

**DIAGRAM OF MUTATIONS IN THE AMYLOID PRECURSOR PROTEIN
CAUSING INTRACEREBRAL HEMORRHAGE AND FAMILIAL ALZHEIMER'S DISEASE**

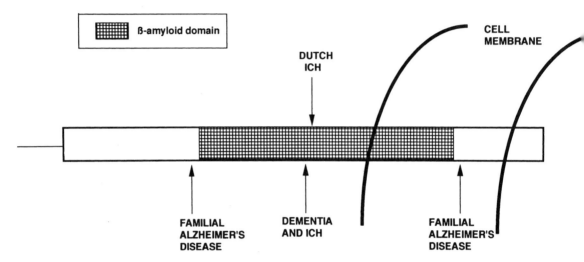

Figure 4. The relative positions of the various described mutations in the amyloid precursor protein molecule and their orientation with respect to the Aβ region.

gous state is a lethal condition.[25] Studies have shown that in HCHWA-D patients heterozygous for the APP[693] mutation, both alleles are expressed in vascular amyloid obtained from leptomeningeal vessels.[52]

Another report described a mutation in codon 692 of APP that substitutes a glycine for an alanine (Figure 4).[53] In a family with this mutation, some members had presenile dementia consistent with AD, and others had an ICH in their early 40s. One patient in that family who had onset of dementia at age 61 did not have the APP[692] mutation. It is not clear if the APP[692] mutation is sufficient and necessary for the development of dementia in this family. Perhaps other polymorphisms or alleles have to be present for the APP[692] mutation to be manifest.

Recent studies using *in situ* hybridization have shown that in some cases of cerebral amyloid angiopathy, APP is produced by vascular myocytes.[54] Since it is known that the PN-2 constituent of APP inhibits fac-

tors IXA and XIA, it is interesting to speculate that in patients with cerebral amyloid angiopathy (either hereditary or sporadic) excess secretion of PN-2 produces a local hypocoagulable state. Furthermore, factor XIA is released by stimulated platelets, which are the first line of defense when vascular integrity is breached. Although this concept may be attractive, it does not fully address some of the known aspects of HCHWA-D. The mutations causing this disease are located within the actual Aβ domain; therefore they would not be expected to affect directly PN-2 function. It is possible that these mutations alter the processing of APP, perhaps leading to excess production and secretion of PN-2, producing increased inhibition of some coagulation factors. Abnormal processing of APP containing the "717" mutations known to cause familial AD has been reported.[55] Further study is needed to determine if similar events occur for APP containing the HCHWA-D mutations.

One line of evidence that may support a role for abnormal processing of APP in some cases (or types) of AD and in HCHWA-D is the finding of reduced soluble APP levels in the cerebral spinal fluid (CSF). A study by Van Nostrand and colleagues[56] compared APP levels in the CSF of patients with HCHWA-D, AD, and age-matched controls. They found a three to fourfold decrease of soluble APP in the CSF in HCHWA-D and AD patients compared to controls.[56] They postulate that this decrease could be due to abnormal processing of APP. One study reported that the processing of APP[717] produced longer Aβ (compared to the processed normal APP molecule) which was less soluble, thereby causing more amyloid fibril deposition.[57] Of course, abnormal processing or deposition of APP or Aβ does not necessarily imply a pathogenic role for either, since other studies have failed to detect any mutations in the APP gene in most cases of familial AD.[58]

Amyloid containing the HCHWA-D common mutation induces several changes in cultured human smooth muscle cells derived from brain vessels. These changes include degeneration as well as an increase in APP levels. More recent studies have demonstrated that Aβ containing the HCHWA-D mutation is able to self-assemble on the surface of cultured human brain smooth muscle cells and forms fibrils.[58a] These findings may explain the selectivity seen for Aβ to involve cerbral vessels and perhaps lead to ICH.

Another unusual feature of HCHWA-D is the lack of amyloid deposits in organs other than the brain.[25] This is of note since studies have shown that APP is produced and deposited in other organs such as kidney, liver, intestine and muscle.[39,59] Assuming that the causative mutations are not limited to the brain (which is unlikely since in some cases the mutation has been found in genomic DNA isolated from lymphocytes), other factors may be controlling either expression of the mutated protein, or the production of a hemorrhagic phenotype. Cerebral vessels have an unusual histology in that the muscular media is typically absent at branch points and an external elastic lam-

ina may also be absent.[60,61] Perhaps this increases the susceptibility of cerebral vessels to hemorrhage in some circumstances.

Icelandic Form

The Icelandic form of amyloid angiopathy, HCHWA-I, is an autosomal dominant disorder that typically causes ICH between the ages of 20 and 30 years.[24,62] The pathology of HCHWA-I is somewhat similar to that of HCHWA-D in that there is deposition of an amyloid protein in small arterioles of the leptomeninges and cortex. Unlike the Dutch form, this amyloid protein has been found to be cystatin C, which is a protease inhibitor that maps to chromosome 20.[63-66] Cystatin C has been found in the spinal fluid and in some tissues outside of the CNS, particularly small arteries and lymph nodes.[67-70]

Detailed genetic analyses have identified a point mutation in exon 2 of cystatin C as the likely cause of HCHWA-I.[71] A single base mutation was found that substitutes a CAG instead of a CTG. This results in a leucine being replaced by a glutamine at this position, and abolishes an *Alu*I restriction site in affected patients (Figure 5).[72]

The pathophysiology of how this mutation causes ICH is unclear. It has been suggested that this mutation may make cystatin C more susceptible to breakdown, thereby accelerating amyloid fibril formation and deposition.[71] Another hypothesis is that the mutated cystatin C cannot be transported out of endothelial cells, resulting in accumulation within the vessel wall.[72] Studies have shown that among patients with the mutation, there is a reduction in the cystatin C level within the CSF.[24,73] It has also been shown that patients with the cystatin C mutation may lack the first 10 amino acids of the protein.[64] Other analyses of cystatin C obtained from various body fluids (i.e., urine, seminal fluid, CSF) have shown heterogeneous amino termini in normal individuals.[71] At this time it is not clear if the mutation results in abnormal function or length of cystatin C.

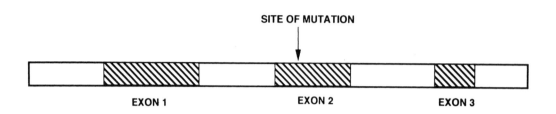

DIAGRAM OF CYSTATIN C GENE AND SITE OF
MUTATION CAUSING ICELANDIC FORM OF
HEREDITARY ICH

SITE OF MUTATION

EXON 1   EXON 2   EXON 3

NORMAL SEQUENCE: GAC GTG GAG CTG GGC CGA

MUTATED SEQUENCE: GAC GTG GAG CAG GGC CGA

Figure 5. A diagram of the exon structure for the cystatin C gene. The location and sequence of the mutation causing Icelandic amyloid angiopathy is indicated.

In summary, the inherited amyloidoses, particularly the Dutch and Icelandic variety, are clear examples of genetic causes of stroke. In both cases, well-defined mutations in specific genes lead to a form of amyloid angiopathy, resulting in ICH, dementia (in some cases) and perhaps cerebral infarctions.

Transthyretin (TTR) is another amyloidogenic protein. Mutations in the transthyretin gene are responsible for type I familial amyloidotic polyneuropathy (FAP), with amyloid fibril deposition in several visceral organs.[74,75] A recent study demonstrated a cerebral amyloid angiopathy due to deposition of mutated TTR involving mainly the leptomeningeal vessels. The patient had FAP, with atypical neurologic features (but no ICH).[76] TTR isolated from the meninges was shown to have a methionine for valine substitution at position 30, which is the same mutation known to cause type I FAP.[74,76] It is not known whether TTR deposition without FAP can produce an isolated amyloid angiopathy.

Spontaneous Amyloid Angiopathy

"Spontaneous" cerebral amyloid angiopathy (CAA) is being recognized with increasing frequency as a cause of ICH in elderly patients.[20,77–79] ICH in the elderly due to CAA has some features that distinguish it from spontaneous ICH. The ICH associated with CAA typically causes a lobar hemorrhage, usually between the junction of the gray matter and white matter. These hemorrhages tend to involve the parietal and temporal lobes, although about 1/3 may also affect the frontal lobes.[20] Cerebellar ICHs can also, on rare occasions, be due to CAA.[78] Patients with CAA often experience more than one ICH.[80–82] Less than 1/3 of CAA-related ICH patients have a history of hypertension.[20,83] In some cases multiple small hemorrhages can be detected by MRI.[84]

The presence of sporadic CAA has been reported and pathologically verified in some elderly patients who develop ICHs after receiving either anticoagulants or thrombolytic agents.[85–87] Studies of patients receiving either thrombolytics or anticoagulants (i.e., coumadin) have reported a clear association between increasing age and the increased risk of ICH.[19,88–91] The incidence of CAA also increases dramatically with age.[92] Although the evidence of an association between these bleeding phenomena and CAA is still somewhat circumstantial, there is mounting evidence to support a causal link.

Small cerebral infarctions have also been described in patients with CAA.[20,93] It is unclear at this time whether this represents the occurrence of two common but causally unrelated events, or if CAA actually can produce, as an independent process, cerebral infarctions. Since many of the risk factors for small infarctions are common in CAA (i.e., hypertension, advanced age), careful case control studies will be needed to assess the significance of this association.

Pathologic examination of cerebral vessels near the ICH site in CAA shows microaneurysm formation and/or fibrinoid necrosis in some cases.[12] Another striking feature of CAA is the presence of clinical and pathologic evidence of Alzheimer's disease (AD) in 40–44% of cases.[20] Since cerebrovascular amyloid deposition is also seen in some cases of AD, it raises the question of whether CAA and some forms of AD may have a common underlying etiology.

As indicated above, several different proteins can be deposited and form "amyloid". In a retrospective study, the Duke group characterized by antibody staining (of brain) the amyloidogenic protein in 19 autopsy-proven cases of CAA. This study showed that 84% of cases had strong staining for Aβ, 37% had strong staining for cystatin C, and 16% stained strongly for apo E (Table 2). In some cases, the sections stained strongly with two or three of the antibodies. In other cases, while staining was present with at least one antibody, it was not striking with any of the antibodies. Note that not every patient with CAA had an ICH. A correlation of CAA staining and clinical characteristics can be seen in Table 2.

The presence of strong staining in some cases with an apo E antibody is intriguing.

Several recent studies have shown a clear association between apo E genotype and the presence of sporadic and late-onset familial AD.[20a] Since AD and CAA show an association in approximately 40% of cases, studies were performed to determine the apo E genotype of cases with documented CAA. A series of reports has shown an increased prevalence of the apo E e4 allele in cases of CAA-associated ICHs and lobar ICHs compared to controls or cases of nonlobar ICHs (Table 1).[93a,93b] This association implies a role for the e4 allele in the causation of CAA-associated ICH, even independent of the development of AD. However, a more recent study found that the e4 genotype was not over-represented in CAA cases without evidence of AD. Rather, this group found an increase in the e2 allele in the CAA-associated ICH cases, particularly when cases with concurrent AD were excluded.[93c] Most of the recent studies had relatively few patients, which increases the risk of drawing inappropriate conclusions about allele frequencies. This is clearly a complex issue that requires further study with much larger numbers of patients.

The mechanism(s) by which deposition of amyloid or cystatin C produces an ICH remains unclear. Perhaps in some patients the e4 allele potentiates the deposition of APP, cystatin C, or other proteins that weaken the vessel wall. It has been suggested that deposition of β-amyloid leads to loss of smooth muscle cells, with weakening and disruption of the vessel wall. This leads eventually to rupture and subsequent ICH, at least in some patients.[93d] Whether the use of concomitant medications (such as aspirin) or the presence of other diseases (i.e., hypertension) hastens the occurrence of an ICH is unknown at present.

The Duke group has analyzed cases of "sporadic" CAA for evidence of gene mutations that can cause ICH. In one case, a white male in his 60's had several ICHs. Analysis of brain DNA from autopsy material showed a heterozygous mutation in exon 2 of cystatin C, which is identical to the mutation seen in HCHWA-I (Figure 6A). Antibody staining of his brain showed

Table 2
Antibody Staining of Brain and Clinical Findings in Cases of Cerebral Amyloid Angiopathy

| Antibody Staining** | ICH | Clinical Diagnosis* | | Other |
		ICH+AD	AD	
Aβ	8++	3++	4++	1++
	2+	1+		
Cystatin C	3++	2++	1++	1++
	6+	2+	3+	
	1−			
Apo E	1++	1++	1++	1−
	7+	3+	2+	
	2−		1−	
Totals	10	4	4	1

*Clinical diagnoses are: ICH, intracerebral hemorrhage (lobar); AD, Alzhemier's disease. One patient in the ICH group had a subarachnoid hemorrhage. The one patient in the "other" group had severe ischemic vascular disease without an ICH or AD.
**Antibody staining abbreviations are: Aβ, amyloid beta-peptide; Apo E, apolipoprotein E. Staining intensities are: ++, moderate or strong staining of all or a majority of vessels within several high-powered fields; +, slight or intermediate staining of some vessels (<50%) per high-powered field; −, staining absent or nondiagnostic.

prominent deposition of cystatin C within the walls of cerebral vessels (Figure 6B).[94] This is an unexpected finding, since this patient was much older than the typical cases of HCHWA-I, was not of Icelandic origin, and had no family history of ICHs.[24,62] It raises the question of whether additional factors (i.e., genetic, environmental) might hasten or retard the expression of the ICH phenotype in individuals with the HCHWA-I mutation of cystatin C.

The identification of one or more genetic markers that can noninvasively identify individuals with CAA, or at high risk for developing CAA, would have obvious medical implications. Such individuals would be candidates to receive strict risk-factor modification to reduce the likelihood of developing an ICH. The administration of powerful anticoagulants (i.e., heparin, coumadin), thrombolytics, and even antiplatelet agents might be reconsidered in such individuals due to the potential for an increased risk of developing an ICH. It is also possible that genetic markers by themselves will not be sensitive enough to identify such individuals. It is possible that other factors such as age will be found to interact with inherited and environ-

mental factors to produce the phenotype now identified as CAA.

Vascular Malformations

Vascular malformations affecting the brain and central nervous system can be classified into four groups: (1) cavernous hemangiomas, (2) arteriovenous malformations, (3) capillary telangiectasias, and (4) venous angiomas.[95] Hemangioblastomas, which some consider a tumor, will be discussed separately. The lesions discussed in this section, while clinically distinct, may at times occur together and in some cases are inherited. The increased use of head CT and MRI has shown, with increasing frequency, a familial association for some types of these vascular malformations.

Cavernous Hemangiomas

Cavernous hemangiomas (CH), sometimes referred to as cavernous angiomas, can occur in any part of the central nervous system, including the brainstem and spinal cord. They are most common in the cerebral

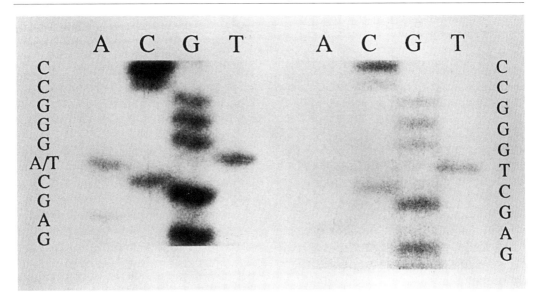

Figure 6A. An autoradiograph of a sequencing gel. The left panel shows a mutation in cystatin C of a patient with a lobar ICH. This mutation is identical to the one seen in cases of Icelandic amyloid angiopathy (see text for details). A control sequence is in the right panel.

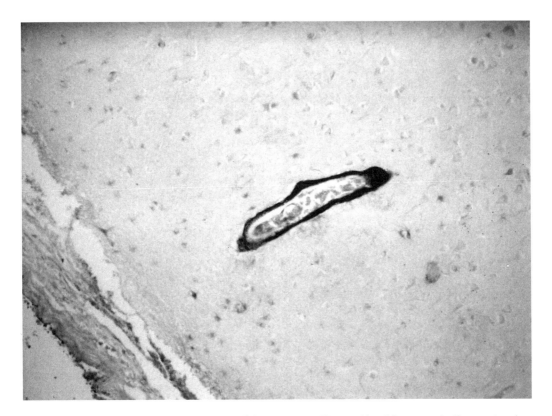

Figure 6B. An antibody stain of the brain of the patient in Figure 6A with a cystatin C mutation (see text for details) causing CAA and an ICH. Many vessels stained strongly positive with an anti-cystatin C antibody.

hemispheres.[96] Pathologically, they typically consist of sinusoidal channels without interspersed brain parenchyma.[97] Unlike arteriovenous malformations, CHs lack large feeding arteries or draining veins.[98] In some cases CHs may be asymptomatic, while in others symptoms such as seizures or focal neurologic deficits (often due to hemorrhage) may occur. CHs can become symptomatic at any time during life. Reports in the literature have shown patients becoming symptomatic even in the prenatal period, although in most studies symptoms first appear in the third or fourth decade of life.[97] These lesions can be detected easily by head CT or MRI scan. In general, CHs are much better visualized by MRI, as compared to CT scans (Figures 7A and 7B).[97–99] Angiography rarely demonstrates a vascular lesion; typical findings include a negative study or an avascular mass.[98]

The vast majority of familial CHs (80%) are supratentorial, while about 20% are infratentorial.[96] There are two reports in the literature of the familial occurrence of isolated brainstem CHs, both of which involved Hispanic families.[100,101] There is a suggestion that infratentorial CHs may be more aggressive in terms of expansion and bleeding frequency.[102] Females may also have a tendency to have more aggressive lesions.[102] This has obvious implications when considering types of therapeutic approaches. On rare occasions CHs can occur in the spinal cord.

The familial occurrence of CH was believed to be quite rare prior to the widespread use of CT and MRI. However, several studies of large pedigrees with familial CH have recently appeared in the literature, supporting the theory that at least some cases are genetic.[99,103,104] Some studies have suggested an increased prevalence for familial CH in Hispanic families, although this may represent ascertainment bias in some cases.[104a] In those cases with familial CH, the pattern of inheritance is consistent with an autosomal dominant disorder with approximately 90% penetrance.[99] There are several cases in the literature of obligate heterozygotes who did not display clinical or radiologic evidence of CH.[99] This suggests that other factors may be necessary for the expression of the CH phenotype. An alternative explanation is that the obligate carriers were not studied serially and that with time they would have developed radiologic or clinical evidence of CH.

A large study of the natural history and genetic features of CH was published by Hayman et al. in 1982.[99] This study was based on a large family of Mexican-American origin, which is of note since several of the CH families reported in the literature were also of Mexican-American or Hispanic origin. The kindred had 122 individuals, of which 22 (11 males, 11 females) had clinical or CT evidence of CH. Not all of the members of the kindred were studied by CT, therefore the actual percentage of affected individuals cannot be determined. In about 50% of cases, multiple CHs were seen on head CT scan.

Several other features emerged from the study of this large kindred. Two of the individuals became progressively demented, one in his 50s and another in her 70s. The 50-year-old patient with dementia had several CHs, some of which had bled. The 70-year-old woman with dementia did not have clinical or radiologic evidence of CHs.[99] No pathologic data are available from these patients to prove whether they had Alzheimer's disease.

Another finding was the progression of the lesions on follow-up studies. In most cases the CHs showed no change in appearance by head CT. Another group of patients showed single lesions that rebled several times although other lesions remained asymptomatic. In a third group of patients, new lesions appeared on follow-up CT in areas that had been normal previously.[99] Another study found that new lesions appeared at a rate of one lesion per decade of age.[105] These findings suggest that CH formation and hemorrhage is a dynamic process, perhaps controlled by local vascular or other factors.

A study performed by Rigamonti et al.

in 1988.[98] focused on 13 patients and their relatives with a positive family history of CH. Seizures were the most common presenting symptom, reported in 69% of symptomatic patients. Among the 44 first-degree relatives of the symptomatic probands, 9 had symptoms and 35 were asymptomatic. MRI scans were performed on 11 of these first-degree relatives and 5 other relatives of the probands. The scans showed CHs in 14 of 16 relatives. This study and others have found a strong tendency for multiple CHs in patients with the familial form of CH, while the sporadic form appears to have a much lower incidence of multiple lesions. It has been stated that the familial form of CH will have multiple lesions in 50–70% of cases, compared to multiple lesions in only 10–13% of "sporadic" cases.[95,96,106] These estimates are certainly susceptible to ascertainment bias, since failure to study (with MRI) all members of an at-risk family may grossly underestimate the familial occurrence.

Some studies have found familial CHs of the brain to be associated with CHs of the retina.[96] It is unclear if familial CH with involvement of the brain and retina is the same disease as familial CH involving just the brain.

The ascertainment of large CH kindreds with many living affecteds has made it possible to perform genetic linkage studies. These studies have identified linkage to a series of markers on chromosome 7, in the q11–22 region (see Table 3 for linked markers).[107,107a] The CH gene appears to be located in an 11cM to 15cM region. Experiments are underway to identify and analyze candidate genes located within this region.

Several conclusions can be drawn about familial CHs: (1) they are inherited as an autosomal dominant trait with near complete penetrance; (2) for unknown reasons they may be more common in families of Hispanic origin; (3) multiple lesions appear to be associated with the familial form more than the sporadic form; and (4) MRI scan is far more sensitive than CT for detecting lesions in the brain.

Based on these findings, we recommend that any symptomatic relative, as well as any asymptomatic first-degree relative of a patient with CH, consider elective MRI for the detection of cranial lesions, particularly if multiple CHs are found in the affected family member(s) and if there is a positive family history. Although many such lesions are asymptomatic and are unlikely to bleed, their presence may result in modification of risk factors, altered use of certain medications, and reconsideration of physical activities that may put the patient at high risk for complications arising from these lesions.

Familial Arteriovenous Malformations

The familial occurrence of arteriovenous malformations (AVM) has been recognized in a handful of case reports. AVMs have been associated with several inherited disorders, most notably Rendu-Osler-Weber syndrome, Sturge Weber disease, neurofibromatosis, hereditary hemorrhagic telangiectasias, and von Hippel Lindau disease.[108–110] In addition, AVMs can also be associated with intracranial aneurysms (IA).[111,112] Although not all cases of IAs have an obvious genetic etiology, the genetics of IA and subarachnoid hemorrhage are discussed in more detail in *Chapter 12*. This section, however, will focus on familial AVMs not associated with these particular syndromes. Cerebrovascular lesions associated with these syndromes will be covered in subsequent sections of this chapter.

There are several significant problems in the study of familial AVMs. One is the incomplete study of at-risk kindreds. With very few exceptions, most of the studies in the literature have reported only affected patients who were symptomatic. There has been a paucity of prospective studies that systematically screen asymptomatic at-risk family members for AVMs. This is unfortunate for two reasons. First, head CT and MRI scans can safely and easily detect clinically silent AVMs that may be amenable to therapy. Although many of these lesions do not become symptomatic or hemorrhage

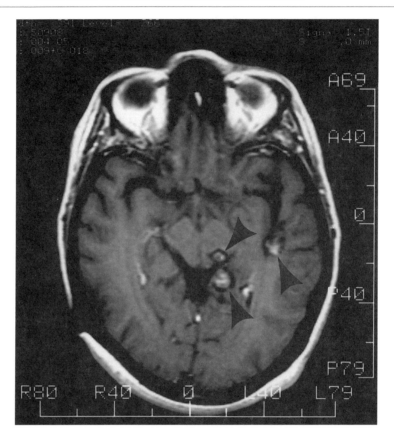

Figure 7A. A contrast-enhanced T$_1$-weighted MRI image. It shows several cavernous hemangiomas (CHs) that slightly enhance in the region of the left midbrain and left temporal lobe (arrowheads).

grossly, in some cases they can produce devastating strokes. The second issue is the mode of inheritance. The studies cited below indicate that familial AVMs are probably inherited as an autosomal dominant trait in some cases. This means that family members may be at considerable risk for inheriting the AVM trait.

Another issue is the proper classification of these lesions. In some cases a large intracerebral hemorrhage can obscure an underlying AVM or tumor, making accurate diagnosis and classification difficult. In other cases, the clinical or pathologic data are not adequate to allow the proper classification of the lesion as an AVM, cavernous hemangioma, or capillary telangiectasia. Since the latter two lesions can also be inherited as part of specific syndromes, proper classification is very important. As noted above, in some

families there may be overlap between AVMs and other vascular lesions.[113]

A report by Allard et al.[109] used MRI to study a kindred with familial cerebral AVMs. In this prospective study, 7 of 17 studied individuals had evidence of cerebral vascular lesions, of whom 3 were asymptomatic. Individuals with lesions ranged in age from 11 to 46 years. Some of the lesions on CT and MRI had a pattern suggestive of an AVM, while others could have been telangiectasias or hemangiomas. Angiography on one "affected" patient did not show any abnormal vessels. The pattern of inheritance was consistent with an autosomal dominant trait.

A study by Boyd and colleagues[114] reported on a father and 4 children who were studied for AVMs. The father was 45 years old when studied, and had a long history of

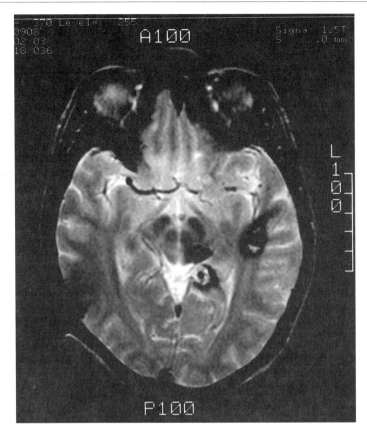

Figure 7B. A T_2-weighted MRI of the patient in Figure 7A. Note the hemosiderin (dark) signal around the CHs. This patient had a positive family history of CHs.

migraine headaches and right facial twitching. A head CT and angiogram confirmed the presence of a large AVM. Four of his children had a head CT scan. Three of the four showed abnormalities consistent with an AVM, which was confirmed by surgery in one case and angiography in another case.

Another study by Snead et al.[115] reported 3 siblings, two by the same father, with AVMs. Two of the patients became symptomatic quite young, at ages 14 and 17 years. The third sibling was studied electively at 11 years of age and had an AVM documented by CT and angiography.

A study by Kidd and Cummings[116] reported on a large pedigree with two cases of documented cerebral hemorrhage, although a vascular lesions was proven in only one case and was not clearly an AVM. This family has numerous other cases of sudden death

of members at a young age, but there is a paucity of radiologic or pathologic evidence to prove that these were due to AVMs.

There is a case report of AVMs occurring in two brothers with myotonic muscular dystrophy.[117] Although this may represent a chance occurrence, it may also indicate a locus for familial AVM on the proximal long arm of chromosome 19 (CH19), where the gene for myotonic dystrophy (DM) is located.[118] An alternative explanation is that, in this particular family, there is some type of abnormality in the region of CH19 (i.e., deletion, duplication, translocation) in or near the DM locus that causes both diseases.

A review of the literature was published recently by Yokoyama and colleagues.[108] They found that cases of familial AVM tended to exhibit symptoms at a somewhat younger age (10–19 years) com-

Table 3
Genetic Markers Linked to Familial
Cavernous Hemangioma*

Marker*
D7S652
D7S558/D7S1789
D7S804

*Adapted from references 107a and 107b.

pared to sporadic cases which tended to have a peak presentation between the ages of 21 and 40.[119,120] The location of the familial AVMs was similar to the sporadic cases, with the vast majority being supratentorial in both circumstances. The cumulative data suggest an autosomal dominant mode of transmission, although other environmental factors may play a role in gene expression. Unfortunately, the reported pedigrees are too small for standard linkage studies, and no genes have been identified. The possibility that familial AVMs will be a genetically heterogeneous group of disorders will further complicate genetic studies.

Based on the data, it seems reasonable to suggest that close relatives of a proband with an AVM who exhibit neurologic symptoms (i.e., seizures, severe headaches, progressive focal deficits, cranial bruits) should be studied noninvasively with a CT or MRI to search for such lesions. This is especially true if there is a positive family history of AVMs or an early age of onset for the proband. Until more extensive prospective studies are performed to determine the frequency of familial AVMs in an unselected population, there is insufficient data to make firm recommendations about screening asymptomatic family members.

Hereditary Hemorrhagic Telangiectasia

Hereditary hemorrhagic telangiectasia (HHT) is also known by the eponym Rendu-Osler-Weber syndrome. It is an autosomal dominant disorder that is characterized by the triad of dermal mucosal telangiectasias, recurrent hemorrhages (particularly nasal and gastrointestinal [GI]), and a familial occurrence.[121,122]

The clinical manifestations of HHT most commonly include epistaxis, which is seen in up to 96% of patients.[123] Mucocutaneous telangiectasias occur most commonly on the palms, nail beds, lips, and tongue. The GI tract is involved in approximately 15% of cases, leading to various types of hemorrhage. Liver involvement can be found in about 8% of patients, manifested mostly by hepatomegaly and cirrhosis.[123,124] Of note to neurologists is the fact that this syndrome can cause cerebral vascular disease by one of several mechanisms.

Involvement of the CNS occurs in 8–12% of affected patients and their family members.[121,125] Pulmonary AVMs, which occur in 4–10% of patients and families with HHT, account for numerous CNS complications.[121,123,126] Pulmonary AVMs can lead to strokes by being a source of air emboli, as well as causing paradoxical emboli. Septic emboli from a pulmonary AVM can produce a cerebral abscess. The most common types of lesions are summarized in Table 4, and include complications secondary to pulmonary AVMs, vascular malformations of the brain (including AVMs), and aneurysms.

Several different CNS vascular malformations can occur in patients with HHT. One report suggested that the most common malformations are cerebral telangiectasias and angiomas, which account for about 17% of all neurologic complications in patients with HHT.[121] However, another study found that cerebral lesions in patients with HHT can have variable and atypical appearances on MRI. Angiography showed that many of these lesions were in fact AVMs that were clearly visible on MRI.[121a] AVMs of the spinal cord have also been reported in these patients. Cerebral saccular aneurysms have been reported, as well as rare instances of carotid-cavernous fistula.[121,127,128] These vascular lesions can lead to subarachnoid, intracerebral, or intraventricular hemorrhage.

Genetic studies of families with HHT

Table 4
CNS Complications of Hereditary Hemorrhagic Telangiectasia

Pulmonary Arteriovenous Fistula	Primary CNS Lesions
Hypoxia	Telangiectasias
Headaches	Arteriovenous malformations
Syncope	Aneurysms
Seizures	Carotid-cavernous fistula
Transient ischemic attacks	Spinal cord AVMs
Cerebral thrombosis	Portal-systemic encephalopathy
Septic emboli	
Paradoxical emboli	
Air embolism	
Brain abscess	

Parts of this table were adapted from reference 121.

have confirmed an autosomal dominant mode of inheritance with penetrance of 97% for at least one manifestation of the disease.[123] It has been stated that a negative family history exists in 15–30% of affected patients, although this result is obviously influenced by the methods of ascertainment and analysis.[129] In general, all patients manifest some symptoms by the age of 40 years.[124]

Two groups have used genetic linkage techniques to identify one locus responsible for HHT. The locus is on the long arm of chromosome 9 in the q33-q34 region.[130,131] A group led by Marchuk at Duke identified endoglin (a transforming growth factor-β [TGF-β] binding protein) as the causative gene/protein in some cases of HHT.[21a] A variety of mutations in this gene have been identified.[21a,131a] It is postulated that defective binding of TGF-β leads to abnormal development and/or repair of endothelial cells and vessels. There is evidence that a homozygous state is lethal.[132]

A second locus for HHT has been mapped to the pericentromeric region of chromosome 12.[132a] Further experiments identified three different mutations in the activin receptor-like kinase 1 (ALK1) gene in patients with HHT.[132b] ALK1 is a cell surface receptor for TGF-β. As with endoglin, it is postulated that defects in ALK1 may lead to defective vascular development or repair. The pathologic analysis of these lesions

typically shows focal dilatation of postcapillary venules, sometimes accompanied by dilated arterioles.[121,133] These lesions become progressively larger with time, and develop fragile walls.[95,110] If the lesions stay relatively small, they form a small tangle of vessels called telangiectasias. However if they enlarge considerably, arterioles and venules may communicate directly, thereby becoming arteriovenous malformations.

In terms of diagnosis, the smaller cerebral telangiectasias probably cannot be well visualized by head CT. Even MRIs may miss smaller lesions, although larger telangiectasias, along with AVMs and even some aneurysms can be detected by MRI.[134] Most of the small telangiectasias will not be seen angiographically, although the AVMs and aneurysms should be visualized in most cases, particularly with a contrast enhanced CT or MRI scan.[134] Symptomatic lesions can be treated surgically if they are accessible.

At-risk family members in kindreds with HHT should probably be screened for pulmonary AVMs, since these lesions can lead to significant neurologic complications. An argument can also be made to screen at-risk family members (with or without neurologic symptoms) with a brain MRI scan to evaluate the presence of a cerebral abscess, AVMs, and other vascular lesions. The identification of such lesions, even if asymptomatic, may have implications for the management of other vascular risk factors, the use of various medications, and for genetic counseling.

Other Disorders

Other vascular malformations in the central nervous system can produce intracerebral hemorrhages. For example, venous angiomas are small lesions that can, on rare occasion, bleed and produce tiny intracerebral hemorrhages. They can be detected in some cases by identification of an enhancing draining vein on CT or MRI.[135,136] These venous angiomas are typically single and sporadic, although in rare cases they can be multifocal and/or familial, with a dominant pattern of inheritance. A recent study identified a single nucleotide change (C-to-T, arginine to tryptophan) at position 2545 in the receptor tyrosine kinase TIE2 gene on chromosome 9p21.[136a] This mutation segregated with venous angiomas in 2 families with an autosomal dominant pattern.

Many different types of tumors (primary and secondary) can bleed and cause an intracerebral hemorrhage. With the exception of hemangioblastomas, we have chosen to forego a discussion of the genetics of other types of neoplasms that may involve the brain. As indicated at the beginning of this chapter, there is significant evidence supporting a genetic etiology for at least a subset of cases of familial subarachnoid hemorrhage due to rupture of an IA. This issue is discussed in detail in *Chapter 12*.

Hemangioblastoma/von Hippel-Lindau Disease

Hemangioblastomas are usually benign cystic tumors that occur within the nervous system, most commonly in the cerebellum.[137,138] Hemangioblastomas are often associated with von Hippel-Lindau disease (VHL) which defines a syndrome that includes retinal angiomas, renal cell carcinoma, pheochromocytoma, and cysts in one

or more organs, as well as hemangioblastomas in the nervous system.[139,140] The relationship of hemangioblastomas to VHL is discussed in more detail below, but is of central importance for understanding this disease.

Hemangioblastomas in the cerebellum may be asymptomatic when small, but eventually will produce headache in the vast majority of patients. Other symptoms such as nausea and vomiting, vertigo, and diplopia.[140] Physical findings that are present in 60–70% of affected patients include nystagmus, ataxia, and papilledema.[140,141] Many of these findings result from increased intracranial pressure produced by obstructive hydrocephalus. Only rarely do these tumors manifest with intracerebral or subarachnoid hemorrhage. Most of the symptoms are referable to mass effect, edema, and obstructive hydrocephalus. They do produce erythropoietin which causes an erythrocytosis in 12–49% of cases.[137,142] The diagnosis of hemangioblastoma can be made easily with brain CT or MRI scans (Figure 8). Small lesions can be detected by enhanced MRI scans. Cerebral angiography is likely to show a hypervascular mass.[141,143]

The genetics of hemangioblastomas is closely related to the genetics of VHL. Von Hippel-Lindau disease is an autosomal dominant syndrome with near complete penetrance. Segregation studies have shown a pattern of inheritance consistent with a single autosomal dominant gene. Most patients will manifest symptoms by the age of 65 years.[139] VHL is defined as follows: for sporadic cases, two or more hemangioblastomas, or, a single hemangioblastoma with visceral evidence of the disease.[139] If there is a family history of hemangioblastomas, only one lesion is required to define VHL. Based on these widely used diagnostic criteria, patients presenting with an isolated hemangioblastoma and a negative family history are defined as not having VHL. Whether these definitions will be supported by ongoing genetic studies remains to be determined.

Detailed clinical studies of VHL have

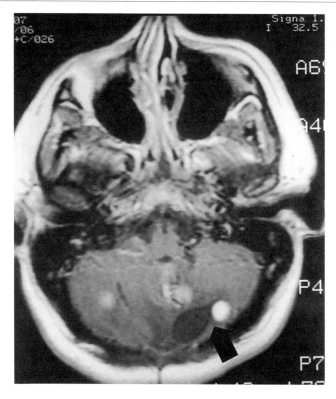

Figure 8. A contrast-enhanced T_1-weighted MRI of a patient with von Hippel-Lindau disease. Note the enhancing nodule and associated cyst in the left cerebellum (arrow). This is a classic hemangioblastoma. There are two other lesions in the cerebellum with slight enhancement, also hemangioblastomas.

shown that posterior fossa hemangioblastomas are the most common CNS lesion, followed by hemangioblastomas in the spinal canal.[142] Retinal and other typical systemic lesions are described above. It is unusual for all but the most severely affected patients to have all of these manifestations of VHL. Obviously, the completeness of diagnosis and ascertainment is dependent on the level of investigation.

Approximately 20% of patients with hemangioblastoma have either multiple lesions or a family history of hemangioblastomas.[140] The proportion of patients presenting with cerebellar hemangioblastomas who are found to have VHL varies between 0% and 40%.[139] In general, patients with sporadic cerebellar hemangioblastomas tend to present later (mean age 42–47 years) than patients with a family history of VHL or clinical evidence of VHL (mean age 30–34 years).[140,142] Sporadic cases of hemangioblastoma almost always have single lesions, while multifocal lesions are strongly associated with VHL. These findings can be summarized as follows: a cerebellar hemangioblastoma is more likely to be familial or to be associated with VHL if it occurs at a younger age or if it is multifocal.

Genetic linkage studies have localized the VHL gene to human chromosome 3p25-p26.[144] Detailed mapping studies have shown the VHL gene to be near the marker D3S601.[145] A study by Latif et al. reported the identification of a putative VHL gene.[146] This gene encodes a cDNA of 1,810 bp. The gene can produce two different proteins of 18 kd and 30 kd, depending on which translation start site is used.[147]

The VHL protein appears to function as

a tumor suppressor gene. A variety of mutations in the VHL gene have been described. Germ line mutations (detected in lymphocytes) tend to cluster in exons 1 and 3, while somatic mutations (found only in the tumor tissue) can occur throughout the gene.[148] Although a great variety of mutations in the VHL gene have been described, there are two basic types; missense mutations (which account for about 60% of the VHL mutations) and truncation mutations (accounting for 40%).[148] Other types of mutations such as substitutions have recently been described.[148a] There does seem to be a correlation between the type of mutation and the disease phenotype. In general, the missense mutations are found in VHL families with pheochromocytomas, whereas mutations causing protein truncation tend to occur in cases without pheochromocytomas.[149]

In many cases of VHL (approximately 50%) there is a loss of heterozygosity (LOH), meaning that one allele is deleted, and the other allele is affected by a mutation.[150] This type of "double-hit" scenario is common with tumor suppressor genes and similar diseases. Two transcriptional factors, Elongin B and Elongin C, have been shown to bind to the VHL protein.[151] In cases of VHL mutations, the VHL protein will not bind normally to Elongin B or C, resulting in excessive elongation and up-regulation of target genes.[152] This can lead to tumorogenesis and increased angiogenesis via a similar mechanism.[148]

Other studies have shown that hypermethylation of the VHL gene can also occur. Hypermethylation leads to inactivation of the gene. If a tumor-suppression gene is inactivated, it can lead to a number of cancers. One study of 53 tumors from 33 patients with VHL found that hypermethylation was present in 33% of tumors without LOH.[150]

The presence of VHL mutations was studied in cases of sporadic cerebellar hemangioblastoma (i.e., without other manifestations of VHL). Somatic mutations were found in 40% of such cases, and LOH was found in one case.[153] Therefore, it appears as

though mutations in the VHL gene can cause sporadic cases of this cerebellar tumor. Characterization of these various mutations has made case-specific genetic counseling possible in some cases.[146]

Conclusions

Numerous types of vascular processes, ranging from a microscopic amyloid angiopathy to arteriovenous malformations can produce the clinical syndrome of an intracerebral hemorrhage. As is true with many other genetic disorders, especially those of late onset or with a complex genetic background, we are just beginning to fully appreciate the importance that genetic factors may play in the etiology of these processes. Understanding the genetics of these disorders and their role in producing an ICH is especially challenging, since in many cases the lesion cannot be detected by simple clinical examination. Some lesions such as amyloid angiopathy currently require microscopic examination for the diagnosis, while others such as AVMs or CHs can be detected by brain MRI scan.

It is likely that we have only begun to scratch the surface of understanding the genetics of these various conditions. One roadblock to progress is the failure to study at-risk but asymptomatic family members of affecteds. The ascertainment of these individuals will be vital for establishing large numbers of affecteds and large families that may then be used for further genetic analyses. The discovery of genetic markers and ultimately the mutations for these disorders will greatly enhance the screening of asymptomatic at-risk individuals. Once asymptomatic gene carriers are identified, risk-factor modification and prophylactic approaches may reduce the occurrence of an ICH in such individuals. Since ICH is among the most deadly of all stroke types, the approach outlined above may make a substantial impact on stroke morbidity and mortality.

References

1. Broderick JP, Brott T, Tomsick T, et al. Intracerebral hemorrhage more than twice as common as subarachnoid hemorrhage. J Neurosurg 1993;78:188–191.
2. Anderson CS, Jamrozik KD, Burvill PW, et al. Determining the incidence of different subtypes of stroke: Results from the Perth Community Stroke Study, 1989–1990. Med J Aust 1993;158:85–89.
3. Broderick JP, Brott T, Tomsick T, et al. The risk of subarachnoid and intracerebral hemorrhages in blacks as compared with whites [see comments]. N Engl J Med 1992;326:733–736.
4. Giroud M, Gras P, Chadan N, et al. Cerebral haemorrhage in a French prospective population study. J Neurol Neurosurg Psychiatry 1991;54:595–598.
5. Schutz H, Bodeker RH, Damian M, et al. Age-related spontaneous intracerebral hematoma in a German community. Stroke 1990;21:1412–1418.
6. Sacco RL, Hauser WA, Mohr JP. Hospitalized stroke in blacks and Hispanics in northern Manhattan. Stroke 1991;22:1491–1496.
7. Wakai S, Nagai M. Histological verification of microaneurysms as a cause of cerebral haemorrhage in surgical specimens. J Neurol Neurosurg Psychiatry 1989;52:595–599.
8. Mohr J, Caplan L, Melski J, et al. The Harvard cooperative stroke registry: A prospective study. Neurology 1978;28:754–762.
9. Furlan A, Whisnant J, Elveback L. The decreasing incidence of primary intracerebral hemorrhage: A population study. Ann Neurol 1979;5:367–373.
10. Feldmann E, Tornabene J. Diagnosis and treatment of cerebral amyloid angiopathy. Clin Geriatr Med 1991;7:617–630.
11. Sillus M, Saeger W, Linke RP, et al. Cerebral amyloid angiopathy: Frequency, significance and immunohistochemistry. Zentralbl Pathol 1993;139:207–215.
12. Vonsattel JP, Myers RH, Hedley WE, et al. Cerebral amyloid angiopathy without and with cerebral hemorrhages: A comparative histological study. Ann Neurol 1991;30:637–649.
13. Wityk RJ, Caplan LR. Hypertensive intracerebral hemorrhage: Epidemiology and clinical pathology. Neurosurg Clin N Am 1992;3:521–532.
14. Molinari GF. Lobar hemorrhages: Where do they come from? How do they get there? [editorial; comment]. Stroke 1993;24: 523–526.
15. Wakai S, Kumakura N, Nagai M. Lobar intracerebral hemorrhage: A clinical, radiographic, and pathological study of 29 consecutive operated cases with negative angiography. J Neurosurg 1992;76:231–238.
16. Brott T, Thalinger K, Hertzberg V. Hypertension as a risk factor for spontaneous intracerebral hemorrhage. Stroke 1986;17: 1078–1083.
17. Kittner SJ, McCarter RJ, Sherwin RW, et al. Black-white differences in stroke risk among young adults. Stroke 1993;24(suppl 12):I13–I15.
18. Ross-Russell R. Observations on intracerebral aneurysms. Brain 1963;86:425–441.
19. Kase C, Mohr J, Caplan L. Intracerbral hemorrhage. In Barnett H, Mohr J, Stein B, et al (eds). Stroke: Pathophysiology, Diagnosis, and Management. New York: Churchill Livingstone, 1992, 561–616.
20. Vinters H. Cerebral amyloid angiopathy. In Barnett H, Mohr J, Stein B, et al (eds). Stroke: Pathophysiology, Diagnosis, and Management. New York: Churchill Livingstone, 1992, 821–858.
20a. Saunders AM, Strittmatter WJ, Schmechel D, et al. Association of apolipoprotein allele epsilon 4 with late onset familial and sporadic Alzheimer's disease. Neurology 43;1993: 1467–1472.
21. Graffagnino C, Herbstreith M, Roses AD, Alberts MJ. Molecular genetic analysis of intracerebral hemorrhage. Arch Neurol 1994; 51:981–984.
21a. McAllister KA, Grogg KM, Johnson DW, et al. Endoglin, a TGF-β binding protein of endothelial cells, is the gene for hereditary haemorrhagic telangiectasia. Nature Genet 1994;8:345–351.
21b. Alberts MJ, Davis JP, Graffagnino C, et al. Endoglin gene polymorphism as a risk factor for sporadic intracerebral hemorrhage. Ann Neurol 1997;41:683–686.
21c. Alberts MJ, Moore SE, Hoffmann KL, et al. Genetic studies of intracerebral hemorrhage. Stroke 1998;29:320. Abstract.
21d. Alberts MJ, Graffagnino C, McClenny C, et al. ApoE genotype and survival from intracerebral hemorrhage. Lancet 1995;346:575.
21e. Alberts MJ, Graffagninno C, McClenny C, et al. Effect of apoE genotype and age on survival after intracerebral hemorrhage. Stroke 1996;27:183.
22. Haan J, Algra PR, Roos RA. Hereditary cerebral hemorrhage with amyloidosis-Dutch type: Clinical and computed tomographic analysis of 24 cases. Arch Neurol 1990;47: 649–653.
23. Haan J, Roos RA, Briet PE, et al. Hereditary cerebral hemorrhage with amyloidosis-Dutch type. Research Group Hereditary

Cerebral Amyloid-Angiopathy. Clin Neurol Neurosurg 1989;91:285–290.

24. Jensson O, Gudmundsson G, Arnason A, et al. Hereditary cystatin C (gamma-trace) amyloid angiopathy of the CNS causing cerebral hemorrhage. Acta Neurol Scand 1987;76:102–114.

25. Luyendijk W, Bots GT, Vegter, et al. Hereditary cerebral haemorrhage caused by cortical amyloid angiopathy. J Neurol Sci 1988;85: 267–280.

26. Roos RA, Haan J, van Broeckhoven C. Hereditary cerebral hemorrhage with amyloidosis-Dutch type: A congophilic angiopathy. An overview. Ann NY Acad Sci 1991;640:155–160.

26a. Bornebroek M, Westendorp RG, Hann J, et al. Mortality from hereditary cerebral hemorrhage with amyloidosis-Dutch type: The impact of sex, parental transmission and year of birth. Brain 1997;120:2243–2249.

27. Haan J, Roos RA, Algra PR, et al. Hereditary cerebral haemorrhage with amyloidosis-Dutch type: Magnetic resonance imaging findings in 7 cases. Brain 1990;1251–1267.

28. Wattendorff A, Bots G, Went L, Endtz L. Familial cerebral amyloid angiopathy presenting as recurrent cerebral haemorrhage. J Neurol Sci 1982;55:121–135.

29. Rozemuller AJ, Roos RA, Bots GT, et al. Distribution of beta/A4 protein and amyloid precursor protein in hereditary cerebral hemorrhage with amyloidosis-Dutch type and Alzheimer's disease. Am J Pathol 1993;142: 1449–1457.

30. Haan J, Bakker E, Jennekens SA, Roos RA. Progressive dementia, without cerebral hemorrhage, in a patient with hereditary cerebral amyloid angiopathy. Clin Neurol Neurosurg 1992;94:317–318.

31. van Duinen S, Castano E, Prelli F, et al. Hereditary cerebral hemorrhage with amyloidosis of Dutch origin is related to Alzheimer disease. Proc Natl Acad Sci 1987; 84:5991–5994.

32. Kang J, Lemaire H, Unterbeck A, et al. The precursor of Alzheimer disease amyloid A4 protein resembles a cell-surface receptor. Nature 1987;325:733–736.

33. Goldgaber D, Lerman M, McBride O, et al. Characterization and chromosomal localization of a cDNA encoding brain amyloid of Alzheimer's disease. Science 1987;235: 877–880.

34. Yoshikai S-i, Sasaki H, Doh-ura K, et al. Genomic organization of the human beta-amyloid precursor gene. Gene 1990;87:257–263.

35. Tanzi R, McClatchey A, Lamperti E, et al. Protease inhibitor domain encoded by an amyloid protein precursor mRNA associated with Alzheimer's disease. Nature 1988;331:528–530.

36. Ponte P, Gonzalez-DeWhitt P, Schilling J, et al. A new A4 amyloid mRNA contains a domain homologous to serine proteinase inhibitors. Nature 1988;331:525–527.

37. Glenner G, Wong C. Alzheimer's disease and Down's syndrome: Sharing a unique cerebrovascular amyloid fibril protein. Biochem Biophys Res Commun 1984;122:1131–1135.

38. Masters CL, Multhaup G, Simms G, et al. Neuronal origin of a cerebral amyloid: Neurofibrillary tangles of Alzheimer's disease contain the same protein as the amyloid of plaque cores and blood vessels. EMBO J 1985;4:2757–2763.

39. Selkoe D, Podlisny M, Joachim C, et al. Beta-amyloid precursor protein of Alzheimer disease occurs at 110- to 135-kilodalton membrane-associated proteins in neural and nonneural tissues. Proc Natl Acad Sci 1988; 85:7341–7345.

40. Koo E, Sisodia S, Archer D, et al. Precursor of amyloid protein in Alzheimer disease undergoes fast antegrade axonal transport. Proc Natl Acad Sci 1990;87:1561–1565.

41. Whitson J, Glabe C, Shintani E, et al. Beta-amyloid protein promotes neuritic branching in hippocampal cultures. Neurosci Lett 1990;110:319–324.

42. Olterdorf T, Fritz L, Schenk D, et al. The secreted form of Alzheimer's amyloid precursor protein with the Kunitz domain is protease nexin-II. Nature 1989;341:144–147.

43. Esch F, Keim P, Beattie E, et al. Cleavage of amyloid β peptide during constitutive processing of its precursor. Science 1990;248: 1122–1124.

44. Smith R, Higuchi D, Broze G. Platelet coagulation factor XIa-inhibitor, a form of Alzheimer amyloid precursor protein. Science 1990;248:1126–1128.

45. Schmaier A, Dahl L, Rozemuller A, et al. Protease nexin-2/amyloid protein precursor: A tight-binding inhibitor of coagulation factor IXa. J Clin Invest 1993;92:2540–2545.

46. Chartier-Harlin M, Crawford F, Houlden H, et al. Early-onset Alzheimer's disease caused by mutations at codon 717of the amyloid precursor protein gene. Nature 1991;353: 844–846.

47. Goate A, Chartier-Harlin M, Mullan M, et al. Segregation of a missense mutation in the amyloid precursor protein gene with familial Alzheimer's disease. Nature 1991;349: 704–706.

48. Murrell J, Farlow M, Ghetti B, et al. A mutation in the amyloid precursor protein associated with hereditary Alzheimer's disease. Science 1991;253:97–99.

49. Mullan M, Crawford F, Axelman K, et al. A pathogenic mutation for probably Alzheimer's disease in the APP gene at the N-terminus of beta-amyloid. Nat Genet 1992; 1:345–347.

50. Levy E, Carman MD, Fernandez-Madrid IJ, et al. Mutation of the Alzheimer's disease amyloid gene in hereditary cerebral hemorrhage, Dutch type. Science 1990;248:1124–1126.

51. Bakker E, van Broeckhoven C, Haan J, et al. DNA diagnosis for hereditary cerebral hemorrhage with amyloidosis (Dutch type). Am J Hum Genet 1991;49:518–521.

52. Prelli F, Levy E, van Duinen SG, et al. Expression of a normal and variant Alzheimer's beta-protein gene in amyloid of hereditary cerebral hemorrhage, Dutch type: DNA and protein diagnostic assays. Biochem Biophys Res Commun 1990;170:301–307.

53. Hendriks L, van Duijn CM, Cras P, et al. Presenile dementia and cerebral haemorrhage linked to a mutation at codon 692 of the beta-amyloid precursor protein gene. Nat Genet 1992;1:218–221.

54. Wisniewski H, Wegiel J, Frackowiak J, et al. Amyloid deposits in leptomeningeal vessels are produced by smooth muscle cells. Neurol 1994;44 (suppl 2):A371.

55. Cai X-D, Golde T, Younkin S. Release of excess amyloid β protein from a mutant amyloid β protein precursor. Science 1993;259: 514–516.

56. Van Nostrand W, Wagner SL, Haan J, et al. Alzheimer's disease and hereditary cerebral hemorrhage with amyloidosis-Dutch type share a decrease in cerebrospinal fluid levels of amyloid beta-protein precursor. Ann Neurol 1992;32:215–218.

57. Suzuki N, Cheung T, Cai X-D, et al. An increased percentage of long amyloid β protein secreted by familial amyloid β protein precursor (βAPP717) mutants. Science 1994;264:1336–1339.

58. Tanzi R, Vaula G, Romano D, et al. Assessment of amyloid β-protein precursor gene mutations in a large set of familial and sporadic Alzheimer disease cases. Am J Hum Genet 1992;51:273–282.

58a. Van Nostrand WE, Melchar JP, Ruffini L. Pathologic amyloid beta-protein cell surface fibril assembly on cultured human cerebrovascular smooth muscle cells. J Neurochem 1998;70:216–223.

59. Joachim CL, Mori H, Selkoe DJ. Amyloid beta-protein deposition in tissues other than brain in Alzheimer's disease. Nature 1989; 341:226–230.

60. Bloom W, Fawcett D. A Textbook of Histology. Philadelphia: WB Saunders, 1975.

61. Glynn L. Medial defects in the circle of Willis and their relation to aneurysm formation. J Pathol Bact 1940;51:213–222.

62. Gudmundsson G, Hallgrimsson J, Jonasson T, Bjarnason O. Hereditary cerebral hemorrhage with amyloidosis. Brain 1972;95:387–404.

63. Ghiso J, Pons EB, Frangione B. Hereditary cerebral amyloid angiopathy: The amyloid fibrils contain a protein which is a variant of cystatin C, an inhibitor of lysosomal cysteine proteases. Biochem Biophys Res Commun 1986;136:548–554.

64. Ghiso J, Jensson O, Frangione B. Amyloid fibrils in hereditary cerebral hemorrhage with amyloidosis of Icelandic type is a variant of gamma-trace basic protein (cystatin C). Proc Natl Acad Sci USA 1986;83: 2974–2978.

65. Schnittger S, Rao VV, Abrahamson M, Hansmann I. Cystatin C (CST3), the candidate gene for hereditary cystatin C amyloid angiopathy (HCCAA), and other members of the cystatin gene family are clustered on chromosome 20p11.2. Genomics 1993;16:50–55.

66. Abrahamson M, Islam MQ, Szpirer J, et al. The human cystatin C gene (CST3), mutated in hereditary cystatin C amyloid angiopathy, is located on chromosome 20. Hum Genet 1989;82:223–226.

67. Lofberg H, Grubb AO, Nilsson EK, et al. Immunohistochemical characterization of the amyloid deposits and quantitation of pertinent cerebrospinal fluid proteins in hereditary cerebral hemorrhage with amyloidosis. Stroke 1987;18:431–440.

68. Abrahamson M. Human cysteine proteinase inhibitors. Isolation, physiological importance, inhibitory mechanism, gene structure and relation to hereditary cerebral hemorrhage. Scand J Clin Lab Invest Suppl 1988; 191:21–31.

69. Blondal H, Guomundsson G, Benedikz E, Johannesson G. Hereditary cerebral hemorrhage: Dementia with cystatin C amyloidosis. Nord Med 1990;105:76–77.

70. Blondal H, Guomundsson G, Benedikz E, Johannesson G. Dementia in hereditary cystatin C amyloidosis. Prog Clin Biol Res 1989;317:157–164.

71. Levy E, Lopez OC, Ghiso J, et al. Stroke in Icelandic patients with hereditary amyloid angiopathy is related to a mutation in the cystatin C gene, an inhibitor of cysteine proteases. J Exp Med 1989;169:1771–1778.

72. Palsdottir A, Abrahamson M, Thorsteinsson L, et al. Mutation in cystatin C gene causes hereditary brain haemorrhage. Lancet 1988; 2:603–604.

73. Shimode K, Fujihara S, Nakamura M, et al. Diagnosis of cerebral amyloid angiopathy by enzyme-linked immunosorbent assay of cys-

tatin C in cerebrospinal fluid. Stroke 1991;22: 860–866.

74. Sakaki Y, Sasaki H, Yoshioka K, Furuya H. Genetic analysis of familial amyloidotic polyneuropathy, an autosomal dominant disease. Clin Chim Acta 1989;185:291–298.

75. Ikeda S, Hanyu N, Hongo M, et al. Hereditary generalized amyloidosis with polyneuropathy. Brain 1987;110:315–337.

76. Kametani F, Ikeda S, Yanagisawa N, et al. Characterization of a transthyretin-related amyloid fibril protein from cerebral amyloid angiopathy in type I familial amyloid polyneuropathy. J Neurol Sci 1992;108: 178–183.

77. Yong WH, Robert ME, Secor DL, et al. Cerebral hemorrhage with biopsy-proved amyloid angiopathy. Arch Neurol 1992;49:51–58.

78. Itoh Y, Yamada M, Hayakawa M, et al. Cerebral amyloid angiopathy: A significant cause of cerebellar as well as lobar cerebral hemorrhage in the elderly. J Neurol Sci 1993;116: 135–141.

79. Andoh Y, Wakai S, Nagai M, et al. Lobar intracerebral hemorrhage secondary to cerebral amyloid angiopathy: A clinicopathologic study of three operated cases. No To Shinkei 1989;41:1217–1223.

80. Mandybur TI. Cerebral amyloid angiopathy: The vascular pathology and complications. J Neuropathol Exp Neurol 1986;45:79–90.

81. Wakui K, Seguchi K, Kuroyanagi T, et al. Multiple intracerebral hemorrhages due to cerebral amyloid angiopathy after head trauma. No Shinkei Geka 1988;16:1287–1291.

82. Inoue A, Sato K, Itagaki S, Nakai O. A case of multiple cerebral hemorrhage related to cerebral amyloid angiopathy. No Shinkei Geka 1988;544–549.

83. Vinters HV. Cerebral amyloid angiopathy. A critical review. Stroke 1987;18:311–324.

84. Hendricks HT, Franke CL, Theunissen PH. Cerebral amyloid angiopathy: Diagnosis by MRI and brain biopsy. Neurol 1990;40:1308–1310.

85. Ramsay DA, Penswick JL, Robertson DM. Fatal streptokinase-induced intracerebral haemorrhage in cerebral amyloid angiopathy. Can J Neurol Sci 1990;17:336–341.

86. Pendlebury WW, Iole ED, Tracy RP, Dill BA. Intracerebral hemorrhage related to cerebral amyloid angiopathy and t-PA treatment. Ann Neurol 1991;29:210–213.

87. Leblanc R, Haddad G, Robitaille Y. Cerebral hemorrhage from amyloid angiopathy and coronary thrombolysis. Neurosurgery 1992; 31:586–590.

88. Wijdicks EF, Jack CJ. Intracerebral hemorrhage after fibrinolytic therapy for acute myocardial infarction. Stroke 1993;24:554–557.

89. Kase CS. Diagnosis and management of intracerebral hemorrhage in elderly patients. Clin Geriatr Med 1991;7:549–567.

90. Anderson JL, Karagounis L, Allen A, et al. Older age and elevated blood pressure are risk factors for intracerebral hemorrhage after thrombolysis. Am J Cardiol 1991;68:166–170.

91. Stroke Prevention in Atrial Fibrillation Investigators. Warfarin versus aspirin for prevention of thromboembolism in atrial fibrillation: Stroke prevention in atrial fibrillation II study. Lancet 1994;343:687–691.

92. Masuda J, Tanaka K, Ueda K, Omae T. Autopsy study of incidence and distribution of cerebral amyloid angiopathy in Hisayama, Japan. Stroke 1988;19:205–210.

93. Kobayashi Y, Suematsu K, Kamada H, et al. Cerebral amyloid angiopathy complicated by multiple cerebral infarcts and intracerebral hemorrhages: Case report. No Shinkei Geka 1987;15:405–408.

93a. Greenberg SM, Rebeck GW, Vonsattel JP, et al. Apolipoprotein E epsilon 4 and cerebral hemorrhage associated with amyloid angiopathy. Ann Neurol 1995;38:254–259.

93b. Greenberg SM, Briggs ME, Hyman BT, et al. Apolipoprotein E e4 is associated with the presence and earlier onset of hemorrhage in cerebral amyloid angiopathy. Stroke 1996; 27:1333–1337.

93c. Nicoll JAR, Burnett C, Love S, et al. High frequency of apolipoprotein E e2 allele in hemorrhage due to cerebral amyloid angiopathy. Ann Neurol 1997;41:716–721.

93d. Greenberg SM, Hyman BT. Cerebral amyloid angiopathy and apolipoprotein E: bad news for the good allele? Ann Neurol 1997;41:701–702.

93e. Hyman BT, Gomez-Isla T, Briggs M, et al. Apolipoprotein E and cognitive change in an elderly population [see comments]. Ann Neurol 1996;40:55–66.

94. Graffagnino C, Herbstreith M, Schmechel D, et al. Cystatin C mutation in a case of sporadic ICH with amyloid angiopathy. Stroke 1995;26:2190–2193.

95. Russell D, Rubinstein L. Pathology of Tumors of the Nervous System. Baltimore: Williams and Wilkins, 1977, 116–145.

96. Dobyns W, Michels V, Groover R, et al. Familial cavernous malformations of the central nervous system and retina. Ann Neurol 1987;21:578–583.

97. Villani R, Arienta C, Caroli M. Cavernous angiomas of the central nervous system. J Neurosurg Sci 1989;33:229–252.

98. Rigamonti D, Hadley M, Drayer B, et al. Cerebral cavernous malformations: Incidence and familial occurrence. N Engl J Med 1988;319:343–347.

99. Hayman L, Evans R, Ferrell R, et al. Familial cavernous angiomas: Natural history and genetic study over a 5-year period. Am J Med Genet 1982;11:147–160.
100. Bicknell J, Carlow T, Kornfeld M, et al. Familial cavernous angiomas. Arch Neurol 1978;35:746–749.
101. Bicknell J. Familial cavernous angioma of the brain stem dominantly inherited in Hispanics. Neurosurgery 1989;24:102–105.
102. Robinson J Jr, Awad I, Magdinec M, Paranandi L. Factors predisposing to clinical disability in patients with cavernous malformations of the brain. Neurosurgery 1993;32:730–735.
103. Pettigrew L, Immken L, Hejtmancik T. Natural history of familial cavernous hemangioma. Stroke 1989;20:139A.
104. Mason I, Aase J, Orrison W, et al. Familial cavernous angiomas of the brain in an Hispanic family. Neurology 1988;38: 324–326.
104a. Gunel M, Awad IA, Finberg K, et al. A founder mutation as a cause of cerebral cavernous malformation in Hispanic Americans. N Engl J Med 1996;334:946–951.
105. Kattapong VJ, Hart BL, Davis LE. Familial cerebral cavernous angiomas: Clinical and radiologic studies. Neurology 1995;45: 492–497.
106. Gangemi M, Maiuri F, Donati P, et al. Familial cerebral cavernous angiomas. Neurol Res 1990;12:131–136.
107. Gil-Nagel A, Dubovsky J, Wilcox KJ, et al. Familial cerebral cavernous angioma: A gene localized to a 15-cM interval on chromosome 7q. Ann Neurol 1996;39:807–810.
107a. Polymeropoulos MH, Hurko O, Hsu F, et al. Linkage of the locus for cerebral cavernous hemangiomas to human chromosome 7q in four families of Mexican-American descent. Neurology 1997;48:752–757.
108. Yokoyama K, Asano Y, Murakawa T, et al. Familial occurrence of arteriovenous malformation of the brain. J Neurosurg 1991;74: 585–589.
109. Allard J, Hochberg F, Franklin P, Carter A. Magnetic resonance imaging in a family with hereditary cerebral arteriovenous malformations. Arch Neurol 1989;46:184–187.
110. King C, Lovrien E, Reiss J. Central nervous sytem arteriovenous malformations in multiple generations of a family with hereditary hemorrhagic telangiectasia. Clin Genet 1977;12:372–381.
111. Fox J. Intracranial Aneurysms. New York: Springer-Verlag, 1983.
112. Takakura K, Saito I, Sasaki T. Special problems associated with subarachnoid hemorrhage. In Youmans J (ed). Neurological Surgery. Philadelphia: WB Saunders, 1990, 1873–1874.
113. Takamiya Y, Takayama H, Kobayashi K, et al. Familial occurrence of multiple vascular malformations of the brain. Neurol Med Chir (Tokyo) 1984;24:271–277.
114. Boyd M, Steinbok P, Paty D. Familial arteriovenous malformations. J Neurosurg 1985; 62:597–599.
115. Snead O, Acker J, Morawetz R. Familial arteriovenous malformation. Ann Neurol 1979;5:585–587.
116. Kidd H, Cumings J. Cerebral angiomata in an Icelandic family. Lancet 1947;31:747–748.
117. Tandan R, Mohire M, Dorwart R. Intracranial arteriovenous malformations in two brothers with myotonic dystrophy. Clin Neurol Neurosurg 1991;93:143–147.
118. Aslandis C, Jansen G, Amemiya C, et al. Cloning of the essential myotonic dystrophy region and mapping of the putative defect. Nature 1992;355:548–551.
119. Moody R, Poppen J. Arteriovenous malformations. J Neurosurg 1970;32:503–511.
120. Perret G, Nishioka H. Report on the cooperative of intracranial aneurysms and subarachnoid hemorrhage. Section VI: Arteriovenous malformations. J Neurosurg 1966; 25:467–490.
121. Romain G, Fisher M, Perl D, Poser C. Neurologic manifestations of hereditary hemorrhagic telangiectasia (Rendu-Osler-Weber disease): Report of 2 cases and review of the literature. Ann Neurol 1978;4: 130–144.
121a. Fulbright RK, Chaloupka JC, Putman CM, et al. MR of hereditary hemorrhagic telangiectasia: Prevalence and spectrum of cerebrovascular malformations. Am J Neuroradiol 1998;19:477–484.
122. Bird R, Jaques W. Vascular lesion of hereditary hemorrhagic telangiectasia. N Engl J Med 1959;260:597–599.
123. Plauchu H, de Chadarevian J-P, Bideau A, Robert J-M. Age-related clinical profile of hereditary hemorrhagic telangiectasia in an epidemiologically recruited population. Am J Med Genet 1989;32:291–297.
124. Porteous M, Burn J, Proctor S. Hereditary haemorrhagic telangiectasia: a clinical analysis. J Med Genet 1992;29:527–530.
125. Hodgson C, Burchell H, Good C, et al. Hereditary hemorrhagic telangiectasia and pulmonary arteriovenous fistula. N Engl J Med 1959;261:625–636.
126. Hodgson C, Kaye R. Pulmonary arteriovenous fistula and hereditary hemorrhagic telangiectasia: A review and report of 35 cases of fistula. Dis Chest 1963;43:449–455.
127. Grollmus J, Hoff J. Multiple aneurysms as-

sociated with Osler-Weber-Rendu disease. Surg Neurol 1973;1:91–93.

128. Djindjian R. Spinal vascular malformations. J Neurosurg 1976;45:727–728.

129. Saunders W. Hereditary hemorrhagic telangiectasia. Arch Otolaryngol 1962;76:245–260.

130. McDonald M, Papenberg K, Ghosh S, et al. A disease locus for hereditary haemorrhagic telangiectasia maps to chromosome 9q33–34. Nature Genet 1994;6:197–204.

131. Shovlin C, Hughes J, Tuddenham E, et al. A gene for hereditary haemorrhagic telangiectasia maps to chromosome 9q3. Nature Genet 1994;6:205–209.

131a. McAllister KA, Baldwin MA, Thukkani AK, et al. Six novel mutations in the endoglin gene in hereditary hemorrhagic telangiectasia type 1 suggest a dominant-negative effect of receptor function. Hum Molec Genet 1995;4:1983–1985.

132. Snyder L, Doan C. Studies in human inheritance: XXV. Is the homozygous form of multiple telangiectasia lethal? J Lab Clin Med 1944;29:1211–1216.

132a. Johnson DW, Berg JN, Gallione CJ, et al. A second locus for hereditary hemorrhagic telangiectasia mapped to chromosome 12. Genome Res 1995;5:21–28.

132b. Johnson DW, Berg JN, Baldwin MA, et al. Mutations in the activin receptor-like kinase 1 gene in hereditary hemorrhagic telangiectasia type 2. Nat Genet 1996;13: 189–195.

133. Hashimoto K, Pritzker M. Hereditary hemorrhagic telangiectasia: An electron microscopic study. Oral Surg 1972;34:751–768.

134. Orrison W. Introduction to Neuroimaging. Boston: Little, Brown, 1989, 127–132.

135. Moritake K, Handa H, Mori K, et al. Venous angioma of the brain. Surg Neurol 1980;14: 95–105.

136. Stein B, Solomon R. Arteriovenous malformations of the brain. In Youmans J, (ed). Neurological Surgery. Philadelphia: WB Saunders, 1990, 1831–1863.

136a. Vikkula M, Boon LM, Carraway KL, et al. Vascular dysmorphogenesis caused by an activating mutation in the receptor tyrosine kinase TIE2. Cell 1996;87:1181–1190.

137. Palmer J. Haemangioblastomas: A review of 81 cases. Acta Neurochir 1972;27:125–148.

138. Bonebrake R, Siqueira E. The familial occurence of solitary hemangioblastoma of the cerebellum. Neurology 1964;14:733–743.

139. Maher ER, Iselius L, Yates JR, et al. von Hippel-Lindau disease: A genetic study. J Med Genet 1991;28:443–447.

140. Jeffreys R. Clinical and surgical aspects of posterior fossa haemangioblastoma. J Neurol Neuorsurg Psychiatry 1975;38:105–111.

141. Cobb CA III, Youmans JR. Sarcomas and neoplasms of blood vessels. In Youmans J (ed). Neurological Surgery. Philadelphia: WB Saunders, 1990, 3152–3170.

142. Neumann HP, Eggert HR, Scheremet R, et al. Central nervous system lesions in von Hippel-Lindau syndrome. J Neurol Neurosurg Psychiatry 1992;55:898–901.

143. Filling-Katz M, Choyke P, Patronas N, et al. Radiologic screening for von Hippel-Lindau disease: The role of enhanced MRI in the central nervous system. J Comput Assist Tomogr 1989;13:743–755.

144. Richards FM, Phipps ME, Latif F, et al. Mapping the von Hippel-Lindau disease tumour suppressor gene: Identification of germline deletions by pulsed field gel electrophoresis. Hum Mol Genet 1993;2:879–882.

145. Richards FM, Maher ER, Latif F, et al. Detailed genetic mapping of the von Hippel-Lindau disease tumour suppressor gene. J Med Genet 1993;30:104–107.

146. Latif F, Tory K, Gnarra J, et al. Identification of the von Hippel-Lindau disease tumor suppressor gene. Science 1993;260:1317–1320.

147. Stackhouse TM, Lerman M, Zbar B. An in vitro analysis of the von Hippel-Lindau tumor suppressor protein. Proc Am Assoc Cancer Res 1995;36:570.

148. Decker HJH, Weidt EJ, Brieger J. The von Hippel-Lindau tumor suppressor gene. Cancer Genet Cytogenet 1997;93:74–83.

148a. Atuk NO, Stolle C, Owen JA, et al. Pheochromocytoma in von Hippel-Lindau disease: Clinical presentation and mutation analysis in a large multigenerational kindred. J Clin Endocrinol Metab 1998;83:117–120.

149. Crossey PA, Richards FM, Foster K, et al. Identification of intragenic mutations in the von Hippel-Lindau disease tumour suppressor gene and correlation with disease phenotype. Hum Mol Genet 1994;3:1303–1308.

150. Prowse AH, Webster AR, Richards FM, et al. Somatic inactivation of the VHL gene in von Hippel-Lindau disease tumors. Am J Hum Genet 1997;60:765–771.

151. Kibel A, Iliopoulos O, DeCaprio JA, et al. Binding of the von Hippel-Lindau tumor suppressor protein to Elongin B and C. Science 1995;269:1444–1446.

152. Pause A, Lee S, Worrell RA, et al. The von Hippel-Lindau tumor-suppressor gene product forms a stable complex with human CUL-2, a member of the Cdc53 family of proteins. Proc Natl Acad Sci USA 1997;94: 2156–2161.

153. Tse JY, Wong JH, Lo KW, et al. Molecular genetic analysis of the von Hippel-Lindau disease tumor suppressor gene in familial and sporadic cerebellar hemangioblastomas. Am J Clin Pathol 1997;107:459–466.

Chapter 12

Subarachnoid Hemorrhage and Intracranial Aneurysms

Mark J. Alberts, MD

Introduction

The diseases discussed in this chapter are not the most common types of stroke, but they are associated with the highest mortalities. For example, aneurysmal subarachnoid hemorrhage (SAH) can have a mortality rate above 50%.[1-3] Patients who survive an SAH are typically left with devastating neurologic deficits. Part of this high mortality is because many SAHs occur without warning. It is somewhat of a paradox that while an SAH can be the most unpredictable and deadly type of stroke, it is potentially one of the most preventable. The investigation of genetic factors that may play a role in SAH and intracranial aneurysm (IA) formation may someday aid in the presymptomatic diagnosis and treatment of IAs.

Several diseases and processes can produce an SAH. Estimates from the Cooperative Study indicate that approximately 80% of nontraumatic SAHs are caused by aneurysmal rupture.[4] This section will focus on the genetics of saccular IAs, also known as "berry" aneurysms, since they are the most common cause of SAH and there is significant evidence supporting the importance of genetic factors in the formation of

these aneurysms. Other types of vascular abnormalities producing SAH such as atherosclerotic aneurysm, mycotic aneurysms, and traumatic aneurysms, which are commonly referred to as arterial dissections, are discussed in *Chapter 10*. Other vascular abnormalities such as arteriovenous malformations, cavernous hemangiomas, and hereditary telangiectases are reviewed in *Chapter 11*.

Genetic Epidemiology

There are several lines of epidemiological evidence supporting the importance of genetic factors in the development of IAs. A questionnaire survey by Norrgard et al.[5] found that 6.6% of first to third-degree relatives of SAH patients also had SAHs. This is slightly above one estimate of a 5.6% risk of SAH that occurs by chance alone.[6] This estimate may be somewhat inaccurate depending on the age of the surveyed population. In addition, a questionnaire survey is likely to grossly underestimate the incidence of familial IAs. This is supported by the results of two other studies. Hachinski et al.(personal communication) found that approximately

From Alberts MJ (ed). *Genetics of Cerebrovascular Disease.* Armonk, NY: Futura Publishing Company, Inc., © 1999.

12% of first-degree relatives of SAH patients had IAs detected by screening with magnetic resonance angiography (MRA). This finding is confirmed by another prospective study by Ronkainen et al.[7] who found that 12% of first-degree relatives of SAH patients had IAs visualized by MRA. Even these studies may underestimate the percentage of SAH cases that are familial, since MRA is not a perfect screening tool, and aneurysm formation is likely to be a dynamic process.[8–10] These two factors are discussed in more detail below.

There have been many case reports in the literature of families with IAs and SAH.[11–17] In most cases the pattern of inheritance supports an autosomal dominant trait, although publication bias may be responsible for the reporting of families with large numbers of affecteds. This issue was recently reviewed by Alberts et al.,[18] who analyzed SAH families with large sibships where asymptomatic siblings had been studied by cerebral angiography. A total of 66 siblings from 7 sibships were available in the literature for analysis. Eighteen were clinically affected with an SAH, and 11 were found to have an IA on elective angiography (Table 1). Twenty-three individuals had normal angiograms. There were a total of 29 known affecteds and 23 known unaffecteds. Data were not available on 13 of the 60 siblings. The total number of affecteds across all the sibships was 56%, which is consistent with an autosomal dominant trait. If ascertainment bias is corrected by excluding the probands from each family, the percent affected drops to 49%, which is still consistent with an autosomal dominant trait.

Some studies have shown that for sibling pairs, IAs tend to occur at the same site or at mirror sites more often than in randomly selected IA patients.[19,20] There may be a tendency for familial cases to have multiple IAs (discussed in more detail below).[5,19,21] There are several reports in the literature of IAs occurring in identical twins.[22–24] Since identical twins by definition have identical genes, this supports the importance of genetic factors in the development of IAs.

In general, females are more likely to

suffer an SAH compared to males, with a ratio of almost 2:1 in some studies.[4,21] A large autopsy study found that females were more likely to have unruptured incidental IAs than males, 63% to 37%.[25] Males may be more likely to have an SAH under 20 years of age, although such cases are rare. Since IAs are more common in women, some investigators have suggested that certain hormones may play a role in IA formation although this remains unproven.[26]

It has been suggested that females are more likely to have multiple IAs, and there may be a predominance of females in cases of familial IA/SAH.[21,27] A detailed study found that multiple aneurysms were present in 22% of all male probands and 26% of all female probands, which was not a significant difference.[28] There was a strong association between multiplicity and a positive family history. Eight of 21 (38%) probands with a positive family history had multiple aneurysms, while only 28 of 144 (19%) without a positive family history had multiple aneurysms. Some of the prior studies of IA may not have fully considered the important role that family history may have on issues such as multiplicity. The overrepresentation of females in all SAH cases will clearly influence most analyses of sex differences, and must be accounted for in such studies.

Age-of-onset data can be important in the study of genetic aspects of some diseases. In general, aneurysmal rupture tends to occur at a mean age between the upper 40's and middle 50's.[4,21] Cases of rupture below the age of 20 or above the age of 65 are unusual. It has been reported that familial IAs tend to rupture at a somewhat younger mean age, 42.3 years, in contrast to all IAs which tend to rupture between 50 and 60 years of age.[4,20,21] Data from the large sibships reviewed by Alberts et al.[18] showed that the peak age of SAH and IA detection was 30–39 years, which is considerably younger than the average age of rupture of 50–54 years for patients with nonfamilial SAH/IAs. The differences in the age-of-onset data for familial and nonfamilial SAH/IA should be viewed with caution. The young age of onset can be a cause as well as a

Table 1

Genetic Aspects of Intracranial Aneurysms and Subarachnoid Hemorrhage in Sibships

	Number of Individuals
No. of sibships	7
No. of persons in sibship	66 (52 with data)
SAH	18 (35%)
IA on Angiogram	11 (21%)
Normal Angiogram	23 (44%)
Number Affected*	29 (56%)

SAH = subarachnoid hemorrhage.

IA = intracranial aneurysm. These individuals had an IA detected by elective angiography.

*Affected refers to individuals with either an SAH or an IA found by elective angiography.

Adapted from data in reference 18.

result of significant ascertainment bias, since such early-onset families will have more members available for study. A large study at Duke University Medical Center, Durham, North Carolina, by Graffagnino and colleagues, did not find any age differences between affecteds in familial cases (mean age 55.7 years) and nonfamilial cases (mean age 51.3 years).[28] The situation may be somewhat analogous to familial Alzheimer's disease (AD), in that the early-onset form of AD, as reported by Pericak-Vance et al., had an obvious genetic pattern because there were more members alive who could be evaluated. However, when large late-onset pedigrees were identified and evaluated, significant genetic factors were also found in that group.[29]

One of the largest studies of familial IAs was reported recently by Ronkainen et al.[29a] They studied 167 familial IA patients who had 215 aneurysms, and compared them to 983 patients with nonfamilial IAs. This study found that the familial cases had a younger age (approximately 5 years) than the nonfamilial cases. More of the familial IAs were small (< 6 mm) and were likely to rupture at a smaller size compared to the nonfamilial cases. Multiple IAs were noted in 25% of the familial cases and 22% of the nonfamilial cases. For siblings with familial IAs, they were more likely to become symptomatic within 10 years of each other than in the nonfamilial cases. For the familial cases, IAs involving the middle cerebral artery

was the most common site, occurring in approximately 50% of such cases.[29a]

Another large familial study in Finland examined the occurrence of IAs in first-degree relatives in families with two or more affected individuals.[29b] Magnetic resonance angiography was used as the initial screening tool, with conventional cerebral angiography used to verify suspected cases with IAs. This study examined 438 individuals from 85 families that did not have a history of polycystic kidney disease. The age-adjusted prevalence of IAs among the screened relatives was approximately 9, which is about four times higher than expected for the general population. These findings further support the concept that familial IA/SAH is a distinct disorder, and that first-degree relatives of affecteds are at a defininite increased risk for IA formation.[29b]

The establishment of a genetic pattern for late-onset IA/SAH has been challenging for several reasons. Many late-onset probands and families will have to be ascertained and studied before any definitive statements can be made. Such late-onset kindreds are often difficult to ascertain due to lack of living affecteds and inadequate medical records on deceased individuals. Despite these challenges, genetic segregation studies of various pedigrees with familial IA/SAH have been performed in an attempt to determine the likely mode of inheritance. One such study found no overall pattern of inheritance that explained all observations, although evidence of an autosomal pattern was observed in some cases.[29c] Another retrospective study by Alberts et al. examined published large sibshihps in which some at-risk siblings had been studied.[18] They found a pattern of inheritance consistent with an autosomal dominant trait in this highly selective case series, although formal segregation studies were not performed on this data set. Based on the data presented above, there is significant evidence that genetic factors are important in causing some cases of early-onset (less than 45 years) familial SAH/IA. It seems likely that genetic factors are also important in at least some cases of late-onset familial SAH/IA cases. Whether early-onset familial SAH/IA differs from late-onset SAH/IA, and to what extent late-onset SAH is genetic, will require further study.

There are several conditions (some inherited), that are associated with the formation of IAs and SAH. These are discussed in the following section, but they also support the hypothesis that SAH and IA formation can be caused by genetic factors. In addition, animal studies have shown a variable genetic susceptibility for aneurysm formation among different strains of rats, further supporting the importance of inherited factors in this disorder.[29d]

Risk Factors

Two major risk factors associated with IA and SAH are hypertension and smoking. Pregnancy has been associated with rupture of IAs, but not necessarily with IA formation.

The role of hypertension in the pathogenesis of IA formation and as a factor in SAH has been the focus of considerable debate. Some studies have shown that hypertension is a risk factor for SAH, while others have shown that it is not overrepresented in SAH patients.[30–34] It appears that hypertension is neither sufficient nor necessary for IA formation or rupture.[35] However, Andrews and Spiegel[21] reported that hypertensive patients less than 55 years of age (male or fe-

male) were more likely than normotensive patients to have multiple aneurysms. Another recent study found that hypertension was a significant risk factor for SAH, with an odds ratio of 2.9 and a relative risk of 2.8.[85a] Hypertension may also be a familial risk for SAH. A study examined the frequency of hypertension among 1290 first-degree and 3588 second-degree relatives of SAH patients. The relative risk attributed to hypertension was 1.8 in these relatives.[85b]

One possible explanation for these disparate results is that in some cases hypertension may hasten the development or expansion of an IA, or may cause rupture of a previously asymptomatic IA. It is likely that certain individuals may be more susceptible to the effects of chronic hypertension in the pathogenesis of IAs and subsequent SAH. The interaction of hypertension with other inherited conditions is discussed in more detail below.

Cigarette smoking has emerged as a significant risk factor for SAH. Numerous epidemiological studies have cited an increased risk for SAH among smokers ranging from 2.6X to 11X control.[32,36,37] This association is of interest since smoking may cause proteolytic disruption of elastin fibers, which may be important in IA formation and rupture (see below).[38]

Associated Conditions

Intracranial aneurysms and SAH have been associated with several other conditions (Table 2). The major inherited conditions that are associated with IAs include polycystic kidney disease, connective tissue disorders, (particularly Ehlers-Danlos syndrome type IV), Marfan's syndrome, coarctation of the aorta, fibromuscular hyperplasia, pseudoxanthoma elasticum, and possibly abdominal aortic aneurysms (AAA). Hypoplasia of an anterior cerebral artery is a fairly common vascular anomaly that is associated with anterior communicating and anterior cerebral artery IAs. Arteriovenous malformations (AVM) and moyamoya disease have also

Table 2
Some Inherited Conditions Associated With Intracranial Aneurysms or Subarachnoid Hemorrhage[1]

Condition	Berry Aneurysms	Fusiform ICA Aneurysms or CCF[2]	Other Lesions
Polycstic kidney disease	++++	0	
Ehlers-Danlos type IV	0/+	++++	dissections; giant aneurysms
Marfan's syndrome	+	+++	aortic aneurysms; giant aneurysms
Fibromuscular dysplasia	++++	0	dissections
Coartation of aorta	++	+	ICH-? etiology
Pseudoxanthoma elasticum	0	++/+++	

[1]Some conditions that are associated with IAs, such as arteriovenous malformations, hypoplasia of anterior cerebral artery, and moyamoya disease have not been included in this table, since it is not clear if these conditions are inherited diseases.

[2]ICA, internal carotid artery; CCF, carotid-cavernous fistula

0, no clear or convincing association

+, some rare or occasional association; may be due to other factors (i.e. hypertension) or a chance occurrence

++, association seen in some, but not all studies; probably a true genetic effect, but cannot ruleout effects of another concomitant process

+++, moderately associated; findings confirmed in many studies; convincing genetic effect

++++, strong association found in most or all studies; striking genetic effect

(Data used for this table are derived from the references cited in the text.)

been associated with IAs.[39,40] Studies have shown that AVMs are found in 1–2% of patients with IAs.[39] Conversely, IAs may be present in 5–16% of patients with AVMs.[39,41] Moyamoya disease and AVMs are discussed in more detail in *Chapters 10 and 11*, respectively.

Polycystic Kidney Disease

Autosomal dominant polycystic kidney disease (ADPKD) is the most common inherited disorder that is associated with IA/SAH.[42–44] This disorder is inherited with essentially complete penetrance. PKD typically affects the kidneys and other organs such as the heart and gastrointestinal tract. Patients usually present between the ages of 30 and 40 with frequent urinary tract infections or renal vascular hypertension.[45]

Genetic linkage studies have found evidence for at least two genetic loci for ADPKD. One is on chromosome 16 (16p13.3).[46] This is termed type I ADPKD (PKD1), and is due to mutations in polycystin. Type II ADPKD (PKD2) has been linked to chromosome 4, and the causative gene has been cloned.[47,48,48a] Both types of PKD are associated with IAs and SAH (see below for details).[42–44,45a] The PKD1 locus is responsible for approximately 85% of cases, and the PKD2 locus responsible for the other 15%.

Polycystin is a large gene which encodes a huge glycoprotein. The gene spans about 52 kb of genomic DNA, and is organized into 46 exons. The transcript is >14,000 bp, and the protein has 4,302 amino acids. It appears to be involved in cell-cell and cell-matrix interactions.[47a,48b] The genetic study of polycystin is difficult due to duplication of part of chromosome 16 (including polycystin), and with the presence of several homologues for parts of the gene.[47b] A variety of mutations has been detected in the 3′ region of the gene. Some of these lead to frameshift mutations and abnormal translation. Others are single base changes or intronic mutations that lead to missplicing.[48c] The pathophysiology of how these mutations lead to IA formation is unclear.

The PKD2 gene on chromosome 4q21–23 encodes a transcript of approximately 5.4 kb with 15 exons. It encodes a protein of 968 amino acids.[48a] The protein has characteristics suggestive of an integral membrane protein. It also has some similarities to ionic voltage channel subunits.[48a] There are some regions of homology between polycystin and the PKD2 protein.

Studies have been performed to identify and characterize mutations in the PKD2 gene. One large analysis examined 35 families thought to have PKD2.[48d] The PKD2 gene was screened using single strand conformational polymorphism (SSCP) techniques. A variety of mutations were identified in these families, including deletions, substitutions, and insertions, leading to frameshifts, splice mutations, and the creation of a new restriction site. Many of the mutations led to frameshifts and premature stop codons. The mutations were scattered throughout the gene, without any evidence for clustering around certain exons or regions.[48d]

The true incidence of IAs in patients with PKD is the subject of controversy. As with many genetic disorders, the method of ascertainment can explain some of the disparate results. A review of cohorts in which all patients were not systematically examined (some did not have an angiogram or an autopsy) showed an incidence of IA/SAH in PKD patients of approximately 10%.[49,50] If one analyzes only those series in which brain autopsies or cerebral angiography were performed, the incidence of IA/SAH increases to 38%.[43] In addition, 22% of PKD cases with IA/SAH had a family history of some type of cerebral hemorrhage.[42] It seems reasonable to conclude that patients with PKD have an increased risk of IA/SAH compared to an age-matched population.

The IAs that arise in patients with PKD appear to have distinct features compared to spontaneous IAs. A critical review of this issue by Lozano and LeBlanc[51] found that 72% of PKD patients with IAs were males, compared to 43% of patients with sporadic IAs. In addition, the mean age of IA rupture in PKD patients was 39.8 years, compared to a mean of 51.4 years for sporadic IA patients.[51] Another study found that about 25% of PKD patients with IAs had multiple aneurysms.[42] These data suggest that IA formation and rupture in PKD patients may be biologically different from IA formation and rupture in patients with sporadic disease.

It has been argued that IA formation in PKD is due primarily to the presence of hypertension (HTN) in affected patients. This concern relates to the larger issue of the role of systemic HTN in the formation and rupture of IA in any setting (as reviewed above, in "Risk Factors"). Some case series have reported an increased incidence of HTN in patients with IAs[31,33] and in patients with IAs associated with PKD.[42,44] Since PKD commonly produces HTN, an association between HTN and IAs is not unexpected. At present, it is reasonable to conclude that some diseases (possibly including PKD) and processes may render the cerebral vasculature more susceptible to the effects of long-term HTN, perhaps leading to more extensive or premature IA formation and rupture.[51] It is possible that HTN is more important for IA formation and rupture in nonfamilial late-onset cases, but less important in the pathogenesis of clearly genetic cases. Detailed research is required to prove these hypotheses.

Ehlers-Danlos Syndrome

Ehlers-Danlos syndrome (EDS) describes a group of heterogeneous disorders that share various connective tissue abnormalities. Among the most common abnormalities are skin changes, joint hypermobility, rupture of large arteries and arterial aneurysms, and rupture of other organs.[52] EDS type IV is one of the most severe forms of the disease, and has been associated with IA formation. However, a detailed review of the literature shows that the IAs in patients with EDS type IV are distinctly different from the berry aneurysms typically associated with most cases of SAH. For example, a study by Mirza et al.[53] found formation of a fusiform aneurysm in the internal carotid artery (ICA) in a patient with EDS. Another study by Krog et al.[54] describes ICA aneurysms in a patient with EDS type IV. This case provides little clinical detail and there is no clear evidence that the patient had a berry aneurysm. A report by Edwards of a vertebral aneurysm in an EDS type IV patient is in fact a false aneurysm that was likely due to a vertebral dissection.[55] Ruby et al.[56] reported a similar case of a pseudoaneurysm involving the ICA in a patient with EDS type IV.

There is little doubt that EDS type IV is associated with formation of carotid-cavernous fistulae (CCF). A report by Fox et al.[57] shows evidence of an intracavernous aneurysm that had ruptured into the cavernous sinus in a patient with EDS. No evidence of berry aneurysms was seen in this study. Another report by Lach et al.[58] provides pathologic confirmation of a dissecting aneurysm and CCF involving the left ICA. In this case, there is clear evidence of a transmural tear involving the left ICA forming a CCF. In the right intracavernous ICA, there is rupture of the internal and external elastic lamina with some lateral aneurysmal dilatation.

There is one convincing case report in the literature by Rubinstein and Cohen of a patient with a true berry aneurysm and EDS.[59] This patient had an unequivocal berry aneurysm at the ICA/posterior communicating artery bifurcation. Fusiform enlargement and two aneurysmal dilatations of the left vertebral artery were also seen. There was abnormal morphology of the vascular wall, particularly collagenous tissue. It is of note that the authors suggest that patients with EDS are likely to have a high frequency of IAs.

From the above studies, it seems reasonable to conclude that there are no overwhelming epidemiological data showing a strong association between EDS type IV and intracranial berry aneurysm formation. EDS type IV is clearly associated with CCF formation, and probably causes diffuse fusiform arterial aneurysms. However, an overabundance of intracranial berry aneurysms in EDS type IV patients has yet to be proven.

There have been several reports describing various mutations in the type III procollagen gene as a cause of EDS type IV.[60–73] These findings have led to speculation that a significant percentage of IAs are due to mutations in the type III procollagen gene. While this is still an area of active study (see candidate genes below), at present it does not appear that a majority of familial IA/SAH cases can necessarily be ascribed to mutations in the type III procollagen gene.[74]

Coarctation of the Aorta

Coarctation of the aorta (COA) has been associated with an increased occurrence of IAs and SAH. An autopsy study by Riffenstein et al.[75] examined 104 cases of COA. Eleven deaths were due to intracranial lesions (not necessarily IAs), and the patients were between the ages of 10 and 30 years at death. The brain was examined in detail in nine of these cases, five of which showed a ruptured IA. Two additional patients had an SAH, although an IA was not specifically identified. In one of these cases, a small unruptured IA was noted.

A review by Campbell[76] examined two large autopsy series with 304 patients with COA. Intracranial hemorrhage was responsible for 11.5% of deaths in this study, at a mean age of 29 years. Another large review by Campbell and Baylis[77] reported that intracranial hemorrhage was responsible for approximately 1 of 8 deaths in patients with COA. An IA was demonstrated by pathologic examination in only half of the cases of intracranial hemorrhage. The inability to identify IAs in the other cases may be due to aneurysmal destruction by the SAH, or perhaps incomplete examination of the brain. Other case reports (many of which are not reviewed in this chapter) and small case series have documented the occurrence of COA and SAH/IA.[78,79] Patients with an intracerebral hemorrhage (ICH) and COA are also described in these reports. It is possible that COA produces an ICH by other mechanisms, in addition to rupture of a saccular aneurysm.

The concept that IA and COA are related has been challenged by Yoshioka et al.[79] They pooled data from five large case series that included more than 1,100 patients with COA. The incidence of IA in this group was 3.7%, which is not dramatically higher than could be seen in an age-matched population examined by autopsy. Viewed from a different perspective, the incidence of COA in patients with an SAH was 0.35%, based on pooled data from eight studies with over 6,000 patients. Yoshioka et al. make the point that the incidence of IA rupture may be elevated in patients with COA, perhaps due to the effects of hypertension and atherosclerosis.

Study by Bobby et al.[80] supports this supposition. The authors followed a large cohort of patients after surgical correction of coarctation. One hundred eighty-two patients were followed for a total of 3,288 patient years. When the series was reported in 1991, there were no deaths from any type of cerebrovascular event. It is possible that early operation of these patients reduced the occurrence and effects of long-term hypertension, including enlargement and rupture of IAs.

The results of the above studies indicate that at present there is some evidence to suggest an increased occurrence of IA and SAH in some patients with COA. However, this association is neither overwhelming nor dramatic. The evidence that this association is due to a common genetic factor is suspect, as there are some data suggesting that the increased occurrence of SAH in these patients may reflect the presence of severe hypertension secondary to the COA. This is clearly an area that requires further study from epidemiological, clinical, and genetic viewpoints.

Fibromuscular Dysplasia

There is a clear and, in some cases, striking association between fibromuscular dysplasia (FMD) and IA. This association is much stronger than the association cited above between IA and COA. Intracranial aneurysms are found with increased frequency in patients with either renal or carotid FMD. A recent literature review by Ouchi et al.[81] cited data from angiographic or autopsy studies. In such studies, the occurrence of IAs in FMD patients ranged from 5% to 51%. This most likely reflects different ascertainment schemes among the various studies.

A large review by Mettinger et al.[82] examined 284 patients with documented aortocranial FMD and found that 21% had IAs. In approximately 1/3 of the affected cases (or 7% of the total) the IAs were multiple. In this cohort of patients, IAs most commonly affected the internal carotid artery, anterior cerebral artery, and middle cerebral artery, in descending order. Another study by the same author found that 19 of 37 FMD patients (51%) had IAs, some of them multiple.[83] The vast majority of IAs occurred in females, but females are more commonly affected by FMD than males.[83,84] A study by So et al.[84] reported SAH in 5 of 32 FMD patients, and asymptomatic IAs in 2 of 32 FMD patients. Females with FMD may have as much as a 60% rate of having multiple aneurysms, mandating full angiographic investigation in such patients.[85]

It appears clear that FMD is associated with an increased risk of IAs. This does not address the issue of how often IA/SAH is associated with FMD. A prospective study by George et al.[85] found angiographic evidence of cerebrovascular FMD in 18 of 87 (21%) cases of ruptured IA. This association seems somewhat higher than that which is seen in most clinical settings. There may be ascertainment bias, or perhaps all angiograms are not interpreted with close attenton to the presence of FMD. Although FMD may be much more common in SAH/IA patients than previously thought, more prospective studies are needed to prove this assertion.

It has been argued that the increased frequency of IAs in FMD patients is due primarily to the presence of severe HTN, particularly in cases of renal FMD. However, in a study by Mettinger and Erickson,[83] only 9 of

19 FMD patients with IAs had concurrent hypertension. In many cases IAs tend to occur primarily on the side of the most affected cervical artery, suggesting that the two processes are linked pathogenically. It is possible that the presence of concurrent HTN (severe in some cases) accelerates IA formation or increases the risk of IA rupture.

In summary, renal or cervical FMD is clearly associated with a significantly increased risk of IA/SAH, which may approach 50% in some series. Concomitant HTN is unlikely to adequately explain the marked increased occurrence of IA/SAH in FMD, suggesting that a common genetic factor is present.

Abdominal Aortic Aneurysm

It is not widely recognized that abdominal aortic aneurysms (AAA) can occur with a familial clustering. There are numerous cases reported in the literature of the familial occurrence of AAAs.[86,87] This has led to a detailed analysis of families containing a proband with an AAA. Blood relatives of an AAA proband had the following increased relative risk of developing an AAA: fathers, 4X; mothers, 4X; brothers, 9X; and sisters, 23X. [88] A study by Majumder et al.[89] performed segregation analysis on first-degree relatives of 91 probands with AAA. This analysis concluded that a single autosomal recessive gene could explain the observed pattern of inheritance for AAA in the families and relatives under study.

A detailed study of the coexistence of IAs and AAAs was reported by Norrgard et al.[90] in 1987. They found that 1.2% (7 of 574) of patients had both AAAs and IAs, diagnosed either clinically or pathologically. Six percent of the 87 AAA patients had blood relatives with IAs. Among 237 IA patients, 8 (3%) had blood relatives with AAAs. Patients with both types of aneurysms were significantly younger (56.8 years vs. 63.5 years) than other AAA patients. However, the overall occurrences of AAAs and IAs in these families is not dramatically higher than what would normally be expected.

Further analysis of the data showed that there were 2 AAA families having 3 relatives with IAs, and 1 family had 4 relatives with IAs. This may suggest that there are subpopulations of individuals with AAAs and IAs in which genetic factors may be important. This concept is supported by detailed genetic studies of the type III procollagen gene in patients with AAA. One study described a mutation in the type III procollagen gene in a family with AAAs.[61] The mutation was a single base change that converted a glycine to an arginine at codon 619. This mutation was found to cause a change in the thermal stability of collagen.

Another study sequenced cDNA from 50 unrelated aortic aneurysm patients looking for mutations in the type III procollagen gene.[91] Three patients had IAs and were apparently included in the study because of family members with aortic aneurysms. Two sequence variations were found among the 54 individuals, causing changes in exons 14 and 30. One of these patients had a dissecting aneurysm of the aorta. Three siblings had the same mutation, but without clinical or ultrasound abnormalities. In the case with a sequence change in exon 30, a similar change was present in several other relatives, including the patient's nephew who had an IA. However 2 of the patient's brothers had the "mutation" but no evidence of an aortic aneurysm. Among the 3 probands with IAs, no mutations in the type III procollagen gene were identified.[91]

The overall results of this study found that only 2% of patients with aortic aneurysms had "mutations" in the type III procollagen gene.[91] This finding could indicate that mutations in the type III procollagen gene are not a common cause of aortic aneurysms. The role of collagen in IA formation is discussed below in more detail.

HLA Antigens

The expression of a specific human leukocyte antigen (HLA) phenotype is under genetic control of the major histocompatibility complex. Association studies of HLA with several other diseases particularly autoimmune diseases, have provided useful information for the study of those entities. There have been numerous studies of HLA antigens and families or sporadic cases of IA/SAH.[17,92–95] A summary of the results of the major studies is provided in Table 3. With the exception of two studies showing an elevation in the B7, Cw2, and DR2 antigens, no consistent results were found. These disparate findings can be explained by several factors including vastly different populations (particularly ethnic background), suboptimal sample sizes, and disease heterogeneity. For the most part, the use of these association studies has been supplanted by modern genetic linkage techniques that are described in *Chapter 2*.

Hypoplasia of the Anterior Cerebral Artery

There is convincing evidence that aneurysms of the anterior communicating artery (ACommA) are associated with hypoplasia of the A1 segment of the anterior cerebral artery (ACA). One study by Kwak and Suzuki[96] found that 80% of patients with ACommA IAs have some anomalies in the anterior portion of the circle of Willis. A follow-up study found hypoplasia of the A1 segment of the ACA in 145 of 213 (68%) of patients with ACommA aneurysms.[97] Another study by Wilson et al.[98] reported that 85% of ACommA IAs were associated with ACA hypoplasia. This was an autopsy study that allowed for detailed examination of the intracranial vasculature.

The risk of developing an ACommA aneurysm due to hypoplasia of the ACA is still unclear. One study reported that 9.1% of hypoplastic ACAs developed aneurysms.[99] Considering the fact that hypoplasia of part of the ACA circulation is a fairly common finding, this represents a significant risk factor for ACommA aneurysms. Studies have also found that hypoplasia of the A1 segment of the ACA is more common in patients with ACommA aneurysms compared to patients with aneurysms in other locations.[98]

The association of ACA hypoplasia with ACommA aneurysms has direct implications when studying the genetics of IA formation. Several studies have suggested that variations in the ACA circulation may result in developmental and structural vascular defects that potentiate IA formation.[99,100] In addition, hypoplasia of the A1 segment of the ACA may result in abnormal hemodynamic effects, exposing the proximal ACA and ACommA to abnormally high blood pressures.[100] These factors may be one explanation for the development of ACommA aneurysms associated with ACA hypoplasia.

If ACommA aneurysms are formed by developmental and hemodynamic factors related to ACA hypoplasia, and if one assumes that ACA hypoplasia is more of a developmental or sporadic entity (as opposed to a genetic phenomenon), then one might expect ACommA aneurysms to be underrepresented in cases of clear familial IA/SAH. In a large retrospective literature review by Alberts et al.,[18] seven large sibships with familial SAH/IA were studied. In these large sibships, ACommA aneurysms were distinctly underrepresented. While most studies have reported that ACommA aneurysms account for between 25–35% of all IAs, they accounted for only 14–15% of IAs in the sibships with familial IAs/SAH. Another report by Sakai et al.[101] found a decreased incidence of ACommA IAs in familial cases. A review of this topic by Andrews also found a reduced incidence of ACommA IAs in familial compared to sporadic cases.[20] These findings, if confirmed by prospective studies, would provide important information for ge-

Table 3
HLA Antigens Associated with SAH/IA

Increased Antigens	Decreased Antigens
A28, **B7**, B21, B27, Bfs, **Cw2**, **DR2**, DR7	B40, Cw3

Bold antigens are those found elevated in more than one study.
Data adapted from references 17, 92–94.

netic studies of familial IAs, and for advising at-risk relatives of probands with ACommA aneurysms. However, at this time the data are too preliminary to make firm recommendations with respect to this issue.

Marfan's Syndrome

Marfan's syndrome is an autosomal dominant disorder affecting the cardiac, vascular, ocular, and skeletal systems.[102] The genetic cause of Marfan's syndrome has been traced to mutations in the fibrillin gene, which is located on chromosome 15q21.1.[103] Fibrillin is a large gene of about 110 kilobases organized into 65 exons.[104] The fibrillin gene produces a 350 kilodalton glycoprotein that is an integral component of microfibrils in both elastic and nonelastic tissue.[103–105] Several different mutations in the fibrillin gene have been described, including a de novo missense mutation, a deletion, and a point mutation. Some of these mutations can produce a truncated fibrillin peptide.[103,106,107]

Marfan's syndrome has been associated with aneurysm formation, particularly affecting the aorta.[108] There are several reports of aneurysms affecting the internal carotid artery in patients with Marfan's syndrome.[109–111] In most of these cases the aneurysms are either extracranial or in the cavernous portion of the internal carotid artery. There is one report by ter Berg et al.[112] of a familial association of IAs in a family with other multiple congenital anomalies thought to represent Marfan's syndrome. Another report cites a 33-year-old woman with Marfan's syndrome who died from septicemia.[113] At autopsy there were two areas of small nodular or early aneurysmal dilatation at arterial forks in the middle cerebral artery. However, a clear berry aneurysm was not identified.

With the exception of these few reports, there is a paucity of convincing evidence that Marfan's syndrome is associated with an increased risk of berry aneurysms. Therefore, at present the association between Marfan's syndrome and berry aneurysms must be con-

sidered very suspect. However, as seen with EDS type IV, Marfan's syndrome is probably associated with extracranial aneurysms as well as intracranial giant aneurysms particularly involving the cavernous carotid.

Pseudoxanthoma Elasticum

Pseudoxanthoma elasticum (PXE) is a rare inherited disorder affecting the elastic fibers of the skin, cardiovascular system, and eyes. The pattern of inheritance is unclear, with reports of autosomal dominant, autosomal recessive, and sporadic cases (see reference 114 for a review). PXE typically results in calcification of the arterial elastic media and intima, with symptoms of premature atherosclerosis (cardiac, peripheral) in some cases.[114,115] Ischemic strokes and gastric hemorrhages can also occur.[116] There are rare case reports of CNS aneurysms, but no reports of intracranial berry aneurysm formation. Three reports described fusiform aneurysms involving the cavernous, parasellar, or subclinoid portion of the ICA[117–119]; one described aneurysms of both common carotid arteries[117]; another described an anterior spinal artery aneurysm.[120] The underlying genetic abnormality in PXE is unknown.

Candidate Genes

The rapid progress in molecular biological research has led to the further understanding of the structure and function of some genes and proteins that may be important in the pathogenesis of IAs. This section will review the potential roles of several candidate genes in IA formation.

Collagen

Collagen defines a family of at least 15 different proteins. In general terms, collagens can be divided into fibrillary and nonfibrillary types. As shown in Table 4, collagen types I, II, and III constitute the major fibrillary collagens. These collagen proteins

Table 4
Major Collagen Types[1]

Collagen Type	Gene Designation(s)	Distribution	Chromosome
Major Fibrillar			
Type I	COL1A1	widespread	17q21-q22
	COL1A2	widespread	7q21-q22
Type II	COL2A1	cartilage+	q12q12-q14
Type III	COL3A1	blood vessels+	2q24-q33
Major Nonfibrillar			
Type IV	COL4A1/COL4A2	Basement membranes	12q33-q34
Minor Fibrillar			
Type V	COl5A2	widespread	2q24-q32
Type XI	COL11A1	cartilage	1q21
	COL11A2	cartilage	6p21.2
Other Nonfibrillar			
Type VI	COL6A1/COL6A2	widespread	21q22.3
	COL6A3	widespread	2q37
Type IX	COL9A1	cartilage, vitreous	6q12-q14
Type X	?	cartilage	?
Type XII	?	associated with type I collagen	?
Type XIII	COL13A1	widespread	10q22

+, indicates that the collagen type has other distributions not listed in the table
[1] Data for this table were adapted and compiled from references 121, 122.

are first synthesized as soluble precursors called procollagens. Through a complex series of post-translational modifications, the procollagens undergo hydroxylation and glycosylation. Disulfide bonds form among the propeptides, and a triple helix folding occurs in a process called nucleated growth. The procollagen peptide then undergoes proteolytic cleavage and further self-assembly and covalent cross-linking.[121]

In the study of cerebral aneurysms, type III collagen has received the most interest. Type III collagen is composed of three identical procollagen chains designated proalpha 1 (III). Type III procollagen is also designated by the abbreviation COL3A1. The COL3A1 gene is 44 kilobases in size, and contains 52 exons. It has been mapped to chromosome 2q24–33.[122] Type III collagen is a major constituent of blood vessels throughout the body, and has therefore received much attention in the pathogenesis of many different types of vascular disease. This is supported by the fact that Ehlers-Danlos syndrome type

IV (EDS IV) can cause vascular abnormalities and in some cases is due to mutations in the COL3A1 gene.[62,66] As discussed above, carotid cavernous sinus fistulae, aortic aneurysms, and various systemic aneurysms (dissecting and fusiform) have been associated with EDS type IV.

The importance of collagen abnormalities as a cause of cerebral aneurysms is better understood and reviewed when the proposed collagen abnormalities are subdivided. A review of the literature shows two broad groups of potential collagen abnormalities: (1) a quantitative protein abnormality (collagen deficiency), and (2) a qualitative protein abnormality (abnormal collagen). These abnormalities could be caused by any of the following mechanisms: (1) mutations in the COL3A1 gene; (2) mutations or other defects in the genes and proteins responsible for the extensive post-translational processing of COL3A1; and (3) environmental or developmental factors. Defects in any of the processing pathways could produce quantitative and qualitative collagen abnormalities. This possibility has obvious implications in research and clinical studies involving cerebral aneurysms.

There have been several reports of collagen deficiency in patients with ruptured IAs. One study by Neal-Dyer et al.[123] found a relative deficiency of type III collagen in 11 of 17 SAH patients. This study measured ratios of secreted collagen obtained from skin and temporal artery biopsies. Another study by Ostergard and Oxlund.[124] found a deficiency of collagen type III in the middle cerebral artery of 14 patients with ruptured IAs. The deficiency of type III collagen did not cause a change in the mechanical strength of the artery, but did increase the extensibility of the MCA. A case report by DePaepe et al.[125] studied a 55-year-old male with multiple IAs who had a 95% reduction in collagen type III levels. This patient had some symptoms that suggested a more generalized connective tissue disorder, but was not considered to have Ehlers-Danlos syndrome. In addition, there was no family history of similar disorders. Deak et al.[126] studied procollagen synthesis in 14 patients with AAA, 8 with an aneurysm at another site and 6 with a first-degree relative with AAA. Two of the 14 patients had markedly abnormal type I/III collagen ratios, but with normal type III collagen mRNA levels. One patient had altered thermal stability of type III collagen. Overall, these studies provide some indirect evidence of quantitative type III collagen changes in some patients with IAs. However, as discussed below, these changes are certainly not universal, nor is it clear whether they are the cause or result of IA formation.

A small study by LeBlanc and colleagues[127] examined collagen synthesis in a case of probable familial IA/SAH. The proband, her mother, and one sibling were found to have IAs. Collagen synthesis was measured in skin fibroblast cell lines from the proband, 2 controls, and a patient with EDS type IV. No changes in type III collagen synthesis were seen in either the proband or control cell lines, while virtual absence of type III collagen was seen in the patient with EDS

type IV. LeBlanc also found normal collagen levels in 4 patients with IAs. It must be emphasized that the presence of normal type III collagen levels does not rule out a mutation in collagen which could affect its strength or thermal stability. These questions can best be answered by direct genetic studies, which are reviewed below. However, the results of the quantitative collagen studies described above do provide some preliminary evidence that collagen deficiency may be present in a subgroup of patients with IAs. Confirmation of this deficiency will require the detailed study of larger number of patients, coupled with genetic studies.

The detection of qualitative changes in type III procollagen has been a focus of several studies. One report by Majamaa et al.[128] found abnormal structural stability of type III procollagen in 2 patients with familial SAH. No changes were found in 4 other patients with familial SAH, or 6 patients with sporadic SAH. Despite these qualitative changes in collagen, no quantitative changes in collagen synthesis were seen in the cell lines of the patients. A study by Pope et al.[129] reported type III collagen mutations producing fragile cerebral arteries in patients with EDS type IV. Most of these patients had CCFs, as opposed to berry aneurysms. (This issue has been discussed above.) It is certainly reasonable to expect that mutations in the collagen gene could cause changes in the strength of selected vessels. However, there are limited data at this time to support vascular fragility on the basis of collagen defects as a cause of most cases of IAs.

The most direct approach for evaluating the presence of COL3A1 mutations is to sequence the gene. A study by Kuivaniemi and colleagues[74] performed extensive cDNA sequencing of COL3A1 in a large cohort of patients with IAs or arterial dissections (half of whom had a positive family history). No clear instances of pathogenic mutations in the COL3A1 cDNA were identified, thereby indicating that mutations in this gene are unlikely to be a common cause of familial or sporadic IAs or cervical dissections.

There is growing evidence to support the presence of mutations in the COL3A1 gene in patients with EDS type IV (as discussed above). Mutations in COL3A1 known to cause EDS type IV include intronic mutations causing splicing defects, single base changes that convert glycine to other amino acids, and large multiexon deletions.[60–62,64–72,130,131] There are several cases of significant mosaicism seen with different types of COL3A1 mutations, indicating that the mutation most likely occurred sometime during early embryonic development.[62,132] Such mosaicism could account for the phenotypic diversity seen in such cases. It also makes predictions of phenotypic involvement from blood leukocyte DNA analysis difficult, since other tissues could have a different percentage of cells with a mutation.

Several recent studies have investigated the occurrence of COL3A1 mutations in individuals with aortic aneurysms, with or without other features of EDS type IV. A study by Tromp et al.[91] sequenced cDNA from 50 unrelated patients with AAAs. Changes in the COL3A1 gene were identified in only 2% of individuals, and it is not immediately apparent whether these changes are true pathogenic mutations. Six of the 50 patients also had IAs, but no COL3A1 mutations were identified in those individuals. Another study by Powell et al.[133] examined genetic variations of COL3A1 in patients with AAAs. They found a statistically significant increase in the frequency of a minor allele (of a *Stu* site) in patients with AAAs compared to controls.

These data support the importance of C0L3A1 mutations in causing EDS type IV and probably some cases of AAAs. However, there is no evidence that mutations in the COL3A1 gene cause cerebral berry aneurysms, either familial or sporadic. As noted above, quantitative or qualitative changes in the collagen protein may be caused by abnormalities in the post-translational processing of collagen. Direct analysis of the collagen gene would not show any abnormalities in such cases. At present, I am not aware of any large genetic studies that have investigated the post-translational processing of type III collagen.

Elastin

Elastin is the major component of elastic fibers that make up part of skin, ligaments, and some blood vessels. It is a highly cross-linked, insoluble protein that is synthesized as a soluble precursor, tropoelastin. The human elastin gene has 34 exons, and has been mapped to chromosome 7q11.1–21.1.[134] The elastin mRNA appears to undergo alternative splicing, with various exons being spliced out; however, the functional significance of this is unclear. These features of elastin are reviewed in reference 135.

Elastin has attracted attention as a candidate gene in IA formation due to the defects in the internal elastic lamina seen in some cases of IAs. Defects in the muscular media occur commonly at the bifurcations of intracranial vessels.[136] This results in the internal elastic lamina being the major source of structural support for these vessels, since cerebral vessels lack an external elastic lamina.[137] Defects in the internal elastic lamina could therefore lead to aneurysm formation.[10,138] Some studies have reported deficiencies and abnormalities of the elastin fibers in cerebral vessels, while others have failed to find such changes.[139–140] Epidemiological studies (reviewed above) have linked cigarette smoking to SAH,[32,37] and some have suggested that smoking may weaken elastin fibers, leading to aneurysm formation.[38] Considering these findings, elastin is certainly worthy of further genetic studies in cases of IA, especially considering its importance in maintaining vessel wall integrity.

Elastase and Elastase Inhibitors

As indicated above, elastin is a critical component of the vessel wall. Elastin can be degraded by a variety of elastases, which are produced by several different cell types, particularly neutrophils. Processes leading to an increase in elastase activity, or reduced activity of inhibitors of elastase, could cause damage to elastin and weaken the vessel wall. Although a variety of elastase inhibitors have been isolated and characterized, the most abundant and potent is alpha-1-antitrypsin (α1AT), which inhibits neutrophil elastase.[141] The a1AT gene is located on chromosome 14q31–32.3 and is 12.2 kb in size, composed of 5 exons and 4 introns.[142] This gene is highly polymorphic, with more than 75 allelic variants having been described, although only a small percentage are clinically relevant.[143] The two parental alleles are codominantly expressed, meaning that the protein from each allele will be produced independently, resulting in a variety of phenotypes.[143]

There is a clear association between a person's genotype and functional activity of the α1AT protein. The correlation between the various allelic variants and functional α1AT is categorized as follows[143]:

- Normal—normal amounts and activity of α1AT.
- Deficient—low serum levels of α1AT; function may be low or normal.
- Dysfunctional—normal amounts of α1AT, but abnormal function.
- Null—no detectable α1AT in serum.

A list of some of the most common and important α1AT allelic variants is provided in Table 5. Variants designated M1, M2, or M3 are normal variants. The two most common abnormal variants are designated S and Z. The S variant has slightly reduced serum α1AT levels, but normal function. The Z variant has significantly reduced serum levels and reduced function.[143]

Patients who are homozygous for the Z variant have the full clinical syndrome of α1AT deficiency, with lung emphysema and liver disease. Patients homozygous for the S variant have reduced α1AT levels, but not low enough to produce clinical problems. Patients who are heterozygotes for both alleles (ZS) may develop clinical symptoms in some cases.[143]

Interest in α1AT and SAH has developed based on several observations. First,

there is a clear association between cigarette smoking, SAH risk, and reduced α1AT activity.[144] Second, studies of patients with SAH have shown reduced α1AT activity (but not serum levels).[145] Other studies have reported reduced α1AT levels in patients with aortic aneurysms.[146] These observations led to studies of the various α1AT genotypes in SAH patients. One study found that the heterozygous deficient genotypes (MZ or MS) are present in 16% of patients with IAs compared to 7% of the general population.[147] Another study found an eightfold increase in the Z allele frequency in a subset of IA patients.[148] Based on these findings, further study of α1AT and IA/SAH is clearly warranted.

Fibrillin

The fibrillin gene was discussed above, in the section on Marfan's syndrome. As with EDS type IV, defects in this gene appear to promote formation of giant, fusiform IAs, not berry IAs. Nonetheless, detailed genetic studies of this gene in cases of familial IA are probably warranted, since other types of mutations in this gene, or perhaps defects in the post-translational processing of fibrillin, may cause formation of berry IAs.

Clinical Evaluations and Genetic Counseling

The recognition of a familial aggregation in some cases of IA, combined with the possibility of a surgical cure for unruptured IAs and the devasting consequences of an SAH, makes consideration of a screening approach mandatory. Since IA formation and growth is a dynamic process, screening with imaging techniques will require serial studies (as opposed to screening for a metabolic disease, which can usually be done only once). Further complicating the issue is the uncertainty surrounding the exact percentage of IA cases that are familial, along with the exact mode of inheritance. It is very

Table 5

Genetic and Phenotypic Variants of Alpha-1-Antitrypsin (α1AT)

Allele Type	Levels of α1AT (mg/dl)	Functional Activity*	Allele Frequency	Comments
M1	150–350	100%	0.64–0.72	normal
M2	150–350	100%	0.14–0.19	normal
M3	150–350	100%	0.10–0.11	normal
Z	15–50	15%	0.01–0.02	α1AT def**
S	100–200	60%	0.02–0.04	normal

*Functional activity if in homozygous state.
**Patients homozygous for the Z allele will develop symptoms of α1AT deficiency. See text for details of other genotypes.
Data adapted from references 143 and 148.

possible that if mass serial screenings were performed, the percentage of cases that are familial would increase substantially above the 12–15% now reported.[7]

Based on the somewhat heterogeneous features seen in cases of familial IA (i.e., different ages of onset, variations in IA location, size, and multiplicity) it is quite possible that familial IA/SAH is genetically heterogeneous, in that different mutations are present in a single gene, or mutations are present in several different genes. It is also possible that familial IA/SAH will be polygenic, requiring specific changes in two or more genes to produce a disease phenotype. These variables offer potential complications for any type of screening paradigm.

The optimal screening tool would be one that was safe, highly sensitive and specific, and inexpensive. A desirable scenario would be the identification of one or more mutations that could be detected easily by screening genomic DNA from peripheral blood. Perhaps such a blood test will become available in the future, but currently it is not available. For now we are left with less desirable screening techniques such as cerebral angiography, MRA, and CT angiography. Although angiography is considered the "gold standard" it can miss some IAs, is expensive as well as invasive, and has a small but definitie risk of serious complications.[149] MRA is relatively safe, but may also miss IAs and cannot be performed on every-one.[9] CT angiography (CTA) is a new technique that uses intravenous contrast to visualize the arteries at the base of the brain through computer reconstruction. Its resolution may be superior to MRA, and it allows three-dimensional reconstruction of the circle of Willis.[150] In my opinion, CTA may supplant MRA as the noninvasive screening method of choice for IAs.

Several published studies have performed detailed decision analyses on the issue of screening strategies for IAs.[6,151] These studies have made certain assumptions about the various risks and other parameters involved in such decisions. Their conclusions are that screening is appropriate in cases of familial IA for at-risk individuals between the ages of (approximately) 35 to 65.[6,151] We would broaden the age range to 25 to 65 years, based on published data showing symptomatic IAs in familial cases below the age of 30 years.[18] We would also define "familial" IA to include an individual with any one of several family characteristics (Table 6). In addition, anyone with an affected relative and symptoms suggestive of IA or a sentinel SAH should be studied by angiography without delay. Physicians and geneticists should also keep in mind that due to the dynamic process of IA formation and growth, a negative neuroimaging screening test at age 30 or 35 does *not* rule out the occurrence of an IA several years later, especially in cases of a strong family

Table 6*
Characteristics Associated with a High Risk of Familial IA/SAH[1]

1. Affected[2] parent and one or more affected siblings.
2. Two or more affected siblings.
3. An affected parent and child.
4. An affected parent and two (or more) affected first degree relatives of the parent.
5. Inherited condition (in family) known to cause IA/SAH plus an affected parent or one (or more) affected siblings.

[1] IA, intracranial aneurysm; SAH, subarachnoid hemorrhage

[2] In all cases, affected refers to the presence of an IA (symptomatic or asymptomatic) or an SAH.

*All relations are stated with respect to the individual at risk.

Note: This table should only be used as a guide for consideration of those at risk for developing an IA/SAH in certain cases with a suggestive family history. Recommendations in specific cases must be based on that particular clinical scenario, weighing the risks vs. benefits of diagnostic studies and treatment.

history. Screening every 3 to 5 years is reasonable in at-risk individuals. In addition, family members at risk for an IA/SAH should be told to avoid behavior such as smoking that may increase the risk of IA formation and rupture.

Any screening for cases of familial IA should be performed by a well-trained, coordinated team consisting of a neurosurgeon and/or neurologist experienced in cerebrovascular disease, and a geneticist who is familiar with the disease and counseling techniques. Since such screening will likely involve a radiologic procedure, a neuroradiologist with expertise in performing and interpreting angiograma, MRAs, and/or CTAs should participate in the process.

Conclusion

There is growing evidence that, in some cases, IA formation and SAH are due to inherited genetic factors. We now have the technology to screen for mutations in specific genes and determine their role in IA formation. Mutations in the type III collagen gene are important in the pathogenesis of EDS type IV and some cerebral aneurysms, but are not the likely cause of berry IAs. Other candidadate genes should be studied in detail in cases of familial IA. It is possible, and even likely, that familial IA will be a genetically heterogeneous disorder. Until a sensi-

tive and specific genetic (blood) marker is identified, clinical criteria will have to be used for developing screening strategies. Considering the devastating consequences of IA rupture, an aggressive screening approach seems warranted for this disease.

References

1. Hijdra A, Braakman R, Van Gijn J, et al. Aneurysmal subarachnoid hemorrhage: Complications and outcome in a hospital population. Stroke 1987;18:1061–1067.
2. Mohr J, Kistler J, Fink M. Intracranial aneurysms. In Barnett H, Mohr J, Stein B, et al (eds). Stroke: Pathophysiology, Diagnosis, and Management. New York: Churchill Livingstone, 1992, 617–643.
3. Mohr J, Caplan L, Melski J, et al. The Harvard Cooperative Stroke Registry. Neurology 1978;28:754–762.
4. Sahs A, Perret G, Locksley H, et al. Intracranial Aneurysms and Subarachnoid Hemorrhage: A Cooperative Study. Philadelphia: Lippincott, 1969.
5. Norrgard 0, Angquist KA, Fodstad H, et al. Intracranial aneurysms and heredity. Neurosurgery 1987;20:236–239.
6. Dippel DW, ter Berg H, Habbema JD. Screening for unruptured familial intracranial aneurysms: A decision analysis. Acta Neurol Scand 1992;86:381–389.
7. Ronkainen A, Hernesniemi J, Ryynanen M. Familial subarachnoid hemorrhage in East Finland, 1977–1990. Neurosurgery 1993;33:787–796.
8. Allcock J, Canham P. Angiographic study of

the growth of intracranial aneurysms. J Neurosurg 1976;45:617–621.

9. Schuierer G, Huk W, Laub G. Magnetic resonance angiography of intracranial aneurysms: Comparison with intra-arterial digital subtraction angiography. Neuroradiology 1992;35:50–54.

10. Sekhar L, Heros R. Origin, growth, and rupture of saccular aneurysms: A review. Neurosurgery 1981;8:248–260.

11. Shinton R, Palsingh J, Williams B. Cerebral haemorrhage and berry aneurysm: Evidence from a family for a pattern of autosomal dominant inheritance. J Neurol Neurosurg Psychiatry 1991;54:838–840.

12. Elshunnar KS, Whittle IR. Familial intracranial aneurysms: Report of five families. Br J Neurosurg 1990;4:181–186.

13. Edelsohn L, Caplan L, Rosenbaum A. Familial aneurysms and infundibular widening. Neurology 1972;22:1056–1060.

14. Evans T, Venning M, Strang F, et al. Dominant inheritance of intracranial berry aneurysm. BMJ 1981;283:824–825.

15. Fox J. Familial intracranial aneurysms. J Neurosurg 1980;57:416–417.

16. Halal F, Mohr G, Toussi T, et al. Intracranial aneurysms: A report of a large pedigree. Am J Med Genetic 1983;15:89–95.

17. Mellergard P, Ljunggren B, Brandt L, et al. HLA-typing in a family with six intracranial aneurysms. Br J Neurosurg 1989;3:479–485.

18. Alberts MJ, Quinones A, Gaffagnino C, et al. Risk of intracranial aneurysms in familial subarachnoid hemorrhage. Can J Neurol Sci 1995;22:121–125.

19. Lozano AM, LeBlanc R. Familial intracranial aneurysms. J Neurosurg 1987;66:522–528.

20. Andrews R. Intracranial aneurysms: Characteristics of aneurysms in siblings. N Engl J Med 1977;297:115.

21. Andrews RJ, Spiegel PK. Intracranial aneurysms: Age, sex, blood pressure, and multiplicity in an unselected series of patients. J Neurosurg 1979;51:27–32.

22. Schon F, Marshall J. Subarachnoid hemorrhage in identical twins. J Neurol Neurosurg Psychiatry 1984;47:81–83.

23. Parekh HC, Gurusinghe NT, Sharma RR. Cerebral berry aneurysms in identical twins: A case report. Surg Neurol 1992;38:277–279.

24. Fairburn B. "Twin" intracranial aneurysms causing subarachnoid hemorrhage in identical twins. BMJ 1973;1:210–211.

25. Inagawa T, Hirano A. Autopsy study of unruptured incidental intracranial aneurysms. Surg Neurol 1990;34:361–365.

26. Suzuki S, Robertson J, White R. Experimental intracranial aneurysms in rats: A gross and microscopic study. J Neurosurg 1980;52:494–500.

27. McKissock W, Richardson A, Walsh L, et al. Multiple intracranial aneurysms. Lancet 1964;1:623–626.

28. Graffagnino C, Maher J, Quinones A, et al. Genetic aspects of intracranial aneurysms and SAH. Stroke 1994;25:257.

29. Pericak-Vance M, Bebout J, Gaskell P, et al. Linkage studies in familial Alzheimer's disease: Evidence for chromosome 19 linkage. Am J Hum Genet 1991;48:1034–1050.

29a. Ronkainen A, Hernesniemi J, Tromp G. Special features of familial intracranial aneurysms: Report of 215 familial aneurysms. Neurosurg 1995;37:43–47.

29b. Ronkainen A, Hernesniemi J, Puranen M, et al. Familial intracranial aneurysms. Lancet 1997;349:380–384.

29c. Schievink WI, Schaid DJ, Rogers HM, et al. On the inheritance of intracranial aneurysms. Stroke 1994;25:2028–2037.

29d. Coutard M, Osborne-Pellegrin M. Genetic susceptibility to experimental cerebral aneurysm formation in the rat. Stroke 1997;28:1035–1042.

30. Hannsen P, Rosengren A, Tsipogianni A, et al. Risk factors for stroke in middle-aged men in Goteborg, Sweden. Stroke 1990;21:223–229.

31. Knekt P, Reunanen A, Aho K, et al. Risk factors for subarachnoid hemorrhage in a longitudinal population study. J Clin Epidemiol 1991;44:933–939.

32. Juvela S, Hillbom M, Numminen H, et al. Cigarette smoking and alcohol consumption as risk factors for aneurysmal subarachnoid hemorrhage. Stroke 1993;24:639–646.

33. Phillips SJ, Whisnant JP. Hypertension and the brain. The National High Blood Pressure Education Program. Arch Int Med 1992;152:938–945.

34. Stehbens WE. Etiology of intracranial berry aneurysms. J Neurosurg 1989;70:823–831.

35. McConnick W, Schmalstieg E. The relationship of arterial hypertension to intracranial aneurysms. Arch Neurol 1977;34:285–287.

36. Kawachi I, Colditz GA, Stampfer MJ, et al. Smoking cessation and decreased risk of stroke in women. JAMA 1993;269:232–236.

37. Longstreth WJ, Nelson LM, Koepsell TD, et al. Cigarette smoking, alcohol use, and subarachnoid hemorrhage. Stroke 1992;23:1242–1249.

38. Cannon D, Read R. Blood elastolytic activity in patients with aortic aneurysm. Ann Thorac Surg 1982;34:10–15.

39. Fox J. Intracranial Aneurysms. New York: Springer-Verlag, 1983.

40. Korishi Y, Kadowaki C, Hara M, et al. Aneurysms associated with moyamoya disease. Neurosurg 1985;16:484–491.

41. Takakura K, Saito I, Sasaki T. Special problems associated with subarachnoid hemorrhage. In Youmans J (ed). Neurological Surgery. Philadelphia: WB Saunders, 1990, 1873–1874.

42. Schievink WI, Torres VE, Piepgras DG, et al. Saccular intracranial aneurysms in autosomal dominant polycystic kidney disease. J Am Soc Nephrol 1992;3:88–95.

43. Fehlings MG, Gentili F. The association between polycystic kidney disease and cerebral aneurysms. Can J Neurol Sci 1991;18: 505–509.

44. Ryu SJ. Intracranial hemorrhage in patients with polycystic kidney disease. Stroke 1990;21:291–294.

45. Report of a meeting of physicians and scientists. Autosomal dominant polycystic kidney disease. Lancet 1992;339:1146–1149.

45a. van Dijk MA, Chang PC, Peters DJ, Breuning MH. Intracranial aneurysms in polycystic kidney disease linked to chromosome 4. J Am Soc Nephrol 1995;6:1670–1673.

46. Reeders S, Breuning M, Davies K, et al. A highly polymorphic DNA marker linked to adult polycystic kidney disease on chromosome 16. Nature 1985;317:542–544.

47. Peters D, Spruit L, Saris J, et al. Chromosome 4 localization of a second gene for autosomal dominant polycystic kidney disease. Nat Genet 1993;5:359–362.

47a. International Polycystic Kidney Disease Consortium. Polycystic kidney disease: The complete structure of the PKD1 gene and its protein. Cell 1995;81:289–298.

47b. European Polycystic Kidney Disease Consortium. The polycystic kidney disease 1 gene encodes a 14 kb transcript and lies within a duplicated region on chromosome 16. Cell 1994;77:881–894.

48. Somlo S, Wirth B, Gennino G, et al. Fine genetic localization of the gene for autosomal dominant polycystic kidney disease (PKD1) with respect to physically mapped markers. Genomics 1992;13:152–158.

48a. Mochizuki T, Wu G, Hayashi T, et al. PKD2, a gene for polycystic kidney disease that encodes an integral membrane protein. Science 1996;272:1339–1342.

48b. Hughes J, Ward CJ, Peral B, et al. The polycystic kidney disease 1 (PKD1) gene encodes a novel protein with multiple cell recognition domains. Nat Genet 1995;10: 151–160.

48c. Peral B, Gamble V, Strong C, et al. Identification of mutations in the duplicated region of the polycystic kidney disease 1 gene (PKD1) by a novel approach. Am J Hum Genet 1997;60:1399–1410.

48d. Veldhuisen B, Saris JJ, de Haij S, et al. A spectrum of mutations in the second gene for autosomal dominant polycystic kidney disease (PKD2). Am J Hum Genet 1997;61: 547–555.

49. Ditlefsen E, Tonjum A. Intracranial aneurysms and polycystic kidneys. Acta Med Scand 1960;168:51–54.

50. Bigelow N. The association of polycystic kidneys with intracranial aneurysms and other related disorders. Am J Med Sci 1953;225: 485–494.

51. Lozano AM, Leblanc R. Cerebral aneurysms and polycystic kidney disease: a critical review. Can J Neurol Sci 1992;19:222–227.

52. Steinmann B, Royce P, Superti-Furga A. The Ehlers-Danlos syndrome. In Royce P, Steinmann B (eds). Connective Tissue and Its Heritable Disorders. New York: Wiley-Liss, 1993, 351–407.

53. Mirza F, Smith P, Lim W. Multiple aneurysms in a patient with Ehlers-Danlos Syndrome: Angiography without sequelae. AJR 1979;132:993–995.

54. Krog M, Almgren B, Eriksson J, et al. Vascular complication in the Ehlers-Danlos syndrome. Acta Chir Scand 1983;149:279–282.

55. Edwards A. Ehlers-Danlos syndrome with vertebral artery aneurysm. Proc Roy Soc Med 1969;62:734–735.

56. Ruby S, Kramer J, Cassidy S, et al. Internal carotid artery aneurysm: A vascular manifestation of type IV Ehlers-Danlos syndrome. Connecticut Med 1989;53:142–144.

57. Fox R, Pope F, Narcisi P, et al. Spontaneous carotid cavernous fistula in Ehlers-Danlos syndrome. J Neurol Neurosurg Psychiatry 1988;51:984–986.

58. Lach B, Nair S, Russell N, et al. Spontaneous carotid-cavernous fistula and multiple arterial dissections in type IV Ehlers-Danlos syndrome. J Neurosurg 1987;66:462–467.

59. Rubinstein M, Cohen N. Ehlers-Danlos syndrome associated with multiple intracranial aneurysms. Neurology 1964;14:125–132.

60. Kontusaari S, Tromp G, Kuivaniemi H, et al. Inheritance of an RNA splicing mutation (G+1 IV520) in the type III procollagen gene (COL3A1) in a family having aortic aneurysms and easy bruisability: Phenotypic overlap between familial arterial aneurysms and Ehlers-Danlos syndrome type IV. Am J Hum Genet 1990;47:112–120.

61. Kontusaari S, Tromp G, Kuivaniemi H, et al. A mutation in the gene for type III procollagen (COL3A1) in a family with aortic aneurysms. J Clin Invest 1990;86:1465–1473.

62. Kontusaari S, Tromp G, Kuivaniemi H, et al. Substitution of aspartate for glycine 1018 in the type III procollagen (COL3A1) gene causes type IV Ehlers-Danlos syndrome: The

mutated allele is present in most blood leukocytes of the asymptomatic and mosaic mother. Am J Hum Genet 1992;51:497–507.

63. Kuivaniemi H, Kontusaari S, Tromp G, et al. Identical G+1 to A mutations in three different introns of the type III procollagen gene (COL3A1) produce different patterns of RNA splicing in three variants of Ehlers-Danlos syndrome IV: An explanation for exon skipping some mutations and not others. JBC 1990;265:12067–12074.

64. Lee B, Vitale E, Superti FA, et al. G to T transversion at position +5 of a splice donor site causes skipping of the preceding exon in the type III procollagen transcripts of a patient with Ehlers-Danlos syndrome type IV. JBC 1991;266:5256–5259.

65. Lee B, D'Alessio M, Vissing H, et al. Characterization of a large deletion associated with a polymorphic block of repeated dinucleotides in the type III procollagen gene (COL3A1) of a patient with Ehlers-Danlos syndrome type IV. Am J Hum Genet 1991;48:511–517.

66. Lloyd J, Narcisi P, Richards A, et al. A T+6 to C+6 mutation in the donor splice site of COL3A1 IV57 causes exon skipping and results in Ehlers-Danlos syndrome type IV. J Med Genet 1993;30:376–380.

67. Richards AJ, Lloyd JC, Ward PN, et al. Characterization of a glycine to valine substitution at amino acid position 910 of the triple helical region of type III collagen in a patient with Ehlers-Danlos syndrome type IV. J Med Genet 1991;28:458–463.

68. Richards AJ, Ward PN, Narcisi P, et al. A single base mutation in the gene for type III collagen (COL3A1) converts glycine 847 to glutamic acid in a family with Ehlers-Danlos syndrome type IV: An unaffected family member is mosaic for the mutation. Hum Genet 1992;89:414–418.

69. Richards AJ, Lloyd JC, Narcisi P, et al. A 27-bp deletion from one allele of the type III collagen gene (COL3A1) in a large family with Ehlers-Danlos syndrome type IV. Hum Genet 1992;88:325–330.

70. Tromp G, Kuivaniemi H, Shikata H, et al. A single base mutation that substitutes serine for glycine 790 of the alpha 1 (III) chain of type III procollagen exposes an arginine and causes Ehlers-Danlos syndrome IV. JBC 1989;264:1349–1352.

71. Tromp G, Kuivaniemi H, Stolle C, et al. Single base mutation in the type III procollagen gene that converts the codon for glycine 883 to aspartate in a mild variant of Ehlers-Danlos syndrome IV. JBC 1989;264:19313–19317.

72. Vissing H, D'Alessio M, Lee B, et al. Multiexon deletion in the procollagen III gene is associated with mild Ehlers-Danlos syndrome type IV. JBC 1991;266:5244–5248.

73. Weil D, D'Alessio M, Ramirez F, et al. A base substitution in the exon of a collagen gene causes alternative splicing and generates a structurally abnormal polypeptide in a patient with Ehlers-Danlos syndrome type VII. EMBO J 1989;8:1705–1710.

74. Kuivaniemi H, Prockop D, Wu Y, et al. Exclusion of mutations in the gene for type III collagen (COL3A1) as a common cause of intracranial aneurysms or cervical artery dissections. Neurology 1993;43:2652–2658.

75. Reifenstein G, Levine S, Gross R. Coarctation of the aorta. Am Heart J 1947;33:146–168.

76. Campbell M. Natural history of coarctation of the aorta. Br Heart J 1970;2:633–640.

77. Campbell M, Baylis J. The course and prognosis of coarctation of the aorta. Br Heart J 1956;18:475–495.

78. Serizawa T, Satoh A, Miyata A, et al. Ruptured cerebral aneurysm associated with coarctation of the aorta: Report of two cases. Neurol Med Chir 1992;3:342–345.

79. Yoshioka S, Kai Y, Uemura S, et al. Ruptured cerebral aneurysm associated with coarctation of the aorta (Japanese.) No To Shinkei Brain & Nerve 1990;42:1055–1060.

80. Bobby J, Emami J, Fanner R, et al. Operative survival and 40 year follow up of surgical repair of aortic coarctation. Br Heart J 1991;65: 271–276.

81. Ouchi Y, Tagawa H, Yamakado M, et al. Clinical significance of cerebral aneurysm in renovascular hypertension due to fibromuscular dysplasia: Two cases in siblings. Angiology 1989;40:581–588.

82. Mettinger K. Fibromuscular dysplasia and the brain. II. Current concept of the disease. Stroke 1982;13:53–58.

83. Mettinger K, Ericson K. Fibromuscular dysplasia and the brain: Observations on angiographic, clinical and genetic characteristics. Stroke 1982;13:46–52.

84. So E, Toole J, Dalal P, et al. Cephalic fibromuscular dysplasia in 32 patients. Arch Neurol 1981;38:619–622.

85. George B, Zerah M, Mourier X, et al. Ruptured intracranial aneurysms: The influence of sex and fibromuscular dysplasia upon prognosis. Acta Neurochir 1989;97:26–30.

85a. Teunissen LL, Rinkel GJ, Algra A, et al. Risk factors for subarachnoid hemorrhage: A systemic review. Stroke 1996;27:544–549.

85b. Bromberg JE, Rinkel GJ, Algra A, et al. Hypertension, stroke, and coronary heart disease in relatives of patients with subarachnoid hemorrhage. Stroke 1996;27:7–9.

86. Johansen K, Koepsell T. Familial tendency for abdominal aortic aneurysms. JAMA 1986;256:1934–1936.

87. Cole C, Barber G, Bouchard A, et al. Abdominal aortic aneurysm: Consequences of posi-

tive family history. Can J Surg 1989;32: 117–120.

88. Webster M, St. Jean P, Steed D, et al. Abdominal aortic aneurysm: Results of a family study. J Vasc Surg 1991;13:366–372.

89. Majumder P, St Jean P, Ferrell R, et al. On the inheritance of abdominal aortic aneurysm. Am J Hum Genet 1991;48:164–170.

90. Norrgard 0, Angqvist K-A, Fodstad H, et al. Co-existence of abdominal aortic aneurysms and intracranial aneurysms. Acta Neurochir (Wien) 1987;87:34–39.

91. Tromp G, Wu Y, Prockop D, et al. Sequencing of cDNA from 50 unrelated patients reveals that mutations in the triple-helical domain of type III procollagen are an infrequent cause of aortic aneurysms. J Clin Invest 1993;91:2539–2545.

92. Norrgard 0, Beckman G, Beckman L, et al. Genetic markers in patients with intracranial aneurysms. Hum Hered 1987;37: 255–259.

93. Ostergaard J, Bruun-Petersen G, Lamm U. HLA antigens and complement types in patients with intracranial saccular aneurysms. Tissue Antigens 1986;28:176–181.

94. Ryba M, Grieb P, Podobinska I, et al. HLA antigens and intracranial aneurysms. Acta Neurochir 1992;116:1–5.

95. Lye RH, Dyer PA. Intracranial aneurysm and HLA-DR2 [letter]. J Neurol Neurosurg Psychiatry 1989;52:291.

96. Kwak R, Suzuki J. Correlation of anterior communicating artery aneurysm with blood circulation at the anterior part of the circle of Willis and its vascular anomalies. Phronesis 1974;11:407–417.

97. Kwak R, Niizuma H, Suzuki J. Hemodynamics in the anterior part of the circle of Willis in patients with intracranial aneurysms: A study by cerebral angiography. Tohoku J Exp Med 1980;132:69–73.

98. Wilson G, Riggs H, Rupp C. The pathologic anatomy of ruptured cerebral aneurysms. J Neurosurg 1954;11:128–134.

99. Marinkovic S, Kovacevic M, Milisavljevic M. Hypoplasia of the proximal segment of the anterior cerebral artery. Anat Anz 1989;168: 145–154.

100. Kirgis H, Fisher W, Llewellyn R, et al. Aneurysms of the anterior communicating artery and gross anomalies of the circle of Willis. J Neurosurg 1966;25:73–78.

101. Sakai N, Sakata K, Yamada H, et al. Familial occurrence of intracranial aneurysms. Surg Neurol 1974;2:25–29.

102. Pyeritz A. Marfan syndrome. In Emery A, Rimoin D (eds). Principles and practices of medical genetics. New York: Churchill Livingstone, 1990, 1047–1063.

103. Dietz HC, Cutting GR, Pyeritz RE, et al. Marfan syndrome caused by a recurrent de

novo missense mutation in the fibrillin gene. Nature 1991;352:337–339.

104. Pereira L, D'Alessio M, Ramirez F, et al. Genomic organization of the sequence coding for fibrillin, the defective gene product in Marfan syndrome. Hum Mol Genet 1993; 2:961–968.

105. Dietz HC, Pyeritz RE, Puffenberger EG, et al. Marfan phenotype variability in a family segregating a missense mutation in the epidermal growth factor-like motif of the fibrillin gene. J Clin Invest 1992;89: 1674–1680.

106. Dietz HC, Valle D, Francomano CA, et al. The skipping of constitutive exons in vivo induced by nonsense mutations. Science 1993;259:680–683.

107. Kainulainen K, Sakai LY, Child A, et al. Two mutations in Marfan syndrome resulting in truncated fibrillin polypeptides. Proc Nat Acad Sci 1992;89:5917–5921.

108. Roberts W, Honig H. The spectrum of cardiovascular disease in the Marfan syndrome: A clinico-morphologic study of 18 necropsy patients and comparison to 151 previously reported necropsy patients. Am Heart J 1982;104:115–135.

109. Latter D, Ricci M, Forbes R, et al. Internal carotid artery aneurysm and Marfan's syndrome. Can J Surg 1989;32:463–466.

110. Ohyama T, Ohara S, Momma F. Aneurysm of the cervical internal carotid artery associated with Marfan's syndrome. Neurol Med Chir (Tokyo) 1992;32:965–968.

111. Finney H, Roberts T, Anderson R. Giant intracranial aneurysms associated with Marfan's syndrome. J Neurosurg 1976;45: 342–347.

112. ter Berg H, Bijlsma J, Veiga Pires J, et al. Familial association of intracranial aneurysms and multiple congenital anomalies. Arch Neurol 1986;43:30–33.

113. Stehbens W, Delahunt B, Hilless A. Early berry aneurysm formation in Marfan's syndrome. Surg Neurol 1989;3 1:200–202.

114. Neldner K. Pseudoxanthoma elasticum. In Royce P, Steinmann B (eds). Connective Tissue and Its Heritable Disorders. New York: Wiley-Liss, 1993, 425–436.

115. Walker E, Fredrickson R, Mayes M. The mineralization of elastic fibers and alterations of extracellular matrix in pseudoxanthoma elasticum: Ultrastructural, immunocytochemistry, and X-ray analysis. Arch Dermatol 1989;125:70–76.

116. Neldner K. Pseudoxanthoma elasticum. Clin Dermatol 1988;6:1–159.

117. Scheie H, Hogan T. Angoid streaks and generalized arterial disease. Arch Ophthamol 1957;57:855–868.

118. Dixon J. Angoid streaks and pseudoxan-

thoma elasticum. Am J Ophthamol 1951;34: 1322–1323.

119. Munyer T, Margulis A. Pseudoxanthoma elasticum with internal carotid artery aneurysm. AJR 1981;136:1023–1024.

120. Kito K, Kobayashi N, Mori N, et al. Ruptured aneurysm of the anterior spinal artery associated with pseudoxanthoma elasticum. J Neurosurg 1983;58:126–128.

121. Kielty C, Hopkinson I, Grant M. Collagen: The collagen family. Structure, assembly, and organization in the extracellular matrix. In Royce P, Steinmann B (eds). Connective Tissue and its Heritable Disorders. New York: Wiley-Liss, 1993, 103–147.

122. Chu M-L, Prockop D. Collagen: Gene structure. In Royce P, Steinmann B (eds). Connective Tissue and its Heritable Disorders. New York: Wiley-Liss, 1993, 149–165.

123. Neil-Dwyer G, Bartlett J, Nicholls A, et al. Collagen deficiency and ruptured cerebral aneurysms: A clinical and biochemical study. J Neurosurg 1983;59:16–20.

124. Ostergaard J, Oxlund H. Collagen type III deficiency in patients with rupture of intracranial saccular aneurysms. J Neurosurg 1987;67:690–696.

125. de Paepe A, van Landegem W, de Keyser F, et al. Association of multiple intracranial aneurysms and collagen type III deficiency. Clin Neurol Neurosurg 1988;90:53–56.

126. Deak S, Ricotta J, Mariani T, et al. Abnormal biosynthesis of type III procollagen in patients with multiple aneurysms. Matrix 1990;10:237.

127. Leblanc R, Lozano A, van der Rest M, et al. Absence of collagen deficiency in familial cerebral aneurysms. J Neurosurg 1989;70: 837–840.

128. Majamaa K, Savolainen ER, Myllyla VV. Synthesis of structurally unstable type III procollagen in patients with cerebral artery aneurysm. Biochim Biophysic Acta 1992; 1138:191–196.

129. Pope FM, Kendall BE, Slapak GI, et al. Type III collagen mutations cause fragile cerebral arteries. Br J Neurosurg 1991;5:551–574.

130. Narcisi P, Wu Y, Tromp G, et al. Single base mutation that substitutes glutamic acid for glycine 1021 in the COL3A1 gene and causes Ehlers-Danlos syndrome type IV. Am J Med Genet 1993;46:278–283.

131. Weil D, D Alessio M, Ramirez F, et al. Structural and functional characterization of a splicing mutation in the pro-alpha 2(I) collagen gene of an Ehlers-Danlos type VII patient. JBC 1990;265:16007–16011.

132. Milewicz D, Witz A, Smith A, et al. Parental somatic and germ-line mosaicism for a mul-

tiexon deletion with unusual endpoints in a type III collagen (COL3A 1) allele produces Ehlers-Danlos syndrome type IV in the heterozygous offspring. Am J Hum Genet 1993;53:62–70.

133. Powell J, Adamson J, MacSweeney S, et al. Genetic variants of collagen III and abdominal aortic aneurysm. Eur J Vasc Surg 1991;5: 145–148.

134. Fazio M, Mattei M-G, Passage E, et al. Human elastin gene: New evidence for localization to the long arm of chromosome 7. Am J Hum Genet 1991;48:696–703.

135. Rosenbloom J. Elastin. In Royce P, Steinmann B (eds). Connective Tissue and its Heritable Disorders. New York: Wiley-Liss, 1993, 167–188.

136. Glynn L. Medial defects in the circle of Willis and their relation to aneurysm formation. J Pathol Bacteriol 1940;51:213–222.

137. Stehbens W. Pathology of the Cerebral Blood Vessels. St Louis: CV Mosby, 1972.

138. Stehbens W. Aneurysms and anatomic variation of cerebral arteries. Arch Pathol 1963; 75:45–64.

139. Ostergaard J, Reske-Nielsen E, Oxlund H. Histological and morphometric observations in the reticular fibers in the arterial beds of patients with ruptured intracranial saccular aneurysms. Neurosurg 1987;20:554–558.

140. Chyatte D, Reilly J, Tilson M. Morphometric analysis of reticular and elastin fibers in the cerebral arteries of patients with intracranial aneurysms. Neurosurgery 1990; 26:939–943.

141. Cox DW. α1-Antitrypsin deficiency. In Royce PM, Steinmann B (eds). Connective Tissue and its Heritable Disorders. New York, Wiley-Liss, 1993, 549–561.

142. Sifers RN, Finegold MJ, Woo SLC. Molecular biology and genetics of alpha1-antitrypsin deficiency. Semin Liver Dis 1992;12:301–310.

143. Brantly M, Nukiwa T, Crystal RG. Molecular basis of alpha-1-antitrypsin deficiency. Am J Med 1988;84(suppl 6A):13–27.

144. Gaetani P, Tartara F, Tancioni F, et al. Activity of alpha1-antitrypsin and cigarette smoking in subarachnoid hemorrhage from ruptured aneurysm. J Neurol Sci 1996;141:33–38.

145. Tartara F, Gaetani P, Tancioni F, et al. Alpha 1-antitrypsin activity in subarachnoid hemorrhage. Life Sciences 1996;59:15–20.

146. Cohen JR, Mandell C, Chang JB, et al. Altered aortic protease and antiprotease activity in patients with ruptured abdominal aortic aneurysms. Surg Gynecol Obstet 1987; 164:355–358.

147. Schievink W, Katzman JA, Piepgras DG, et al. Alpha-1-antitrypsin phenotypes among

patients with intracranial aneurysms. J Neurosurg 1996;84:781–784.

148. St. Jean P, Hart B, Webster M, et al. Alpha-1-antitrypsin deficiency in aneurysmal disease. Hum Hered 1996;46:92–97.

149. Dion J, Gates P, Fox A, et al. Clinical events following neuroangiography: A prospective study. Stroke 1987;18:997–999.

150. Gray L, Kallmes D, Chotas HC, et al. Volume-rendered 3D CT of the circle of Willis for detection of aneurysms in subarachnoid hemorrhage. Radiology 1993;189(suppl): 193C. Abstract.

151. ter Berg H, Dippel D, Limburg M, et al. Familial intracranial aneurysms: A review. Stroke 1992;23:1024–1030

Chapter 13

Genetic Causes of Pediatric Stroke

P. Ian Andrews, MBBS, FRACP, Monique M. Ryan, MBBS,
Raymond S. Kandt, MD

Introduction

This chapter will review inherited etiologies of pediatric stroke, concentrating on those with Mendelian inheritance (autosomal dominant, autosomal recessive, and X-linked), and mitochondrial inheritance. Relatively common causes of pediatric stroke for which genetic predisposition is one of multiple risk factors will also be briefly discussed. Among the many papers reviewed, the works of Roach and Riela,[1] Edwards and Hoffman[2] and Natowicz and Kelley[3] were comprehensive reference sources.

Stroke is uncommon in childhood. An incidence of 2.52/100,000/year in North American children up to 14 years of age[4] is comparable to the incidence of 2.1/ 100,000/year in Swedish children of the same age group observed during a similar period,[5] although markedly less than the 13.02/100,000/year reported recently from France.[6] The 10-year population-based survey of pediatric stroke in Rochester, Minnesota identified one ischemic stroke and three cerebral hemorrhages in a population of 15,384 children less than 15 years of age, to give an incidence of 0.63/100,000/year ischemic strokes and 1.89/100,000/year cerebral hemorrhages.[4] The Mayo Clinic hospital-based survey during the same period

identified 69 stroke patients less than 15 years of age; 55% had ischemic strokes and 45% had cerebral hemorrhages, which is similar to the balance of ischemic (61%) and hemorrhagic strokes (39%) in the French series.[6] In the large Swedish hospital-based series a weighting of 44% ischemic and 56% hemorrhagic strokes was seen.[5] A large proportion of those with ischemic strokes (41%) had preexisting heart disease, thereby implying an embolic etiology for these strokes. Unlike stroke in adulthood, atherosclerosis is an infrequent risk factor for stroke in the pediatric age group; instead, a wide variety of cardiac, hematologic, metabolic and other systemic diseases predispose children to stroke. In many cases an etiologic diagnosis remains elusive (29% in the Mayo clinic series[4]). Accurate diagnosis is required to direct therapy, to prevent further strokes, to identify relatives at risk, and to provide an opportunity for early intervention.

Since clinical and neuroimaging features often distinguish among different types of strokes, i.e., ischemic strokes of arterial origin (embolic and thrombotic), strokes due to venous occlusion, and cerebral hemorrhage, this chapter is organized to group etiologies of pediatric strokes under these headings. Many etiologies, however, are associated with more than one type of stroke, as shown

From Alberts MJ (ed). *Genetics of Cerebrovascular Disease.* Armonk, NY: Futura Publishing Company, Inc., © 1999.

in Table 1. The final section, a collection of miscellaneous etiologies of neonatal and pediatric stroke, and disorders with stroke-like episodes, includes those with a less clear genetic role and those not classified by the groupings outlined above.

The clinical presentation of older children and adolescents with stroke is similar to that in adults. Infants and neonates, however, often present with less easily identified or interpreted symptoms and signs. Focal or generalized seizures often occur at the onset of stroke, especially in younger children,[7–11] in which case differentiation from Todd's paresis is important. Clinical features are related to the size and location of the stroke and the underlying etiology. Frequently, the systemic features of the disorder provide the best clinical information with which to identify the etiology of the stroke. Table 2 lists some of the peripheral features which aid in the clinical diagnosis of some inherited disorders associated with childhood stroke.

Neuroimaging with computerized tomography (CT) or magnetic resonance imaging (MRI) aids in localization and classification of the stroke (ischemic versus hemorrhagic), and is especially useful in infants in whom the neurologic examination is less precise. Neuroimaging may also provide specific information about etiology; for example, MRI may reveal the diminished carotid flow void and the basal vascular network of collateral channels of moyamoya,[12] or CT may show basal ganglia calcification suggestive of mitochondrial encephalopathy, lactic acidosis and stroke-like episodes (MELAS).[13] Cerebral angiography is important in the diagnosis and management of some stroke syndromes, but is occasionally associated with complications.[14,15] Therefore, angiography is generally restricted to the investigation of spontaneous intracerebral hemorrhage and ischemic stroke of uncertain etiology. Other specific laboratory investigations are directed by the clinical and neuroimaging features of the stroke. The frequent occurrence of further strokes and potential involvement of other family members prompts a relatively low threshold for investigation of the inherited stroke syndromes. Interpretation of laboratory investigations should always include consideration of the different age-dependent values normally found in neonates and children.

The prognosis for many of the inherited pediatric stroke syndromes is often unfavorable, since strokes may be recurrent, and underlying brain or systemic dysfunction may be integral parts of the inherited condi-

---

Table 1
Inherited Syndromes with Multiple Stroke Mechanisms

Antiphospholipid antibodies	VST, AT, AE
ATIII deficiency	VST, AT, AE, ICH
Ehlers-Danlos syndrome	AE, ICH, SAH
Fabry's disease	AT, AE, ICH, SAH
Fibromuscular dysplasia	AT, ICH, SAH
Homocystinuria	VST, AT
Kohlmeier-Degos disease	AT, ICH, VST
Marfan's syndrome	AE, AT, SAH
MELAS	nonarterial stroke-like episodes, AT
Moyamoya	AT, AE, ICH, SAH
Neurofibromatosis type 1	AT, ICH
Protein C deficiency	VST, AT, AE, ICH
Protein S deficiency	VST, AT, AE, ICH
Reistance to activated protein C	VST, AT, AE, IVH
Rendu-Osler-Weber Syndrome	septic microemboli, paradoxical emboli, air emboli, ICH, SAH
Sickle cell disease	AT, watershed infarction, ICH, VST, fat embolism
Tuberous sclerosis complex	AT, AE, ICH, SAH

AT, arterial thrombosis; AE, arterial embolism; ICH, intracranial hemorrhage; SAH, subarachnoid hemorrhage; VST, venous sinus thrombosis.

Table 2
Clinical Clues to the Cause of Some Inherited Stroke Syndromes

Central nervous system
Headache syndromes or migraine: moyamoya, MELAS, antiphospholipid antibodies, AVM, Rendu-Osler-Weber syndrome, intracerebral aneurysm, familial cavernous hemangioma. Kohlmeier-Degos disease
Alternating hemiplegia: moyamoya, homocystinuria, MELAS, organic acidurias
Mental retardation: homocystinuria, MELAS, TSC, NF1
Deafness: MELAS
Dementia: MELAS, sickle cell disease, homocystinuria
Brain tumor: TSC, NF1
Macrocephaly: NF1, Bannayan-Zonona syndrome
Ophthalmoplegia: Fabry disease, mitochondrial encephalomyopathy

Ophthalmologic
Corneal opacity: Fabry's disease
Lisch nodules: NF1
Ectopia lentis: homocystinuria, Marfan's syndrome
Cataract, whorl keratopathy: Fabry's disease
Retinal infaction: antiphospholipid antibodies
Retinopathy: overlap syndromes with MELAS
Retinal hemorrhage: ROW
Retinal hamartoma: TSC

Peripheral nervous system
Peripheral painful dysesthesia: Fabry's disease

Skin
Angiokeratoma, hypohidrosis, oedema, decreased body hair: Fabry disease
Livedo reticularis: Sneddon's syndrome (antiphospholipid antibodies)
Ulceration/gangrene: antiphospholipid antibodies, MELAS
Purpura fulminans: protein S deficiency, protein S deficiency
Vascular lesions: ROW, Bannayan-Zonana syndrome
Multiple lipomas: Bannayan-Zonana syndrome
Reticular pigmentation, nail dystrophy, leucokeratosis: dyskeratosis congenita
Bruising: coagulopathies
Café au lait spots, neurofibromas, plexiform neuromas: NF1
Hypopigmented patches, angiofibromas, shagreen patches subungual fibromas: TSC
Xanthelasma: dyslipoproteinemias
Umbilicated papules with atrophic centers: Kohlmeier-Degos disease
Hyperelastic skin: Ehlers-Danlos, pseudoxanthoma elasticum

Body habitus
Growth failure: progeria, Fabry's disease, MELAS, Bannayan-Zonana syndrome, sickle cell disease
Marfanoid habitus: Marfan's syndrome, homocystinuria

Pulmonary disease
Hemoptysis: ROW

Cardiac disease
Conduction defects: MELAS, Fabry's disease
Ischemic heart disease: premature aging syndromes, dyslipoproteinemias
Rhabdomyoma: TSC
Left ventricular failure, valvular disease: Fabry's disease
Coarctation of the aorta, primary pulmonary hypertension: moyamoya
Aneurysm of ascending aorta, mitral valve prolapse: Marfan's syndrome

Table 2 (*continued*)

Reproductive system
Spontaneous abortion: antiphospholipid antibodies
Delayed puberty, priapism: Fabry's disease

Previous venous thromboses
Protein C, protein S, antithrombin III deficiency
Antiphospholipid antibodies

Myopathy
Mitochondrial myopathy: MELAS

Renal disease
Hypertension: Fabry's disease, FMD, NF1, TSC
Progressive renal failure: Fabry's disease

Gastrointestinal disease
Bleeding: ROW, Kohlmeier-Degos disease, bleeding disorders
Diarrhea: Fabry's disease

MELAS, mitochondrial encephalomyelopathy, lactic acidosis, and stroke-like episodes; AVM, arteriovenous malformation; TSC, tuberous sclerosis complex; NF1, neurofibromatosis type 1: ROW, Rendu-Osler-Weber disease; FMD, fibromuscular dysplasia.

tion. This poor prognosis contrasts with the relatively good prognosis of sporadic stroke in childhood.

In addition to supportive treatment, therapy is directed toward elimination of provocative factors and amelioration of the underlying disorder. In the acute phase, maintenance of airway and circulation may be required; large ischemic or hemorrhagic strokes may be associated not only with dysfunction of the injured region, but also with edema and raised intracranial pressure, requiring intensive therapy. Provocative factors such as dehydration and infection should be treated appropriately. Specific therapy is available for particular inherited stroke syndromes and short-term and long-term anticoagulation is also appropriate for some syndromes. Rehabilitation is guided by specialized therapists, educators, and medical professionals. Fortunately children have a great propensity to learn, explore, and develop, and the many activities of everyday life within a supportive family provide a natural rehabilitation. Acute and chronic care require involvement, education, and support of the entire family and oftentimes the community. In this chapter, only the therapies specific to each etiology of inherited stroke are discussed. General and specific therapeu-

tic measures may be found in the texts by Roach and Riela[1] and Edwards and Hoffman,[2] as well as many of the references cited.

Arterial Strokes

Ischemic stroke, or cerebral infarction, may be subclassified into two broad categories based on the mechanism of arterial occlusion: embolic or thrombotic. Although clinical features depend upon the region infarcted, acute hemiparesis is the presenting feature in more than 90% of reported childhood ischemic strokes.

An etiologic diagnosis for ischemic stroke is commonly elusive (34% in the Mayo Clinic series and at least 57% in the Japanese series[4,10]). Lacunar strokes in childhood are uncommon. They have been attributed to thrombotic and embolic disease, but, even with modern diagnostic techniques, their cause is typically unknown.[8,10,16]

Embolic Stroke

The pathophysiology of embolic stroke has been discussed in *Chapter 8*. Hemorrhagic transformation is more common in larger infarcts and those affecting primarily the cerebral cortex, but we have found it to be relatively rare in childhood.

Cerebral embolism typically has a sudden onset, with maximal clinical deficit apparent soon after the onset of symptoms. This distinctive history, however, may not be provided for many infants with embolic stroke. Size and location of the ischemic region distal to the embolus determine the clinical deficit. Cerebral emboli most commonly originate in the heart,[16] with various congenital heart diseases being associated with an increased incidence of cerebrovascular accidents in childhood. In these patients, cerebral emboli may arise from multiple sources: mural thrombi, prosthetic heart valves, cardiac catheterization, surgical intervention, bacterial endocarditis, and paradoxical emboli in patients with right-to-left shunts. Thrombotic stroke may also be more common in patients with congenital heart disease, especially in patients with cyanotic disorders with high hematocrit and iron deficiency. Many forms of congenital heart disease have an increased incidence in families, with a presumed genetic component.[17] In addition, both Mendelian inheritance of primary cardiac defects and cardiac defects in the context of multisystem inherited disorders have been associated with embolic stroke in young people, as shown in Table 3.

Tuberous Sclerosis Complex (TSC)

TSC is a multisystem disorder. It is inherited as an autosomal dominant trait, but sporadic occurrence is frequent. Although Vogt described the classic triad of mental retardation, seizures and adenoma sebaceum (now termed angiofibroma[18]), the clinical features are variable and may involve most organ systems.[19] TSC has been linked in most kindreds to the loci TSC1 and TSC2, on chromosomes 9 and 16 respectively.[20] The TSC2 gene maps to the same region as that of the PKD1 gene responsible for adult-onset polycystic kidney disease. An etiologic link between the two has been suggested, which may account for their shared characteristics of cerebral aneurysms and multiple renal cysts.[21]

Strokes are uncommon in TSC and have been associated with moyamoya[22] and cerebral aneurysms. With regard to cerebral embolism, it has been hypothesized that either tumor fragments from cardiac rhabdomyoma, or, more likely, thrombus particles created by turbulence of intracardiac blood flow secondary to rhabdomyoma, have embolized to the brain.[23,24] It has also been suggested, however, that these cases with stroke-like symptomatology have been due to local hamartomas (cortical tubers), and that the abnormal neuroimaging was the result of continuing myelination in the lesion.[25]

MRI or angiography may show moyamoya, and surgery may be indicated for such patients (as discussed below). Cardiac examination and echocardiography may reveal evidence of rhabdomyoma. If cardiogenic cerebral embolism associated with rhabdomyoma is suspected, treatment with anticoagulant or antiplatelet agents may be preferable to surgical removal of the rhabdomyoma, since many rhabdomyomas spontaneously become smaller or disappear.[24,26,27]

Thrombotic Strokes

The neurologic deficit that develops due to cerebral thrombosis often has a slow onset and may be observed upon awakening.[10,28] The imprecise history available for younger patients, however, may make distinction between thrombotic and embolic stroke difficult. Any cerebral vessel may be affected, but the supraclinoid internal carotid and middle cerebral arteries are most commonly involved.[29] In contrast to embolic stroke, rapid spontaneous resolution is not expected. Febrile illnesses have frequently preceded thrombotic stroke in reported childhood cases.[10,11]

Sickle Cell Disease

Sickle cell disease is one of many clinical disorders due to defects of the beta hemoglobin gene,[30] located on chromosome 11. Inheritance follows an autosomal recessive

Table 3
Inherited Primary Cardiac Defects Associated with Stroke in Young People

Inherited Primary Cardiac Defects
Mitral valve prolapse
Familial atrial myxoma
Hypertrophic obstructive cardiomyopathy and other cardiomyopathies
Familial sinus node disease and other inherited dysrhythmias

Inherited Syndromes with Cardiac Defects
Mitral valve prolapse
   Ehlers-Danlos syndrome, Marfan's syndrome, Duchenne muscular dystrophy, Becker muscular
   dystrophy, myotonic dystrophy, von Willebrand disease, Fragile X syndrome, pseudoxanthoma
   elasticum,[484] osteogenesis imperfecta[3]
Cardiac rhabdomyoma
   tuberous sclerosis
Atrial myxoma
   Myxoma, spotty pigmentation and endocrine overactivity syndrome (Carney's complex)[485]

Based on the review of Natowicz and Kelley.[3]

pattern. The molecular defect is a single base mutation of thymidine to adenosine, resulting in substitution of valine for glutamic acid at the sixth amino acid position of the beta hemoglobin molecule.[9] Patients homozygous for hemoglobin S have hemoglobin molecules which contain two normal alpha chains and two abnormal beta chains. Eighty to ninety-eight percent of their hemoglobin is in the form of hemoglobin S, the remainder being HbF (fetal hemoglobin).[14] In conditions of low oxygen tension and acidosis, these abnormal hemoglobin molecules polymerize, causing increased cell rigidity and the characteristic sickled shape. This rigidity promotes sludging in small vessels and mechanical injury with subsequent hemolysis. Approximately 0.16% of the black population of the USA is homozygous for hemoglobin S.[31] Patients with this condition suffer multisystem disease, with 90% of patients symptomatic by the age of 5 years.[32]

Ischemic or hemorrhagic stroke occurs in 5.5–17% of North American pediatric sickle cell patients, increasing to 32% by age 40 years.[33] These figures, however, may be underestimations, as MRI suggests that silent cerebral infarcts may occur.[34]

Approximately 75% of strokes seen in patients are ischemic and 25% are hemorrhagic.[33,35] Cerebral venous thrombosis is rare.[36] The majority of first ischemic strokes occur between 1 and 15 years of age, typically between 5 and 10 years.[35,37] Hemiplegia is the most common presenting feature, but symptoms related to lesions in all regions of the CNS may occur, including cortex, deep white matter, basal ganglia, brainstem and spinal cord.[34,35,37] Transient ischemic attacks are relatively common (12% of patients with stroke), and may be the dominant feature.[37] Focal or generalized seizures may accompany stroke in these patients.[9,35] Recurrent strokes occur in 67–90% of sickle cell patients, usually within the first 3 years, although this incidence can be decreased with therapy.[33,35,38,39]

Subarachnoid or intraparenchymal cerebral hemorrhage secondary to rupture of arterial aneurysms or rupture of intraparenchymal dilated vessels tends to occur in older patients.[9,33,39,40] Presentation of intracranial hemorrhage varies from acute headache and meningismus to collapse, coma, and death according to the site and magnitude of hemorrhage.

A slowly progressive encephalopathy (which may be due to small vessel disease in a manner similar to multi-infarct dementia), and fat embolism to the brain from bone marrow infarction have also been reported in children with sickle cell disease.[9,33] Headache and central nervous system bacterial infection (particularly with *Streptococcus pneumoniae* in younger children) are also increased in frequency.

Factors that may precipitate a stroke or are associated with an increased incidence of stroke in sickle cell disease include dehydration, acidosis, pregnancy,[33,37] a past history of bacterial meningitis, aplastic crises or frequent painful crises, low hematocrit or increased reticulocyte count, leucocytosis, abnormal liver function tests, cardio-megaly, hyperventilation for EEG, cerebral angiography, and a family history of stroke.[39,41] Additionally, occasional patients have suffered stroke immediately following blood transfusion. An abnormal baseline brain MRI has also been claimed to identify a population at increased risk of clinical cerebrovascular disease.[42] Relatively increased levels of hemoglobin F (above 10%) may be partially protective against cerebrovascular complications.[43] Patients with alpha-thalassemia trait also seem relatively protected from the complications of sickle cell disease.[44]

Brain CT and MRI scanning typically show evidence of large-vessel disease in the form of watershed infarction (including the deep white matter watershed regions) or infarcts in the territories of the major cerebral arteries.[34,45] Small subcortical or cortical infarcts are less common. Occlusion, stenosis or moyamoya pattern of the large cerebral arteries can often be identified by MRI. Angiography in patients with stroke reveals large-vessel occlusive disease in 60–95% of patients, typically of the intracranial internal carotid and the proximal middle and anterior cerebral arteries.[9,39] Because of reports of stroke following intravenous radiographic contrast, it has been suggested that angiograms be performed only in those patients pretreated with blood transfusions,[14,46] or by magnetic resonance angiography, which carries no risk of induction of sickling. Doppler ultrasound studies of the extracerebral carotid arteries and the large intracerebral arteries with transcranial techniques also provide evidence of large-vessel disease. This technique may identify patients at increased risk of cerebrovascular complications of sickle cell disease,[47] although concerns persist regarding the sensitivity of this form of imaging.[48]

Neuropathologic studies of patients with stroke commonly show endothelial and smooth muscle proliferation with organizing and recanalizing thrombi of the distal internal carotid and proximal intracranial divisions of the internal carotid arteries.[49,50] Other vascular abnormalities observed include proliferation, dilatation, congestion and thrombosis of small intracranial collateral vessels, a moyamoya pattern of collaterals, and dural sinus thromboses.[49–51] These pathologic studies show good correlation with the abnormalities detected by CT and MRI scanning.

The pathogenesis of ischemic and hemorrhagic stroke with large-and small-vessel disease is multifaceted. A unifying hypothesis has been proposed based on observations of increased cerebral blood flow (which is inversely proportional to hematocrit) in sickle cell patients without stroke, relatively decreased cerebral blood flow in patients with stroke, and severely blunted responsiveness to increased inspired carbon dioxide.[9,52–54] The suggestion is that, in order to maintain brain oxygenation, sickle cell patients maximally dilate cerebral blood vessels and increase cerebral blood flow. Abnormal red blood cells and turbulent blood flow, however, predispose to endothelial hyperplasia and thrombus formation in large vessels. This leads to compromised cerebral blood flow, especially to distal arterial fields. The underlying red cell defect causes sludging in these distal regions (i.e., watershed regions), with lower oxygen tensions and lower pH. Because of previous maximal compensatory cerebral vasodilation, this compromised brain perfusion is not redeemable via autoregulation, and hence results in infarctions in the boundary regions between cerebral arteries.

Large vessel occlusion leads to significant infarctions typically confined to single-vessel territories, possibly associated with occlusion of smaller vessels resulting from stasis and the sickling phenomenon itself. In addition, cerebral vasodilation, endothelial damage, and hypervolemia may predispose to rupture of both intraparenchymal vessels and aneurysms.

Patients with sickle cell anemia occasionally have other abnormalities of platelet function and coagulation, which may add to the prothrombotic tendency seen in sickle cell disease. These include increased platelet thromboxane production,[55] enhanced thrombin generation,[56] and increased tissue factor activity,[57–59] possibly related to reduced levels of proteins C and S,[60] elevated plasma levels of cytokines interleukin 1 and tumor necrosis factor, and activation of the coagulation system as a result of deformation of red cell membranes in sickling.[57]

Diagnosis of sickle cell disease depends upon identification of hemoglobin S with hemoglobin electrophoresis. Anemia, polychromasia, reticulocytosis, and sickled red cells may be seen on the peripheral blood film. Prenatal diagnosis by hemoglobin electrophoresis and DNA analysis is well established.[61]

Management of the acute ischemic stroke is supportive. It is important to treat or prevent factors which may lead to sickling and ischemic stroke such as infection, hypoxia, acidosis, hypovolemia, and hypothermia. Exchange transfusion aimed at reducing hemoglobin S levels to less than 20% also reduces the risk of ongoing sickling, although it has not been proven to limit evolution of the acute stroke.[37] It has even been linked, in some cases, to neurologic deterioration, prompting recommendations that simple transfusion constitutes the best initial intervention.[62] Chronic transfusion therapy with packed red cells or exchange transfusion, aimed at reducing hemoglobin S levels to 10–30%, has been effective in preventing ischemic stroke and, possibly, hemorrhagic stroke in almost all patients.[9,37,39,63,64] It also prevents progression of arteriographic abnormalities of cerebral vessels. Long-term transfusion carries significant risks including iron overload, viral infections and development of erythrocyte alloantibodies, prompting trials of less intensive transfusion regimens with variable results.[65,66] In those patients without demonstrable large-vessel lesions on imaging, and presumed microvascular disease, the value of transfusion therapy in prevention of recurrent strokes is less clear.[62]

Heparin, coumadin, and aspirin have not been proven effective in preventing recurrent strokes.[9,67] Hydroxyurea offers potential therapeutic effect by increasing hemoglobin F, reducing anemia, and increasing erythropoietin,[68,69] although it has significant side effects and potential long-term toxicity. Butyrate may also rapidly increase hemoglobin F, at least in the short term, which may limit cerebrovascular complications.[70] Bone marrow transplantation (BMT) has been used to successfully treat 16 of 17 patients reviewed by Kirkpatrick et al.,[71] although its use is limited by availability of histocompatible siblings, consequences of BMT (sepsis, hemorrhage, graft-versus-host disease, endocrine dysfunction, secondary malignancy), and cost. Concerns have also been raised regarding an increased incidence of neurologic complications after BMT for sickle cell disease.[72] Cord blood transplants, intrauterine BMT, and gene therapy may become useful in the future. Avoidance or treatment of other risk factors associated with cerebrovascular disease, such as smoking, hypertension, and the oral contraceptive pill, seem to be sensible precautions.

Sickle Cell Trait

Patients with sickle cell trait have one normal beta hemoglobin gene and one with the sickle mutation. Children with sickle cell trait have infrequently suffered cerebrovascular accidents.[73–75] Two children with dural sinus thromboses and venous infarction following anesthesia have been reported, as well as one case in which a posterior circulation infarct was seen in a child with variant anatomy and probable dehydration.[76–78] This is consistent with adult data which suggest that neurologic complications in patients with sickle cell trait are often associated with other complicating factors such as acidosis, anesthesia, strenuous exercise, oral contraceptives, and possibly migraine. Sickle trait may also be identified by hemoglobin electrophoresis.

Hemoglobin SC Disease

Patients with hemoglobin SC have one beta hemoglobin gene with the sickle cell mutation and the other beta hemoglobin gene with a different point mutation (hemoglobin C). Their clinical manifestations are usually milder and of later onset than those seen in sickle cell disease. Stroke in childhood has been reported, but details are limited.[39,75]

Homocystinuria

Homozygous homocystinuria is an autosomal recessive disorder of homocysteine and methionine metabolism. It was first recognized in 1962[79] and is the most commonly cited genetic disease affecting the cerebral vasculature and increasing the risk for stroke.[3] Several biochemical disturbances produce similar biochemical and clinical features. The most common defect is in the enzyme cystathionine beta-synthase, which catalyzes the conversion of homocysteine and serine to cystathionine, using pyridoxal phosphate as a cofactor. The gene for cystathionine beta-synthase has been localized to the subtelomeric region of chromosome 21q22:3 and multiple defects have been linked to the clinical disease.[80] Individuals with cystathionine beta-synthase deficiency have elevated levels of homocysteine, homocystine, and methionine in blood and urine. Homozygous deficiency of this enzyme occurs in 1 in 60,000 to 1 in 200,000 individuals.[81] Several defects in the 5-methyltetrahydrofolate-dependent homocysteine pathway also result in elevated blood and urine levels of homocysteine and homocystine, but not methionine. These defects include: inability to convert cobalamin to its active form, abnormal formation of methylcobalamin but normal adenylcobalamin, 5,10 methylenetetrahydrofolate reductase deficiency, and Immerslund syndrome (defective vitamin B12 absorption despite normal intrinsic factor).[82]

Clinical manifestations of homocystinuria show considerable variability between and within families.[83] Thrombotic and thromboembolic strokes occur throughout life and have been reported as early as the first year.[84] Thirty percent of patients have suffered thromboembolic events by the age of 20 years, and 60% by age 40.[81] Cerebral venous sinus thrombosis has occasionally been reported in children with homocystinuria.[85] Other typical clinical features include: ectopia lentis (downwardly dislocated lens, usually before age 10), Marfanoid habitus, osteoporosis, mental retardation in two-thirds of the cases (usually evident in infancy even before strokes occur[86]), generalized or focal seizures in 10–15%, and a possibly increased incidence of psychiatric illnesses in near relatives (reviewed by Mudd[82]). Skeletal abnormalities may include genu valgum, pes cavus, and pectus excavatum, while cutaneous manifestations may include fair coloring, a malar flush, and livedo reticularis. Stroke may occur with normal phenotype and intelligence.[85,87] The vascular pathology predisposing to stroke is found in large, medium, and small arteries and veins throughout the body and may be present from early infancy.[82] The typical arterial pathology comprises diffuse or patchy fibrous thickening of the intima, frayed elastic and muscle fibers, and increased collagen within the media. Veins may be normal or show fibrous endophlebitis.[82] Pathogenesis is attributed to the toxic effects of homocystine and homocysteine on vascular endothelial cells,[88] with subsequent induction of smooth muscle proliferation, potentiation of auto-oxidation of low-density lipoprotein cholesterol,[89] increased incorporation of lipoprotein(a) into fibrin,[90] and promotion of thrombosis.[91–94] It has been suggested that contributing factors, such as coexistent resistance to activated protein C, may be necessary for thromboembolism to occur in those with only modestly raised plasma levels of homocysteine,[95,96] but this has been refuted by other researchers.[97] Angiography reveals arterial wall irregularities with arterial narrowing and aneurysm formation throughout the body, including the carotid arteries,[98] but may be hazardous because intravenous contrast agents may precipitate thrombosis.

Quantification of amino acids in urine and plasma is usually reliable for diagnosis of all forms of homocystinuria, though even small doses of exogenous pyridoxine may give false-negative results in pyridoxine-sensitive patients.[82] Identification of excess sulphur-containing amino acids (i.e., homocystine) in the urine with the cyanide-nitroprusside test will indicate patients with cystathionine beta-synthase deficiency, but is not reliable for patients with defects in the 5-methyltetrahydrofolate-dependent homocysteine pathway.[86] False-negatives may occur if the urine is dilute or has been standing too long.[99] Cystathionine beta-synthase activity may be measured in liver biopsy specimens, lymphocytes stimulated by phytohemagglutinin and cultured skin fibroblasts.[82] Antenatal diagnosis of cystathionine beta-synthase deficiency and methylenetetrahydrofolate reductase deficiency is possible.[100,101]

Management is based on control of the biochemical abnormality and supportive measures. Low methionine diets appear to be of value and some children have had normal development with the institution of such dietary treatment from the first days of life.[102] Choline, betaine, folate, cystine, and vitamin B12 therapy have also been used effectively, as reviewed by Mudd et al.[82] In the 40% of patients with cystathionine beta-synthase deficiency who are pyridoxine-sensitive, treatment with large doses of pyridoxine, usually 150 mg/day (25 to 1200 mg/day) may prevent complications.[82] Replete folate stores and adjunctive betaine[103,104] are also beneficial in these patients. Supportive therapy includes dipyrimadole and aspirin, avoidance of immobility and other risk factors for stroke, and selective use of heparin and coumadin.

Hyperhomocysteinemia

In 1976, Wilcken et al.[105] first linked pathological homocysteine accumulation with premature ischemic heart disease. Abnormal homocysteine and methionine metabolism, demonstrated by methionine loading, has been reported in about one-third of patients less than 50 or 55 years of age with peripheral, coronary, and cerebral vascular disease, and is an independent risk factor, unrelated to hyperlipidemia, hypertension, diabetes, and smoking.[106] It has been attributed to deficiencies of folate and of activity of cystathionine beta-synthase and methylenetetrahydrofolate reductase.[107–109] The published prevalence of hyperhomocysteinemia is well in excess of that of heterozygous homocysteinuria, indicating that factors other than mutations of the genes for cystathionine beta-synthase and methylenetetrahydrofolate reductase are important.[106] Plasma homocysteine concentrations have been demonstrated, in several series, to be significantly and negatively correlated to those of blood or serum folate and/or serum B12 in patients with vascular and renal disease but not in controls; this suggests a genetic or acquired increased dependency on these vitamins in these patients.[110] Treatment with folate or pyridoxine,[104] or choline and betaine,[104] reduces plasma homocysteine levels and improves response to methionine loading, but has not as yet been proven to decrease vascular risk. Hyperhomocysteinemia has been causally linked with symptomatic cerebrovascular disease in only one pediatric patient.[104]

Fabry's Disease

Fabry's disease is an X-linked recessive disorder of neutral glycosphingolipid catabolism, due to diminished activity of the lysosomal enzyme alpha-galactosidase A.[111] The gene for alpha-galactosidase A has been localized to the long arm of the X chromosome and cloned.[112] More than forty mutations of the gene have been identified, including partial duplication, partial deletions, and point mutations, with certain mutations predicting milder phenotypes.[112] A patient with Fabry's disease in which exon 6 is deleted from the mRNA is the first mammalian example of a genomic splice-site mutation preventing transcription of the preceding exon.[113,114]

Disease incidence is approximately one in 40,000.[112] In the hemizygous affected male, clinical disease onset usually occurs in childhood or adolescence. Either neurologic or dermatologic manifestations are the first complaints. Periodic crises of temperature-sensitive, severe burning or lightning-like pains in the palms or soles are characteristic, but 10–20% of patients deny painful crises or acroparesthesias.[111] Small, punctate, dark red flat or slightly raised angiectases (angiokeratomas) may also be an early sign. These typically occur in clusters between the umbilicus and the knees, increasing in size and number with age, and may occur on the oral mucosa and the conjunctivae. Hypohidrosis is an early and almost constant finding.[111] Delayed puberty, retarded growth, sparse body hair, and an acromegalic-like face have been reported.[111] With increasing age additional features may develop: a characteristic corneal opacity[115]; posterior or anterior lenticular opacities[116]; priapism[117]; left ventricular enlargement, valvular dysfunction and cardiac conduction abnormalities; progressive renal impairment (which causes death in the majority of patients with Fabry's disease); chronic lung disease; lymphedema of the legs; episodic diarrhea; anemia; mental retardation; seizures; personality changes; and a characteristic extension deformity of the terminal interphalangeal finger joint.[111] Rare atypical cases are characterized by isolated progressive cardiomyopathy, typically manifesting in middle age.[113] Due to preservation of some alpha-galactosidase activity, a number of affected males may have mild, slowly progressive disease. Some heterozygous females may manifest clinical features of the disease late in life.[111,118]

Although Fabry's disease is a frequently cited cause of pediatric stroke, in fact, stroke and other vascular complications rarely occur before adulthood.[119] Symptoms depend upon the size and location of the occluded artery. Hemispheric strokes are the most common,[120,121] but transient ischemic attacks, brainstem lesions,[122,123] cerebral aneurysms,[124] and intraparenchymal hemorrhage with and without hypertension of renal disease[125] have been reported. Involvement of the vertebral and basilar arteries may result in prominent posterior circulation symptoms such as vertigo, diplopia, dysarthria, nystagmus and ataxia.[126] Presentation with headache and an aseptic meningeal reaction has been described in one patient.[127] Recurrent cerebrovascular events are seen in the majority of cases.[125] Incidence of stroke in Fabry's disease has not been demonstrated to correlate with level of enzyme activity.[128] Cardiovascular involvement in Fabry's disease can include myocardial infarction secondary to coronary artery occlusion, valvular prolapse due to infiltrative changes,[129] and hypertrophic cardiomyopathy.[130] All can result in cardiogenic embolism to the cerebrovascular circulation.

Neuroimaging reveals multifocal white matter and periventricular infarcts, possibly in association with ischemic or atrophic changes in the cerebellum and brainstem.[126] Tortuous basal vessels, possibly with fusiform aneurysmal dilatation, may be visualized on MRI or angiography.[126] Pathologic findings in Fabry's disease are due to the diffuse deposition of the accumulated glycosphingolipids, predominantly in the lysosomes of endothelial, perithelial and smooth-muscle cells of blood vessels.[111] This may be associated with endothelial proliferation producing narrowing of the lumen and subsequent thrombosis and ischemia in distal tissues, exacerbated by hypertension and by the dolichoectasia of vessels seen in some postmortem series,[126] which presumably arises as a result of loss of vascular smooth-muscle elasticity and strength secondary to glycosphingolipid deposition. Two processes, glycosphingolipid deposition in perineurium and neurons of somatic and autonomic nerves and loss of peripheral sensory fibers, may be important in the autonomic and peripheral nervous system dysfunction.[111] Abnormal storage has been reported in many regions of the central nervous system.[111,131,132]

Diagnosis may be confirmed by determination of the elevated trihexosylceramide content of plasma or urine sediment, or assay of alpha-galactosidase A in plasma, leucocytes, tears, biopsied tissue or cultured skin fibroblasts.[111] Skin and renal biopsies demonstrate typical lipid deposits.[133] Classically affected hemizygotes have no detectable alpha-galactosidase function; atypical hemizygous males with mild, slowly progressive disease have up to 35% normal alpha-galactosidase function, and heterozygous females have enzyme activity that ranges from undetectable to normal, depending upon the age and sex of the patient, the source of the material, and random X inactivation.[111,118]

Treatment is supportive. The pain of Fabry's disease has been treated with phenytoin,[134] carbamazepine,[135] phenoxybenzamine,[136] neurotropin,[137] and prednisolone,[127] with varying efficacy. Renal failure may require symptomatic therapy with dialysis and renal transplantation,[111] and successful renal transplantation may occasionally improve systemic disease features.[138] Antiplatelet therapy has not been demonstrated to prevent thromboembolic complications of vascular endothelial disease, but anticoagulant therapy may be indicated in those with conditions predisposing to cardiogenic embolism. Most attempts at enzyme replacement have been unsuccessful, however, administration of purified alpha-galactosidase from human plasma has reduced circulating globotriaosylceramide levels without eliciting an immune response[139] and offers hope for future trials with enzyme produced by recombinant DNA technology. Identification of the genetic defects enables carrier detection and prenatal diagnosis.[140,141] Prenatal diagnosis can be also achieved by assay of alpha-galactosidase activity in chorionic villi or amniotic cells. Genetic counseling, screening of possible heterozygotes, and prenatal diagnostic tests may help some affected families "prevent" this disease.

Cerebral Autosomal Dominant Arteriopathy with Subcortical Infarcts and Leucoencephalopathy (CADASIL)

This condition, first described by Sourander and Walinder in 1977,[142] is a hereditary multi-infarct dementia clinically characterized by recurrent subcortical ischemic strokes, psychiatric disturbances, pseudobulbar palsy, and dementia. Identification of eight large European pedigrees has enabled identification of CADASIL as a distinct clinical entity, representing a subpopulation of patients previously classified as affected by hereditary dementia or Binswanger's disease. Inheritance is autosomal dominant, with delayed penetrance. Defects in the notch3 gene, a large transmembrane receptor localized to chromosome 19p13:1, underlie the disorder.[143] This gene is near, but probably not allelic to, the gene for familial hemiplegic migraine.[144]

Onset of symptoms is usually in the fourth or fifth decade, typically as recurrent transient ischemic attacks or strokes in the setting of a strong family history and absence of vascular risk factors. One pedigree has been reported, however, which has had a clinically distinct phenotype characterized by relatively early onset of symptoms, typically in adolescence, with TIAs or frequent migraines progressing to strokes and psychotic mood disorders.[145]

Neuroimaging reveals small areas of well-demarcated hypodensity without contrast enhancement, scattered throughout the basal ganglia and adjacent white matter, as well as coalescent white matter lesions of diffuse hypodensity. These changes are symmetrical, particularly marked in the regions of the external capsule and anterotemporal regions, and spare the cerebral cortex. In all families described, asymptomatic family members have been identified who have had typical abnormalities on CT or MRI.[146] Angiography has been essentially normal in most individuals studied. Pathologic examination reveals diffuse demyelination and a widespread vasculopathy of the small arteries penetrating the white matter and basal ganglia, with associated multiple small infarcts and atrophy. Arteries have thickened walls with nonamyloid eosinophilic deposits in the media, duplication of the internal elastic lamina, and obliteration of the lumen.[146,147]

No effective therapy is available.

Dyslipoproteinemias

Hereditary dyslipoproteinemias have been associated with pediatric stroke.[3,148,149] Typically, stroke due to this group of disorders affects adolescents and young adults, but stroke has occurred in infants as young as

15 and 19 months.[148] Clinical features are consistent with thrombotic stroke, and a family history of premature stroke and heart disease is common. When performed, angiograms have revealed focal or multifocal regions of irregular narrowing of large and medium intracranial arteries, consistent with atherosclerosis,[148,149] with a normal extracranial vasculature. Disorders of lipid metabolism are discussed in detail in *Chapter 5*.

Neurofibromatosis Type 1 (NF1)

NF1 is a multisystem disorder affecting at least 1 in every 3000 people.[150] It is inherited as an autosomal dominant trait, but sporadic occurrence is frequent. It is associated with deletion or inactivation of a huge gene located on the short arm of chromosome 17 (17q11.2).[151] The gene contains 50 exons, and two alternatively-spliced mRNAs have been identified.[151] Recent studies have suggested that the functions of the gene product (neurofibromin) include regulation of cellular proto-oncogenes (*ras*) important in growth and differentiation pathways, involvement in intracellular signal transduction pathways through protein kinase phosphorylation cascades, and possibly association with cytoplasmic microtubules. There is little correlation between the various abnormal genotypes (large and small deletions, translocations, insertions, and point mutations), and phenotype, as would be expected in a condition manifesting marked phenotypic variability within affected members of single families.[150] It is not clear how alteration or expression of neurofibromin, the product of the gene, is related to strokes in NF1.[151] Pathologic examination of affected large- and medium-sized arteries reveals intimal smooth muscle proliferation, medial thinning, and elastic tissue fragmentation.[152]

Clinical features of NF1 are protean,[150] and there is marked variability of expression. Diagnosis is based on the criteria of the National Institutes of Health Consensus Development Conference.[153] Stroke in childhood may rarely be the presenting feature of NF1.[154] Moyamoya is the typical vascular anomaly linking childhood stroke with NF1, but distal arterial occlusion with ischemic stroke, (most commonly involving the supraclinoid carotid artery) has also been reported.[155,156] Neurofibromatosis can present as a systemic vasculopathy, with involvement of the intracranial vasculature described in one child.[157] Hemiplegia has also resulted, in a single case, from a spontaneous vertebral arteriovenous fistula.[158] Cerebral hemorrhage secondary to systemic hypertension or intracranial aneurysm[159] may also occur. Treatment for the stroke is supportive and surgical intervention as described below may be useful in patients with moyamoya.

Premature Aging Syndromes

Progeria syndrome (Hutchinson-Gilford syndrome) is characterized by premature aging, growth deficiency, and dysmorphism with premature atherosclerosis.[30,160] Average life expectancy is 14 years.[30,160] Inheritance is usually described as autosomal recessive, but autosomal dominant inheritance associated with advanced paternal age has also been proposed.[30] Abnormal regulation of nucleotide excision repair before DNA synthesis has been demonstrated in cultured progeriod skin fibroblasts exposed to ultraviolet radiation, possibly accounting for the accelerated degenerative changes seen in progeria,[161] while identified abnormalities of hyaluronic acid[162] and decorin[163] excretion could indicate an underlying connective tissue defect. Although cerebrovascular disease is rare in this syndrome, several cases of childhood stroke due to atherosclerotic cardiovascular and carotid artery disease have been reported.[164,165] Therapy is supportive only.

Vascular Anomalies Associated with Ischemic Stroke

Moyamoya

Moyamoya is a Japanese expression meaning "something hazy like a puff of smoke drifting in the air" and was coined by

Suzuki and Takaku[166] to describe the angiographic blush of basal telangiectatic vessels adjacent to stenosed or occluded internal carotid arteries. This appearance is not specific to a single disease entity, rather it is a finding associated with carotid stenosis or occlusion of many etiologies. The vast majority of reports come from Japan, but moyamoya is well recognized in other ethnic groups.[167] The majority of patients do not have recognized predisposing factors[168,169]; however, a hereditary basis for the disease is supported by the observations of racial preponderance, familial occurrence in 7% of cases,[170] the association with particular HLA phenotypes,[169,171] and association with several inherited disorders (Table 4). In addition, moyamoya has been associated with other vascular abnormalities such as coarctation of the aorta,[172] renal, hepatic, coronary and splenic arteriopathy,[15,173,174] and primary pulmonary hypertension,[175] perhaps indicating a common vascular abnormality underlying these conditions. Several acquired diseases have also been associated with moyamoya, notably radiation therapy to the head and neck, tuberculous meningitis, lep-

tospiral meningitis, "tonsillitis,"[171] and acquired anti-Ro/SS-A and anti-La/SS-B antibodies.[176]

Moyamoya was associated with 4.4% of childhood strokes in one North American series.[4] Children typically present with recurrent strokes associated with hemiplegia or other significant deficit and/or transient ischemic attacks. Progression may be relentless,[166,174] resulting in intellectual deterioration, although some patients exhibit a relatively benign course. Episodes may be precipitated by crying, coughing, or straining. Some patients may develop alternating hemiplegia, chorea,[177] torticollis[178] or other movement disorders.[168] Focal or generalized seizures occur in 40% of cases.[168] Headache may occur and occasionally precedes other features, especially in older children. Adults with moyamoya more commonly present with subarachnoid or intraparenchymal hemorrhage.[179]

Angiography provides the diagnosis. It typically reveals stenosis or complete occlusion of the intracranial portion of one (in 40% of cases[168]; see Figure 1) or both internal carotid arteries, or anterior and middle cerebral arteries. Less often, distal branches of the anterior and middle cerebral arteries are affected. The posterior circulation and external carotid arteries may be similarly affected. A basal network of telangiectatic collateral vessels, giving rise to the name "moyamoya," is apparent. Progression,[179] stability,[180] and, in one case, resolution[181] of the angiographic lesions have been reported. The observation of telangiectatic cerebral vessels is associated with a worse prognosis.[182]

MRI may reveal flow void in the telangiectatic collateral channels and abnormal signal in the large vessels as well as evidence of previous cerebrovascular accidents[12] (Figure 2). Comparison of conventional and magnetic resonance angiography has revealed the latter to be an acceptable alternative, albeit slightly less reliable for identification of collateral ves-

Table 4
Inherited Conditions Associated with Moyamoya

Antiphospholipid syndromes[467]
Apert's syndrome
Down syndrome
Fanconi anemia
Hypomelanosis of Ito[486]
Immunodeficiency syndromes
Incontinentia pigmenti[487]
Marfan's syndrome
NADH-CoQ reductase deficiency
Neurofibromatosis
Polycystic kidney disease
Pseudoxanthoma elasticum
Sickle cell disease
Sturge-Weber syndrome
Tuberous sclerosis complex
Type 1 glycogenosis

Based on the review of Roach and Riela.[1]

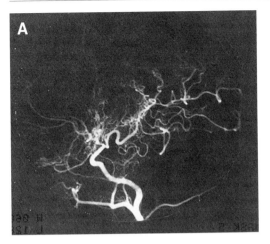
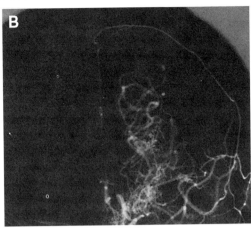

Figure 1. These lateral and AP left carotid angiograms show moyamoya in a 3 year old boy with Trisomy 21 and recurrent strokes from age 2 years. Note the "puff of smoke" appearance in both films due to proliferation of collateral channels near the origin of the middle and anterior cerebral arteries (MCA & ACA). Both show diminished flow in the MCA and ACA. Diffuse stenosis of the MCA is best seen in the AP view (B).

sels and assessment of degree of arterial stenosis.[183] CT may reveal previous strokes, atrophy, and telangiectatic vessels, but may be normal in 13% of patients at presentation.[168] Intracranial aneurysms[184] and arteriovenous malformations[185] have also been described in patients with moyamoya disease. Electroencephalography typically reveals bilateral slowing of background activity and may show epileptiform activity.[186] A characteristic buildup of diffuse high-voltage slow waves after cessation of hyperventilation has been described.[186]

The underlying pathology is the gradual occlusion of the large intracerebral arteries, with secondary profusion of collateral dilated perforating arteries between the intracranial carotid and other large vessels at the base of the brain and the leptomeningeal vessels.[187] Histopathology reveals early intimal thickening and edema, due to proliferation of smooth muscle cells with some infiltration by macrophages and T cells. Later, intimal fibrosis, buckling and splitting of the internal elastic lamina, and atrophy or necrosis of the media are seen.[168,188] Some arteries may show scat-

tered lymphocytes in the adventitia.[168] Mural thrombi in abnormal vessels and distal platelet and fibrin microthromboemboli have been described, but are usually absent. These findings may suggest an etiologic role for a chronic inflammatory stimulus, possibly producing increased angiogenic factors,[189] in the development of vasoocclusive lesions and collateral vessels.

Medical therapy has not generally been useful in moyamoya,[190] although two patients with progressive disease despite surgical intervention have been reported to have stabilized when treated with nicardipine, a calcium channel blocker,[191] and oxygen therapy has been suggested for acute management of TIAs and strokes.[192] Various surgical approaches have been tried, with the aim of revascularization of the ischemic brain through the creation of anastomoses between the intracranial and extracranial vasculature.[168] Anastomosis of the superficial temporal artery to a branch of the middle cerebral artery has been helpful in some cases but is technically difficult in small children.[193] Surgical placement of a section of the uninterrupted superficial tem-

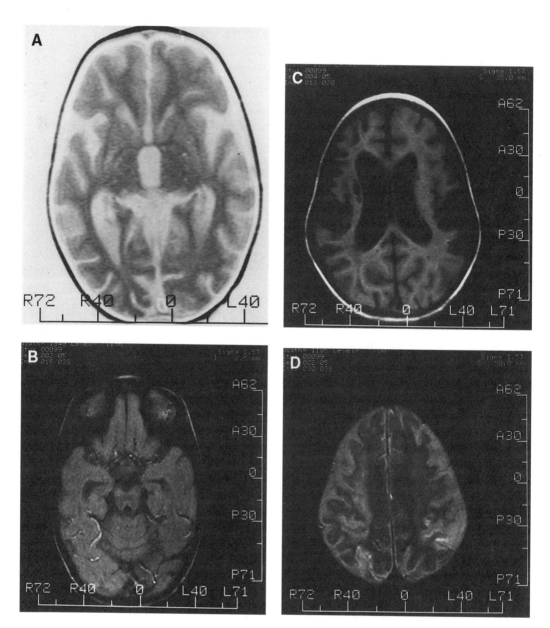

Figure 2. MRI films of the same patient as in Figure 1. **A:** A T_2 image shows diffuse cerebral atrophy and signal voids of collateral channels in the substance of the basal ganglia. **B:** Discontinuity of the signal voids at the origins of the ACA and MCA bilaterally is seen, suggesting interruption of flow in these vessels. Multiple surrounding signal voids represent the hypertrophied collateral channels. **C and D:** Higher T_1 and T_2 images show diffuse cerebral atrophy and a deep right hemispheric infarction, which is likely a watershed infarct.

poral artery against the arachnoid or pial surface of the underlying, relatively ischemic brain (encephaloduroarteriosynangiosis) has also been successful in several studies, as has encephalomyosynangiosis, in which temporal muscle is sutured to surgically created dural defects, approximating muscle to the surface of the brain.[194] Evidence is accumulating as to the efficacy of these procedures, both alone and in combination, but questions remain regarding the optimal form and timing of surgical intervention. Prognosis is variable but a recent series identified high morbidity and mortality, particularly in those symptomatic from an early age.[195]

Fibromuscular Dysplasia

Fibromuscular dysplasia (FMD) is an uncommon, segmental, nonatheromatous angiopathy of uncertain etiology, which mainly affects young women.[196] Small-to-medium sized arteries throughout the body may be involved.[197] Intracranial saccular aneurysms are also commonly identified in affected adults, although rarely reported in children. Stroke and central nervous system symptoms typically result from involvement of the cervical carotid artery (bilaterally in up to 86% of cases[199]). The intracranial carotid artery and its branches are commonly involved in symptomatic children[198] and involvement of the vertebrobasilar system has also been reported in children.[197–199,200] Autosomal dominant inheritance with variable expression and reduced penetrance in males has been postulated.[201] FMD has also been linked with Turner's syndrome.[202,203]

Patients may present with transient ischemic events or cerebral infarction due to thrombosis of, or embolism from, the diseased artery or arteries. Spontaneous or traumatic carotid dissection may be the presenting problem.[199,200,204] Adults with FMD have presented with symptomatic intracranial aneurysms[205] and carotid-cavernous

fistula.[206] Hypertension secondary to renal artery involvement may increase the risk of hemorrhage from cerebral aneurysms associated with FMD.[207] Fibromuscular dysplasia is also an incidental finding at autopsy or during the investigation of other disorders.[196]

Diagnosis depends upon angiography, which typically shows the appearance of a "string of beads" in the affected artery.[11] A tubular stenosis may occasionally be observed.[196] These arterial lesions may be unifocal or multifocal. The segmental arterial abnormalities result from alternating regions of thickening, with medial hypertrophy and fibrosis or sometimes hyperplasia of intima or adventitia, adjacent to regions of thinned media or deteriorated elastic tissue.[196] Intimal disease is most common in symptomatic affected children. Progression or stability of existing lesions and the development of new lesions have been reported with serial angiography.[196]

Endarterectomy, surgical resection, arterial dilatation, bypass surgery, and antiplatelet and anticoagulant medications have been used therapeutically, but efficacy is difficult to judge in view of the variable, often benign natural history and small number of patients studied.[196] Three modes of therapy have been recommended based on clinical status: conservative management for asymptomatic individuals; antiplatelet and/or anticoagulant medications for those with embolic disease; and surgical treatment for those with distal hemodynamic insufficiency.[196]

Kohlmeier-Degos Disease (Malignant Atrophic Papulosis)

Kohlmeier-Degos disease is a rare, progressive vasculopathy of uncertain etiology. It is characterized by a recurring maculopapular rash with pathologic findings of progressive noninflammatory occlusion of small- and middle-sized arteries and veins, causing ischemic infarction of the skin and viscera.[207,208] In most cases this is a sporadic disorder, but two families with an affected mother and child are reported.[209,210] Presentation is typically with nonspecific symptoms such as anorexia, fever and malaise, and subsequent development of cutaneous, gastrointestinal and neurologic manifestations. Multiple intestinal infarcts and secondary perforation cause death in most patients. CNS involvement occurs in at least 20% of patients, most of whom are young adults,[211] but 11 children with neurologic involvement have been reported.[212,213] Their neurologic course was characterized by seizures, headache, and progressive global neurologic deterioration over years, often in a stepwise fashion, but with marked variability in clinical course and severity of ultimate outcome. CT scans of the affected children showed cerebral atrophy and multiple focal lesions consistent with small infarcts. Hemorrhagic strokes,[213] choreoathetosis, ataxia, sensory loss, and venous sinus thrombosis have been reported in affected adults.[207] Cerebrospinal fluid protein is often elevated and the EEG frequently is nonspecifically abnormal.[211] Englert et al. identified anticardiolipin antibodies in one adult and speculated that these may be causally related to the syndrome.[214] Pathologic examinations reveal vascular endothelial swelling, intimal proliferation and small artery and arteriolar occlusion,[207] with widespread foci of ischemic and occasionally hemorrhagic infarction throughout the cerebral hemispheres, brainstem, and spinal cord.[208,211]

There is no specific treatment. Central nervous system dysfunction has not been prevented by treatment with aspirin, persantin, heparin or coumadin, or immunosuppression with corticosteroids, azathioprine, methotrexate or cyclophosphamide.[207] One patient reportedly improved with aspirin and dipyridamole,[215] but a subsequent patient did not.[216]

Familial Arterial Dissection

In series of patients with a history of spontaneous cervical (internal carotid or vertebral) arterial dissection, a family history can be elucidated in approximately 5% of individuals and has been demonstrated to identify patients at an increased risk of recurrent dissection.[217] Marfan's syndrome, FMD, Ehlers-Danlos syndrome, and polycystic kidney disease have all been linked to arterial dissection. In addition, a recent report described an abnormality of type I collagen in a patient with multiple cervical dissections.[218] Cystic medial necrosis has been described in several pedigrees,[219] abnormalities of elastin have been identified in one family,[220] and a neural crest disorder has been suspected in a series of five families.[221] Cervical artery dissections have also been linked to intracranial aneurysms,[222,223] with a possible etiologic link being an underlying deficiency of alpha-1-antitrypsin.[224]

Venous Strokes

Symptomatology associated with thrombosis of the cortical veins or dural sinuses is frequently less dramatic than that seen in arterial occlusion. Overall, 10–20% of cerebrovascular accidents in childhood are secondary to venous occlusion. These may be associated with sinusitis/otitis, hypercoagulable states, congestive cardiac failure, dehydration, and polycythemia, with the period of greatest risk being the first year of life.[225] While disorders of hypercoagulation have typically been associated with venous infarcts,[226] in recent years these defects have also been increasingly linked to arterial thrombosis as shown in Table 1 and discussed below. It has become increasingly evident that venous thrombosis is frequently a polygenic disease, with severe thrombophilia the result of carriage of multiple gene defects, frequently coupled with environmental factors predisposing to thrombosis.[227–230] In addition to the disorders described below, hereditary hypercoagulability has been described in heparin cofactor II deficiency,[231,232] congenital plasminogen deficiency,[233] congenital deficiency of tissue plasminogen activator,[234] factor XII deficiency,[235] and congenital dysfibrinogenemia.[236] Management of these disorders has been reviewed by Bauer[237] and Andrew.[238]

Protein C Deficiency

Protein C is a vitamin K-dependent protein synthesized in the liver. It is activated by thrombin in the presence of calcium. Activation is accelerated by thrombomodulin, which is produced by endothelial cells. Protein C normally acts as an anticoagulant by inactivating coagulation factors Va and VIIIa and plasminogen activator 1. The gene for protein C has been localized to human chromosome 2q13–14,[239] and multiple substitutions, point mutations, in-frame deletions or insertions, and alterations in splice and promoter sequences have been described.[240–243] Homozygosity, heterozygosity, and compound heterozygosity have all been linked to an increased risk of thrombosis. Most patients with protein C deficiency have both reduced immunologic and biological activity of protein C (type I). Some patients have a normal absolute concentration of the protein with reduced functional activity (type II).[244] Both forms of protein C deficiency are rare, autosomal dominant disorders. Diagnosis may be difficult in the young child, as levels of protein C are low in the first year and do not approach adult values until the fourth year of life.[245] Prevalence has been estimated at 1 in 200 to 300.[246] In addition, acquired protein C deficiency may result from severe liver or cardiac disease, sepsis, bone marrow transplantation, vitamin C deficiency, coumadin and L-asparaginase therapy, and consumptive coagulopathies.

Most symptomatic patients are young adults with heterozygous protein C deficiency who present with venous thromboses,[247–249] including cerebral venous thromboses. Several series report children with cerebral venous stroke and cerebral arterial strokes associated with heterozygous protein C deficiency.[248–252] In patients with heterozygous protein C deficiency, initiation of coumadin therapy may lead to skin and fat necrosis.[253,254] Therefore, initial anticoagulation with heparin during gradual introduction of long-term therapy with coumadin has been recommended.

Individuals with homozygous protein C deficiency, or double heterozygosity for types I and II, present in the newborn period, typically within the first 12 hours after birth,[255,256] with neonatal purpura fulminans (massive bleeding into the skin, accompanied by widespread thrombosis of capillaries, venules, and occasionally larger vessels together with necrosis). Thrombosis of larger vessels in other organs may occur.[257] Seizures, hemiparesis, or hydrocephalus may ensue. Neuroimaging or autopsy findings include intracerebral or intraventricular bleeding, cortical infarction, subarachnoid and subdural hemorrhage, and cerebral venous thromboses. Mental retardation and delayed psychomotor development have been observed with long-term follow-up.[255–257] These neuropathologic abnormalities may be due to cerebral venous sinus thromboses, secondary to the hypercoagulable state of these infants. In some infants CNS thrombotic events have been of intrauterine origin.[256] All patients have a profound reduction of protein C activity (mostly undetectable, all less than 6% of normal activity). Other laboratory studies are consistent with disseminated intravascular coagulation. In the majority of cases both parents prove to be heterozygous for protein C deficiency without clinical manifestations of thrombotic disease.

A similar clinical syndrome of purpura fulminans in the absence of inherited protein C deficiency may be seen following viral or bacterial infection in infancy or childhood. Affected children often improve with heparin, in contrast to those cases seen in neonates with homozygous protein C deficiency.[256]

Recommendations for treatment of homozygous protein C deficiency include initial treatment with intravenous fresh frozen plasma (FFP), with ongoing management including serial FFP transfusion or institution of replacement therapy with concentrated protein C until all lesions are healed (which may be 4–8 weeks).[258] Slow introduction of coumadin during phased withdrawal of FFP can achieve chronic anticoagulation while avoiding redevelopment of skin lesions.[259,260] Recurrence of purpura fulminans may be treated with reinstitution of intravenous FFP infusions. Family members should be screened for protein C deficiency,[237] and consideration given, in high-risk situations, to prophylactic therapy for asymptomatic deficient relatives of symptomatic patients.[261] Liver transplantation has been used successfully in the treatment of homozygous protein C deficiency.[260]

Protein S Deficiency

Protein S is a vitamin-K dependent protein which acts as cofactor for the anticoagulant activity of protein C. Two genes exist for protein S. They have been localized to the centromeric region of chromosome 3.[262] One of these genes is active (designated protein S alpha or PROS1[263]) and several partial deletions (and likely other mutations) are associated with protein S deficiency.[264,265] The other protein S gene (protein S beta or PROS2) shows 97% homology in its coding sequences but is probably a pseudogene.[2,43,263]

Activity of protein S corresponds with the plasma concentration of its free form, which amounts to up to 40% of total protein S. The remainder circulates bound to a complement cofactor, the C4b binding protein. Protein S deficiencies are diagnosed on the basis of low protein S activity and/or low free or total protein S antigen.[266] These diagnoses can be difficult as there may be marked physiologic and pathologic variation in the amount of circulating free protein S. C4b binding protein is an acute phase reactant,[267] which may be elevated during acute inflammation, increasing the unavailable fraction of protein S and contributing to an increased risk of thrombosis.[268] The oral

contraceptive[269] and cigarette smoking also decrease plasma levels of protein S and acquired deficiency of protein S has been associated with oral anticoagulant therapy,[270] pregnancy, nephrotic syndrome, liver disease and disseminated intravascular coagulation. In addition APC resistance (see below) has, in the past, been misdiagnosed, in some cases, as deficiency of protein S.[271]

Type I deficiency of protein S is characterized by low total and free protein S antigen levels. In contrast, type IIa has a low free protein S but normal total protein S antigen levels, whereas in type IIb only protein S activity is decreased (classification as in Comp 1990).[266] Zoller et al. have proposed that type IIa and type I deficiencies are phenotypic variants of the same disease expressed differently due to variation in C4b binding protein levels.[268] All forms of protein S deficiency follow autosomal dominant inheritance and 19 causal mutations have thus far been described, the majority in association with a type I deficiency.[243] Diagnosis using PCR techniques to amplify the coding sequence of the protein S gene must implement selective amplification of the exons and flanking sequences of Psalpha to avoid confusion with the pseudogene.[266]

Like protein C deficiency, protein S deficiency has been associated with embolic, thrombotic, venous, and hemorrhagic infarction in adolescents and young adults.[267,270,272] Initial treatment with heparin and chronic anticoagulation with coumadin prevent thromboses. The risk of coumadin-induced skin necrosis without prior heparinization is less with protein S deficiency than with protein C deficiency.[273]

A single neonate with purpura fulminans has been reported.[274] Treatment with cryoprecipitate and coumadin was beneficial.

Antithrombin III (ATIII) Deficiency

ATIII deficiency is an autosomal dominant disorder manifested by recurrent thrombotic episodes due to dysfunctional natural anticoagulation. The incidence is between 1 in 2,000 and 1 in 5,000 people.[275] ATIII deficiency is responsible for approximately 4% of venous thrombosis in young patients.[276] ATIII is a major physiologic inhibitor of activated serine proteases, including thrombin and the activated forms of factors IX, X, XI, and XII. The gene encoding ATIII is localized to chromosome 1q23-q25.[277] Deficiency of ATIII has been associated with partial deletions, insertions, and single base-pair substitutions, resulting in various subtypes of ATIII deficiency associated with quantitative (type 1) or qualitative (type 2) abnormalities.[277,278]

Clinical expression is very variable, influenced by heterozygosity or homozygosity for type 1 or 2 mutations, as well as by other genetic and environmental factors.[276] The typical presentation is with deep venous thrombosis in young adulthood, with approximately two-thirds of published cases of antithrombin deficiency having a thrombotic event between the ages of 10 and 35.[279] In childhood, ATIII deficiency has also been associated with cerebral venous infarction[280] and embolic arterial infarction.[281–283] In addition, deficiency of this fibrinolytic agent may also be acquired, secondary to renal or gastrointestinal protein-losing syndromes (which may themselves be inherited), or to treatment with L-asparaginase.

Treatment of thrombotic events consists of high-dose heparin or ATIII concentrate (prepared from pooled human plasma[284]) in the acute phase, and coumadin as long-term therapy.[280] Levels of ATIII are normally lower in infants up to 6 or 12 months than in older children or adults.[285]

Resistance to Activated Protein C

This condition, first described by Dahlback et al. in 1993,[286] is thought to account for 60% of venous thrombosis.[287] Protein C is a key part of a natural anticoagulant pathway regulating the activity of the procoagulant proteins factor Va (FVa) and factor VIIIa (FVIIIa).[286–288] Vascular damage activates blood coagulation by exposing tissue factor, which binds FVIIa, with inception of a cascade resulting in the generation of thrombin, activated cofactors FVa and FVIIIa, and FIXa. In vivo control of the coagulation cascade is achieved by the dual capacity of factor V to act in balancing procoagulant and anticoagulant activity.[286–288] Active thrombin reaching intact vascular endothelium attaches to a specific receptor, thrombomodulin, and thereby loses its procoagulant activity, becoming an activator of the anticoagulant protein C. Activated protein C (APC) inhibits coagulation by degrading factors VIIIa and Va, with free protein S[289] and intact factor V[290] acting as cofactors to APC.

APC inactivates FVa by cleavage of the protein at three sites, and a single point mutation in the gene for FVa at nucleotide 1691, causing replacement of arginine (R) with glutamine (Q) at position 506 of the FVa heavy chain, is the underlying defect in more than 90% of cases of resistance to APC.[290–293] This mutation does not affect the activation of FV to FVa, and procoagulant activity is normal. It does alter one of the APC cleavage sites, resulting in less efficient inactivation of FVa, and, hence, a prothrombotic state.[294–296] Heterozygosity for the FV:Q506 allele is seen in between 2% and 15% of the general population,[297] with marked variation between different ethnic groups.[298] Family studies suggest autosomal dominant inheritance.[299] Thrombosis is more common, and tends to occur at a younger age, in homozygotes as compared to heterozygotes. The risk is lifelong and becomes greater with increasing age.[300,301]

Clinically, resistance to APC manifests primarily as venous thrombosis, commonly in association with an environmental risk factor such as pregnancy, trauma, surgery, malignancy, or usage of the oral contraceptive pill, but occasionally occurring in the absence of identifiable triggers for thrombosis.[302,303] Arterial thromboembolism also occurs as a result of APC resistance, with evidence that the risk may be higher in pediatric as opposed to adult populations, particularly in the neonatal period.[304,305] Resistance to APC has been causative in a single case of neonatal purpura fulminans.[306]

Diagnosis of resistance to APC can be made by an APTT-based test comparing clotting times in the APTT reaction in the presence and absence of added APC, with a sensitivity and specificity of over 85%.[291,297] This investigation must be modified in those receiving anticoagulant medications, or those with other coagulation deficiencies.[288,299] Factor V mutation analysis can also be performed, using a variety of techniques amplifying the nucleotide region of interest from genomic DNA or mRNA.[303]

Recommended therapy of resistance to APC, at present, follows similar guidelines to those of the other deficiencies of natural anticoagulation.[238] Avoidance of risk factors for thrombosis, prophylactic therapy in high-risk situations, and chronic anticoagulation for recurrent thrombosis are the cornerstones of prophylaxis.[239] Acute management consists of heparinization, followed by long-term coumadin therapy. Screening of family members of children with APC resistance, to identify those who warrant prophylactic anticoagulation in high-risk situations, may also be appropriate.[239]

Combined deficiency of proteins C and S and antithrombin III has recently been described in two siblings with Cohen syndrome.[307] Clearly this cannot be the result of a contiguous gene syndrome, but may represent an abnormality of connective tissue, aberrant carboxylation, or a common inhibitor.[307]

Cerebral Hemorrhages

Large intraparenchymal and subarachnoid hemorrhages typically present with sudden-onset severe headache, associated with vomiting and loss of consciousness. Smaller lesions, especially those not involving motor cortex or brainstem, can produce much less significant symptomatology especially in infants, in whom irritability, vomiting, or seizures, possibly associated with photophobia or low-grade fever, may be the cause of presentation.[308]

Vascular Malformations

Vascular malformations are abnormalities of development of the fetal vasculature. Four main types are described: (1) arteriovenous, (2) cavernous, (3) venous, and (4) capillary (telangiectasis). These malformations are the most common causes of spontaneous intracranial hemorrhage in children,[4] although aneurysms are more common in adolescence.[309]

Arteriovenous Malformations

Arteriovenous malformations (AVMs) are the type of vascular malformation most likely to produce symptoms. They account for 40% of intracranial hemorrhages and 20% of strokes of all forms in children less than 15 years of age.[4] About 90% of brain AVMs are located above the tentorium.[310] The incidence of rupture is about 1 per 100,000 children per year.[4]

AVMs are usually sporadic, but reports of families with isolated CNS AVMs and childhood intracerebral hemorrhage suggest some instances of autosomal dominant inheritance.[311–313] Some familial cases of spinal AVM[314] and brain AVM[315] associated with other cutaneous vascular malformations likely represent cases of Rendu-Osler-Weber syndrome (see below).

Symptoms of AVMs may be secondary to rupture, to "steal" phenomenon causing local ischemia, or to venous hypertension.[310] Rupture with hemorrhage is the most common mode of presentation, and is associated with a 10% mortality (higher if located in the posterior fossa) and significant neurologic deficit in 30–50% of survivors.[310] About 15% of children present with seizures, which are typically focal with or without secondary generalization,[310,316] and are disproportionately common with lesions of the temporal lobe. Other symptoms include a slowly progressive, focal neurologic deficit,[310] hydrocephalus, or precocious puberty.[317] Neonates may present with congestive cardiac failure.

Symptoms depend on the localization of the AVM and the magnitude of the hemorrhage. They vary from coma and death, to subtle, recurrent minor bleeding episodes manifest primarily by headache. Hemorrhage may be intraventricular, intraparenchymal, subarachnoid or some combination thereof. Bleeding is relatively more common with infratentorial AVMs[318] and with smaller lesions.[310] The lifetime risk of rebleeding with AVMs is greater when the first bleed occurs in childhood than if the first hemorrhage occurs in adulthood.[310] Recurrent hemorrhage occurs in 6–7% of patients in the first year after the initial hemorrhage and 2–3% in subsequent years, as reviewed by Martin and Edwards.[310] Mortality rate increases with each subsequent hemorrhage.[319] The risk of subsequent bleeding is lower in those who present with seizure (2% per year) than in those who present after an initial hemorrhagic episode.[310]

CT may show focal calcification and often dramatic enhancement of the lesion with intravenous contrast. A minority of AVMs will not show enhancement.[320] Intraparenchymal and intraventricular hemorrhage are well demonstrated by CT, and subarachnoid hemorrhage may also be seen. MRI may reveal old hemorrhage and evidence of the abnormal vessels and may be particularly useful in identifying the rare angiographically inapparent (occult) AVMs.[310] Cerebral arteriography usually confirms and delineates the AVM. Both carotid and vertebral arteries should be studied since AVMs may be multiple and aneurysms accompany about 8% of AVMs in adults (less in children).[310] Even angiographically occult AVMs may bleed and their natural history is qualitatively similar to those which are visible angiographically.[310]

Pathologically, AVMs are characterized by the direct anastomosis of thin-walled, dilated arterial and venous vessels in the absence of a capillary bed. Degenerative changes are seen both in the blood vessels themselves (calcification, generally more prominent in veins and shunting vessels), and in the adjacent brain parenchyma (gliosis, atrophy, and cortical necrosis).[321,322]

Because of the high incidence of rebleeding, surgical excision of the AVM is generally recommended.[322] Surgery is generally delayed following an acute hemorrhage, but may be required emergently for unrelenting neurologic deterioration.[310] Complete excision is an effective cure, but incomplete excision leaves considerable risk of rebleeding.[322] Surgical mortality in children operated on for brain AVMs ranges from 0–3% and morbidity from 0–8%, which is lower than the risk of nonintervention.[310] Very large, high-flow lesions may not be amenable to one-stage surgical excision, and deeply located lesions may not be resectable. Embolization has been advocated for lesions not amenable to surgery, but may result in incomplete obliteration with subsequent risk of rebleeding.[310,323] In some patients a combined approach with embolization followed by surgery is preferable, while the use of coils to induce thrombosis of lesions can improve prognosis in those too unstable for open resection.[323,324] Focused radiation, using gamma-beam radiation or heavy charged-particle Bragg peak radiosurgery,[325] has been used successfully in children. Potential problems with these forms of treatment are the long time interval until therapeutic effect is observed, and the risk of radiation injury to nearby brain tissue.

Cavernous Malformations

These vascular lesions, consisting of thin-walled abnormal vessels with no direct communication with the arterial system, are encountered less commonly than AVMs.[326] Presentation may be with seizures, sudden onset of neurological deficit, headache, or slowly progressive deficit, due to hemorrhage or mass effect. Most are supratentorial.[326] Familial occurrence of the angiomas is well recognized and follows an autosomal dominant pattern with variable expression.[327] Penetrance depends upon the method of identification, but may approach

100% in some families studied by MRI. One responsible gene has been mapped to chromosome 7q11–22.[327] A single family has been reported in which typical cerebral cavernous angiomas were associated with angiomas of the retina and liver.[328] Cavernous malformations are typically inapparent on angiogram, but may be detected with CT or MRI and are frequently multiple. Indications for surgical resection include hemorrhage and intractable seizures, while management of asymptomatic lesions remains unclear. Ongoing symptoms are seen in a significant proportion of patients treated conservatively.[329]

Hereditary Hemorrhagic Telangiectasia (Rendu-Osler-Weber Disease)

This is an autosomal dominant disorder characterized by skin, mucosal, pulmonary, gastrointestinal, genitourinary, and central nervous system telangiectasias, which commonly bleed. Expression is highly variable. Incidence is 1–2 per 100,000,[30] but it is likely that a significant proportion of cases are undiagnosed. Multiple loci for Rendu-Osler-Weber disease have now been identified. The first, ORW1, involves the gene (mapped to chromosome 9q33–34[330,331]) for endoglin, an endothelial and placental receptor for transforming growth factor beta. Mutations at two separate sites on chromosome 12[332,333] have also been linked to HHT, with the most recently described locus involving mutations in the activin receptor-like kinase 1 gene.[333] Mutations linked to chromosome 9 may predispose to a higher incidence of pulmonary arteriovenous malformations than mutations at other sites.[333]

Symptoms typically begin in adolescence or young adulthood, with cutaneous and conjunctival telangiectasias increasing in size and number after the first decade. Some affected individuals may have no external vascular lesions.[334] Up to 36% of patients have neurologic involvement secondary to brain abscess, vascular malformations of the brain, meninges or spinal cord, or porto-systemic encephalopathy.[335] Intraparenchymal and subarachnoid hemorrhage have been described in conjunction with arteriovenous malformations, arterial aneurysms and carotid-cavernous fistulae, telangiectasias, and angiomas.[335,336] Ischemic stroke in these patients may occur due to septic and nonseptic emboli and air emboli from arteriovenous communication in the lung, paradoxical emboli through pulmonary arteriovenous communications, and thrombosis secondary to polycythemia.[335,337,338] Treatment is largely symptomatic and may include iron replacement therapy or periodic transfusion. Resection or occlusion of pulmonary arteriovenous fistulae is recommended for those causing neurologic complications.[338] Little information is available as yet regarding the relative benefits of conservative management or surgical intervention for cerebral lesions, but screening of family members has been recommended with an aim of early identification and treatment of pulmonary AVMs. Children require repeated examinations due to the natural history of additional symptomatology with increasing age.[338,339]

Wyburn-Mason Syndrome

Wyburn-Mason syndrome and Bonnet-Dechaume-Blanc syndrome are likely the same clinical entity,[340] characterized by unilateral facial angioma (50%) with retinal and intracranial arteriovenous malformations. The intracranial AVM involves the ipsilateral optic nerve, chiasm, and midbrain and may be demonstrated by CT, MRI, and angiography.[339,340] Progressive or sudden visual loss is common in childhood. These patients may also suffer subarachnoid or intraparenchymal hemorrhage or seizures. Most cases are single, but the original report described an affected mother and child.[341]

Bannayan-Zonana Syndrome

This is an autosomal dominant syndrome associated with macrocephaly, postnatal growth failure, mesodermal lipomas and hemangiomas, and variable developmental delay.[342] One affected boy suffered multiple cerebral hemorrhages secondary to cryptic vascular malformations.[342]

Aneurysms

Symptomatic intracranial arterial aneurysms in children account for 0.6% of patients with cerebral aneurysms and subarachnoid hemorrhage.[343] Aneurysms account for 36% of subarachnoid hemorrhages in patients up to 20 years of age.[309] A bimodal distribution is described, with increased incidence before 2 years and after 10 years.[344] Symptomatic aneurysms in children are generally saccular. They are usually located at the bifurcations of arteries of, or near, the circle of Willis,[345] though a higher proportion are situated more distally, or in the posterior circulation, in children than in adults.[345,346] Intraparenchymal hemorrhage occurs in about 50% of children with symptomatic aneurysms, often combined with subarachnoid hemorrhage.[347] Presentation may also be with cranial nerve or brainstem dysfunction secondary to mass effect.[345] Mortality in cases presenting with intracranial hemorrhage is reported as 54%, but, in general, surgical treatment is more successful than in adults.[345] Cerebral aneurysms are described further in *Chapter 12*.

Several inherited disorders have been associated with symptomatic cerebral aneurysms in childhood (Table 5). The cerebral aneurysms associated with alpha-glucosidase deficiency (type II glycogenosis) have been fusiform and located in the vertebrobasilar system, with or without carotid involvement.[348,349] Coarctation of the aorta also has an association with intracranial aneurysms.[349] An association between Marfan's syndrome and intracranial aneurysm has long been proposed, with many single case reports in the literature, but was unproven in a recent cohort study.[350]

Isolated familial cerebral aneurysms have also been reported, and may account for as many as 20% of all subarachnoid hemorrhages.[351] These have a tendency to occur

Table 5
Inherited Disorders Associated with Intracranial Aneurysm

Ehlers-Danlos syndrome
Marfan's syndrome
Moyamoya
Pseudoxanthoma elasticum
Adult polycystic kidney disease
Infantile polycystic kidney disease
Alpha-glucosidase deficiency (type II glycogenosis)
Fibromuscular dysplasia
Sickle cell disease
Tuberous sclerosis
Alpha-1-antitrypsin deficiency
Alkaptonuria
Hereditary hemorrhagic telangiectasia
Fabry's disease
Neurofibromatosis type 1
3M syndrome
Noonan syndrome

in similar sites within a kindred (most frequently involving the middle cerebral artery), and to become symptomatic at a relatively young age, although presentation in childhood is seen in a minority only.[352] Autosomal dominant inheritance is probably the most common means of inheritance, but genetic heterogeneity is likely.[351,353]

Abnormalities of type III collagen have been identified in half of a series of patients with intracranial aneurysm in the absence of a systemic connective tissue disorder,[353–355] as well as in Ehlers-Danlos syndrome (type IV) and in Marfan's syndrome, and have been suggested as a potential molecular basis for the increased incidence of systemic and intracranial aneurysms in polycystic kidney disease. Despite evidence of decreased type III collagen and increased sensitivity to proteases of type III collagen in several studies on patients with spontaneous intracranial aneurysm or cervical artery dissection,[354,356] a recent series in which sequence analysis was performed on 55 unrelated patients with such a history failed to identify causative mutations in the gene for type III collagen (COL3A1).[357]

An increased incidence of intracranial

aneurysms in heterozygous and homozygous alpha-1-antitrypsin deficiency has been identified,[358] suggesting that aneurysm formation may occur as a result of a variety of abnormalities of extracellular matrix proteins and proteases.

Bleeding Disorders

Intracranial bleeding can occur at any age in individuals with inherited disorders that cause a bleeding tendency.[3] The severity of the bleeding disorder rather than the specific deficiency determines the likelihood of intracranial hemorrhage. Hypertension or an associated cerebrovascular malformation may further predispose these patients to intracranial hemorrhage.[3] Many patients have a preceding history of trauma, although the intervening period may be several days. Using hemophilia as an example, intracranial hemorrhage may occur at birth, is uncommon in the first year of life, and becomes more frequent with the increasing activity of childhood.[359] Rapid diagnosis and treatment is important for improvement of outcome. Studies reporting improved outcome with early treatment, or initiation of treatment prophylactically in those with severe hemophilia and a history of trauma,[360] have led to the formation of guidelines recommending replacement to 30% to 50% of the normal factor VIII level in such patients.[360] Treatment may also include surgical evacuation of hemorrhages after appropriate factor replacement, with best effect seen with subdural, epidural, and cerebellar collections. Despite advances in diagnosis and therapy, mortality is high (16% in hemophiliac patients in a recent study[361]). Morbidity is also very significant, especially in those with intraparenchymal hemorrhage.[361] Inherited disorders of the coagulation cascade and platelet function associated with cerebral hemorrhage in childhood are listed in Table 6. Acquired bleeding disorders due to inherited abnormalities in other systems, for example vitamin K deficiency secondary to cystic fibrosis, abetalipoproteinemia, and alpha-1-anti-

Table 6
Bleeding Disorders Associated with Cerebral Hemorrhage in Children

Coagulation factor deficiencies
  Factor V deficiency, factor VII deficiency, factor VIII deficiency (hemophilia A), factor IX deficiency (hemophilia B), factor X deficiency, factor XI deficiency, factor XIII deficiency, afibrinogenemia, heparin cofactor II deficiency.
Fanconi anemia
Thrombocytopenia associated with inherited metabolic disease: hyperglycinemia, methylmalonic acidemia, isovaleric acidemia
Glanzmann thrombasthenia
Inherited thrombocytopenias

trypsin deficiency may also be associated with intracranial hemorrhage in childhood. These disorders are discussed in detail in *Chapters 11 and 12.*

Hemoglobinopathies

A syndrome of hypertension, convulsions, and intracerebral or subarachnoid hemorrhage occurring several days after blood transfusion has been reported in a small number of children with homozygous beta-thalassemia and other inherited hemoglobinopathies.[362] In two such autopsied cases intraparenchymal microaneurysms were observed. Sinniah et al. described two similar cases in children with beta-thalassemia which occurred 4 to 5 days after blood transfusion,[363] and later proposed a circulating prothrombinase inhibitor as a cause.[364] In some cases the etiology of cerebral dysfunction after transfusion for beta-thalassemia was not clarified.[365]

Miscellaneous Genetic Etiologies of Pediatric Stroke

Mitochondrial Encephalo-myelopathy, Lactic Acidosis and Stroke-like Episodes (MELAS)

MELAS is one of several multisystem syndromes attributed to mutations in mitochondrial DNA (mtDNA). The major diagnostic features include stroke-like episodes, lactic acidosis, ragged red fibers or other abnormalities of mitochondrial structure on muscle biopsy, seizures (generalized or focal), dementia, recurrent headache and vomiting.[366–368] Other clinical features include short stature, cortical blindness, sensorineural hearing loss, and weakness. Ophthalmoplegia, retinal degeneration, heart block, and elevated cerebrospinal fluid protein above 100 mg/dl are usually absent, which distinguishes these patients from those with Kearns-Sayre syndrome.[367] Patients are usually normal at birth with normal early development.[13,368] Symptoms typically begin in childhood or adolescence, but may begin in infancy or young adulthood.[367,369] Since mtDNA is almost exclusively derived from the maternal oocyte, maternal transmission of MELAS (and several other mtDNA disorders) is observed.[368]

Although the stroke-like episodes are not usually the first symptom, they do typically occur in childhood or young adulthood.[368] The neurologic deficits seen may evolve in the setting of a febrile illness with vomiting, seizures, and lethargy and may be either transient and fluctuating or permanent.[13,370] Typically a progressive encephalopathy with accumulating multifocal neurologic deficits and worsening seizures and dementia culminates in death in early to middle adulthood.[370] Involvement of the occipital lobe is common (Figure 3), and may result in cortical blindness.[13]

Neuroimaging reveals focal lesions, which may be transient or permanent. The lesions typically involve the cortex and may or may not conform to recognized vascular distributions.[13,371] Multifocal, hyperintense T_2-weighted signal confined to the cortex of cerebrum or cerebellum (especially the deeper layers) and adjacent white matter may be relatively specific for MELAS.[371] In addition, CT may reveal basal ganglia calcification, basal ganglia lucencies, and diffuse cerebral atrophy.[13,372] Cerebral angiography may be normal or show large-vessel occlusion[13] (though autopsy studies do not report large-vessel occlusions[13,373]), or occasionally demonstrates increased caliber of large vessels or a capillary blush with early venous filling.[372]

Neuropathology reveals multiple focal areas of necrosis, accentuated in the occipital cortex, but also present in other sections of cortex, subcortical white matter, basal ganglia, thalamus and substantia nigra, and associated with cerebral atrophy and gliosis.[373,374] Abnormal mitochondria have been observed in the smooth muscle and endothelial cells of small cerebral and systemic arteries, as well as in the substance of the brain and other organs, and have been implicated as a cause of the multifocal ischemic lesions of patients with MELAS. Localized cellular dysfunction secondary to inadequate oxidative phosphorylation in the cerebral vasculature or parenchyma, or intermittent vasospasm due to mitochondrial angiopathy, have been postulated as the causative mechanisms of cerebral infarction.[375,376,378] Single-photon emission computed tomography (SPECT) studies have supported metabolic dysfunction rather than hypoperfusion.[376]

An A to G mutation in the tRNALeu at position 3243 of the mtDNA has been found in muscle or blood from approximately 80% of patients with MELAS.[368,379] Asymptomatic or mildly symptomatic relatives of the MELAS patients may have the same tRNA mutation in muscle or blood; however, the proportion of mutant mitochondrial genomes is greater in the patients with MELAS than in the healthy relatives with the same tRNA mutation.[380] In addition, the proportion of mutant mitochondrial genomes was greater in the mildly symptomatic relatives than in the asymptomatic relatives.[380] This suggests that the proportion of mutant mitochondrial genomes in affected individuals is important in determining pathogenesis, with a threshold effect seen in many cases.[368] A subgroup of patients with the nucleotide 3243 MELAS mutation have progressive extraocular ophthalmoplegia but no strokes.[381] The most common biochemical abnormality in clinically defined patients was initially reported to be complex I deficiency,[382] but in patients with both clinical features and the specific

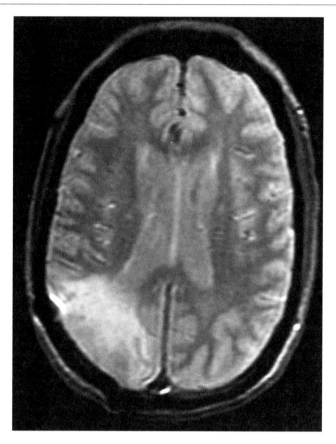

Figure 3. Proton density-weighted image of a 36-year-old man with MELAS. A recent right occipital watershed infarct is shown. The patient had a 20-year history of recurrent stroke-like episodes, sensorineural deafness, and seizures. (Provided by Vincent P. Matthews, MD, Bowman Gray School of Medicine, Winston-Salem, NC.)

mtDNA defect, multiple defects of respiratory chain enzymes may occur.[368]

A separate mutation in the same tRNA at position 3271 has also been identified[383] which has been reported to cause the MELAS phenotype in under 10% of patients. A third mutation at nucleotide pair 3291 was reported in a single patient[384] and a single patient with typical MELAS, and two asymptomatic relatives had a mutation involving substitution of G for A at position 11084.[385] No mutation has been identified in approximately 10% of patients with a clinical diagnosis of MELAS.

Patients with overlapping features of MELAS and MERRF (myoclonus epilepsy and ragged red fibers) or Kearns-Sayre syndrome have suffered stroke-like episodes.[13,368] Patients with many clinical features of MELAS, including the tRNA mutation but without lactic acidosis or ragged red fibers on muscle biopsy, have also suffered stroke-like episodes.[368] Alternating hemiplegia of childhood has also been linked with mitochondrial disease[386] through the demonstration in some patients of abnormal phosphorus magnetic resonance spectra on scanning,[387] although this may represent a secondary phenomenon.

Treatment for MELAS has been reviewed by Shoffner and Wallace,[388] who report improvements with coenzyme Q, succinate, prednisone, and prednisolone in isolated cases. They observed difficulty in assessing efficacy of therapy because of variability of clinical manifestations throughout the patients' lives. In one patient long-term treatment with riboflavin (300 mg/day) and nicotinamide (4 gm/day) reduced recurrent encephalopathic episodes and myoclonus, improved sural nerve conduction velocities, and was associated with improvement of several parameters of muscle energy metabolism.[389] Management should also include examination, genetic screening, and counseling of relatives when appropriate.

Other Metabolic Disorders with Stroke-like Features

Several metabolic abnormalities are associated with symptoms or neuroradiologic features that may suggest ischemic stroke. These disorders are not associated with focal ischemic cerebral damage due to arterial occlusion, but are likely related to local intracellular biochemical dysfunction, which may manifest as focal or diffuse cerebral dysfunction. Commonly, patients manifest other features of the underlying disease in addition to the stroke.

Aicardi et al. drew attention to a group of inherited metabolic disorders characteristically associated with lucency of the basal ganglia on CT scan, including Wilson's disease, Leigh's disease, and mitochondrial encephalopathies.[390] Similarly, methylmalonic acidemia,[391,392] propionic acidemia,[393–395] isovaleric acidemia, and glutaric aciduria types I and II[396,397] may be associated with recurrent focal neurologic symptoms and basal ganglia lucencies on CT scan,[3] as well as cerebellar hemorrhages.[394] Strokes may occur simultaneously with, or independently of, metabolic decompensations.[397]

Female carriers of the mutant X-linked gene for ornithine transcarbamylase (OTC) deficiency may present with paroxysmal neurologic dysfunction related to protein intake, with symptoms possibly including seizures and development of hemiplegia.[398,399,400] Another urea cycle defect, deficiency of carbamoylphosphate synthetase (CPS), has been associated with recurrent stroke-like episodes, seen in the context of focal cortical atrophy and episodic encephalopathy.[401] CPS deficiency is inherited in an autosomal recessive pattern with the gene located on chromosome 2.[401]

Sulfite oxidase deficiency is seen in isolation or as a general deficiency of the enzyme group utilizing the molybdenum cofactor.[30] Isolated sulphite oxidase deficiency is transmitted in an autosomal recessive fashion and is extremely rare. Presentation is with refractory neonatal seizures and feeding difficulties, with subsequent development of profound mental retardation, seizure disorders, late-onset lens dislocation, and spasticity.[402] Neuroimaging and pathologic examination reveal cerebral lesions of differing ages resulting in cerebral atrophy, loss of volume in the white matter and basal ganglia, cystic degeneration, and calcification, with no abnormalities of the vasculature.[402] A similar clinical picture is seen with molybdenum cofactor deficiencies.[403] Therapy with cysteamine has been suggested as potentially efficacious because trials of penicillamine and captopril have been unsuccessful. Prenatal diagnosis is available.[404]

Phosphoglycerate kinase deficiency is a rare X-linked recessive disorder characterized by chronic nonspherocytic hemolytic anemia, neonatal jaundice, recurrent hemolytic crises, and occasional rhabdomyolysis.[405,406] Mild to moderate neurologic deterioration has been reported in some affected males, usually consisting of speech delay and slowly progressive dementia, and sometimes including seizures and extrapyramidal movement disorder.[405–407] In addition, recurrent episodes of coma, focal neurologic dysfunction, and focal or generalized seizures have been observed in some affected males.[407] These episodes typically occur with febrile illnesses and acute hemolytic crises. Focal neurologic disability may be transient or permanent. Focal brain abnormalities including edema, increased radioisotope uptake, and T_2 hyperintense signal abnormality in patches not conforming to a vascular distribution have been detected by neuroimaging in several patients during these episodes.[407] Other patients had normal brain isotope scans, CT scans, and MRI between and during episodes of neurologic dysfunction. Angiography and postmortem examinations have not revealed arterial occlusion. Disordered energy metabolism has been proposed to account for these neurologic findings, but the precise mechanism is unknown.[407] Aggressive supportive therapy for hemolytic crises and, possibly, splenectomy may be helpful.[404]

Carbohydrate-deficient glycoprotein syndrome (CDG) is a recently described group of inherited metabolic diseases characterized by deficiency of carbohydrate moieties normally found on many serum glycoproteins, glycopeptides and lipoglycoproteins. It is inherited in an autosomal recessive fashion.[408] Major clinical features in infancy include mental retardation, failure to thrive, liver dysfunction with subsequent ascites and pericardial effusion, stroke-like episodes, and inverted nipples.[409,410] In later childhood and adolescence, acquired microcephaly, ataxia, peripheral neuropathy, pigmentary retinopathy, absence of secondary sexual development (in females only), and truncal skeletal abnormalities become obvious.[409] Neuroimaging demonstrates pontocerebellar hypoplasia with or without cortical atrophy,[409] while biochemical investigations typically reveal low thyroid-binding globulin, cholesterol and beta-lipoproteins,[408] and an abnormal immune isoelectric focusing pattern of transferrin is typical and provides a useful screening test.[408]

Stroke-like episodes may occur spontaneously or in the setting of a febrile illness and are usually characterized by transient coma, variable hemiplegia, and seizures.[409] Permanent hemiplegia associated with an ischemic parieto-occipital stroke was observed in one child,[411] and blindness lasting months in two others.[412] The precise biochemical and genetic defects are not known.[409] No therapy is currently available. Indirect prenatal diagnosis may be possible by analysis of isoforms of glycoproteins in the amniotic fluid or fetal blood.[413]

Noonan Syndrome

Noonan syndrome, a relatively common syndrome first described in 1963, is characterized by short stature, webbed neck, hypertelorism, ptosis, wide-spaced nipples, cubitus valgus, and frequent cardiovascular anomalies.[414] It is probably inherited in an autosomal dominant fashion. Noonan syndrome has been linked to intracranial aneurysm,[415] arteriovenous malformation,[416] and thromboembolic infarcts.[417] The genetic defect has not as yet been identified.

Neonatal Stroke

The majority of infants with ischemic stroke have normal prenatal and perinatal histories.[7,418] Clinical manifestations of stroke in the newborn are nonspecific and focal neurologic findings are often absent or subtle. Coker et al. reported seizures in 80% (12/15) of neonates with ischemic stroke.[7] Neonatal stroke is usually idiopathic, and often attributed to placental emboli.

Resistance to APC has recently been recognized as an imported inherited cause (see above).

Other inherited diseases associated with stroke in the child or adolescent may also be associated with stroke in the newborn. In particular, the inherited bleeding disorders may present in the prenatal or neonatal periods. The fulminant characteristics of homozygous protein C or S deficiency (neonatal purpura fulminans) are described above. Caution needs to be applied to interpretation of assays for proteins C and S and for ATIII, however, since levels of all are low in the neonatal period and increase with age.[238,419] Although uncommon, intracranial hemorrhage secondary to hemophilia and other factor deficiencies in the neonatal period has been described.[420] Thrombocytopenia has also been implicated in neonatal strokes, with etiologies including maternal ITP, isoimmune platelet group incompatibility,[421] X-linked and autosomal recessive inherited thrombocytopenia, and inherited metabolic disorders such as glycinemia, methylmalonic and isovaleric acidemias. Stroke has also been reported in neonates born to mothers with antiphospholipid antibodies,[422] putatively as a result of transplacental passage of IgG antibodies conferring a thrombophilic tendency, while a similar mechanism has been held to account for a case of intracranial hemorrhage in the infant of a mother with acquired factor VIII:C inhibitor.[423]

Porencephaly is a cavitary hemispheric lesion that typically extends from ventricular to pial surfaces. Porencephalic lesions arising before the 25th week of gestation are commonly associated with local abnormalities of gyral development.[424] They are sometimes called schizencephalies, and commonly involve the insula and adjoining precentral and postcentral gyri bilaterally. Later lesions are encephaloclastic without disruption of gyral development.[424] Both forms have been attributed to vascular lesions in some instances, and to toxins and prenatal infections in others.[424] Cases are typically sporadic, but familial cases, identified by CT scanning, have been reported.[425,426] These patients had contralateral hemiplegia from infancy and some suffered seizures, mental retardation, learning difficulties, and microcephaly. The pattern of inheritance was consistent with an autosomal dominant trait, but also potentially consistent with alternative etiologies, i.e., in utero infection.[427] Treatment is supportive.

Two unrelated families with multiple siblings manifesting an unusual syndrome of hydranencephaly with a distinct proliferative vasculopathy without evidence of an acquired cause may also represent an inherited disorder.[428,429]

Aplasia Cutis Congenita

Focal congenital lesions of the scalp resulting in partial or full thickness deficiency of skin usually occur sporadically, but can be associated with an autosomal dominant or recessive[430,431] pattern of inheritance, or with various genetic syndromes as reviewed by Singman et al.[432] Aplasia cutis congenita has occurred in association with various anomalies of cerebral and cerebrovascular genesis, including arteriovenous malformation,[432] superior sagittal sinus thrombosis,[433] and congenital midline porencephaly,[434] prompting recommenda-

tions that the presence of ACC should indicate examination of family members and investigation of affected individuals.[432]

Migraine

Migraine is frequently familial. Various hypotheses regarding the pattern of inheritance have been proposed, including autosomal dominant,[435] autosomal recessive,[436] and polygenic.[437] The relationship between stroke and migraine in childhood is controversial, although several authors cautiously suggest a slightly increased incidence of stroke in children with migraine,[438,439] and several reports describe small numbers of migrainous children with ischemic strokes of otherwise unexplained etiology.[438-441] Patients whose migraines are complicated by the development of neurologic signs seem to be at higher risk for stroke than those with common migraine.[442] The focal neurologic deficit reported with migrainous stroke typically begins with or before the headache and persists longer than the usual neurologic deficit associated with previous migraines reported by some of these patients. CT or MRI scans may reveal a region of injury. Angiography, when performed acutely, has revealed transient segmental narrowing.[438,439] Strokes associated with childhood migraine tend to have a good prognosis, with minor residual deficit and little risk of recurrence.[438,439]

Caution should be exercised in attributing stroke to migraine. Other identifiable causes of stroke should be excluded. Migraine is a relatively common disorder in childhood (ranging from 2.5% to 10% prevalence depending upon age and sex[443]) and, therefore, coincidental occurrence of stroke and migraine may occur in the absence of etiologic association. Migrainous headaches may be symptoms of several disorders which predispose to stroke, such as AVMs,[438,444] cerebral tumors,[438] antiphospholipid antibody syndrome (see below), MELAS, atrial myxoma, and the CADASIL syndrome (see above). In addition, migraine may be an additional risk factor for stroke in those already predisposed, i.e., in those with a history of fibromuscular dysplasia or mitral valve prolapse.[438]

Rarely, childhood stroke is seen in the context of a history of familial hemiplegic migraine, an autosomal dominant syndrome of transient hemiparesis and paresthesia followed by migrainous headache, which has recently been mapped to chromosome 19 in the same region as CADASIL (see above).[145] A recent report describes a single family in which an autosomal dominant pattern of inheritance was associated with migraine attacks (with aura in the majority) and white-matter abnormalities on neuroimaging, in the absence of stroke but with possible progression to subcortical dementia, also probably linked to chromosome 19.[445] The relationships between this condition, CADASIL, and familial hemiplegic migraine have as yet to be defined.

Autoimmune Diseases

Autoimmune diseases such as systemic lupus erythematosus, mixed connective tissue disease, Takayasu's arteritis, periarteritis nodosa, rheumatoid arthritis, dermatomyositis, inflammatory bowel disease, Kawasaki disease, Behçet's disease, Wegener's granulomatosis, immune thrombocytopenic purpura, and possibly granulomatous arteritis are associated with childhood ischemic and hemorrhagic vascular accidents.[11,446,447] This may be in the context of the transient or permanent presence of antiphospholipid antibodies.[448] There is a well-recognized trend towards increased frequency of autoimmune diseases within families, although different family members may suffer different autoimmune diseases.

Human Leucocyte (HLA) Antigens and Childhood Stroke

Mintz et al. reported four patients with unexplained stroke, all of whom were heterozygous for the class I histocompatibility antigen HLA-B51.[449] The patients were aged 7 months to 8 years at the time of ischemic stroke. Occlusions of single large, medium, and small arteries were demonstrated by angiogram in three of the patients. Since the incidence of HLA-B51 was relatively low in the community (2.7–13.1%), they suggested that these results indicate a common immunogenic marker associated with a genetic predisposition to childhood stroke.[450] In addition, HLA-B51 is associated with Behcet's disease (an autoimmune vascuitic disease), Kawasaki disease (a systemic vasculitis of childhood),[451] and moyamoya disease (see above),[174] all of which have been linked to stroke.

An association of saccular aneurysms with the major histocompatibility complex, notably HLA-DR2, has also been reported,[452] which similarly supports the concept of a genetic predisposition to otherwise idiopathic cerebral arterial aneurysms.

Antiphospholipid Antibodies

Antiphospholipid antibodies (APL-Abs) are a heterogeneous group of antibodies which are increasingly associated with cerebrovascular disease in young patients. The lupus anticoagulant and anticardiolipin antibodies are the most commonly assayed subset of these APL-Abs.[453,454] Since 1963, numerous reports have linked circulating APL-Abs to 4–76% of patients with premature stroke in the absence of other risk factors.[455-459] While predominantly a disorder of young adults, children and adolescents have also been affected.[456,460,461] The disorder is usually considered to be an acquired, immune-mediated hypercoagulable state. Antibody-induced inhibition of the protein C-protein S-thrombomodulin complex and inhibition of endothelial factors have been implicated in disease pathogenesis. There have been several reports, however, of familial occurrence of premature strokes associated with APL-Abs,[459,462] as well as familial occurrence of APL-Abs without stroke.[463] Development of APL-Abs may be related to inheritance of HLA-DR antigens

DR7 and DRw35, as well as some complement deficiency alleles, as reviewed by Mc-Neil et al.[464,465]

APL-Abs have been associated with several disorders of hypercoagulation including thrombosis and thromboembolism of both arterial and venous systems, and recurrent spontaneous abortion.[465] Venous thrombosis most commonly affects the iliofemoral veins but has also been reported in the cerebral venous sinuses, while arterial thrombosis most commonly manifests as cerebrovascular disease.[466] Neurologic syndromes associated with APL-Abs consist of four main patterns of involvement: (1) Acute ischemic encephalopathy, frequently preceded by transient ischemic attacks. (2) Recurrent brain infarctions. These are typically small and cortical, but brainstem infarcts, large hemispheric infarcts, lacunar infarcts, venous sinus thromboses, and transient ischemic attacks have been reported. Multi-infarct dementia may occur. The risk for recurrent stroke is high (70%) with a rate of 13% per year and mortality of about 10% per year. Many patients have had other risk factors for stroke, such as smoking, ingestion of the oral contraceptive pill, pregnancy, factor XII deficiency, hypertension, hypertriglyceridemia, polycythemia, or diabetes mellitus. Intracranial hypertension secondary to recurrent venous thrombosis has been reported. (3) Chronic migrainous headache. (4) Ophthalmologic complications, including amaurosis fugax, retinal infarction, optic neuritis, and acute ischemic retinopathy.

Other neurologic complications associated with APL-Abs include chorea,[466] epilepsy, moyamoya syndrome,[467] and myelitis. In general, higher APL Ab titers are associated with more neurologic symptoms,[456,466] with low levels of APL-Abs frequently detectable in healthy controls, possibly reflecting damage to circulatory phospholipid surfaces.

Three cases have been reported in recent years which would appear consistent with complications in the neonate of maternal anticardiolipin antibodies, presumably as a result of transplacental passage of IgG antibodies.[468]

Cerebral angiography in patients with neurologic symptoms and APL-Abs has produced variable results: internal carotid artery occlusion, thrombosis, or stenosis; middle cerebral branch or trunk occlusion; vertebral artery stenosis; aneurysm; vasculitis; or no abnormality.[457,458] Cerebrospinal fluid studies in patients with neurologic symptoms are usually normal (including IgG index, oligoclonal banding, and myelin basic protein), but some show mild lymphocytic pleocytosis and elevated protein.[458] Histopathology has revealed large- and medium-sized arteries occluded by thrombus, without evidence of vasculitis.[458]

Tests for APL-Abs include identification of elevated anticardiolipin antibodies (IgG, IgM, or both) and prolongation of the phospholipid-dependent coagulation tests (tissue thromboplastin inhibition test or activated partial thromboplastin test [aPTT]) which detect the so-called lupus anticoagulant. Abnormal laboratory findings may also include Coombs positive hemolytic anemia, thrombocytopenia, positive antinuclear antibodies, falsely positive VDRL, antimitochondrial antibodies type 45, and reduced serum complement.

No controlled studies of treatment strategies for stroke associated with antiphospholipid antibodies are reported and efficacy of treatment is difficult to judge without more information on the natural history. Treatment of all coexisting risk factors for stroke is important.[466] Because the risk of recurrent stroke is high[458,469] and because recurrent strokes have occurred despite therapy with aspirin and dipyrimadole,[457,458] coumadin therapy has been recommended for those at high risk, i.e., those with previous stroke, Sneddon's syndrome, other connective tissue disorders, cardiac lesions, or high anticardiolipin antibody levels. Moderate- to high-dose coumadin conferred better protection against recurrent thrombotic events than low-dose coumarin or aspirin therapy in one retrospective study.[469] The addition of prednisone to coumadin therapy for those "failing" with coumadin has been suggested,[470] as has long-term therapy with subcutaneous heparin.[471,472] Prednisone may suppress lupus anticoagulant activity but typically has little effect on anticardiolipin antibody levels,[466] and some patients treated with corticosteroids have suffered recurrent strokes.[458] Immunosuppression and plasmapheresis appears to have been beneficial in some cases of acute encephalopathy,[458] and prednisone combined with antiplatelet activity has sometimes been beneficial in preventing fetal loss.[470,471] One uncontrolled study described the use of immunoglobulin infusions in four children, with subsequent improvement in one, but no controlled trials have been conducted for this form of therapy.[467] The combination of aspirin with calcium antagonists has sometimes been effective for patients with recurrent headache.[466]

Ehlers-Danlos Syndrome

At least 10 subtypes of this syndrome have been described, with common features of hyperextensible joints, fragile or hyperelastic skin, easy bruising, excessive scarring, and vascular lesions, possibly associated with abnormal platelet function and cardiac defects.[473] Cerebrovascular manifestations are most commonly seen in Ehlers-Danlos type IV, which is caused by decreased or abnormal synthesis of type III procollagen, secondary to various defects of the COL3A1 gene on chromosome 2, and inherited in an autosomal dominant fashion with frequent new mutations.[473] Several reports have described asymptomatic parental mosaicism which has explained sibling recurrence in the absence of clinical parental disease.[474,475] Phenotypic features vary, but in type IV Ehlers-Danlos syndrome joint and skin changes are not usually prominent. Therefore, complications such as spontaneous rupture of the colon, gravid uterus, or arterial rupture commonly lead to presentation and diagnosis. Significant neurologic complications were seen in 19 of 201 patients in a recent review, most frequently spontaneous intracranial arterial dissection, intracranial aneurysm, and development of carotid-cavernous fistula, indicating that Ehlers-Danlos syndrome, despite its rarity, can be an important cause of stroke in young people.[476]

Intracranial aneurysm most commonly affects the internal carotid artery, typically in or adjacent to the cavernous sinus, and presents after spontaneous or posttraumatic rupture as subarachnoid hemorrhage or carotid-cavernous fistula.[473] Arterial dissection can affect any of the intracranial or extracranial arteries, and can present directly as a result of vessel occlusion or as embolic infarction.[477] Diagnosis is best performed by angiography, but both conventional angiography and intraarterial therapeutic intervention are rendered hazardous by excessive vascular fragility. Pathologic examination reveals fragmentation of the internal elastic lamina and microscopic ruptures of the media and adventitia associated with markedly reduced total collagen content.[478]

No treatment exists and life expectancy is significantly reduced, the mean age of death being in the early 30s for women, and slightly older for men.[479] Prenatal diagnosis is available.

Pseudoxanthoma Elasticum

This rare disease has been classified by Pope in four subtypes; two are inherited in an autosomal dominant and two in an autosomal recessive fashion.[480] Clinical features include yellowish cutaneous plaques affecting primarily the trunk and flexures, abnormal laxity of the skin, arterial occlusive disease resulting in symptomatic premature peripheral vascular disease, and hemorrhage from the gastrointestinal and other mucous membranes. Characteristic ocular lesions are seen in approximately 85% of patients with pseudoxanthoma elasticum, consisting of angioid streaks gray or red lines extending peripherally from the optic disc, and possibly resulting in visual loss secondary to macular degeneration or retinal hemorrhages.[481,482] Rare cerebrovascular complications in childhood typically manifest as stroke secondary to accelerated atherosclerosis and arterial calcification, with vessel occlusion[483] or occasional dissection,[484] or aneurysm formation[483] and acute rupture. No effective therapy exists, but avoidance of cardiovascular risk factors and genetic counseling is advisable.

References

1. Roach ES, Riela AR. Pediatric Cerebrovascular Disorders. 2nd ed. Armonk, NY: Futura Publishing Co, 1995.
2. Edwards MSB, Hoffman HJ. Cerebral Vascular Disease in Children and Adolescents. Baltimore, MD: Williams and Wilkins, 1989, 229–238.
3. Natowicz M, Kelley RI. Mendelian aetiologies of stroke. Ann Neurol 1987;22:175–192.
4. Schoenburg BS, Mellinger JF, Schoenberg DG. Cerebrovascular disease in infants and children: A study of incidence, clinical features, and survival. Neurology 1978;28: 763–768.
5. Eeg-Olofsson O, Ringheim Y. Stroke in children: Clinical characteristics and prognosis. Acta Neurol Scand 1983;72:391–395.
6. Giroud M, Lemesle M, Gouyon J-B, et al. Cerebrovascular disease in children under 16 years of age in the city of Dijon, France: A study of incidence and clinical features from 1985 to 1993. J Clin Epidemiol 1995;48 :1343–1348.
7. Coker SB, Beltran RS, Myers TF, Hmura L. Neonatal stroke: Description of patients and investigation into pathogenesis. Paediatr Neurol 1988;4:219–223.
8. Kappelle LJ, Willemse J, Ramos LMP, van Gijn J. Ischaemic stroke in the basal ganglia and internal capsule in childhood. Brain Dev 1989;11:283–292.
9. Pavlakis SG, Prohovnik I, Piomelli S, DeVivo DC. Neurologic complications of sickle cell disease. Adv Pediatr 1989;36:247–276.
10. Satoh S, Shirane R, Yoshimoto T. Clinical survey of ischemic cerebrovascular disease in children in a district in Japan. Stroke 1991;22:586–589.
11. Trescher WH. Ischemic stroke syndromes in childhood. Pediatr Ann 1992;21:374–383.
12. Brooks BS, El Gammal T, Adams RJ. MR imaging of moyamoya in neurofibromatosis. AJNR 1987;8:178.
13. Pavlakis SG, Phillips PC, DiMauro S. Mitochondrial myopathy, encephalopathy, lactic acidosis and strokelike episodes: A distinctive clinical syndrome. Ann Neurol 1984;16:481–488.
14. Grotta JC, Manner C, Pettigrew LC. Red blood cell disorders and stroke. Stroke 1986;17:811–817.
15. Jansen JN, Donker AJM, Luth WJ, Smit LME. Moyamoya disease associated with renovascular hypertension. Neuropediatrics 1990; 21:44–47.
16. Roach ES, Riela AR. Cerebral embolism. In Pediatric Cerebrovascular Disorders. 2nd ed. Armonk, NY: Futura Publishing Co, 1995, 51–68.
17. Nora JJ. Etiologic aspects of heart disease. In Pine JW (ed). Heart Disease in Infants, Children and Adolescents. Baltimore: Williams and Wilkens, 1989:15–23.
18. Vogt H. Zur diagnostik der tuberosen sklerose. Z Erforsch Behandl jugendl Schwachsinns 1908:2.
19. Gomez MR. Tuberous Sclerosis. 2nd ed. New York: Raven Press, 1988.
20. Kwiatkowski DJ, Armour J, Bale AE, et al. Report and abstracts of the Second International Workshop on the Human Chromosome 9 Mapping 1993. Cytogenet Cell Genet 1993;64:93–121.
21. Kandt RS, Haines JL, Smith M, et al. Linkage of an important gene locus for tuberous sclerosis to a chromosome 16 marker for polycystic kidney disease. Nature Genet 1992;2: 37–41.
22. Hilal SK, Solomon GE, Gold AP, Carter S. Primary cerebral occlusive disease in children. Parts I and II. Radiology 1971; 99:71–94.
23. Harvey FH, Alvord EC. Juvenile cerebral arteriosclerosis and other cerebral arteriopathies of childhood: Six autopsied cases. Acta Neurol Scand 1972;48:479–509.
24. Kandt RS, Gebarski SS, Goetting MG. Tuberous sclerosis with cardiogenic cerebral emboli: Magnetic resonance imaging. Neurology 1985;35:1223–1225.
25. Gomez MR. Strokes in tuberous sclerosis: Are rhabdomyomas a cause? Brain Dev 1989; 11:14–19.
26. Watson GH. Cardiac rhabdomyomas in tuberous sclerosis. Ann NY Acad Sci 1991; 615:50–57.
27. Wallace G, Smith HC, Watson GH, et al. Tuberous sclerosis presenting with fetal and

neonatal cardiac tumours. Arch Dis Child 1990;65:377–379.

28. Roach ES, Riela AR. Cerebrovascular syndromes. In Pediatric Cerebrovascular Disorders. 2nd ed. Armonk, NY: Futura Publishing Co, 1995, 35–50.

29. Gold AP, Chancellor YB, Gilles FH. Strokes in children. Part 2. Stroke 1973;4:1009.

30. McKusick VA. Mendelian Inheritance in Man. 9th ed. Baltimore: Johns Hopkins University Press, 1990.

31. Motulsky AG. Frequency of sickling disorders in US blacks. N Eng J Med 1973;288:31–33.

32. Bainbridge R, Higgs DR, Maude GH. Clinical presentation of homozygous sickle cell disease. J Paediatr 1985;106:881.

33. Powars DR, Wilson B, Imbus C. The natural history of stroke in sickle cell disease. Am J Med 1978;65:461–471.

34. Pavlakis SG, Bello J, Prohovnik I. Brain infarction in sickle cell anaemia: A magnetic resonance imaging correlates. Ann Neurol 1988;23:125–130.

35. Wood DH. Cerebrovascular complications of sickle cell anemia. Stroke 1978;9:73–75.

36. Schenk EA. Sickle cell trait and superior longitudinal sinus thrombosis. Ann Intern Med 1964;60:465–470.

37. Ohene-Frempong K. Stroke in sickle cell disease: Demographic, clinical and therapeutic considerations. Semin Haematol 1991;28: 202–208.

38. Balkaran B, Char G, Morris JS, et al. Stroke in a cohort of patients with homozygous sickle cell disease. J Paeds 1992;120:3 360–366.

39. Russell MO, Goldberg HI, Hodson A. Effect of transfusion therapy on arteriographic abnormalities and on recurrence of stroke in sickle cell disease. Blood 1984;63:162–169.

40. van Hof J, Ritchy K, Shaywitz BA. Intracranial haemorrhage in children with sickle cell disease. Am J Dis Child 1985;139: 1120–1123.

41. De Montalembert M, Beauvais P, Bachir D, et al. Cerebrovascular accidents in sickle cell disease: Risk factors and blood transfusion therapy. Eur J Paediatr 1993;152:201–204. 42. Kugler S, Anderson B, Cross D, et al. Abnormal cranial MRI scans in sickle-cell disease. Arch Neurol 1993;50:629–635.

43. Powars DR, Weiss JN, Chan LS, Schroeder WA. Is there a threshold level of fetal hemoglobin that ameliorates morbidity in sickle cell anemia? Blood 1984;63:921–926.

44. Adams RJ, Kutlar A, McKie V, et al. Alpha thalassemia and stroke risk in sickle cell anemia. Am J Hematol 1994;45:279–282.

45. El Gammal T, Adams RJ, Nichols FT. MR and CT investigation of cerebrovascular disease in sickle cell patients. AJNR 1986;7:1043–1049.

46. Roach ES, Riela AR. Hematologic and neoplastic disorders. In Pediatric Cerebrovascular Disorders. 2nd ed. Armonk, NY: Futura Publishing Co, 1995, 85–108.

47. Adams R, McKie V, Nichols F, et al. The use of transcranial ultrasonography to predict stroke in sickle cell disease. N Engl J Med 1992;326:605–610.

48. DeBaun MR, Glauser TA, Siegel M, et al. Noninvasive central nervous system imaging in sickle cell anemia. J Ped Hematol Oncol 1995;17:29–33.

49. Merkel KHH, Ginsberg PL, Parker JC, et al. Cerebrovascular disease in sickle cell anemia: A clinical, pathological and radiological correlation. Stroke 1978;9:45–52.

50. Rothman SM, Fulling KH, Nelson JS. Sickle cell anemia and central nervous system infarction: A neuropathologic study. Ann Neurol 1986;20: 684–690.

51. Hughes JG, Diggs LW, Gillespie CE. The involvement of the central nervous system in sickle cell anemia. J Pediatr 1940;17:166–181.

52. Huttenlocher PR, Moohr JW, Johns L, Brown FD. Cerebral blood flow in sickle cell cerebrovascular disease. Pediatrics 1984;73: 615–621.

53. Herold S, Brozovic M, Gibbs J, et al. Measurement of regional blood flow, blood volume and cerebral metabolism in patients with sickle cell disease using positron emission tomography. Stroke 1986;17:692–698.

54. Prohovnik I, Pavlakis SG, Piomelli S, et al. Cerebral hyperemia, stroke, and transfusion in sickle cell disease. Neurology 1989;39: 344–348.

55. Foulon I, Bachir D, Galacteros F, Maclouf J. Increased in vivo production of thromboxane in patients with sickle cell disease is accompanied by an impairment of platelet functions to the thromboxane A2 agonist U46619. Arterios Thromb 1993;13:421–426.

56. Peters M, Plaat BEC, Ten Cate H, et al. Enhanced thrombin generation in children with sickle cell disease. Throb Haemost 1994;71: 169–172.

57. Onyemelukwe GC, Jibril JB. Anti-thrombin III deficiency in Nigerian children with sickle cell disease. Tropic Geograph Med 1992; 44:37–41.

58. Helley D, Eldor A, Girot R, et al. Increased procoagulant activity of red blood cells from patients with homozygous sickle cell disease and beta-thalassemia. Thromb Haemost 1996;76:322–327.

59. Kurantsin-Mills J, Ofosu FA, Safa TK, et al. Plasma factor VII and thrombin-antithrombin III levels indicate increased tissue factor activity in sickle cell patients. Br J Haematol 1992;81:539–544.

60. Karayalcin G, Lanzkowsky P. Plasma protein C levels in children with sickle cell disease. Am J Ped Hem Onc 1989;11:320–323.

61. Posey YF, Shah D, Ulm JE, et al. Prenatal diagnosis of sickle cell anemia: Hemoglobin electrophoresis versus DNA analysis. Am J Clin Pathol 1989;92:347–351.

62. Pavlakis SG, Prohovnik I, Pionelli S, DeVivo DC. Neurologic complications of sickle cell disease. Adv Pediatr 1989;36:247–276.

63. Russell MO, Goldberg HI, Reis L. Transfusion therapy for cerebrovascular abnormalities in sickle cell disease. J Pediatr 1976;88:382–387.

64. Williams J, Goff JR, Anderson HR Jr, et al. Efficacy of transfusion therapy for one to two years in patients with sickle cell disease and cerebrovascular accidents. J Pediatr 1980;96:205–208.

65. Miller ST, Jensen D, Rao SP. Less intensive long-term transfusion therapy for sickle cell anemia and cerebrovascular accident. J Paeds 1992;120:54–57.

66. Cohen AR, Martin MB, Silber JH, et al. A modified transfusion program for prevention of stroke in sickle cell disease. Blood 1992;79:1657–1661.

67. Greenberg J, Ohene-Frempong K, Halus J. Trial of low doses of aspirin as prophylaxis in sickle cell disease. J Pediatr 1983;102:781.

68. Charache S, Dover GJ, Moore RD, et al. Hydroxyurea: Effects on hemoglobin F production in patients with sickle cell anemia. Blood 1992;79:2555–2565.

69. Rodgers GP, Dover GL, Uyesaka N, et al. Augmentation by erythropoietin of the fetal-hemoglobin response to hydroxyurea in sickle cell disease. N Eng J Med 1993;328:81–86.

70. Perrine SP, Ginder GD, Faller DV, et al. A short-term trial of butyrate to stimulate fetal-globin-gene expression in the beta-globin disorders. N Eng J Med 1993;328:81–86.

71. Kirkpatrick DV, Barrios NJ, Humbert JM. Bone marrow transplantation for sickle cell anemia. Semin Hematol 1991;28:240–243.

72. Walters MC, Sullivan KM, Bernaudin F, et al. Neurologic complications after allogeneic marrow transplantation for sickle cell anemia. Blood 1995;85:879–884.

73. Cooper MR, Toole JF. Sickle cell trait: Benign or malignant? Ann Intern Med 1972;77:997.

74. Greenberg J, Massey EW. Cerebral infarction in sickle trait. Ann Neurol 1985;18:354–355.

75. Portnoy BA, Herion JC. Neurologic manifestations in sickle-cell disease. Neurology 972;12:643–652.

76. Schenk EA. Sickle cell trait and superior longitudinal sinus thrombosis. Ann Intern Med 1964;60:465–470.

77. Dalal FY, Schmidt GB, Bennett EJ, Ramamurthy S. Sickle-cell trait. Br J Anaesth 1974;46:387–388.

78. Partington MD, Aronyk KE, Byrd SE. Sickle cell trait and stroke in children. Pediatr Neurosurg 1994;20:148–151.

79. Carson NA, Neill DW. Metabolic abnormalities detected in a survey of mentally backward individuals in Ireland. Arch Dis Child 1962;47:505–513.

80. Kluijtmans LAJ, Boers GHJ, Stevens EMB, et al. Defective cystathionine beta-synthase regulation by S-adenosylmethionine in a partially pyridoxine responsive homocystinuria patient. J Clin Invest 1996;98:285–289.

81. Mudd SH, Skovby F, Levy HL. The natural history of homocystinuria due to cystathionine beta-synthase deficiency. Am J Hum Genet 1985;37:1–31.

82. Mudd SH, Levy HL, Skovby F. Disorders of transsulfuration. In Scriver CR, et al (eds). The Metabolic Basis of Inherited Disease. New York: McGraw-Hill, 1989, 693–734.

83. Kraus JP. Molecular basis of phenotype expression in homocystinuria. J Inher Metab Dis 1994;17:383–390.

84. Caronna JJ. Homozygous homocystinuria presenting with superior sagittal sinus thrombosis. Neurol Alert 1992;10:68.

85. Cochran FB, Packman S. Homocystinuria presenting as sagittal sinus thrombosis. Eur Neurol 1992;32:1–3.

86. Swaiman KF. Aminoacidopathies resulting from deficiency of enzyme activity. In Swaiman KF (ed). Pediatric Neurology: Principles and Practice. St Louis: Mosby, 1989, 944–948.

87. Newman G, Mitchell JR. Homocystinuria presenting as multiple arterial occlusions. Q J Med 1984;210:251–258.

88. Harker LA, Ross R, Slichter SJ, Scott CR. Homocysteine-induced arteriosclerosis: The role of endothelial cell injury and platelet response in its genesis. J Clin Invest 1976;58:731–741.

89. Parthasaranthy S. Oxidation of low density lipoprotein by thiol compounds leads to its recognition by the acetyl LDL receptor. Biochem Biophysics Acta 1987;917:337–340.

90. Harpel PC, Chang VT, Borth W. Homocysteine and other sulfhydryl compounds enhance the binding of lipoprotein(a) to fibrin: A potential biochemical link between thrombosis, atherogenesis and sulfhydryl compound metabolism. Proc Natl Acad Sci USA 1992;89:10193–10197.

91. Graeber JE, Slott JH, Ulane RE, et al. Effect of homocysteine and homocystine on platelet and vascular arachadonic acid metabolism. Pediatr Res 1982;16:490–493.

92. Rodgers GM, Kane WH. Activation of endogenous factor V by homocysteine-induced vascular endothelial cell activator. J Clin Invest 1986;77:1909–1916.

93. Panganamala RV, Karpen CW, Merola AJ. Peroxide mediated effects of homocysteine on arterial prostacyclin synthesis. Prostaglandins Leukot Med 1986;22:349–356.

94. McCully KS, Carvalho AC. Homocysteine thiolactone, N-homocysteine thiolactonyl retinamide and platelet aggregation. Res Commun Chem Pathol Pharmacol 1987;56:349–360.

95. Rodgers GM, Conn MT. Homocysteine, an atherogenic stimulus, reduces protein C activation by arterial and venous endothelial cells. Blood 1990;75:895–901.

96. Mandel H, Brenner B, Berant M, et al. Coexistence of hereditary homocystinuria and factor V Leiden-effect on thrombosis. N Engl J Med 1996;334:363–368.

97. D'Angelo A, Fermo I, D'Angelo SV. Thrombophilia, homocystinuria and mutation of the factor V gene. N Eng J Med 1996;335:289.

98. Wicherink-Bol HF, Boers GH, Drayer JI, Rosenbusch G. Angiographic findings in homocystinuria. Cardiovasc Intervent Radiol 1983;6:125.

99. Cohn RM, Roth KS. Pediatric Clinical Chemistry. Philadelphia: WB Saunders, 1984, 657–680.

100. Fleisher LD. Investigation of cystathionine synthase in cultured fetal cells and the prenatal determination of genetic status. J Pediatr 1974;85:677.

101. Christensen E, Brandt NJ. Prenatal diagnosis of 5,10 methylenetetrahydrofolate reductase deficiency. N Eng J Med 1985: 313:50.

102. Perry TL, Hanson S, Love DL. Serum carnosinase deficiency in carnosinaemia. Lancet 1968;1:1229.

103. Wilcken DE, Dudman NP, Tyrrel PA. Homocystinuria due to cystathionine beta-synthase deficiency: The effect of betaine treatment in pyridoxine-responsive patients. Metabolism 1985;34:1115.

104. Dudman NPB, Wilcken DEL, Wang J, et al. Disordered methionine/homocysteine metabolism in premature vascular disease: Its occurrence, cofactor therapy, and enzymology. Arterioscler Thromb 1993;13:1253–1260.

105. Wilcken DE, Wilcken B. The pathogenesis of coronary artery disease: A possible role for methionine metabolism. J Clin Invest 1976;57: 1079–1082.

106. Dudman NPB, Guo X-W, Gordon RB, et al. Human homocysteine catabolism: three major pathways and their relevance to development of arterial occlusive disease. J Nutr 1996;126:1295S-1300S.

107. Dudman NPB, Kim MH, Wang J, et al. Thermolabile methylenetetrahydrofolate reductase causes mild hyperhomocyst(e)inemia in patients with vascular disease. Am J Hum Genet 1993;53(suppl):899. Abstract.

108. Boers GHJ, Smals AGH, Tribels FJM. Heterozygosity for homocystinuria in premature peripheral and cerebral occlusive disease. N Eng J Med 1985;313:709–715.

109. Clarke R, Daly L, Robinson K. Hyperhomocysteinemia: An independent risk factor for vascular disease. N Eng J Med 1991;324:1149–1155.

110. Brattstrom L, Lindgren A, Israelsson B, et al. Hyperhomocystinaemia in stroke: Prevalence, cause, and relationships to type of stroke and stroke risk factors. Eur J Clin Invest 1992;22:214–221.

111. Desnick RJ, Bishop DF. Fabry disease and Schindler disease. In Scriver CR et al (eds). The Metabolic Basis of Inherited Disease. New York: McGraw-Hill, 1989, 1751–1796

112. Bishop DF, Kornreich R, Desnick RJ. Structural organisation of the alpha-galactosidase A gene: Further evidence for the absence of a 3' untranslated region. Proc Natl Acad Sci USA 1988;85:3903.

113. Eng CM, Resnick-Silverman LA, Niehaus DJ, et al. Nature and frequency of mutations in the alpha-galactosidase A gene that cause Fabry disease. Am J Hum Genet 1993;53:1186–1187.

114. Sakuraba H, Eng CM, Desnick RJ, Bishop DF. Invariant exon skipping in the human alpha-galactosidase A gene pre-mRNA:Ag+1 to t substitution in a 5''-splice site causing Fabry disease. Genomics 1992;12:643–650.

115. Franceschetti AT. La cornea verticallata (Gruber) et ses relations avec la maladie de Fabry (Angiokeratoma corporis diffusum). Opthalmologica 1968;74:232.

116. Scher NA, Latson RD, Desnick RJ. The ocular manifestations in Fabry's disease. Arch Ophthalmol 1979;74:760.

117. Wilson SK, Klionosky BL, Rhamy RK. A new etiology of priapism: Fabry's disease. J Urol 1973;109:646.

118. Nagao Y, Nakashima H, Fukuhara Y. Hypertrophic cardiomyopathy in late-onset variant of Fabry disease with high residual activity of alpha-galactosidase A. Clin Genet 1991;39:233–237.

119. Zeluff GW, Caskey CT, Jackson D. Heart attack or stroke in a young man? Think Fabry's disease. Heart and Lung 1978;7:1056.

120. Duperrat B. L'angiokeratome diffus de Fabry. Presse Medicale-Paris 1959;67:1814.

121. Bethune JE, Landrigan PL, Chipman CD. Angiokeratoma corporis diffusum univer-

sale (Fabry's disease) in two brothers. N Eng J Med 1961;264:1280–1285.

122. Lou HOC, Reske-Nielson E. The central nervous system in Fabry's disease. Arch Neurol 1971;25:351.

123. Kahn P. Anderson-Fabry disease: A histopathologic study of three cases with observations on the mechanism of production of pain. J Neurol Neurosurg Psychiatry 1973;36:1053.

124. Maisey DN, Cosh JA. Basilar artery aneurysm and Anderson-Fabry disease. J Neurol Neurosurg Psych 1980;43:85–87.

125. Wise D, Wallace HJ, Jelliniek EH. Angiokeratoma corporis diffusum: A clinical stdy of eight affected families. Q J Med 1962;31:177.

126. Mitsias P, Levine SR. Neurological complications of Fabry's disease. Ann Neurol 1996;40:8–17.

127. Uyama E, Ueno N, Uchino M, et al. Headache associated with aseptic meningeal reaction as clinical onset of Fabry's disease. Headache 1995;35:498–501.

128. Grewal RP. Stroke in Fabry's disease. J Neurol 1994;241:153–156.

129. Sakuraba H, Yanagawa Y, Igarasha T, et al. Cardiovascular manifestations in Fabry's disease: A high incidence of mitral valve prolapse in hemizygotes and heterozygotes. Clin Genet 1986;29:276–283.

130. Colucci WS, Lorell BH, Schoen FJ, et al. Hypertrophic obstructive cardiomyopathy due to Fabry's disease. N Engl J Med 1982;307:926–928.

131. DeVeber GA, Scwarting GA, Kolodny EH, Kowall NW. Fabry disease: Immunocytochemical characterization of neuronal involvement. Ann Neurol 1992;31:409–415.

132. Sung JH, Hayano M, Mastri AR, Desnick RJ. Neuropathology and neural glycosphingolipid deposition in Fabry's disease. Exerpta Medica International Congress Series 1975;1:267.

133. Adams RD, Lyon G. Neurology of Hereditary Metabolic Diseases of Children. Washington: Hemisphere Publishing Corp, 1982, 272–275.

134. Lockman LA, Hunninhake DB, Krivit W, Desnick RJ. Relief of pain in Fabry's disease by diphenylhydantoin. Neurology 1973;23:871.

135. Lenoir G, Rivron M, Gubler MC. La maladie de Fabry. Archives Francaises de Pediatrie 1977;34:704–716.

136. Funderburk SJ, Philippart M, Dale G. Priapism after phenoxybenzamine in a patient with Fabry's disease. N Eng J Med 1974;290:646.

137. Inagaki M, Ohno K, Ohta S, Sakuraba H. Relief of chronic burning pain in Fabry disease with neurotropin. Pediatric Neurology 1990;6:211–213.

138. Clement M, McGonigle RJS, Monkhouse PM, et al. Renal transplantation in Anderson-Fabry disease. In Desnick RJ (ed). Enzyme Therapy in Genetic Diseases. 2nd ed. New York: AR Liss, 1980, 393.

139. Desnick RJ, Dean KJ, Grabowski GA. Enzyme Therapy XVII: Metabolic and immunologic evaluation of alpha-galactosidase A gene and its application to gene diagnosis of heterozygotes. Ann Neurol 1991;29:560–564.

140. Ishii S, Sakuraba H, Shimmoto M. Fabry disease: Detection of 13-bp deletion in alpha-galactosidase A gene and its application to gene diagnosis of heterozygotes. Ann Neurol 1991;29:560–564.

141. Kirkilionis AJ, Riddell DC, Spence MW, Fenwick RG. Fabry disease in a large Nova Scotia kindred: Carrier detection using leucocyte alpha-galactosidase activity and an Ncol polymorphism detected by an alpha-galactosidase cDNA clone. J Med Genet 1991;28:232–240.

142. Sourander P, Walinder J. Hereditary multi-infarct dementia: Morphological and clinical studies of a new disease. Acta Neuropathol (Berl). 1977;39:24–254.

143. Joutel A, Vahedi K, Corpechat C, et al. Strong clustering and stereotyped nature of *Notch3* mutations in CADASIL patients. Lancet 1997;350:1511–1515.

144. Dichgans M, Mayer M, Muller-Myhsok B, et al. Identification of a key recombinant narrows the CADASIL gene region to 8cM and argues against allelism of CADASIL and familial hemiplegic migraine. Genomics 1996;32:151–154.

145. Verin M, Rolland Y, Landgraf F, et al. New phenotype of the cerebral autosomal dominant arteriopathy mapped to chromosome 19: Migraine as the prominent clinical feature. J Neurol Neurosurg Psych 1995;59:579–585.

146. Bousser M-G, Tournier-Lasserve E. Summary of the proceedings of the first international workshop on CADASIL. Paris May 19–21 1993. Stroke 1994;25:704–707.

147. Baudrimont M, Dubas F, Joutel A, et al. Autosomal dominant leukoencephalopathy and subcortical ischaemic stroke: A clinico-pathological study. Stroke 1993;24:122–125.

148. Daniels SR, Bates S, Lukin RR, et al. Cerebral arteriopathy (arteriosclerosis) and ischemic childhood stroke. Stroke 1982;13:360–365.

149. Gleuck CJ, Daniels SR, Bates S, et al. Pediatric victims of unexplained stroke and their families: Familial lipid and lipoprotein abnormalities. Pediatrics 1982;69:308–316.

150. Listernick R, Charrow J. Neurofibromatosis type 1 in childhood. J Pediatr 1990;116:845–853.

151. Gutmann DH, Collins FS. The neurofibromatosis type 1 gene and its protein product, neurofibromin. Neuron 1993;10:335–343.

152. Slayer WR, Slayer DC. The vascular lesions of neurofibromatosis. Angiology 1974;25:501–519.

153. National Institutes of Health Consensus Development Conference. Neurofibromatosis. Conference Statement. Arch Neurol 1988;45: 575–578.

154. Hornstein L, Borchers D. Stroke in an infant prior to the development of manifestations of neurofibromatosis. Neurofibromatosis 1989;2:116–120.

155. Tomsick TA, Lukin RR, Chambers AA, Benton C. Neurofibromatosis and intracranial arterial occlusive disease. Neuroradiology 1976;11:229–234.

156. Gorelick MH, Powell CM, Rosenbaum KN, et al. Progressive occlusive cerebrovascular disease in a patient with neurofibromatosis type 1. Clin Pediatr 1992;31:313–315.

157. Lehrnbecher T, Gassel AM, Rauh V, et al. Neurofibromatosis presenting as a severe vasculopathy. Eur J Pediatr 1994;153: 107–109.

158. Wada K, Ohtsuka K, Terayama K, et al. Neurofibromatosis with spinal paralysis due to arteriovenous fistula. Arch Orthop Trauma Surg 1989;108:322–328.

159. Muhonen MG, Godersky JC, Van Gilder JC. Cerebral aneurysms associated with neurofibromatosis. Surg Neurol 1991;36: 470–475.

160. Smith DW. Progeria syndrome. In Smith DW (ed). Recognizable Patterns of Human Malformation. Philadelphia: WB Saunders, 1982, 112–113.

161. Lipman JM, Applegate-Stevens A, Soyka LA, Hart RW. Cell-cycle defect of DNA repair in progeria skin fibroblasts. Mutation Res 1989;219:273–281.

162. Sweeney KJ, Weiss AS. Hyaluronic acid in progeria and the aged phenotype? Gerontology 1992;38:139–152.

163. Beavan LA, Quentin-Hoffmann E, Schonherr E, et al. Deficient expression of decorin in infantile progeroid patients. J Biol Chem 1993;268:9856–9862.

164. Naganuma Y, Konishi T, Hongou K. A case of progeria syndrome with cerebral infarction. No To Hattatsu 1990;22:71–76.

165. Wagle WA, Haller JS, Cousins JP. Cerebral infarction in progeria. Pediatr Neurol 1992;8:476–477.

166. Suzuki J, Takaku A. Cerebrovascular 'moyamoya' disease. Arch Neurol 1969;20:288.

167. Nishimoto A. Moyamoya disease. Neurol Med Chir 1979;19:221–228.

168. Hoffman HJ, Griebel RW. Moyamoya syndrome in children. In Edwards MSB, Hoffman HJ (eds). Cerebral Vascular Disease in Children and Adolescents. Baltimore: Williams and Wilkins 1989:229–238.

169. Kitahara T, Okumura K, Semba A. Genetic and immunologic analysis on moyamoya. J Neurol Neurosurg Psychiatry 1982;1048–1052.

170. Kitahara T, Ariga N, Yamamura A. Familial occurrence of moyamoya: A report of three Japanese families. J Neurol Neurosurg Psychiatry 1979;42:208–214.

171. Aoyagi M, Ogami K, Matsushima Y, et al. Human leukocyte antigen in patients with moyamoya disease. Stroke 1995;26:415–417.

172. Baltaxe HA, Bloch S, Mooring PK. Coarctation of the thoracic aorta associated with cerebral arterial occlusive disease. AJNR 1982;3:577–580.

173. Kawakita Y, Abe K, Miyata Y, Horikoshi S. Spontaneous thrombosis of the internal carotid artery in children. Folia Psychiatry Neurol Jpn 1965;19:245–255.

174. Halonen H, Halonen V, Donner M, et al. Occlusive disease of intracranial main arteries with collateral networks in children. Neuropaediatrie 1973;4:187–206.

175. Kapusta L, Daniels O, Renier WO. Moyamoya syndrome and primary pulmonary hypertension in childhood. Neuropediatrics 1990;21:162–163.

176. Provost TT, Moses H, Morris EL. Cerebral vasculopathy associated with collateralisation resembling moyamoya phenomenon and with anti-Ro/SS-A and anti-La/SS-B antibodies. Arthritis Rheum 1991;34:1052–1055.

177. Garaizer C, Prats JM, Zuazo E. Chorea of acute onset due to moyamoya disease. Acta Neuropediatr 1994;1:59–63.

178. Yasumoto K, Hashimoto T, Miyazaki M, Kuroda Y. Recurrent torticollis as a presentation of moyamoya disease. J Child Neurol 1993;8:187–188.

179. Handa J, Handa H. Progressive cerebral arterial occlusive disease: Analysis of 27 cases. Neuroradiology 1972;3:119.

180. Leone RG, Schatzki SC, Wolpow ER. Neurofibromatosis with extensive intracranial arterial occlusive disease. AJNR 1982;3:572.

181. Taveras JM. Multiple progressive intracranial occlusions: A syndrome of children and young adults. Am J Roentgenol Radium Ther Nucl Med 1969;106:235–268.

182. Kurokawa T, Chen YJ, Tomita S. Cerebrovascular occlusive disease with and without the moyamoya vascular network in children. Neuropediatrics 1985;16:29.

183. Houkin K, Aoki T, Takahashi A, Abe H. Diagnosis of moyamoya disease with magnetic resonance angiography. Stroke 1994; 25:2159–2164.

184. Kwak R, Ito S, Yamamoto N, Kadoya S. Significance of intracranial aneurysms associated with moyamoya disease. Part 1. Differences between intracranial aneurysms associated with moyamoya disease and usual saccular aneurysms -review of the literature. Neurol Med Chir 1984;24:97–103.

185. Schmit BP, Burrows PE, Kuban K, et al. Acquired cerebral arteriovenous malformation in a child with moyamoya disease. J Neurosurg 1996;84:677–680.

186. Kodama N, Aoki Y, Hiraga H. Electroencephalographic findings in children with moyamoya disease. Arch Neurol 1979;36: 16–19.

187. Yamashita M, Oka T, Tanaka K. Cervicocephalic arterial thrombi and thromboemboli in moyamoya disease-possible correlation with progressive intimal thickening in the intracranial major arteries. Stroke 1984; 15:264.

188. Matsuda J, Ogata J, Yutani C. Smooth muscle cell proliferation and localization of macrophages and T cells in the occlusive intracranial major arteries in moyamoya disease. Stroke 1993;24:1960–1967.

189. Takahashi A, Sawamura Y, Houkin K, et al. The cerebrospinal fluid in patients with moyamoya disease (spontaneous occlusion of the circle of Willis) contains high level of basic fibroblast growth factor. Neurosci Lett 1993;160:214–216.

190. Scott RM. Surgical treatment of moyamoya disease in children. Concepts Pediatr Neurosurg 1985;6:198–221.

191. Hosain SA, Hughes JT, Forem SL, et al. Use of a calcium channel blocker (nicardipine HCl) in the treatment of childhood moyamoya disease. J Child Neurol 1994;9:378–380.

192. Fujiwara J, Nakahara S, Enomoto T, et al. The effectiveness of O2 administration for transient ischaemic attacks in moyamoya disease in children. Childs Nerv Syst 1996;12:69–75.

193. Matsushima Y, Aoyagi M, Kuomo Y, et al. Effects of encephalo-duro-arterio-syangiosis on childhood moyamoya patients: Swift disappearance of ischemic attacks and maintenance of mental capacity. Neurol Med Chir 1991;31:708–714.

194. Matsushima T, Inoue T, Suzuki SO, et al. Surgical treatment of moyamoya disease in pediatric patients: Comparison between the results of indirect and direct revascularisation procedures. Neurosurgery 1992;31: 401–405.

195. Ezura M, Yoshimoto T, Fujiwara S, et al. Clinical and angiographic follow-up of childhood-onset moyamoya disease. Childs Nerv Syst 1995;11:591–594.

196. Sandok BA. Fibromuscular dysplasia of the cephalic arterial system. In Toole JF (ed). Vascular Diseases, Part III. Elsevier Science Publications, 1989;283–292.

197. So EL, Toole JF, Dalal P, Moody DM. Cephalic fibromuscular dysplasia in 32 patients: Clinical findings and radiological features. Arch Neurol 1981;38:619–622.

198. Shields WD, Ziter FA, Osborn AG. Fibromuscular dysplasia as a cause of stroke in infancy and childhood. Pediatrics 1977;59: 899–901.

199. Perez-Higueras A, Alvarez-Ruiz F, Martinez-Bermejo A. Cerebellar infarction from fibromuscular dysplasia and dissecting aneurysm of the vertebral artery: Report of a child. Stroke 1988;19:521–524.

200. Vles JS, Hendriks JJ, Lodder J, Janevski B. Multiple vertebro-basilar infarctions from fibromuscular dysplasia related dissecting aneurysm of the vertebral artery in a child. Neuropediatrics 1990;21:104–105.

201. Rushton AR. The genetics of fibromuscular dysplasia. Arch Intern Med 1980;140:233–236.

202. Komori H, Matsuishi T, Abe T, et al. Turner syndrome and occlusion of the internal carotid artery. J Child Neurol 1993;8:412–415.

203. Miyake S, Arai J, Hayashi M. A case of Turner syndrome with renal hypertension: Fibromuscular dysplasia in cerebro-renal arteries. No To Hattatsu 1985;17:438–442.

204. Qvarfordt PG, Ehrenfield WK. Spontaneous dissection of the extracranial internal carotid artery. In Cerebral Vascular Disease in Children and Adolescents. Baltimore: Williams and Wilkins, 1989, 203–213.

205. George B, Mourier KL, Gelbert F. Vascular abnormalities in the neck associated with intracranial aneurysms. Neurosurgery 1989; 24:499–508.

206. Zimmerman R, Leeds NE, Naidich TP. Carotid-cavernous fistula associated with intracranial fibromuscular dysplasia. Radiology 1977;122:725–726.

207. Roach ES, Riela AR. Vasculopathies of the central nervous system. In Pediatric Cerebrovascular Disorders. 2nd ed. Armonk, NY: Futura Publishing Co, 1995:163–180.

208. Rosemberg S, Lopes MBS, Sotto MN, Graudenz MS. Childhood Degos disease with prominent neurological symptoms: Report of a clinicopathological case. J Child Neurol 1987;2:42–46.

209. Hall-Smith P. Malignant atrophic papulosis (Degos disease): two cases occurring in the same family. Br J Dermatol 1969;81:817–822.

210. Moulin G, Barrut D, Franc MP, Pierson A.

Papulose atrophiante de Degos familiale (mer-fille). Ann Dermatol Venereol 1984; 111:149–155.

211. Petit WA, Soso MJ, Higman H. Degos disease: Neurologic complications and cerebral angiography. Neurology 1982;32:1305–1309.

212. Sotrel A, Lacson AG, Huff KR. Childhood Kohlmeier-Degos disease with atypical skin lesions. Neurology 1983;33:1146–1151.

213. Subbiah P, Wijdicks E, Muenter M, et al. Skin lesion with a fatal neurologic outcome (Degos disease). Neurology 1996;46:636–640.

214. Englert HJ, Hawkes CH, Boey ML, et al. Degos disease: Association with anticardiolipin antibodies and the lupus anticoagulant. Br Med J 1984;289:576.

215. Stahl D, Thomsen K, Hou-Jensen K. Malignant atrophic papulosis treatment with aspirin and dipyrimadole. Arch Dermatol 1978;114:1687.

216. Pallesen RM, Rasmussen NR. Malignant atrophic papulosis: Degos' syndrome. Acta Chir Scand 1979;145:279.

217. Schievink WI, Mokri B, Piepgras DG, Kuiper JD. Recurrent spontaneous arterial dissections: Risk in familial versus nonfamilial disease. Stroke 1996;27:622–624.

218. Mayer SA, Rubin BS, Starman BJ, Byers PH. Spontaneous multivessel cervical artery dissection in a patient with a substitution of alanine for glycine (G13A) in the alpha 1 (l) chain of type 1 collagen. Neurology 1997;47: 552–556.

219. Nicod P, Bloor C, Godfrey M, et al. Familial aortic dissecting aneurysm. J Am Coll Cardiol 1985;55:236–238.

220. Mokri B, Roche PC, O'Brien JF, et al. Abnormalities of elastin in spontaneous internal carotid and vertebral artery dissections. Ann Neurol 1994;36:263. Abstract.

221. Schievink WI, Michels VV, Mokri B, et al. A familial syndrome of arterial dissections with lentiginosis. N Engl J Med 1995;332: 576–579.

222. Schievink WI, Mokri B, Michels VV, Piepgras DG. Familial association of intracranial aneurysms and cervical artery dissections. Stroke 1991;22:1426–1430.

223. Majamaa K, Portimojarvi H, Sotaniemi KA, Myllyla VV. Familial aggregation of cervical artery dissection and cerebral aneurysm. Stroke 1994;25:1704–1705.

224. Schievink WI, Prakash UBS, Piepgras DG, Mokiri B. Alpha-1-antitrypsin deficiency in intracranial aneurysms and cervical artery dissection. Lancet 1994;343:452–453.

225. David M, Andrew M. Venous thromboembolic complications in children: A critical review of the literature. J Pediatr 1993;123: 337–346.

226. Martinez HR, Rangel-Guerra A, Marfil LJ. Ischemic stroke due to deficiency of coagulation inhibitors. Stroke 1993;24:19–25.

227. Zoller B, Berntsdotter A, Garcia de Frutos P, Dahlback B. Resistance to activated protein C as an additional genetic risk factor in hereditary deficiency of protein S. Blood 1995;85:3518–3523.

228. Koeleman BPC, Reitsma PH, Allaart RC, Bettina RM. Activated protein C resistance as an additional risk factor for thrombosis in protein-C deficient families. Blood 1994;84: 1031–1035.

229. Pabinger I, Mustafa S, Rintelen C, et al. Factor V (FV) Leiden mutation (APC-resistance) increases the risk for venous thromboembolism in patients with a gene defect of the protein C (PC) or protein S (PS) gene. Thromb Haemost 1995;73:1361. Abstract.

230. Radtke KP, Lane DA, Greengard JS, et al. Combined hereditary APC resistance and ATIII type II deficiency associated with venous thrombosis. Thromb Haemost 1995;73: 1377. Abstract.

231. Tollefsen DM. Laboratory diagnosis of antithrombin and heparin cofactor II deficiency. Semin Thromb Haemost 1009;16: 162.

232. Bertina RM, Van der Linden IK, Engesser L, et al. Hereditary heparin cofactor II deficiency and the risk of development of thrombosis. Thromb Haemost 1987;56:196.

233. Petaja M, Rasi V, Myllala G. Familial hypofibrinolysis and venous thrombosis. Br J Haematol 1989;71:393.

234. Bick RL, Pegram M. Syndromes of hypercoagulabilty and thrombosis: A review. Semin Thromb and Hemost 1994;20:109–132.

235. Lammle B, Wuillemin W, Huber I, et al. Thromboembolism and bleeding tendency in congenital factor XII deficiency: A study on 74 subjects from 14 Swiss families. Thromb Haemost 1991;65:117.

236. Bithell TC. Hereditary dysfibrinogenemia. Clin Chem 1985;31:509.

237. Bauer KA. Management of patients with hereditary defects predisposing to thrombosis including pregnant women. Thromb Haemost 1995;74:94–100.

238. Andrew M. Developmental hemostasis: Relevance to thromboembolic complications in pediatric patients. Thromb Haemost 1995;74:415–425.

239. Patracchini P, Aiello V, Palazzi P, et al. Sublocalization of the human protein C gene on chromosome 2q13-q14. Human Genet 1989;81:191–192.

240. Romeo G, Hassan HJ, Staempfli S, et al. Hereditary thrombophilia: Identification of nonsense and missense mutations in the

protein C gene. Proc Natl Acad Sci USA 1987;84:2829–2832.

241. Bovill EG, Tomczak JA, Grant B, et al. Protein C Vermont: Symptomatic type II protein C deficiency associated with two GLA domain mutations. Blood 1992;79:1456–1465.

242. Sugahar Y, Miura O, Yuen P, Aoki N. Protein C deficiency Hong Kong 1 and 2: Hereditary protein C deficiency caused by two mutant alleles, a 5-nucleotide deletion and a missense mutation. Blood 1992;80:126–133.

243. Aiach M, Gandrille S, Emmerich J. A review of mutations causing deficiencies of antithrombin, protein C and protein S. Thromb Haemost 1995;74:81–89.

244. Strickland DK, Kessler CM. Biochemical and functional properties of protein C and S. Clin Chim Acta 1987;170:1–24.

245. Nardi M, Karpatkin M. Prothrombin and protein C in early childhood: Normal levels are not achieved until the fourth year of life. J Pediatr 1986;109:843–845.

246. Miletich JP, Sherman L, Broze GJ Jr. Absence of thrombosis in subjects with heterozygous protein C deficiency. N Eng J Med 1987;317:991–996.

247. Camerlingo M, Finazzi G, Casto L, et al. Inherited protein C deficiency and non-hemorrhagic arterial stroke in young adults. Neurology 1991;41:1371–1373.

248. Van Kuijck MAP, Rotteveel JJ, van Oostrom CG, Novakova I. Neurological complications in children with protein C deficiency. Neuropediatrics 1994;25:16–19.

249. Uysal S, Anlar B, Altay C, Kirazli S. Role of protein C in childhood cerebrovascular accidents. Eur J Pediatr 1989;149:216–218.

250. Israels SJ, Seshia SS. Childhood stroke associated with Protein C or S deficiency. J Pediatr 1987;111:562–564.

251. Camerlingo M, Finazzi G, Casto L. Inherited protein C deficiency and nonhemorrhagic arterial stroke in young adults. Neurology 1991;41:1371–1373.

252. Kohler J, Kasper J, Witt I, von Reutern GM. Ischaemic stroke due to protein C deficiency. Stroke 1990;21:1077–1080.

253. Broekmans AW, Bertina RM, Loelinger EA. Protein C and the development of skin necrosis during anticoagulant therapy. Thromb Haemost 1983;49:244.

254. Comp PC. Hereditary disorders predisposing to thrombosis. Prog Hemost Thromb 1986;8:71.

255. Tarras S, Gadia C, Meister L. Homozygous protein C deficiency in a newborn: Clinicopathologic correlation. Arch Neurol 1988;45: 214–216.

256. Marlar RA, Montgomery RR, Broekmans AW. Diagnosis and treatment of homozy-

gous protein C deficiency: Report of the Working Party on homozygous protein C deficiency of the Subcommittee on Protein C and Protein S, International Committee on Haemostasis and Thrombosis. J Pediatr 1989;114:528–534.

257. Marciniak E, Wilson HD, Marlar RA. Neonatal purpura fulminans: A genetic disorder related to the absence of protein C in blood. Blood 1985;65:15–20.

258. Dreyfus M, Magny JF, Bridey F, et al. Treatment of homozygous protein C deficiency and neonatal purpura fulminans with a purified protein concentrate. N Eng J Med 1991;325:1565–1568.

259. Hartman KR, Manco-Johnson M, Rawlings JS, et al. Homozygous protein C deficiency: early treatment with warfarin. Am J Pediatr Hematol Oncol 1989;11:395–401.

260. Peters C, Casella JF, Marlar RA, et al. Homozygous protein C deficiency: Observations on the nature of the metabolic abnormality and the effectiveness of warfarin therapy. Pediatrics 1988;81:272–276.

261. Pabinger I, Kyrle PA, Heistinger M, et al. The risk of thromboembolism in asymptomatic patients with protein C and protein S deficiency: A prospective cohort study. Thromb Haemost 1994;71:441–445.

262. Ploos van Amstel HK, van der Zanden AL, Bakker E, et al. Two genes homologous with human protein S cDNA are located on chromosome 3. Thromb Haemost 1987;58: 982–987.

263. Schmidel DK, Tatro AV, Phelps LG, et al. Organisation of the human protein S genes. Biochemistry 1990;29:7845–7852.

264. Ploos van Amstel HK, Huisman MV, Reitsma PH, et al. Partial protein S gene deletion in a family with hereditary thrombophilia. Blood 1989;73:479.

265. Schmidel DK, Nelson RM, Broxson EH Jr, et al. A 5.3 kb deletion including exon XIII of the protein S alpha gene occurs in two protein S-deficient families. Blood 1991;77:551–559.

266. Comp PC. Laboratory evaluation of protein S status. Semin Thromb Hemost 1990;16: 177–181.

267. Prats JM, Garaizar C, Zuazo E, et al. Superior sagittal sinus thrombosis in a child with protein S deficiency. Neurology 1992;42: 2303–2305.

268. Zoller B, de Frutos PG, Dahlback B. Evaluation of the relationship between protein S and C4b-binding protein isoforms in hereditary protein S deficiency demonstrating type I and type III deficiencies to be phenotypic variants of the same genetic disease. Blood 1995;12:3524–3531.

269. Boerger LM, Morris PC, Thurnau GR. Oral

contraceptives and gender affect protein S status. Blood 1987;69:692–694.

270. Koelman JHTM, Bakker CM, Plandsoen WCG. Hereditary protein S deficiency presenting with cerebral sinus thrombosis in an adolescent girl. J Neurol 1992;239:105–106.

271. Faioni EM, Franchi F, Asti D, et al. Resistance to activated protein C in nine thrombophilic families: Interference in a protein S functional assay. Thromb Haemost 1993;70:1067.

272. Sacco RL, Owen J, Mohr JP. Free protein S deficiency: Clinical manifestations. Ann Intern Med 1987;106:677–682.

273. Engesser L, Broekmans AW, Briet E, et al. Hereditary protein S deficiency: Clinical manifestations. Ann Intern Med 1987;106:677–682.

274. Mahasandana C, Suvatte V, Marlar RA. Neonatal purpura fulminans associated with homozygous protein S deficiency. Lancet 1990;355:61–62.

275. Abildgaard U. Antithrombin and related inhibitors of anticoagulation. In Poller L (ed). Recent Advances in Blood Coagulation. Vol 3. Edinburgh: Churchill Livingstone, 1981, 151–173.

276. Hirsh J, Piovella F, Pini M. Congenital antithrombin III deficiency. Incidence and clinical features. Am J Med 1989;87 supp 3B:34S-38S.

277. Lane DA, Kunz G, Olds RJ, Thein SL. Molecular genetics of antithrombin deficiency. Blood Reviews 1996;10:59–74.

278. Lane DA, Olds RR, Thein SL. Antithrombin III and its deficiency states. Blood Coag Fibrinolysis 1992;3:315–341.

279. Thaler E, Lechner K. Antithrombin III deficiency and thromboembolism. Clin Haematol 1981;10:369–390.

280. Ambruso DR, Jacobson LJ, Hathaway WE. Inherited antithrombin III deficiency and cerebral thrombosis in a child. Pediatrics 1980;65;125–131.

281. Vomberg PP, Breederveld C. Cerebral thromboembolism due to antithrombin III deficiency in two children. Neuropediatrics 1987;18:42–44.

282. Johnson EJ, Prentice CR, Parapia LA. Premature arterial disease associated with familial antithrombin III deficiency. Thromb Haemost 1990;63:13–15.

283. Chowdhury V, Lane DA, Auberger K, et al. Homozygous antithrombin III deficiency: Report of two new cases (99LeutoPhe) associated with arterial and venous thrombosis. Thromb Haemost 1994;72:166–329.

284. Tengborn L, Frohm B, Nilsson L-E. Catabolism and coagulation properties of heat-treated antithrombin III concentrate. Thromb Res 1987;48:701–711.

285. Catrine A, Nilsson T. Antithrombin in infancy and childhood. Acta Paediatr Scan 1975;64:624–628.

286. Dahlback B, Hildebrand B. Inherited resistance to activated protein C is corrected by anticoagulant cofactor activity found to be a property of factor V. Proc Natl Acad Sci USA 1993;90:1004–1008.

287. Griffin JH, Evatt BL, Wideman C, Fernandez JA. Anticoagulant protein C pathway defective in a majority of thrombophilic patients. Blood 1993;82:1989–1999.

288. Marlar RA, Kleiss AJ, Griffin JH. Human protein C: Inactivation of factors V and VIII in plasma by the activated molecule. Ann N Y Acad Sci 1981;370:303–310.

289. Bertina RM, van Wijngaarden A, Reinalda-Poot J, et al. Determination of plasma protein S: The plasma cofactor of activated protein C. Thromb Haemost 1985;53:268–272.

290. Shen L, Dahlback B. Factor V and protein S as synergistic cofactors to activated protein C in degradation of factor VIIIa. J Biol Chem 1994;269(18):735–738.

291. Zoller B, Svensson PJ, He X, Dahlback B. Identification of the same factor V gene mutation in 47 out of 50 thrombosis-prone families with inherited resistance to activated protein C. J Clin Invest 1994;94:2521–2524.

292. Vorberg J, Roelse J, Koopman R, et al. Association of idiopathic thromboembolism with single point mutation at Arg506 of factor V. Lancet 1994;343:1535–1536.

293. Greengard JS, Sun X, Xu X, et al. Activated protein C resistance caused by Arg506 Gl mutation in factor Va. Lancet 1994;343:1361–1362.

294. Aparicio C, Dahlback B. Molecular mechanisms of activated protein C resistance: Properties of factor V isolated from an individual with homozygosity for the Arg 506 to Gln mutation in the factor V gene. Biochem J 1996;313:467–472.

295. Heeb MJ, Kojima Y, Greengard JS, Griffen JH. Activated protein C resistance: Molecular mechanisms based on studies using purified Gln506-factor V. Blood 1995;85:3405–3411.

296. Nicolaes GAF, Tans G, Thomassen MCLGD, et al. Peptide bond cleavages and loss of functional activity during inactivation of factor Va in factor VaR506Q by activated protein C. J Biol Chem 1995;270(21):158–166. @R3:297.   Bertina RM, Koeleman BPC, Koster T, et al. Mutation in blood coagulation factor V associated with resistance to activated protein C. Nature 1994;369:64–67.

298. Rees DC, Cox M, Clegg JB. World distribution of factor V Leiden. Lancet 1995;346:1133–1134.

299. Svensson PJ, Dahlback B. Resistance to acti-

vated protein C as a basis for venous thrombosis. N Eng J Med 1994;330:517–521.

300. Ridker PM, Hennekens CH, Lindpaintner K, et al. Mutation in the gene coding for coagulation factor V and the risk of myocardial infarction, stroke, and venous thrombosis in apparently healthy men. N Engl J Med 1995;332: 912–917.

301. Rosendaal FR, Koster T, Vandenbroucke JP, Reitsma PH. High risk of thrombosis in patients homozygous for factor V Leiden (activated protein C resistance). Blood 1995;85: 1504–1508.

302. Andrew M. Developmental hemostasis: Relevance to thromboembolic complications in pediatric patients. Thromb Haemost 1995;74:415–425.

303. Dahlback B, Hillarp A, Rosen S, Zoller B. Resistance to activated protein C, the FV:Q506 allele, and venous thrombosis. Ann Hematol 1996;72:166–176.

304. Simioni P, de Ronde H, Prandoni P, et al. Ischaemic stroke in young patients with activated protein C resistance. Stroke 1995;26: 885–890.

305. Ganesan V, Kelsey H, Cookson J, et al. Activated protein C resistance in childhood stroke. Lancet 1996;347:260.

306. Pipe SW, Schmaler AH, Nichols WC, et al. Neonatal purpura in association with factor V R506Q mutation. J Pediatr 1996;128: 706–709.

307. Schlichtemeier TL, Tomlinson GE, Kamen BA, et al. Multiple coagulation defects and the Cohen syndrome. Clin Genet 1994;45: 212–216.

308. Roach ES, Riela AR. Intracranial hemorrhage. In Pediatric Cerebrovascular Disorders. 2nd ed. Armonk, NY: Futura Publishing Co, 1995, 69–84.

309. Sedzimir BS, Robinson J. Intracranial haemorrhage in children and adolescents. J Neurosurgery 1973;38:269–281.

310. Martin NA, Edwards MSB. Supratentorial arteriovenous malformations. In Edwards MSB, Hoffman HJ (eds). Cerebral Vascular Disease in Children and Adolescents. Baltimore: Williams and Wilkens, 1989, 283–308.

311. Snead OC III, Acher JD, Morawetz R. Familial arteriovenous malformation. Ann Neurol 1979;5:585–587.

312. Allard JC, Hochberg FH, Franklin PD, Carter AP. Magnetic resonance imaging in a family with hereditary cerebral arteriovenous malformations. Arch Neurol 1989;46: 184–187.

313. Yokoyama K, Asano Y, Murakawa T, et al. Familial occurrence of arteriovenous malformation of the brain. J Neurosurg 1991;74: 585–589.

314. Kaplan P, Hollenburg RD, Frasher FC. A spinal arteriovenous malformation with hereditary cutaneous haemangiomas. Am J Dis Child 1976;130:1329–1331.

315. Zaremba J, Sepien M, Jelowicka M, Ostrowska D. Hereditary neurocutaneous angioma: A new genetic entity? J Med Genet 1979;16:443–447.

316. Celli P, Ferrante L, Palma L. Cerebral arteriovenous malformations in children: Clinical features and outcome in children and adults. Surg Neurol 1984;22:43–49.

317. Ventureya ECG, Herder S. Arteriovenous malformations of the brain in children. Childs Nerv Syst 1987;3:12–18.

318. Humphreys RP. Infratentorial arteriovenous malformations. In Edwards MSB, Hoffman HJ (eds). Cerebral Vascular Disease in Children and Adolescents. Baltimore: Williams and Wilkins, 1989:309–320.

319. Perret G, Nishioka H. Report on the co-operative study of intracranial aneurysms and subarachnoid haemorrhage: Arteriovenous malformations. An analysis of 545 cases of craniocerebral arteriovenous malformations and fistulae reported to the co-operative study. J Neurosurgery 1966;25:467–490.

320. Norman D. Computerized tomography of cerebrovascular malformations. In Wilson CB, Stein BM (eds). Intracranial Arteriovenous Malformations. Baltimore: Williams and Wilkins, 1984, 105–120.

321. Willinsky RA, Lasjaunias P, Terbrugge K, Burrows P. Multiple cerebral arteriovenous malformations (AVMs). Neuroradiology 1990;32:207–210.

322. Drake CG. Cerebral arteriovenous malformations: considerations for and experience with surgical treatment of 166 cases. Clin Neurol Neurosurg 1978;26:145–208.

323. Ciricillo SF, Edwards MSB, Schmidt GS, et al. Interventional neuroradiological management of vein of Galen malformations in the neonate. Neurosurgery 1990;27:22–28.

324. Steiner L, Lindquist C, Steiner M. Radiosurgery with focussed gamma-beam irradiation in children. In Edwards MSB, Hoffman HJ (eds). Cerebral Vascular Disease in Children and Adolescents. Baltimore: Williams and Wilkins, 1989, 367–388.

325. Steinberg GK, Fabrikant JI, Marks MP, et al. Stereotactic heavy-charged-particle Bragg peak radiation for intracranial arteriovenous malformations. N Engl J Med 1990; 323:96–101.

326. Scott RM, Barnes P, Kupsky W, Adelman LS. Cavernous angioma of the central nervous system in children. J Neurosurg 1992;76:38–46.

327. Gil-Nagel A, Dubovsky J, Wilcox KJ, et al.

Familial cerebral cavernous angioma: A gene localised to a 15cM interval on chromosome 7q. Ann Neurol 1996;39:807–810.

328. Drigo P, Mammi I, Battistella PA, et al. Familial cerebral, hepatic and retinal cavernous angioma: a new syndrome. Childs Nerv Syst 1994;10:205–209.

329. Robinson JR, Little JR, Awad I. Natural history of cavernous angioma. Stroke 1990;21:171.

330. McDonald MT, Pepenberg KA, Ghosh S, et al. A disease locus for hereditary haemorrhagic telangiectasia maps to 9q33–34. Nature Genet 1994;6:197–204.

331. McAllister KA, Grogg KM, Johnson DW, et al. Endoglin, a TGF-beta binding protein of endothelial cells, is the gene for hereditary hemorrhagic telangiectasia type 1. Nat Genet 1994;8:345–351.

332. Vincent P, Plauchu H, Hazan J, et al. A third locus for hereditary haemorrhagic telangiectasia maps to chromosome 12q. Hum Mol Genet 1995;4:945–949.

333. McAllister KA, Lennon E, Bowles-Biesecker B, et al. Genetic heterogeneity in hereditary haemorrhagic telangiectasia: Possible correlation with clinical phenotype. J Med Genet 1994;31:927–932.

334. Fitz-Hugh T Jr. The importance of atavism in the diagnosis of hereditary haemorrhagic telangiectasia (Rendu-Osler-Weber disease): Rationale of treatment and report of an additional family. Am J Med Sci 1923;166:884–893.

335. Roman G, Fisher M, Perl DP, Poser CM. Neurologic manifestations of hereditary haemorrhagic telangiectasia (Rendu-Osler-Weber disease): Report of 2 cases and review of the literature. Ann Neurol 1978;4:130–144.

336. McCue M, Harlenberg M, Nance WE. Pulmonary arteriovenous malformations related to Rendu-Osler-Weber syndrome. Am J Med Genetics 1984;19:19–27.

337. Harkonen M. Hereditary hemorrhagic telangiectasia (Osler-WeberRendu disease) complicated by pulmonary arteriovenous fistula and brain abscess. Acta Med Scand 1981;209:137–139.

338. Haitjema T, Westermann CJJ, Overtoom TCT, et al. Hereditary hemorrhagic telangiectasia (Osler-Weber-Rendu Disease): New insights in pathogenesis, complications and treatment. Arch Intern Med 1996;156:714–719.

339. Roach ES, Riela AR. Vascular malformations. In Pediatric Cerebrovascular Disorders. 2nd ed. Armonk, NY: Futura Publishing Co, 1995, 201–218.

340. Berg BO. Neurocutaneous syndromes: Phakomatoses and allied conditions. In Swaiman KF (ed). Pediatric Neurology: Principles and Practice. St Louis: CV Mosby, 1989:795–817.

341. Bonnet P, Dechaume J, Blanc E. L'aneurisme Cirsoide de la retine (aneurisme racemeux) ses relations avec l'aneurisme cirsoid du cerveau. Bul Soc Fr Opthal 1938;51:521–524.

342. Miles JH, Zonana J, McFarlane J. Macrocephaly with hamartomas: Bannayan-Zonana syndrome. Am J Med Genet 1984;19:225–234.

343. Locksley HB. Report on the co-operative study of the intracranial aneurysms and subarachnoid haemorrhage. Section V, part 1: Natural history of subarachnoid haemorrhage, intracranial aneurysms and arteriovenous malformations. Based on 6,368 cases in the co-operative study. Neurosurgery 1966;25:219–239.

344. Orozco M, Triueros F, Quintana F. Intracranial aneurysms in early childhood. Surg Neurol 1978;9:247–252.

345. Humphreys RP. Intracranial arterial aneurysms. In Edwards MSB, Hoffman HJ (eds). Cerebral Vascular Disease in Children and Adolescents. Baltimore: Williams and Wilkins, 1989, 247–254.

346. Amacher AL, Drake CG. The results of operating on cerebral aneurysms and angiomas in children and adolescents: Cerebral aneurysms. Childs Brain 1979;5:151–165.

347. Shucart WA, Wolpert SM. Intracranial aneurysms in childhood. Am J Dis Child 1974;127:288–293.

348. Makos MM, McComb RD, Hart MN, Bennett DR. Alpha-glucosidase deficiency and basilar artery aneurysm: Report of a sibship. Ann Neurol 1987;22:629–633.

349. Braunsdorf WE. Fusiform aneurysm of basilar artery and ectatic internal carotid arteries associated with glycogenosis type 2 (Pompe's disease). Neurosurgery 1987;21:748–749.

350. Van den Berg JSP, Limburg M, Hennekam RCM. Is Marfan syndrome associated with symptomatic intracranial aneurysms? Stroke 196:27:10–12.

351. Schievink WI, Schaid DJ, Rogers HM, et al. On the inheritance of intracranial aneurysm. Stroke 1994;25:2028–2037.

352. Ter Berg HWM, Biljsma JB, Willemse J. Familial occurrence of intracranial aneurysms in childhood: A case report and review of the literature. Neuropediatrics 1987;18:227–230.

353. Bromberg JEC, Rinkel GJE, Algra A, et al. Familial subarachnoid hemorrhage: Distinctive features and patterns of inheritance. Ann Neurol 1995;38:929–934.

354. Ostergaard JR, Oxlund H. Collagen type III deficiency in patients with rupture of in-

tracranial saccular aneurysms. J Neurosurg 1987;67:690–696.

355. De Paepe A, van Landegem W, de Keyser F, de Reuck J. Association of multiple intracranial aneurysms and collagen type III deficiency. Clin Neurol Neurosurg 1988;90:53–56.

356. Majamaa K, Savolainen E-R, Myllala VV. Synthesis of structurally unstable type III procollagen in patients with cerebral artery aneurysm. Biochim Biophys Acta 1992;1138:191–196.

357. Kuivaniemi H, Prockop DJ, Wu Y, Madhatheri SL, et al. Exclusion of mutations in the gene for type III collagen (COL3A1) as a common cause of intracranial aneurysms or cervical artery dissections. Neurology 1993;43:2652–2658.

358. Schievink WI, Katzmann JA, Piepgras DG, Schaid DJ. Alpha-1-antitrypsin phenotypes among patients with intracranial aneurysms. J Neurosurg 1996;84:781–784.

359. Eyster ME, Gill FM, Blatt PM. Central nervous system bleeding in haemophiliacs. Blood 1978;51:1179.

360. Andes WA, Wulff K, Smith WB. Head trauma in haemophilia: A prospective study. Arch Intern Med 1984;144:1981–1983.

361. de Tezanos Pinto M, Fernandez J, Perez Bianco PR. Update on 156 episodes of central nervous sytem bleeding in haemophiliacs. Haemostasis 1992;22:259–267.

362. Wasi P, Na-Nakorn S, Pootrakul P, et al. A syndrome of hypertension, convulsion and cerebral haemorrhage in thalassaemic patients after multiple blood transfusions. Lancet 1978;September 16:603–604.

363. Sinniah D, Vignaedra V, Ahmed K. Neurologic complications of beta-thalassaemia major. Arch Dis Child 1977;52:977–978.

364. Sinniah D, Ekert H, Bosco J, Nathan L, Koe SL. Intracranial haemorrhage and circulating coagulation in beta-thalassaemia major. J Pediatr 1981;99:700–703.

365. Logothetis J, Constantoulakis M, Economidou J, et al. Thalassaemia major (homozygous beta-thalassaemia). Neurology 1972;22:294–304.

366. Gonatas NK, Shy GM. Childhood myopathies with abnormal mitochondria. Exerpta Medica International Congress Series 1965;100:606.

367. DiMauro S, Bonilla E, Zeviani M, et al. Mitochondrial myopathies. Ann Neurol 1985;17:521–538.

368. Ciafaloni E, Ricci E, Shanske S, et al. MELAS: Clinical features, biochemistry, and molecular genetics. Ann Neurol 1992;31:391–398.

369. Fujii T, Okuno T, Ito M. MELAS of infantile onset: Mitochondrial angiopathy or cytopathy? J Neurol Sci 1991;103:37–41.

370. Hirano M, Pavlakis SG. Mitochondrial myopathy, encephalopathy, lactic acidosis, and stroke-like episodes (MELAS): Current concepts. J Child Neurol 1994;9:4–13.

371. Matthews PM, Tampieri D, Berkovic SF, et al. Magnetic resonance imaging shows specific abnormalities in the MELAS syndrome. Neurology 1991;41:1043–1046.

372. Hasuo K, Tamura S, Yasumori K, et al. Computed tomography and angiography in MELAS (mitochondrial myopathy, encephalopathy, lactic acidosis and stroke-like episodes): Report of three cases. Neuroradiology 1987;29:393–397.

373. Ohama E, Ohara S, Ikuta F. Mitochondrial angiopathy in cerebral blood vessels of mitochondrial encephalopathy. Acta Neuropathol (Berl) 1987;74:226–233.

374. Sakuta R, Nonaka I. Vascular involvement in mitochondrial myopathy. Ann Neurol 1989;25:594–601.

375. Forster C, Hubner G, Muller-Hocker J, et al. Mitochondrial angiopathy in a family with MELAS. Neuropaediatrics 1992;23:165–168.

376. Morita K, Ono S, Fukunaga M, et al. Increased accumulation of N-isopropyl-p-(123I)-iodoamphetamine in two cases with mitochondrial encephalomyopathy with lactic acidosis and stroke-like episodes (MELAS). Neuroradiology 1989;31:358–361.

377. Seyama K, Suzuki K, Mizuno Y, et al. Mitochondrial encephalomyopathy with lactic acidosis and stroke-like episodes with special reference to the mechanism of cerebral manifestations. Acta Neurol Scand 1989;80:561–568.

378. Gropen TI, Prohovnik I, Tatemichi TK, Hirano M. Cerebral hyperemia in MELAS. Stroke 1994;25:1873–1876.

379. Goto Y, Nonaka I, Horai S. A mutation in the tRNALeu gene associated with the MELAS subgroup of mitochondrial encephalomyopathies. Nature 1990;348:651–653.

380. Goto Y, Horai S, Matsuoka T, et al. Mitochondrial myopathy, encephalopathy, lactic acidosis and stroke-like episodes (MELAS): A correlative study of the clinical features and mitochondrial DNA mutation. Neurology 1992;42:545–550.

381. Moraes CT, Ciacci F, Silvestri G, et al. Atypical clinical presentations associated with the MELAS mutation at position 3243 of human mitochondrial DNA. Neuromusc Disord 1993;3:43–50.

382. Koga Y, Nonaka I, Kobayashi M. Findings in muscle complex I (NADH coenzyme Q reductase) deficiency. Ann Neurol 1988;24:749–756.

383. Tokunaga M, Mita S, Sakuta R, et al. Increased mitochondrial DNA in blood vessels and ragged-red fibres in mitochondrial myopathy, encephalopathy, lactic acidosis, and stroke-like episodes (MELAS). Ann Neurol 1993;33:275–280.

384. Goto Y, Tsugane K, Tanabe Y, et al. A new point mutation at nucleotide pair 3291 of the mitochondrial tRNALeu gene in a patient with mitochondrial myopathy, encephalopathy, lactic acidosis, and stroke-like episodes (MELAS). Biochem Biophys Res Comm 1994;202:1624–1630.

385. Lertrit P, Noer AS, Jean-Francois MJB, et al. A new disease-related mutation for mitochondrial encephalopathy, lactic acidosis and stroke-like episodes (MELAS) syndrome affects the ND4 subunit of the respiratory complex 1. Am J Hum Genet 1992;51: 457–468.

386. Bourgeois M, Aicardi J, Goutieres F. Alternating hemiplegia of childhood. J Pediatr 1993;122:673–679.

387. Arnold DL, Silver K, Andermann SK. Evidence for mitochondrial dysfunction in patients with alternating hemiplegia of childhood. Ann Neurol 1993;33:604–6–7.

388. Shoffner JM, Wallace DC. Oxidative phosphorylation diseases: disorders of two genomes. In Harris H, Hirschhorn K (eds). Advances in Human Genetics. New York: Plenum Press, 1990, 267–330.

389. Penn AMW, Lee JWK, Thuiller P, et al. MELAS syndrome with mitochondrial tRNA[Leu(UUR)] mutation: Correlation of clinical state, nerve conduction, and muscle 31P magnetic resonance spectroscopy during treatment with nicotinamide and riboflavin. Neurology 1992;42:2147–2152.

390. Aicardi J, Gordon N, Hagberg B. Holes in the brain. Dev Med Child Neurol 1985;27: 249–260.

391. Korf B, Wallman JK, Levy HL. Bilateral lucency of the globus pallidus complicating methylmalonic acidaemia. Ann Neurol 1986;20:364–366.

392. Stockler S, Slavc I, Ebner F, Baumgartner R. Asymptomatic lesions of the basal ganglia in a patient with methylmalonic acidemia. Eur J Paed 1992;151:920.

393. Gebarski SS, Gabrielsen TO, Knake JE, Latack JT. Cerebral CT findings in methylmalonic and propionic acidaemias. AJNR 1983;4:955–957.

394. Dave P, Curless RG, Steinman L. Cerebellar haemorrhage complicating methylmalonic acidemia and propionic acidemia. Arch Neurol 1984;41:1293–1296.

395. Haas RH, Marsden DL, Capistrano-Estrada S, et al. Acute basal ganglia infarction in propionic acidemia. J Child Neurol 1995;10: 18–22.

396. Matinez-Lage JF, Casas C, Fernandez MA, et al. Macrocephaly, dystonia, and bilateral temporal arachnoid cysts: Glutaric aciduria type 1. Childs Nerv Syst 1994;10:198–203.

397. Vallee L, Fontaine M, Nuyts J-P, et al. Stroke, hemiparesis and deficient mitochondrial beta-oxidation. Eur J Pediatr 1994;153: 598– 603.

398. De Grauw TJ, Smit LME, Brockstedt M, et al. Acute hemiparesis as the presenting sign in a heterozygote for ornithine transcarbamylase deficiency. Neuropediatrics 1993;21:133–135.

399. Christodoulou J, Qureshi IA, McInnes RR, Clarke JTR. Ornithine transcarbamylase deficiency presenting with strokelike episodes. J Pediatr 1993;122:423–425.

400. Bajaj SK, Kurlemann G, Schuierer G, Peters PE. CT and MRI in a girl with late-onset ornithine transcarbamylase deficiency: Case report. Neuroradiology 1996;38:796–799.

401. Batshaw M, Brusilow S, Walser M. Treatment of carbamyl phosphate synthetase deficiency with keto analogues of essential amino acids. N Engl J Med 1975;292:1085–1090.

402. Brown GK, Scholem RD, Croll HB, et al. Sulfite oxidase deficiency: Clinical, neuroradiologic, and biochemical features in two new patients. Neurology 1989;39:252–257.

403. Johnson JL, Rajagopalan KV, Lanman JT, et al. Prenatal diagnosis of molybdenum cofactor deficiency by assay of sulphite oxidase activity in chorionic villus samples. J Inher Metab Dis 1991;14:932–937.

404. Kamoun P, Tardy P. Therapeutic attempts in sulfite oxidase deficiency. Eur J Pediatr 1990;149:594–596.

405. Valentine WN, Hsieh HS, Paglia DE, et al. Hereditary haemolytic anaemia associated with phosphoglycerate kinase deficiency in erythrocytes and leucocytes. N Engl J Med 1969;280:528–534.

406. Konrad PN, McCarthy DJ, Mauer AM, et al. Erythrocyte and leucocyte phosphoglycerate kinase deficiency with neurologic disease. J Pediatr 1973;82:456–460.

407. Wilson M, Vowels M, Wise G. The neurologic manifestations of phosphoglycerate kinase deficiency. Australian Paediatric Neurology Meeting 1988. Abstract.

408. Stibler H, Jaeken J, Kristiansson B. Biochemical characteristics and diagnosis of the carbohydrate-deficient glycoprotein syndrome. Acta Paediatr Scan 1991(suppl); 375: 21–31.

409. Jaeken J, Hagberg B, Stromme P. Clinical presentation and natural history of the carbohydrate-deficient glycoprotein syndrome. Acta Paediatr Scan 1991(suppl);375:6–13.

410. Petersen MB, Brostrom K, Stibler H, Skovby F. Early manifestations of the carbohydrate-deficient glycoprotein syndrome. J Pediatr 1993;122:66–70.
411. Stromme P, Maehlen J, Strom EH. Post-mortem findings in two patients with the carbohydrate-deficient glycoprotein syndrome. Acta Paediatr Scan 1991(suppl);375:55–62.
412. Eeg-Olofsson KE, Wahlstrom J. Genetic and epidemiologic aspects of the carbohydrate-deficient glycoprotein syndrome. Acta Paediatr Scan 1991(suppl);375:63–65.
413. Jaeken J, Stibler H, Hagberg B. General summary. Acta Paediatr Scan 1991(suppl)375:66–67.
414. Noonan JA, Ehmke DA. Associated non-cardiac malformations in children with congenital heart disease. J Pediatr 1983;63:468–470.
415. McAnena O, Padilla JR, Buckley TF. Intracranial aneurysm in association with Noonan's syndrome. Ir Med J 1984;77:140–141.
416. Schon F, Bowler J, Baraitser M. Cerebral arteriovenous malformation in Noonan's syndrome. Postgrad Med J 1992;68:37–40.
417. Hinnant CA. Noonan syndrome associated with thromboembolic brain infarcts and posterior circulation abnormalities. Am HJ Med Genet 1995;56:241–244.
418. Levy SR, Abroms IF, Marshall PC, Rosquete EE. Seizures and cerebral infarction in the full-term newborn. Ann Neurol 1986;17:366–370.
419. Lao TT, Yin JA, Yuen PMP. Coagulation and anticoagulation systems in newborns: Correlation with their mothers at delivery. Gynecol Obstet Invest 1990;29:181–184.
420. Olson TA, Alving BM, Chesiel JL, et al. Intracerebral and subdural haemorrhage in a neonate with haemophilia A. J Paed Haematol/Oncol 1985;7:384–387.
421. Morales WJ, Stroup M. Intracranial haemorrhage in-utero due to isoimmune neonatal thrombocytopaenia. Obstet Gynecol 1985;65:205–215.
422. Silver RK, MacGregor SN, Pasternak JF, Nelley SE. Fetal stroke associated with elevated maternal anticardiolipin antibodies. Obstet Gynecol 1992;80:497–499.
423. Ries M, Wolfel D, Maier-Brandt B. Severe intracranial haemorrhage in a newborn infant with transplacental transfer of an acquired factor VIII:C inhibitor. J Pediatr 1995;127: 649–650.
424. Friede RL. Porencephaly: Hydranencephaly, multicystic encephalopathy. In Development Neuropathology. 2nd ed. Berlin: Springer-Verlag, 1989, 28–43.
425. Berg RA, Aleck KA, Kaplan AM. Familial porencephaly. Arch Neurol 1983;40:567–569.
426. Zonana J, Adornato BT, Glass ST, Webb MJ. Familial porencephaly and congenital hemiplegia. J Pediatr 1986;109:671–674.
427. Al-Shahwan SA, Singh B. Familial congenital hemiparesis. J Child Neurol 1995;10:413–414.
428. Fowler M, Dow R, White TA, Greer CH. Congenital hydrocephalus-hydranencephaly in five siblings with autopsy studies: A new disease. Dev Med Child Neurol 1972;14: 173–188.
429. Harper C, Hockey A. Proliferative vasculopathy and an hydranencephalic-hydrocephalic syndrome: A neuropathologic study of two siblings. Dev Med Child Neurol 1983;25:232–244.
430. Freiden J. Aplasia cutis congenita: A clinical review and proposal for clasification. J Am Acad Dermatol 1986;14:646–660.
431. McMurray BR, Martin LW, Dignan SJP, Fogelson MH. Hereditary aplasia cutis congenita and associated defects. Clin Pediatr 1977;16:610–614.
432. Singman R, Asaikar S, Hotson G, Prose NS. Aplasia cutis congenita and arteriovenous fistula. Arch Neurol 1990;47:1255–1258.
433. Lavine D, Lehman JA Jr, Thomas R. Congenital scalp defect with thrombosis of the sagittal sinus. Plast Reconstr Surg 1978;61:599–602.
434. Yokota A, Matsukado Y. Congenital midline porencephaly: A new brain malformation associated with scalp anomaly. Childs Brain 1979;5:380–397.
435. Dalsgaard-Nielson T. Migraine and heredity. Acta Neurol Scand 1965;41:287–300.
436. Goodell H, Lewontin R, Wolff HG. Familial occurrence of migraine headache. Arch Neurol Psychiatry 1954;72:325–334.
437. Baier WK. Genetics of migraine and migraine accompagnee: A study of eighty-one children and their families. Neuropediatrics 1985;16:84–91.
438. Barlow CF. Migraine with seizures, stroke and syncope. Clin Dev Med 1984;91: 126–154.
439. Rossi LN, Penzian JM, Deonna T, et al. Does migraine-related stroke occur in childhood? Dev Med Child Neurol 1990;32:1016–1021.
440. Castaldo JE, Anderson M, Reeves AG. Middle cerebral artery occlusions with migraine. Stroke 1982;13:308–311.
441. Ferguson KS, Robinson SS. Life-threatening migraine. Arch Neurol 1982;39:374–376.
442. Rothrock J, North J, Madden K, et al. Migraine and migrainous stroke: Risk factors and prognosis. Neurology 1993;43: 2472–2476.
443. Fenichel GM. Clinical Pediatric Neurology: A Signs and Symptoms Approach. 2nd ed. Philadelphia: WB Saunders, 1992, 74–87.
444. Welch KMA, Levine SR. Migraine-related

stroke in the context of the International Headache Society classification of head pain. Arch Neurol 1990;47:458–462.

445. Chabriat H, Tournier-Lasserve E, Vahedi K, et al. Autosomal dominant migraine with MRI white-matter abnormalities mapping to the CADASIL locus. Neurology 1995;45:1086–1091.

446. Roach ES, Riela AR. Inflammatory vascular disorders. In Pediatric Cerebrovascular Disorders. 2nd ed. Armonk, NY: Futura Publishing Co, 1995, 121–140.

447. Graf WD, Milstein JM, Sherry DD. Stroke and mixed connective tissue disease. J Child Neurol 1993;8:256–259.

448. Love PE, Santoro SA. Antiphospholipid antibodies: Anticardiolipin and the lupus anticoagulant in systemic lupus erythematosus (SLE) and in non-SLE disorders. Ann Intern Med 1990;112:682.

449. Mintz M, Epstein LG, Koenigsberger MR. Idiopathic childhood stroke is associated with human leucocyte antigen (HLA)-B51. Ann Neurol 1992;31:675–677.

450. Baricordi OR, Sensi A, Pivetti-Pezzi P. Behcet's disease associated with HLA-B51 and DRw52 antigens in Italians. Hum Immunol 1986;17:297–301.

451. Krensky AM, Grady S, Shanley K, Yunis EJ. HLA antigens in mucocutaneous lymph node syndrome in New England. Pediatrics 1981;67:741–743.

452. Ostergaard JR, Bruun-Peterson G, Lamm LU. HLA antigens and complement types in patients with intracranial saccular aneurysms. Tissue Antigens 1986;28:176–181.

453. Harris EN. The Second International Anticardiolipin Standardization workshop: The Kingston anti-phospholipid antibody (KAPS) group. Am J Clin Pathol 1983;25:232–244.

454. Loizou S, McCrea JD, Rudge AC. Measurement of anticardiolipin antibodies by an enzyme-linked immunosorbent assay (ELISA): Standardization and quantitation of results. Clin Exp Immunol 1985;62:738–745.

455. Nencini P, Baruffi MC, Abbate R, et al. Lupus anticoagulant and anticardiolipin antibodies in young adults with cerebral ischaemia. Stroke 1992;23:189.

456. Briley DP, Coull BM, Goodnight SH. Neurologic disease associated with antiphospholipid antibodies. Ann Neurol 1989;25:221–227.

457. Brey RL, Hart RG, Sherman DG, Tegeler CH. Antiphospholipid antibodies and cerebral ischaemia in young people. Neurology 1990;40:1190–1196.

458. Levine SR, Deegan MJ, Futrll N, Welch KMA. Cerebrovascular and neurologic disease associated with antiphospholipid antibodies: 48 cases. Neurology 1990;40:1181–1189.

459. Ford PM, Brunet D, Lillicrap DP, Ford SE. Premature stroke in a family with lupus anticoagulant and antiphospholipid antibodies. Stroke 1990;21:66–71.

460. Mueh JR, Herbst KD, Rapaport SI. Thrombosis in patients with the lupus anticoagulant. Ann Intern Med 1980;92:156.

461. Boey ML, Colaco CB, Gharavi AE. Thrombosis in systemic erythematosus: Striking association with the presence of circulating lupus anticoagulant. Br Med J 1985;313:709–715.

462. Jacobson DM, Lewis JH, Bontempo FA. Recurrent cerebral infarction in two brothers with antiphospholipid antibodies that block coagulation reactions. Stroke 1986;17:98–102.

463. Exner T, Barber S, Kronenberg H, Rickard KA. Familial association of the lupus anticoagulant. Br J Haematol 1980;45:89–96.

464. Smith HR, Auld SL, Stein RL. Familial association of antiphospholipid antibodies. Arthritis Rheum 1991;34(suppl):S42. Abstract.

465. McNeil HP, Chesterman CN, Krilis SA. Immunology and clinical importance of antiphospholipid antibodies. Adv Immunol 1991;49:193–280.

466. Coull BM, Levine SR, Brey RL. The role of antiphospholipid antibodies in stroke. Neurol Clin 1992;10:125–143.

467. Schoning M, Klein R, Krageloh-Mann I, et al. Antiphospholipid antibodies in cerebrovascular ischaemia and stroke in childhood. Neuropediatrics 1994;25:8–14L.

468. Contractor S, Hiatt M, Kosmin M, Kim HC. Neonatal thrombosis with anticardiolipin antibody in baby and mother. Am J Perinatol 1992;9:409–410.

469. Rosove MH, Brewer PM. Antiphospholipid thrombosis: Clinical course after the first thrombotic event in 70 patients. Ann Intern Med 1992;117:303–308.

470. Lubbe WF, Pattison NS. Corticosteroid treatment of pregnant women with antiphospholipid antibodies and previous fetal loss. Am J Obstet Gynecol 1990;162:1341–1342.

471. Bick RL. The antiphospholipid thrombosis (APL-T) syndromes: Characteristics and recommendations for classification and treatment. Am J Clin Pathol 1991;96:424.

472. Babikian VL, Levine SR. Therapeutic considerations for stroke patients with antiphospholipid antibodies. Stroke 1992;23:I-33–I-37.

473. Byers PH. Disorders of collagen biosynthesis and structure. In Scriver CR et al (eds). The Metabolic and Molecular Basis of Inherited Disease. 7th ed. New York: McGraw-Hill, 1995, 4054–4056.

474. Kontusaari S, Tromp G, Kuivaniemi H, et al. Substitution of aspartate for glycine 1018 in type III procollagen (COL3A1) causes type IV Ehlers-Danlos syndrome: The mutated allele is present in most blood leucocytes of the asymptomatic and mutated mother. Am J Hum Genet 1992;51:497.

475. Mileciwz DMcG, Witz AM, Smith ACM, et al. Parental somatic and germline mosaicism for a multi-exon deletion with unusual endpoints in a type III collagen (COL3A1) allele produces Ehlers-Danlos syndrome type IV in the heterozygous offspring. Am J Hum Genet 1993;53:62.

476. North KN, Whiteman DAH, Pepin MG, Byers PH. Cerebrovascular complications in Ehlers-Danlos syndrome type IV. Ann Neurol 1995;38:960–964.

477. Schievink WI, Limburg M, Oorthuys JWE, et al. Cerebrovascular disease in Ehlers-Danlos syndrome type IV. Stroke 1990;21:626–632.

478. Krog M, Almgren B, Eriksson I, Nordstrom S. Vascular complications in the Ehlers-Danlos syndrome. Acta Chir Scand 1983;149: 279–282.

479. Pepin MG, Superti-Furga A, Byers PH. Natural history of Ehlers-Danlos syndrome type IV (EDS type IV): Review of 137 cases (abstr). Am J Hum Genet 1992;51:A44.

480. Pope FM. Historical evidence for the genetic heterogeneity of pseudoxanthoma elasticum. Br J Dermatol 1975;92:493–509.

481. Yap E-Y, Gleaton MS, Buettner H. Visual loss associated with pseudoxanthoma elasticum. Retina 1992;12:315–319.

482. Iqbal A, Alter M, Lee SH. Pseudoxanthoma elasticum: A review of neurological complications. Ann Neurol 1978;4:18–20.

483. Josien E. Extracranial vertebral artery dissection: Nine cases. J Neurol 1992;239:327–330.

484. Lebwohl M, Distefano D, Prioleau PG, et al. Pseudoxanthoma elasticum and mitral-valve prolapse. N Engl J Med 1982;307:228–231.

485. Carney JA, Hruska LS, Beauchamp GD, Gordon H. Dominant inheritance of the complex of myxomas, spotty pigmentation and endocrine overactivity. Mayo Clin Proc 1986;61:165–172.

486. Echenne BP, Leboucq N, Humbertclaude V. Ito hypomelanosis and moyamoya disease. Pediatr Neurol 1995;13:169–171.

487. Pellegrino RJ, Shah AJ. Vascular occlusion associated with incontinentia pigmenti. Pediatr Neurol 1994;10:73–74.

Chapter 14

Inherited Systemic Disorders that Cause Stroke

Mark J. Alberts, MD

Introduction

This chapter will review some inherited systemic disorders that cause stroke or are commonly associated with stroke. Stroke has been associated with many of the thousands of inherited disorders that are currently known. Typically, stroke will not be the only, or even the most common, manifestation of these various diseases. Three major characteristics were used to identify the disorders to be reviewed in this chapter: (1) the disease is fairly common; (2) stroke is an important manifestation of the process; or (3) the disease process is of interest from a scientific perspective. Unless otherwise noted, the actual clinical manifestations of the stroke will not be described in detail because it is assumed that the reader is familiar with the clinical presentation of most common stroke syndromes. However, for those diseases where the stroke syndrome is unique in terms of location or symptoms, further clinical details are provided.

Although an effort was made to avoid redundancy, some of the diseases reviewed here may have been discussed in other chapters. Many of the common risk factors for stroke (i.e., hypertension, hyperlipi-demia) have been reviewed extensively in prior chapters and will not be included below. For other diseases discussed in earlier chapters, this chapter will offer more of a systemic perspective.

Hemoglobinopathies

Inherited disorders of hemoglobin are probably the most common single gene condition affecting humans on a worldwide basis.[1] The high prevalence of this disease, combined with the fact that stroke is a common manifestation and complication of some hemoglobinopathies, makes this group of disorders an important genetic etiology for stroke.

Any discussion of hemoglobinopathies first requires an understanding of normal hemoglobin synthesis. The hemoglobin protein is a tetramer composed of two pairs of peptide chains. These peptide chains are also referred to as globin units; the the most common are designated alpha, beta, gamma, and lambda.[1] Each of these globin subunits is coded for by a pair of genes (one from each parent).

The type of hemoglobin that an individ-

From Alberts MJ (ed). *Genetics of Cerebrovascular Disease*. Armonk, NY: Futura Publishing Company, Inc., © 1999.

ual has is defined by the combination of globin units that makes up the hemoglobin tetramer. Hemoglobin A has a composition of alpha 2 beta 2, and comprises about 97% of total hemoglobin in a normal individual. Hemoglobin A2 is composed of alpha 2 lambda 2 and comprises 3% of normal hemoglobin. The other common normal hemoglobin variant is fetal hemoglobin, designated Hgb, and has a composition of alpha 2 gamma 2.

The genes for the beta, gamma, and lambda globin subunits are located on adjacent loci on chromosome 11, while the alpha subunit genes are located on chromosome 16.[2-4] The inheritance of these genes follows classic mendelian genetics. However, the marked variations that can be seen in each of these genes can produce a myriad of hemoglobin subtypes in any given individual.

Hemoglobinopathies can be subdivided into two major types of disorders. In one type, there is a *structural abnormality* in one of the globin chains. This is the cause of most of the hemoglobin variants and the sickling syndromes such as sickle cell anemia and sickle cell trait. The disorders causing a reduced *rate of synthesis* of a globin gene are defined as the thalassemias. It should be emphasized that, in some cases, there is no sharp distinction between the structural variants and the thalassemias, since some abnormal hemoglobins may be produced at a reduced rate. In addition, in certain populations, structural variants and thalassemias may coexist.

Sickle Cell Anemia

Sickle cell anemia (SSA) occurs when an individual is homozygous for hemoglobin S. Hemoglobin S contains an adenine to thymine substitution in the betar-globin gene, which causes a substitution of valine for glutamic acid.[5] The cellular pathophysiology of how these changes produce cell sickling and resulting anemia will not be reviewed here. Almost every organ system in the body can be affected with sickle cell anemia, and once symptoms are produced, the

illness is designated sickle cell disease (SSD).[6] From a vascular and neurologic perspective, the most significant phenomena are the vasoocclusive events that complicate this disease, as well as the occasional vascular ruptures that can produce hemorrhages.

Patients with SSD can develop a variety of cerebrovascular abnormalities including arterial ischemic strokes, cerebral hemorrhages, and venous thromboses.[7,8] The overall prevalence of neurologic events in these patients ranges from 5% to 17%, and the incidence of stroke can range from 10% to 17%.[7-10] Studies using MRI have found a 10% incidence of asymptomatic strokes in these patients.[11] Strokes may occur in all parts of the central nervous system, including the spinal cord and retina. Intracerebral hemorrhage tends to occur more frequently in adults, while cerebral infarction predominates in children, although either type of stroke can occur in any age group.[7,8]

In the past, it was thought that arterial cerebral thrombosis causing stroke in SSA patients was due to sludging of blood vessels leading to focal vascular occlusion. The sludging was presumed to be due to the sickling of red blood cells.[1,12,13] This phenomenon certainly happens, but is probably responsible for more small-vessel strokes than large-vessel strokes in patients with SSD. Studies using a variety of imaging techniques (i.e., MRI, MR angiography, and cerebral angiography), as well as pathologic evaluations have demonstrated the presence of a large-vessel vasculopathy particularly affecting the intracranial portion of the internal carotid arteries and proximal portion of the other large intracranial vessels.[13-15] This vasculopathy appears to be due to intimal thickening and hyerplasia affecting the large intracranial vessels and in some cases smaller vessels. The etiology of this intimal thickening is unknown, although some have suggested that it is due to chronic intimal damage from the impact of high-velocity rigid erythrocytes, while others have suggested that it is due to chronic hypoxia affecting the vessel wall (for a review, see references 1 and 16). However, it is now clear that the intimal thickening involving proximal cerebral vessels is the major vascular process in the production of large-vessel cerebral infarctions in patients with SSD.[12,13,17,18]

The presence of the large-vessel vasculopathy and thickening in the distal ICA may lead to the formation of moyamoya vessels, which in some cases may rupture, producing an intracerebral hemorrhage or subarachnoid hemorrhage particularly in older patients.[19] Sickle cell disease can also be complicated by venous thrombosis, possibly due to sludging of erythrocytes particularly when an individual is dehydrated or infected. This type of venous thrombosis can affect the sagittal sinus as well as other venous channels, producing a venous infarction with hemorrhage and occasional seizures.[20]

Patients with SSD may have other types of cerebral hemorrhages such as subdural hematomas and aneurysmal SAH. Case reviews suggest that aneurysmal SAH occurs in up to 3% of SSD patients.[21,22] Multiple aneurysms can be seen in up to 30% of such cases, mandating a thorough work-up for unruptured aneurysms. The SAH may occur at a younger age than seen in a population of SAH patients without SSD. [21]

Many SSA patients will have elevated levels of fetal hemoglobin (HgbF), since this variety of hemoglobin does not have the β-globin gene that contains the sickle cell mutation. Several studies have suggested that individuals with high levels of HgbF (typically $\geq$ 8%–10%) are associated with a reduced risk of clinical complications.[23,24] Other studies have found a relationship between the presence of alpha-thalassemia (described below) and the risk of stroke. Those studies indicated that SSA patients with alpha-thalassemia had a significantly reduced occurrence of stroke.[25] This effect was apparent even after controlling for higher HgbF levels in patients with SSA who did not have a stroke. The mechanism(s) underlying this protective effect are unclear, but may relate partially to improved erythorcyte physiology, less sickling, and improved rheology in patients with both SSA and alpha-thalassemia.[26]

Some studies have suggested that the risk of stroke in patients with sickle cell anemia increases when the level of fetal hemoglobin falls below 8% to 10%.[23,24] Other studies have used transcranial Dopplers (TCD) as a screening device. Normally, a TCD will show elevated velocities in the major intracranial vessels in patients with sickle cell anemia. The elevated velocities are most likely due to the anemia. However, if the velocities are markedly elevated (greater than 150 cms per second for the middle cerebral artery, for example) this indicates a higher risk for stroke in those individuals.[27-30] This is most likely an indication of the large-vessel vasculopathy described above. MR imaging can also detect the large vessel vasculopathy.[31]

Current treatment recommendations emphasize aggressive transfusion therapy to prevent recurrent strokes once a stroke has occurred. Most protocols endeavor to keep the HgbS concentration less than 30%. One study found a dramatic decrease in clinical recurrences of stroke following transfusion therapy.[17] In addition, prolonged transfusion therapy appeared to greatly attenuate the progressive stenosis found in the large intracranial vessels.[17] Some investigators have reported that less aggressive transfusion protocols with target HgbS concentrations of up to 50% were well tolerated with few vascular complications.[32]

As indicated above, the risk of vascular complications decreases as the fetal Hgb concentration increases. Many SSA patients have elevated levels of fetal hemoglobin, which does not contain the beta-globin chain that harbors the SSA mutation.[1] This is a compensatory response and may lessen the clinical manifestations of SSA in some patients. One method to increase HgbF levels is with hydroxyurea. A recent large, double-blind, randomized study of hydroxyurea in almost 300 adult patients with SSA found that such treatment reduced the rate of painful crises by almost half.[33] Whether hydroxyurea will prove effective in reducing the rate of neurovascular events remains unclear at present.[34]

Sickle Cell Trait

Sickle cell trait affects approximately 8% of African-Americans, and is due to heterozygosity for hemoglobin S.[20] These individuals have approximately 30% hemoglobin S and 70% hemoglobin A. Typically, individuals with sickle cell trait do not develop any significant medical problems. However, during periods of severe hypoxia or dehydration, sickling may occur, producing a painful crisis.[1] Infarctions involving major organs, particularly stroke, are extremely rare in individuals with sickle cell trait, although there are several well-documented reports of stroke in such individuals.[35,36] In general, sickle cell trait patients should be evaluated for other causes of stroke before it is attributed to the hemoglobinopathy.

Hemoglobin SC Disease

This disorder is a double heterozygous state, with patients having both hemoglobin S and hemoglobin C.[37] Hemoglobin C results from the substitution of lysine for glutamic acid in the beta-chain. There is an increased tendency for patients with SC disease to experience sickle phenomena, leading to painful crises as well as other organ involvement.[1] Stroke has been described in SC patients, with an incidence of 2% to 5%, which is less than in patients with SSD.[20]

Thalassemias

A great variety of mutations can produce either alpha- or beta-thalassemia, including large deletions, point mutations, insertions, defective RNA processing, and others (see reference 1 for a review). Both alpha- and beta-thalassemia can be further subdivided depending on whether there is no production of the particular globin chain, or only a partial reduction in production of one chain. These differences relate largely to the type and severity of the mutation present. In addition, it is fairly common for one form of thalassemia to coexist with one of the structural hemoglobin variants. The various combinations lead to a myriad of possible pathologic and clinical syndromes (see reference 1 for a complete review).

The key pathophysiologic step in thalassemia is the imbalance in production of one of the globin chains. In beta-thalassemia, the excess alpha-globin chains cannot form stable hemoglobin tetramers. These chains precipitate in erythrocyte precursors, leading to abnormal cell proliferation and destruction.[37] These defects can be partially compensated by persistent synthesis of fetal hemoglobin in these patients.[1] The homozygous form of the disease leads to more symptoms compared with the heterozygous variety.[38]

Alpha-thalassemia occurs in up to 30% of African-Americans.[39] In cases of alpha-thalassemia, there is a relative overproduction of beta- or gamma-globin chains. These chains can form homotetramers and are relatively soluble; therefore the symptoms in cases of alpha-thalassemia related to ineffective erythropoeisis tend to be less severe than is seen in cases of beta-thalassemia.[40] However, the resulting hemoglobin in the alpha-thalassemias is an inefficient carrier of oxygen, leading to tissue hypoxia.[1]. As with the beta-thalassemias, patients who are homozygous have more severe clinical symptoms.

Cerebrovascular disease appears to be an uncommon isolated complication of the thalassemias. Other systemic symptoms related to bone marrow expansion and iron overload tend to predominate. However, as indicated above, the thalassemias can have significant modifying effects on the cerebrovascular manifestations of SSA.

Coagulation Disorders

Most of the major inherited coagulation disorders have been reviewed in *Chapter 7* and will not be reiterated here. However, the increased recognition and, perhaps, prevalence of the antiphospholipid antibody (APA) syndromes as an etiology for various types of ischemic stroke necessitates a further discussion of these syndromes.

The APA syndromes describe a somewhat heterogeneous group of disorders. Included in this category are idiopathic lupus anticoagulants, idiopathic anticardiolipin antibodies, coagulation disorders seen in cases of lupus, (and other related autoimmune and collagen-vascular disorders) and Sneddon's syndrome (Table 1).[20] There is clear overlap among many of these cases in some individuals and families, while in others they appear as distinct disorders. Antiphospholipid antibodies can be seen in patients with other systemic disorders, including Sjogren syndrome, Behcets syndrome, rheumatoid arthritis, and other connective tissue diseases, as well as in pregnancy and due to ingestion of certain drugs such as neuroleptics.[20,41,42]

The APA syndromes are thought to be a more common cause of stroke among young adults, partially because recent studies and clinical practices have emphasized their importance in this population.[43] It has been suggested that APAs are present in approximately 10% of all stroke patients, although their causative role may be greater or less depending on the individual's age and other clinical factors.[20] This author has seen several patients older than 60 years of age with no other identifiable etiology of stroke except the presence of APAs (typically anticardiolipin antibodies).

Patients with any APA or with Sneddon's syndrome can experience large-vessel as well as small-vessel strokes. These strokes can occur throughout the cerebrum and brainstem and can be single or multiple. Atherothrombotic and thromboembolic mechanisms have been described in affected patients. Retinal arterial and cerebral venous thrombosis can also occur.[20,44,45] Thrombosis of cerebral veins is fairly common with these disorders, leading to venous infarctions, increased intracranial pressure, and seizures in some cases. Migraine headaches are a fairly common complaint. Tranisient ischemic attacks, personality change, seizures, and vascular dementia can also occur in some patients with an APA syndrome or Sneddon's syndrome.[46,47]

Although one of the APAs, namely the lupus anticoagulant, might be expected to cause cerebral hemorrhages, this a faulty assumption. The designation of a lupus "anticoagulant" is really a misnomer, since the "anticoagulant" effect is an *in vitro*, not *in vivo*, phenomenon. Cases of ICH or SAH in patients with a lupus anticoagulant are exceedingly rare and most likely unrelated. Further information related to testing for these antibodies is provided in *Chapter 7*.

An increasing number of reports show that, in some cases, an APA syndrome may occur within families. One large kindred reported by Smith et al. showed positive APAs in 13 of 22 members from three generations.[48] In another family described by Matthey et al.,

Table 1
Antiphospholipid Antibody Syndromes

Lupus anticoagulant and/or anticardiolipin antibody syndrome (sporadic or hereditary)
- idiopathic
- related to autoimmune or connective tissue disorders
- drug related
- secondary to infections
- other (pregnancy)

Sneddon's syndrome (sporadic or hereditary)
- idiopathic
- associated with autoimmune or connective tissue disorders
- associated with lupus anticoagulants or anticardiolipin antibodies
- associated with activated protein C resistance/factor V Leiden mutation

all 4 members (both parents and 2 children) had anticardiolipin antibodies, and 2 had a lupus anticoagulant.[49] Several reports have found a familial association for lupus anticoagulants. One study by Exner et al. found positive lupus anticoagulants in 2 pairs of siblings, both of whom had other symptoms such as Raynaud's phenomenon, arthralgias, or lupus.[50] Other studies have found a high incidence of lupus anticoagulants or anticardiolipin antibodies in relatives of patients with lupus.[51,52] In most cases, relatives with positive antiphospholipid antibodies had other clinical or laboratory abnormalities such as Raynaud's phenomenon, positive antinuclear antibody tests, and the like. At present, there are no reports of genetic linkage for these disorders.

A disorder receiving increased recognition is Sneddon's syndrome. This syndrome is characterized by livedo reticularis (a lattice type of rash) and ischemic cerebrovascular events.[53] In addition to stroke, patients may also experience TIAs, dementia, and seizures. The dementia may occur due to multiple and recurrent strokes in some cases. Other symptoms such as Raynaud's phenomenon, headaches, and other connective tissue disorders may also occur.[54] It is seen more commonly in females than in males, and can occur in either sporadic or hereditary forms.[54] The hereditary form appears to be most consistent with an autosomal dominant pattern of transmission.[47,55] In some families, affected relatives may experience only a rash or vasospastic phenomena without strokes.[56] No genetic linkage studies have been reported to date on the familial form of this disease. Pathologically, Sneddon's syndrome is characterized by a progressive arteriopathy that is noninflammatory and tends to involve medium-sized vessels both in the brain and in the periphery (particularly the digits).[54] In approximately half of the cases, the livedo skin lesions precede the neurologic symptoms.[57]

Recent studies have shown that many (but not all) patients with Sneddon's syndrome have APAs (either an anticardiolipin antibody or lupus anticoagulant).[47,58,59] In addition, relatives with livedo reticularis only may also have APAs.[60] In rare cases, unaffected relatives in families with Sneddon's syndrome may also have positive APAs.[60] It seems apparent that, at least in some cases, these antibodies play a role in the pathogenesis of some of the features of Sneddon's syndrome. A small study also found the presence of activated protein C resistance due to the factor V Leiden mutation in some patients with Sneddon's syndrome.[61]

It is interesting to speculate that a hereditary predisposition to the production of APAs may in some way mediate or cause some cases of Sneddon's syndrome, as well as strokes in individuals without Sneddon's syndrome. One cannot exclude that the APAs may be a result of Sneddon's syndrome, instead of a primary cause. However, recent large epidemiological studies showing the presence of APAs in young patients with no other etiology for stroke does imply a significant etiologic role for these antibodies in causing cerebrovascular disease.[62]

The optimal treatment for patients with symptomatic APAs is most likely coumadin. Retrospective studies have shown that patients treated with coumadin have a much lower risk of recurrent ischemic events compared with patients treated with aspirin or with no treatment.[63,64] The treatment of patients with Sneddon's syndrome, but without APAs, remains problematic. The use of steroids in patients with APAs or Sneddon's syndrome has been suggested, but there is a paucity of data supporting its efficacy.[20] Calcium channel blockers (i.e., Verapamil, Nifedipine) may be helpful in reducing the occurrence of migraine headaches, amaurosis fugax, and Raynaud's phenomenon in patients with an APA syndrome or Sneddon's syndrome.[47,65]

In summary, it appears that the APA syndromes, either alone or as part of larger syndromes (Sneddon's syndrome and lupus) can be hereditary diseases. Similar to many other conditions, as more symptomatic or asymptomatic relatives of affected patients are carefully studied and tested, the percentage of cases that appear to be familial is likely to increase. The identification of the causative gene(s) for these diseases will add important information about the pathogenesis, testing, and treatment of these disorders.

Mitochondrial Disorders

Over the past 10 years, there has been a plethora of clinical reports and laboratory research related to the inherited mitochondrial disorders. These disorders have many systemic manifestations, including stroke, and are quite unique from a genetic perspective. This section will review the genetics of mitochondria and the inherited mitochondrial diseases that cause strokes.

Each cell has many mitochondria, and each mitochondria has its own complement of DNA.[66] The mitochondrial DNA (mt-DNA) is a circular molecule with 16,569 bases. The 2 strands (designated heavy and light) are differentially transcribed and encode subunits for 13 proteins and 24 RNAs (22 transfer RNAs and 2 ribosomal RNAs).[67] The four major enzymes encoded by the mt-DNA include NADH dehydrogenase, ATP synthase, cytochrome C-oxidase, and ubiquinol:cytochrome C oxidoreductase (Figure 1).[68] Although mt-DNA is relatively independent of nuclear DNA (chromosomal DNA), some of the proteins encoded by each type of DNA are interdependent.[69] For example, nuclear DNA encodes some subunits of proteins for oxidative phosphorylation, and many of these protein complexes have other subunits encoded by mt-DNA.

Mitochondrial DNA has other unique and important characteristics. It is maternally inherited, meaning that sons and daughters inherit all of their mt-DNA from their mothers, since ovum have mitochondria but sperm do not.[70] Each mitochondria contains 2 to 10 mt-DNA molecules, but mt-DNA does not recombine (unlike nuclear DNA). In addition, mt-DNA appears to be 10 times more likely than chromosomal DNA to undergo mutations.[71] Since mt-DNA has no introns, these frequent mutations are likely to affect coding sequences.[68]

All of these features lead to an accumulation of mt-DNA mutations through successive generations.

Two additional important aspects of mutated mt-DNA are the concepts of heteroplasmy and homoplasmy. Heteroplasmy is seen when mutated and normal mt-DNA coexist within the same cell. Homoplasmy is when all of the mt-DNA in a cell is either normal or mutated. As cells divide, the proportion of normal and mutated mt-DNA will vary in each daughter cell. There is most likely a cellular and organ threshold for mutations above which clinical symptoms are likely to occur. However, this threshold may vary depending on many factors, including age, oxidative stress, protective mechanisms, concomitant diseases, and the type of mutation present.[68,72] Because of these factors, one might expect a significant degree of genetic and clinical heterogeneity among the inherited mitochondrial disorders, which is exactly what is seen.[72]

Mutations in mt-DNA can produce a variety of systemic mitochondrial diseases (Table 2). These diseases may have clinical and pathological overlap in some cases. In terms of cerebrovascular disease, the MELAS syndrome is the most important. The acronym stands for **m**itochondrial **e**ncephalomyelopathy, **l**actic **a**cidosis, and **s**troke-like episodes.[73] Although the disease is maternally inherited, not all relatives will have symptoms due to the heterogeneity noted above. The vast majority of patients (approximately 75% to 80%) have a point mutation at nucleotide position 3243 in the gene for the transfer RNA for leucine.[74,75] The actual mutation is an adenine to guanine transition.[75] Ten percent of MELAS patients will have a different mutation at nucleotide position 3271.[76,77] Another mutation has been described at nucleotide position 11,084.[78] This point mutation affects the ND-4 subunit of NADH dehydrogenase. In patients with MELAS symptoms and the 3243 mutation, an analysis of mt-DNA from muscle biopsies has shown that mutant genomes exist in 55% to 95% of the mitochondria.[73]

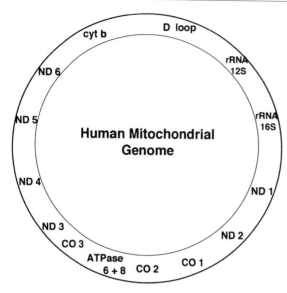

Figure 1. A schematic of the human mitochondrial genome, showing the major genes. The transfer RNA genes (tRNA) are scattered throughout the mitochondrial genome and are not shown in this figure. The D-loop region is involved with DNA replication. Most of the genes coding for metabolic proteins are transcribed in a clockwise direction, with the exception of ND 6, which is transcribed in a counterclockwise direction. For the tRNA genes, about two-thirds are transcribed clockwise, and one-third counterclockwise. CO, cytochrome oxidase; cyt, cytochrome; ND, NADH dehydrogenase; rRNA, ribosomal RNA.

Patients with MELAS commonly have strokes in the parietal-occipital region, often producing visual field defects or blindness, although other vascular territories can certainly be involved.[73] In many cases, these strokes do not respect typical vascular boundaries.[79] Other common symptoms include seizures, myoclonus, dementia, and ataxia.[73] Patients may have a prodrome of severe migraine headaches, although this does not occur in all individuals. The first stroke typically occurs before age 40, although later ages of onset can also occur. Changes can be easily detected on T2-weighted MRI images. Areas of increased signal on T2-weighted MRI images can be detected in deeper cortical layers and superficial regions of white matter. Deep white-matter regions do not usually appear abnormal on MRI.[80]

MELAS patients may have an angiopathy affecting the brain and muscle. Studies have shown abnormal endothelium particularly in brain capillaries, although the precise pathophysiology of the strokes is unclear at this time.[81] Cerebral angiography often fails to demonstrate vascular occlusions.[79,81] Several mitochondrial protein complexes show abnormally decreased activity, including ND-1 (part of NADH dehydrogenase), cytochrome oxidase 2, and cytochrome oxidase C.[73,81,82] It is appealing to postulate that regional areas of relative cerebral hypoxia or mild ischemia are sufficient to cause symptoms in MELAS patients who would otherwise be asymptomatic without such mutations.

The diagnosis of MELAS might be suggested by the onset of typical symptoms in a young individual without another etiology for the symptoms. Individuals may have a positive family history, although this may be obscured in some cases. Elevated serum lactate levels are common in MELAS patients and should suggest the diagnosis.[73] The mitochondrial mutations can be easily detected using peripheral blood (leukocytes) for analysis with PCR techniques.[83] Molecular genetic tests are available in selected laboratories to confirm the diagnosis in cases with a high index of suspicion.

The treatment of MELAS patients remains problematic. There are case reports of treatment with coenzyme Q resulting in biochemical improvements, if not clinical changes.[84]

Table 2
Inherited Mitochondrial Diseases*

Chronic progressive external ophthalmoplegia
MELAS syndrome
Myoclonic epilepsy with ragged red fibers
Leber's hereditary optic neuropathy
Kearns-Sayre syndrome
Leigh disease

*Adapted from reference 68.

Disorders of Homocysteine Metabolism

Several studies during the past decade have shown that disorders of homocysteine metabolism, which result in homocystinuria or hyperhomocysteinemia, are common and significant genetic risk factors for the development of arterial thrombo-occlusive disease.[85,86] This section will review the biochemical, genetic, and clinical aspects of these disorders of homocysteine metabolism.

The conversion of homocysteine (an amino acid) to cystathionine is catalyzed by cystathionine β-synthase (CBS), which is a pyridoxine dependent enzyme. Another related reaction is catalyzed by 5, 10-methylene tetrahydrofolate reductase (MTHFR), which catalyzes the formation of homocysteine to methionine (Figure 2).[87] Therefore, functional deficiencies in either CBS or MTHFR may result in elevated levels of homocysteine. Lastly, defects in the synthesis of methylcobalamin, which is a vitamin B12 coenzyme required by MTHFR, as well as dietary deficiencies in vitamins B6, B12, and folate can also affect the conversion of homocysteine to methionine.[88–90]

The classic form of homocystinuria is due to reduced activity of the CBS enzyme. This type of classic homocystinuria is inherited as an autosomal recessive trait and is characterized by dislocated optic lenses (present in at least 95% of patients), as well as mental retardation, osteoporosis, skeletal abnormalities, and arterial occlusive phenomena producing stroke, myocardial infarction, renal ischemia, and venous occlusions.[91]

The CBS gene has been mapped to CH21q22.3.[92] More than a dozen different mutations in the CBS gene have been described in patients and families with homocystinuria. Most of the CBS mutations identified to date are located in exons 2 through 10, with the majority in exons 3 and 8.[93] In addition, an intronic mutation has also been identified. In the Irish population, a G to A transition (designated G307S) appears to be the most common cause of homo-

cystinuria, and is associated with B6 unresponsiveness.[93] This mutation is in exon 8, and probably accounts for 50% to 60% of all cases of homocystinuria.[94] A different missense mutation, also in exon 8, appears to be responsible for approximately 40% of cases that are B6 responsive.[95] The majority of patients with homocystinuria appear to be compound heterozygotes, meaning that each allele for the CBS gene contains a different mutation.[93] This is not unexpected, since each allele would be inherited from a different parent, and it would be statistically unlikely that each parent of an affected individual would have the same mutation.

Clinical studies have shown that patients with fully developed homocystinuria are at increased risk for premature atherothrombotic vascular disease as well as venous thrombosis.[96] Untreated patients with homocystinuria have a 33% to 50% prevalence of thrombotic events by age 25.[87,96] Cerebrovascular events (arterial or venous) account for approximately one-third of all thrombotic complications, while peripheral thromboses account for half of such events. Myocardial infarction is less common, present in only 4–10% of patients with these complications.[87] There is anecdotal evidence that patients with homocystinuria are at increased risk for vascular events (i.e., thrombosis) following angiographic procedures.[97] Whether this is true when using newer contrast agents is unclear at this time.

Natural history studies of patients with homocystinuria have shown that more than half do not experience thrombotic events.[96,98] The reason for this heterogeneity had been unclear until a recent study by Mandel et al. This group examined the presence of factor V Leiden mutation in patients with homocyteinuria. This mutation is the most common cause of activated protein C resistance, the most common familial cause of thrombosis.[99] They found that all of the patients with homocystinuria who had thrombotic events were either homozygous or heterozygous for the factor V Leiden mutation.[98] They hy-

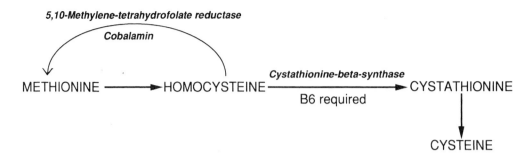

Figure 2. A depiction of some of the key steps in the metabolism of homocysteine. Note that homocysteine can be metabolized to cystathionine via cystathionine-beta-synthase, which is the defective enzyme in classic homocystinuria. Homocysteine can also be metabolized to methionine via the methylene-tetrahydrofolate reductase (MTHFR) pathway. Recently described mutations in the MTHFR gene and enzyme support the importance of this pathway in homocysteine metabolism (see text for details).

pothesized that, in some patients, the additive effect of homocystinuria and the factor V Leiden mutation (causing resistance to activated protein C) is needed to produce thrombosis, although other stressors (ie., surgery, infection) may also act as triggers for thrombosis in some cases.

The exact pathophysiology of these vascular lesions is not known. Pathologically, the lesions appear to be premature atherosclerotic plaques. The primary mechanism and triggering process remain unclear. Several hypotheses have been advanced, including abnormal oxidation and aggregation of low-density lipoproteins, as well as a direct toxic effect of homocysteine on the vascular endothelium.[86,100] Homocystinuria also appears to cause increased platelet activation and platelet aggregation, both of which may accelerate atherosclerosis.[101] A variety of other mechanisms have also been advanced (see reference 86 for a review). Most of these conclusions are based on in vitro experiments, and their applicability to the in vivo process of premature atherosclerosis is not certain.

Over the past decade, there has been an increasing interest in the identification of patients with vascular disease who have hyperhomocysteinemia, but not the full syndrome of homocystinuria, presumably on the basis of heterozygous CBS deficiency.[102] It is estimated that 1.5%–2.0% of the general population may be heterozygous for CBS deficiency.[102,103] In general, individuals with plasma homcysteine levels above 14 uM have an increased risk of vascular disease.[104] Two large studies, one in 1985 and one in 1991, found that hyperhomocysteinemia due to apparent partial deficiencies in CBS was a significant independent risk factor for the development of premature vascular disease.[102,103] These studies evaluated homocysteine levels in patients with premature cerebrovascular disease, peripheral vascular disease, or coronary artery disease. A methionine-loading test was used to detect elevated levels of homocysteine in these patients. These studies also examined CBS activity but not the presence of mutations in the gene. They found elevated homocysteine levels following a methionine load, and reduced CBS activity in some patients.[102,103]

The assertion that the hyperhomocysteinemia in these patients was caused by abnormal CBS activity has recently been challenged because another laboratory has not been able to detect such deficiencies.[105] Furthermore, genetic analyses of these patients for a common mutation in the CBS gene failed to detect this mutation.[106] These data have called into question the concept that heterozygous mutations in the CBS gene are responsible for cases of hyperhomocysteinemia and premature atherothrombotic events. Of course, the CBS gene has not yet been analyzed for all known heterozygous mutations. Therefore, the possibility that some heterozygous mutations in the CBS gene are present cannot be ruled out at this time. However, the current data have led to the search for other inherited disorders that may cause hyperhomocysteinemia.

Another large case-control study evaluated the relationship between plasma homocysteine levels and the occurrence of vascular disease (including stroke).[106a] This study included 750 cases and 800 controls. They found that patients with fasting homocysteine levels in the top fifth had a 2.2 relative risk for vascular disease when compared to the bottom four-fifths of the controls. The overall magnitude of the effect was determined to be similar to that found for smoking or hyperlipidemia.[106a] This large study further supports the importance of elevated homocysteine levels as a common and significant risk factor for vascular disease.

As indicated above, methylene tetrahydrofolate reductase (MTHFR) is important in the metabolism of homocysteine to methionine. The gene for MTHFR has been identified and is localized to CH1p36.3.[107] At least nine different mutations, mostly missense mutations, have been described in this gene.[108] Total or severe deficiencies in this enzyme can cause homocysteinemia and homocystinuria. Depending on the type of mutation, different degrees of enzyme activity will be present. Those mutations causing the lowest levels of enzymatic activity are associated with an earlier age of onset, typically within the first year of life. Such patients will have symptoms in infancy, including mental retardation and seizures. Premature thrombotic events can also occur, including, in rare cases, familial strokes presenting in adulthood.[108–110]

Severe MTHFR deficiency due to the inherited mutations described above appears to be an infrequent cause of homocysteinemia. In 1988, a much more common thermolabile variant of MTHFR was described, with activity less than 50% of controls. This deficiency appears to be inherited as an autosomal recessive trait.[111] Genetic analysis has found a relatively common single base transition in the MTHFR gene that changes an alanine to a valine and produces reduced activity of the enzyme along with elevated levels of homocysteine and increased thermal lability.[112]

A recent study found that individuals with premature vascular disease who are homozygous for the above described mutation have reduced activities of MTHFR, along with almost a 100% increase in plasma homocysteine levels following a methionine load.[112] Another related study screened 60 patients with premature cardiovascular disease (32 with stroke) and 100 controls for the presence of the MTHFR mutation. Fifteen percent of diseased patients were found to be homozygous for the mutation, compared to 5.4% of controls.[106] The identification of this fairly common mutation in the MTHFR gene, and its correlation with hyperhomocysteinemia in patients with premature vascular disease, suggest that it may be a common cause of such diseases.

In terms of therapy, early use of methionine restriction may be extremely beneficial in some patients with CBS deficiency. Some patients appear to be pyridoxine responsive, but this appears to be associated with specific mutations in this gene. Other mutations are strongly associated with lack of pyridoxine responsiveness.[93] Most patients require pyridoxine doses of 200 mg to 1,000 mg/day.[87] Patients who are B6 unresponsive may have an earlier age of onset for vascular events. Adding folate in doses of 5 mg/day may enhance the effects of B6 therapy. Betaine, a methyl-donor, may also reduce cysteine levels and some complications. Doses of 100 mg/kg/day are recommended.[87] Despite the fairly wide use of antiplatelet agents and/or antithrombotic agents, their efficacy remains unproven in this disorder. Patients who are homozygous for the MTHFR mutation may be responsive to high plasma folate levels, which could suggest one line of therapy. However, this has not been tested in prospective studies.

Connective Tissue Diseases

There are a number of inherited connective tissue disorders that may cause various types of stroke (Table 3). In some cases, this is due to direct involvement of the cerebral vessels, while in other cases the pathophysiology is either indirect or more obscure.

Marfan's syndrome is an autosomal dominant trait with systemic connective tissue and vascular manifestations.[113] As many as a third of cases have a negative family history, suggesting that new mutations are the cause in many cases of Marfan's.[113] Marfan's syndrome is due to mutations in the fibrillin gene, which has been cloned and localized to CH15q21.1.[114] Fibrillin is a large gene, with a cDNA of more than 10 kB.[115,116] A variety of abnormalties in the fibrillin gene have been identified, including missense mutations, insertions, and deletions.[116–118] These mutations appear to be fairly evenly distributed over the entire coding sequence, although several involve epidermal growth factor-like motifs.[117] Fibrillin is a component of microfibrils that are present in many different types of connective tissue.[119] The mutations in fibrillin may lead to production of a defective protein unable to undergo enzymatic processing, hydroxylation, or binding of calcium.[117]

Aortic dissection is one of the more common neurovascular complications of Marfan's syndrome.[117] It can result in stroke by secondary occlusion of the carotid and vertebral arteries, and produces obvious cardiac and systemic complications.[117] Less commonly, dissections of the common carotid artery, internal carotid artery, or vertebral artery have been reported to cause strokes.[120,121] Aneurysms involving the intracranial or extracranial vasculature have been described in patients with Marfan's.[120] Intracranially, the most common location is the cavernous portion of the internal carotid artery.[122] Oftentimes, these intracranial aneurysms may be large and produce mass effects as opposed to subarachnoid hemorrhage.[120,123] In addition to these complications, connective tissue throughout the body can be affected, producing abnormalities in the eyes (ectopia lentis), skin, body habitus

Table 3
Inherited Connective Tissue Disorders that Cause Stroke

Disease	Cerebrovascular Complications	Frequency
Ehlers-Danlos type 4	carotid cavernous fistulae	common
	intracranial aneurysms	
	(cavernous carotid)	common
	arterial dissections	moderate
	cervical arterial aneurysms	rare
Marfan's syndrome	dissection of aorta	common
	carotid dissection	moderate
	aneurysms (intra and extracranial)	moderate
Pseudoxanthoma elasticum	vaso-occlusive disease of	
	carotid or vertebral	common
	moyamoya	moderate
	intracranial aneurysms	moderate
	arterial dissections	rare
Neurofibromatosis type 1	intracranial occlusive disease	
	with moyamoya	common
	arterial venous fistula/aneurysms	
	of extracranial arteries	moderate
	intracranial aneurysms	moderate
	extracranial occlusion of vessels	rare

(Marfanoid apearance), heart valves,and joints.[117]

The Ehlers-Danlos syndromes (EDS) describe a somewhat complex group of connective tissue abnormalities with wide ranging systemic effects that are highly variable.[124] There are ten different types of EDS, although type 4 is most commonly recognized as being associated with cerebrovascular events.[120,125] EDS type 4 is inherited as an autosomal dominant trait. It is now known to be due to abnormalities in type III collagen resulting from a variety of mutations in the type III procollagen gene.[126–128]

The most common cerebrovascular abnormality associated with EDS type 4 is a spontaneous carotid cavernous fistula (CCF).[120,129,130] The occurrence of bilateral carotid cavernous fistulae with a positive family history of the same should suggest EDS type 4.[120]Another common complication of EDS type 4 is the development of intracranial aneurysms, which may be involved in the formation of the CCFs described above.[120,125,131,132] This association is discussed in more detail in *Chapter 9*. Dissections of the internal carotid artery or vertebral artery can occur, as well as the formation of cervical extracranial aneurysms.[120,131,133,134] Vessels affected by EDS type 4 may be difficult to repair surgically, and in some cases, angiography may lead to further dissections and bleeding.[135]

Pseudoxanthoma elasticum (PXE) affects elastic fibers of the skin, eyes, and vascular system.[136] The skin lesions are circular and consist of raised yellow orange lesions. Angiod streaks in the eyes occur in about 85% of affected patients, and retinal hemorrhage may accompany up to one-third of these cases.[137,138] Pathologic studies have shown abnormal elastic fibers of the skin and blood vessels, but it is unclear if these changes are primary or secondary to another underlying defect. There is mineralization of elastic fibers with increased production of these fibers in some skin areas.[136]

Two modes of inheritance have been described for PXE: autosomal dominant and autosomal recessive. The molecular genetic etiology of PXE is unknown. Two candidate genes/proteins have been studied—elastin and fibrillin. One study examined the elastin gene for gross mutations in one family with PXE but none were detected.[139] Another study used immunofluorescence to study fibrillin in 16 patients with PXE, but no convincing abnormalities were found.[140]

The most common cerebrovascular abnormalities in PXE are stenotic or occlusive lesions involving the carotid and vertebral arteries. Lesions in the extracranial or intracranial portion of these arteries have also been described.[141,142] Spontaneous arterial dissections and intracranial aneurysms have been reported, but these are considerably less common.[120]

Another common genetic disorder is neurofibromatosis (NF). NF is essentially two diseases, classified as NF-1 and NF-2. NF-1 affects multiple organ systems and typically has vascular involvement. NF-2 is defined by bilateral acoustic neuromas and other tumors.[143] For the most part, NF-2 does not have significant cerebrovascular manifestations.

NF-1 is inherited as an autosomal dominant trait, with new mutations quite common. It is due to a variety of mutations in neurofibromin, which is a tumor suppressor gene found on CH17.[144] This gene may also be involved in the regulation of the growth and development of connective tissues. In terms of cerebrovascular disease, a common abnormality is stenosis or occlusion of the supraclinoid internal carotid artery.[145,146,146a] This often leads to a moyamoya pattern of collaterals which may rupture, producing an ICH or SAH. The underlying pathology of the stenotic lesions appears to be due to proliferation of smooth muscle cells.[146a] Arteriovenous fistulae of the extracranial cerebral vessels, as well as intracranial aneurysms, have also been reported.[147,148]

In patients with findings as described above, a search for other systemic manifestations of NF-1 is certainly reasonable. Typical lesions to look for include neurofibromas of the skin, cafe-au-lait spots, Lisch nodules, axillary or inguinal freckling, renal artery stenosis, and a variety of nervous system tumors including gliomas and schwannomas.[143]

Other Disorders

Fabry's disease is an X-linked dominant disorder. Males are affected more than females, but females carrying one mutated gene also manifest symptoms. The disease is caused by deficiency of the lysosomal emzyme alpha-galactosidase A, leading to accumulation of ceramide trihexoside throughout the body. Organ systems that are predominantly affected include the kidneys, cornea, vasculature, and peripheral nerves. Systemic complications include renal failure, rash (angiokeratoma) myocardial infarction, valvular dysfunction, and painful crises.[149]

The cerebrovascular disorders include ischemic strokes most commonly, although ICH and SAH can occur. Infarctions tend to involve the vertebrobasilar territory in about two-thirds of cases.[150] Three major underlying mechanisms appear to be responsible. One is the elongation and tortuosity of large vessels, leading to thrombosis of these vessels, which may be caused by defects in the internal elastic lamina and smooth muscle cells. The second mechanism involves infarction in the territory of small vessels and may be caused by stenosis of these vessels due to the deposition of glycosphingolipids in the vessel wall. Dolicoectasia of large vessels may also lead to compression and pinching at the origin of some penetrating vessels. A third mechanism is cardiac embolism due to involvement of the heart valves.[150] Rare cases of ICH have also been reported.[150]

Treatment of Fabry's disease remains problematic. Renal transplantation may alleviate some of the enzyme deficiency.[150] Direct enzymatic replacement has also been attempted with varying degrees of short-term success.[151] Treatment with antiplatelet agents has been suggested, although large prospective studies have not been performed due to the rarity of the disorder.[152]

Although this volume has focused on strokes in humans, there has been important relevant research using animal models of stroke that may have applications for understanding strokes in humans. One of the most common models is the spontaneously hypertensive stroke prone rat (SHRSP). As implied by the name, this strain is hypertensive and has a high rate of stroke (although in most cases the stroke is surgically produced by ligating the middle cerebral artery). By breeding large numbers of animals, it is possible to perform genetic linkage studies to determine which loci may be responsible for the high rate of stroke. A recent study used this type of analysis and found a locus with an LOD score > 16, indicating very strong linkage.[153] Further analysis showed that the genetic marker at this locus (on chromosome 5) was within the gene for atrial natriuretic factor (ANF), and very close to the gene for brain natriuretic factor (BNF). The analysis was performed based on the severity of the ischemic stroke, rather than the presence of hypertension. The precise mechanism underlying this relationship is unclear. Both ANF and BNF can affect vasoreactivity, as well as modulate vessel growth.[153] Since impaired vasoreactivity has been shown to present in the SHRSP strain, these findings may suggest one mechanism of stroke susceptibility.[154] This area of research, combining genetic approaches with informative animal models, may prove useful for identifying other loci and genes having a role in stroke pathogenesis.

Summary

This chapter has provided a review of some of the more common inherited systemic disorders that are associated with stroke. As discussed, a variety of these can cause stroke. These range from hemoglobinopathies to coagulation and mitochondrial disorders, and include connective tissue disorders and abnormalities in homocysteine metabolism. In evaluating patients with inherited stroke and, in some cases, patients with spontaneous stroke, a careful search for clinical and laboratory evidence of these systemic disorders is prudent. The identification of these disorders may not only establish a diagnosis for the stroke, but also may indicate potential therapies and provide valuable information for genetic counseling of relatives of the affected patient.

References

1. Weatherall D, Clegg J, Higgs D, et al. The hemoglobinopathies. In Scriver C, et al (eds). The Metabolic and Molecular Basis of Inherited Disease. New York: McGraw-Hill, 1995, 3417–3471.
2. Nicholls R, Jonasson J, McGee J, et al. High resolution mapping of the human alpha-globin locus. J Med Genet 1987;24:39–46.
3. Koeffler H, Sparkers R, Stang H, et al. Regional assignment of genes for human alpha globin and phosphoglycerate phosphatase to the short arm of chromosome 16. Proc Natl Acad Sci USA 1981;78:7015–7018
4. Sanders-Haigh L, Anderson W, Francke U. The beta-globin gene is on the short arm of chromosome 11. Nature 1980;283:683–686.
5. Ingram V. Specific chemical differences between the globins of normal human and sickle-cell anaemia haemoglobin. Nature 1956;178:792–794.
6. Serjeant GR. The clinical features of sickle cell disease. Baillieres Clin Haematol 1993;6:93–115.
7. Mohr J. Sickle cell anemia, stroke, and transcranial studies. N Engl J Med 1992;326:637–639.
8. Powars D, Wilson B, Imbus C, et al. The natural history of stroke in sickle cell disease. Am J Med 1978;65:461–471.
9. Grotta J, Manner C, Pettigrew C, et al. Red blood cell disorders and stroke. Stroke 1986;17:811–817.
10. Ohene-Frempong K. Stroke in sickle cell disease: Demographic, clinical, and therapeutic considerations. Semin Hematol 1991;28:213–219.
11. Zimmerman R, Gill F, Goldberg H, et al. MRI of sickle cell cerebral infarction. Neuroradiology 1987;29:232–237.
12. Pavlakis S, Bello J, Prohovnik I, et al. Brain infarction in sickle cell anemia: Magnetic resonance imaging correlates. Ann Neurol 1988;23:125–130.
13. Rothman S, Fulling K, Nelson J. Sickle cell anemia and central nervous system infarction: A neuropathological study. Ann Neurol 1986;20:684–690.
14. Francis RB. Large-vessel occlusion in sickle cell disease: Pathogenesis, clinical consequences, and therapeutic implications. Med Hypotheses 1991;35:88–95.
15. Merkel K, Ginsberg P, Parker J, et al. Cerebrovascular disease in sickle cell anemia: A clinical, pathological and radiographic correlation. Stroke 1978;9:45–52.
16. Nagel R. Severity, pathobiology, epistatic effects, and genetic markers in sickle cell anemia. Semin Hematol 1991;28:180–201.
17. Russell M, Goldberg H, Hodson A, et al. Effect of transfusion therapy on arteriographic abnormalities and on recurrence of stroke in sickle cell disease. Blood 1984;63:162–169.
18. Adams R, Nichols F, McKie K, et al. Cerebral infarction in sickle cell anemia: Mechanism based on CT and MRI. Neurology 1988;38:1012–1017.
19. Vernant J, Delaporte J, Buisson G, et al. Cerebrovascular complications of sickle-cell anemia. Rev Neurol 1988;144:465–473.
20. Coull B, Goodnight S. Antiphospholipid antibodies and coagulation disorders in ischemic stroke. In Barnett H, Mohr J, Stein B, Yatsu F, (eds). Stroke: Pathophysiology, Diagnosis, and Management. New York: Churchill Livingstone, 1992, 859–873.
21. Anson J, Koshy M, Ferguson L, et al. Subarachnoid hemorrhage in sickle-cell disease. J Neurosurg 1991;75:552–558.
22. Balkaran B, Char G, Morris JS, et al. Stroke in a cohort of patients with homozygous sickle cell disease. J Pediatr 1992;120:360–366.
23. Bordin J, Kerbauy J, Lourenco D, et al. Level of fetal hemoglobin as an indicator of clinical complications in sickle cell anemia. Braz J Med Biol Res 1989;22:1347–1353.
24. Powars D, Weiss J, Chan L, et al. Is there a threshold level of fetal hemoglobin that ameliorates morbidity in sickle cell anemia? Blood 1984;63:921–926.
25. Adams RJ, Kutlar A, McKie V, et al. Alpha thalassemia and stroke risk in sickle cell anemia. Am J Hematol 1994;45:279–282.
26. Steinberg M, Rosenstock W, Coleman M, et al. Effects of thalassemia and microcytosis on the hematologic and vasoocclusive severity of sickle cell anemia. Blood 1984;63:1353–1360.
27. Adams RJ, Nichols FT, Aaslid R, et al. Cerebral vessel stenosis in sickle cell disease: Criteria for detection by transcranial Doppler. Am J Pediatr Hematol Oncol 1990;12:277–282.
28. Adams R, McKie V, Nichols F, et al. The use of transcranial ultrasonography to predict

stroke in sickle cell disease [see comments]. N Engl J Med 1992;326:605–610.

29. Adams RJ, Nichols FT, Figueroa R, et al. Transcranial Doppler correlation with cerebral angiography in sickle cell disease. Stroke 1992;23:1073–1077.

30. Seibert JJ, Miller SF, Kirby RS, et al. Cerebrovascular disease in symptomatic and asymptomatic patients with sickle cell anemia: Screening with duplex transcranial Doppler US correlation with MR imaging and MR angiography. Radiology 1993;189: 457–466.

31. Wiznitzer M, Ruggieri PM, Masaryk TJ, et al. Diagnosis of cerebrovascular disease in sickle cell anemia by magnetic resonance angiography. J Pediatr 1990;117:551–555.

32. Cohen AR, Martin MB, Silber JH, et al. A modified transfusion program for prevention of stroke in sickle cell disease. Blood 1992;79:1657–1661.

33. Charache S, Terrin ML, Moore RD, et al. Effect of hydroxyurea on the frequency of painful crises in sickle cell anemia: Investigators of the Multicenter Study of Hydroxyurea in Sickle Cell Anemia. N Engl J Med 1995;332:1317–1322.

34. Vichinsky E, Lubin B. A cautionary note regarding hydroxyurea in sickle cell disease. Blood 1994;83:1124–1128.

35. Radhakrishnan K, Thacker A, Maloo J, et al. Sickle cell trait and stroke in the young adult. Postgrad Med J 1990;66:1078–1080.

36. Reyes M. Subcortical cerebral infarctions in sickle cell trait. J Neurol Neurosurg Psychiatry 1989;52:516–518.

37. Bunn H, Forget B. Hemoglobin: Molecular, Genetic and Clinical Aspects. Philadelphia: WB Saunders, 1986.

38. Chalevelakis G, Clegg J, Weatherall D. Imbalanced globin chain synthesis in heterozygous β-thalassemic bone marrow. PNAS 1975;72:3853–3857.

39. Dozy A, Kan Y, Embury S, et al. Alpha-globin gene organization in blacks precludes the severe form of alpha-thalassemia. Nature 1979;280:605–607.

40. Higgs D, Vickers M, Wilkie A, et al. A review of the molecular genetics of the human alpha-globin gene cluster. Blood 1989;73:1081–1104.

41. Manoussakis M, Gharavi A, Drosos A, et al. Anticardiolipin antibodies in unselected autoimmune rheumatic disease patients. Clin Immunol Immunopathol 1987;44:297–307.

42. Love P, Santoro S. Antiphospholipid antibodies: Anticardiolipin and the lupus anticoagulant in systemic lupus erythematosus (SLE) and in non-SLE disorders. Ann Int Med 1990;112:682–698.

43. Brey R, Hart R, Sherman D, et al. Antiphos-

pholipid antibodies and cerebral ischemia in young adults. Neurology 1990;40:1190–1196.

44. Levine S, Deegan M, Futrell N, et al. Cerebrovascular and neurologic disease associated with antiphospholipid antibodies. Neurology 1990;40:1181–1189.

45. Gastineau D, Kazmer F, Nichols W, et al. Lupus anticoagulant: An analysis of the clinical and laboratory features of 219 cases. Am J Hematol 1985;19:265–275.

46. Levine S, Welch K. The spectrum of neurologic disease associated with antiphospholipid antibodies: Lupus anticoagulants and anticardiolipin antibodies. Arch Neurol 1987;44:876–883.

47. Pettee A, Wasserman B, Adams N, et al. Familial Sneddon's syndrome. Neurology 1994;44:399–405.

48. Smith H, Auld S, Stein R. Familial association of antiphospholipid antibodies. Arthritis Rheum 1991;34:S42.

49. Matthey F, Walshe K, Mackie I, et al. Familial occurrence of the antiphospholipid syndrome. J Clin Pathol 1989;42:495–497.

50. Exner T, Barber S, Kronenberg H, et al. Familial assocation of the lupus anticoagulant. Br J Haematol 1980;45:89–96.

51. Mackie I, Colaco C, Machin S. Familial lupus anticoagulant. Br J Haematol 1987;67:359–363.

52. Mackworth-Young C, Chan J, Harris N, et al. High incidence of anticardiolipin antibodies in relatives of patients with systemic lupus erythematosus. J Rheumatol 1987;14:723–726.

53. Sneddon I. Cerebrovascular lesions and levido reticularis. Br J Dermatol 1965;77:180–185.

54. Rebello M, Val J, Gaijo F, et al. Livedo reticularis and cerebrovascular lesions (Sneddon's syndrome). Brain 1983;106:965–979.

55. Berciano J. Sneddon syndrome: Another Mendelian etiology of stroke. Ann Neurol 1988;24:586–587.

56. Lossos A, Ben-Hur T, Ben-Nariah Z, et al. Familial Sneddon's syndrome. J Neurol 1995; 242:164–168.

57. Rumpl E, Neuhofer J, Pallua A, et al. Cerebrovascular lesions and levido reticularis (Sneddon's syndrome): A progressive cerebrovascular disorder? J Neurol 1985;231: 324–330.

58. Levine S, Langer S, Albers J, et al. Sneddon's syndrome: An antiphospholipid antibody syndrome? Neurology 1988;38:798–800.

59. Vargas J, Yebra M, Pascual M, et al. Antiphospholipid antibodies and Sneddon's syndrome. Am J Med 1989;87:597.

60. Lousa M, Sastre JL, Cancelas JA, et al. Study of antiphospholipid antibodies in a patient with Sneddon's syndrome and her family. Stroke 1994;25:1071–1074.

61. Biousse V, Frances C, Piette J-C, et al. Acti-

vated protein C resistance caused by factor V Arg 506→Gln mutation in 14 patients with Sneddon's syndrome. Neurology 1996;46: A479.

62. Hess D, Krauss J, Adams R, et al. Anticardiolipin antibodies: A study of frequency in TIA and stroke. Neurology 1991;41:525–528.

63. Rosove M, Brewer P. Antiphopholipid thrombosis: Clinical course after the first thrombotic event in 70 patients. Ann Int Med 1992;117:303–308.

64. Khamashta M, Cuadrado M, Mujic F, et al. The management of thrombosis in the antiphospholipid-antibody syndrome. N Engl J Med 1995;332:993–997.

65. Winterkorn J, Kupersmith M, Wirtschafter J, et al. Brief Report: Treatment of vasospastic amaurosis fugax with calcium-channel blockers. N Engl J Med 1993;329:396–398.

66. Attardi G, Schatz G. Biogenesis of mitochondria. Annu Rev Cell Biol 1988;4:289–333.

67. Anderson S, Bankier A, Barrell B, et al. Sequence and organization of the human mitochondrial genome. Nature 1981;290:457–465.

68. Johns D. Mitochondrial DNA and disease. N Engl J Med 1995;333:638–644.

69. Clayton D. Structure and function of the mitochondrial genome. J Inherit Metab Dis 1992;15:439–447.

70. Giles R, Blanc H, Cann H, et al. Maternal inheritance of human mitochondrial DNA. Proc Natl Acad Sci 1980;77:6715–6719.

71. Wallace D, Ye J, Neckelmann S, et al. Sequence analysis of cDNAs for the human and bovine ATP synthase β-subunit: Mitochondrial DNA genes sustain seventeen times more mutations. Curr Genet 1987;12:81–90.

72. Wallace D. Diseases of the mitochondrial DNA. Annu Rev Biochem 1992;61:1175–1212.

73. Ciafaloni E, Ricci E, Shanske S, et al. MELAS: Clinical features, biochemistry, and molecular genetics. Ann Neurol 1992;31:391–398.

74. Tanaka M, Ino H, Ohno K, et al. Mitochondrial DNA mutations in mitochondrial myopathy, encephalopathy, lactic acidosis, and stroke-like episodes (MELAS). Biochem Biophys Res Commun 1991;174:861–868.

75. Goto Y, Nonaka I, Horai S. A mutation in the tRNALeu(UUR) gene associated with the MELAS subgroup of mitochondrial encephalomyopathies. Nature 1991;346: 651–653.

76. Sakuta R, Goto Y, Horai S, et al. Mitochondrial DNA mutations at nucleotide posotions 3243 and 3271 in mitochondrial myopathy, encephalopathy, lactic acidosis, and stroke-like episodes: A comparative study. J Neurol Sci 1993;115:158–160.

77. Marie SK, Goto Y, Passos-Bueno MR, et al. A Caucasian family with the 3271 mutation in mitochondrial DNA. Biochem Med Metab Biol 1994;52:136–139.

78. Lertrit P, Noer AS, Jean-Francois MJ, et al. A new disease-related mutation for mitochondrial encephalopathy lactic acidosis and strokelike episodes (MELAS) syndrome affects the ND4 subunit of the respiratory complex I [see comments]. Am J Hum Genet 1992;51:457–468.

79. Pavlakis S, Phillips P, DiMauro S, et al. Mitochondrial myopathy, encephalopathy, lactic acidosis, and strokelike episodes: A distinctive clinical syndrome. Ann Neurol 1984;16: 481–488.

80. Matthews PM, Tampieri D, Berkovic SF, et al. Magnetic resonance imaging shows specific abnormalities in the MELAS syndrome. Neurology 1991;41:1043–1046.

81. Forster C, Hubner G, Muller-Hocker J, et al. Mitochondrial angiopathy in a family with MELAS. Neuropediatrics 1992;23:165–168.

82. Moraes CT, Ricci E, Bonilla E, et al. The mitochondrial tRNA(Leu(UUR)) mutation in mitochondrial encephalomyopathy, lactic acidosis, and strokelike episodes (MELAS): Genetic, biochemical, and morphological correlations in skeletal muscle. Am J Hum Genet 1992;50:934–949.

83. Hammans SR, Sweeney MG, Brockington M, et al. Mitochondrial encephalopathies: Molecular genetic diagnosis from blood samples [see comments]. Lancet 1991;337:1311–1313.

84. Abe K, Fujimura H, Nishikawa Y, et al. Marked reduction in CSF lactate and pyruvate levels after CoQ therapy in a patient with mitochondrial myopathy, encephalo-pathy, lactic acidosis and stroke-like episodes (MELAS). Acta Neurol Scand 1991;83:356–359.

85. Motulsky A. Nutritional ecogenetics: Homocysteine-related arteriosclerotic vascular disease, neural tube defects, and folic acid. Am J Hum Genet 1996;58:17–20.

86. McCully K. Homocysteine and vascular disease. Nat Med 1996;2:386–389.

87. Skovby F. The homocystinurias. In Royce P, Steinmann B, (eds). Connective Tissue and its Heritable Disorders. New York: Wiley-Liss, 1993, 469–486.

88. Selhub J, Jacques P, Wilson P, et al. Vitamin status and intake as primary determinants of homocystinemia in an elderly population. JAMA 1993;270:2693–2698.

89. Rosenblatt D. Inherited disorders of folate transport and metabolism. In Scriver C, Beaudet A, Sly W, et al, (eds). The Metabolic and Molecular Basis of Inherited Disease. New York: McGraw-Hill, 1995, 3111–3128.

90. Fenton W, Rosenberg L. Inherited disorders of cobalamin transport and metabolism. In Scriver C, Beaudet A, Sly W, et al (eds). The

Metabolic and Molecular Basis of Inherited Disease. New York: McGraw-Hill, 1995, 3129–3149.

91. Mudd S, Levy H, Skovby F. Disorders of transsulfuration. In Scriver C, Beaudet A, Sly W, et al (eds). The Metabolic Basis of Inherited Disease. New York: McGraw-Hill, 1989, 693–734.

92. Munke M, Kraus J, Ohura T, et al. The gene for cystathione beta-synthase (CBS) maps to the subtelomeric region on human chromosome 21q and to the proximal mouse chromosome 17. Am J Hum Genet 1988;42:550–559.

93. Kraus J. Molecular basis of phenotype expression in homocystinuria. J Inherit Metab Dis 1994;17:383–390.

94. Hu F, Gu Z, Kozich V, et al. Molecular basis of cystathionine beta-synthase deficiency in pyridoxine responsive and nonresponsive homocystinuria. Hum Mol Genet 1993;2: 1857–1860.

95. Shih V, Fringer J, Mandell R, et al. A missense mutation (I278T) in the cystathionine beta-synthase gene prevalent in pyridoxine-responsive homocystinuria and associated with mild clinical phenotype. Am J Hum Genet 1995;57:34–39.

96. Mudd S, Skovby F, Levy H, et al. The natural history of homocystinuria due to cystathionine β-synthase deficiency. Am J Hum Genet 1985;37:1–31.

97. Morrells C, Fletcher B, Weilbaecher R, et al. The roentgenographic features of homocystinuria. Radiology 1968;90:1150–1158.

98. Mandel H, Brenner B, Berant M, et al. Coexistence of hereditary homocystinuria and factor V Leiden-effect on thrombosis. N Engl J Med 1996;334:763–768.

99. Bertina R, Koelman B, Koster T, et al. Mutation in blood coagulation factor V associated with resistance to activated protein C. Nature 1994;369:64–67.

100. Naruszewicz M, Mirkiewicz E, Olszewski A, et al. Thiolation of low-density lipoprotein by homocysteine thiolactone causes increased aggregation and altered interaction with cultured macrophages. Nutr Metab Cardiovasc Dis 1994;4:70–77.

101. Harker L, Slichter S, Scott C, et al. Homocystinemia: Vascular injury and arterial thrombosis. N Engl J Med 1974;291:537–543.

102. Clarke R, Daly L, Robinson K, et al. Hyperhomocysteinemia: An independent risk factor for vascular disease. N Engl J Med 1991;324:1149–1155.

103. Boers G, Smals A, Trijbels F, et al. Heterozygosity for homocystinuria in premature peripheral and cerebral occlusive arterial disease. N Engl J Med 1985;313:709–715.

104. Boushey C, Beresford S, Omenn G, et al. A quantitative assessment of plasma homocysteine as a risk factor for vascular disease. JAMA 1995;274:1049–1057.

105. Engbersen A, Franken D, Boers G, et al. Thermolabile 5,10-methylenetetrahydrofolate reductase as a cause of mild hyperhomocysteinemia. Am J Hum Genet 1995;56:142–150.

106. Kluijtmans L, van den Heuvel L, Boers G, et al. Molecular genetic analysis of mild hyperhomocysteinemia: A common mutation in the methylenetetrahydrofolate reductase gene is a genetic risk factor for cardiovascular disease. Am J Hum Genet 1996;58:35–41.

106a.Graham IM, Daly LE, Refsum HM, et al. Plasma homocysteine as a risk factor for vascular disease. The European Concerted Action project. JAMA 1997;277:1775–1781.

107. Goyette P, Sumner J, Milos R, et al. Human methylenetetrahydrofolate reductase: Isolation of cDNA, mapping and mutation identification. Nat Genet 1994;7:195–200.

108. Goyette P, Frosst P, Rosenblatt D, et al. Seven novel mutations in the methylenetetrahydrofolate reductase gene and genotype/phenotype correlations in severe methylenetetrahydrofolate reductase deficiency. Am J Hum Genet 1995;56:1052–1059.

109. Rosenblatt D. Inherited disorders of folate transport and metabolism. In Scriver C, Beaudet A, Sly W, et al (eds). The Metabolic Basis of Inherited Disease. New York: McGraw-Hill, 1989, 2049–2064.

110. Visy JM, Le Coz P, Chadefaux B, et al. Homocystinuria due to 5,10-methylenetetrahydrofolate reductase deficiency revealed by stroke in adult siblings [see comments]. Neurology 1991;41:1313–1315.

111. Kang S-S, Wong P, Susmano A, et al. Thermolabile methylenetetrahydrofolate reductase: An inherited risk factor for coronary artery disease. Am J Hum Genet 1991;48: 536–545.

112. Frosst P, Blom H, Milos R, et al. A candidate genetic risk factor for vascular disease: A common mutation in methylenetetrahydrofolate reductase. Nat Genet 1995;10:111–113.

113. Godfrey M. The Marfan Syndrome. In Beighton P (ed). McKusick's Heritable Disorders of Connective Tissue. St. Louis: CV Mosby Co, 1993, 51–135.

114. Kainulainen K, Pulkkinen L, Savolainen A, et al. Location on chromosome 15 of the gene defect causing Marfan syndrome. N Engl J Med 1990;323:935–939.

115. Maslan C, Corson G, Maddox B, et al. Partial sequencing of a candidate gene for the Marfan syndrome. Nature 1991;353:334–337.

116. Tynan K, Comeau K, Pearson M, et al. Mutation screening of complete fibrillin-1 cod-

ing sequence: Report of five new mutations, including two in 8-cysteine domains. Hum Mol Genet 1993;2:1813–1821.

117. Pyeritz R. The Marfan Syndrome. In Royce P, Steinmann B, (eds). Connective Tissue and its Heritable Disorders. New York: Wiley-Liss, 1993, 437–468.

118. Dietz H, Cutting G, Pyeritz R, et al. Defects in the fibrillin gene cause the Marfan syndrome: Linkage evidence and identification of a missense mutation. Nature 1991;352:337–339.

119. Sakai L, Keene D, Engvall E. Fibrillin, a new 350-kD glycoprotein, is a component of extracellular microfibrils. J Cell Biol 1986;103:2499–2509.

120. Schievink W, Michels V, Piepgras D. Neurovascular manifestations of heritable connective tissue disorders. Stroke 1994;25: 889–903.

121. Youl B, Coutellier A, Dubois B, et al. Three cases of spontaneous extracranial vertebral artery dissection. Stroke 1990;21:618–625.

122. Matsuda M, Matsuda I, Handa H, et al. Intracavernous giant aneurysm associated with Marfan's syndrome. Surg Neurol 1979;12:119–121.

123. Finney H, Roberts T, Anderson R. Giant intracranial aneurysm associated with Marfan's syndrome: Case report. J Neurosurg 1976;45:342–347.

124. Beighton P. The Ehlers-Danlos syndromes. In Beighton P (ed). McKusick's Heritable Disorders of Connective Tissue. St. Louis: CV Mosby, 1993, 189–251.

125. Krog M, Almgren B, Eriksson I, et al. Vascular complications of the Ehlers-Danlos syndrome. Acta Chir Scand 1983;149:279–282.

126. Kontusaari S, Tromp G, Kuivaniemi H, et al. Inheritance of an RNA splicing mutation (G+1 IVS20) in the type III procollagen gene (COL3A1) in a family having aortic aneurysms and easy bruisability: Phenotypic overlap between familial arterial aneurysms and Ehlers-Danlos syndrome type IV. Am J Hum Genet 1990;47:112–120.

127. Lee B, Vitale E, Superti FA, et al. G to T transversion at position +5 of a splice donor site causes skipping of the preceding exon in the type III procollagen transcripts of a patient with Ehlers-Danlos syndrome type IV. J Biol Chem 1991;266:5256–5259.

128. Kuivaniemi H, Kontusaari S, Tromp G, et al. Identical G+1 to A mutations in three different introns of the type III procollagen gene (COL3A1) produce different patterns of RNA splicing in three variants of Ehlers-Danlos syndrome. IV. An explanation for exon skipping some mutations and not others. J Biol Chem 1990;265:12067–12074.

129. Halbach V, Higashida R, Dowd C, et al.

Treatment of carotid-cavernous fistuals assocation with Ehlers-Danlos syndrome. Neurosurgery 1990;26:1021–1027.

130. Schievink W, Piepgras D, Earnest F, et al. Spontaneous carotid-cavernous fistulae in Ehlers-Danlos syndrome: Case report. J Neurosurg 1991;74:991–998.

131. Schievink W, Limburg M, Oorthuys J, et al. Cerebrovascular disease in Ehlers-Danlos syndrome type IV. Stroke 1990;21:626–632.

132. Bannerman R, Ingall G, Graf C. The familial occurrence of intracranial aneurysms. Neurology 1970;20:283–292.

133. Brodribb A. Vertebral aneurysm in a case of Ehlers-Danlos syndrome. Br J Surg 1970;57: 148–151.

134. Ruby S, Kramer J, Cassidy S, et al. Internal carotid artery aneurysm: A vascular manifestation of type IV Ehlers-Danlos syndrome. Conn Med 1989;53:142–144.

135. Cikrit D, Miles J, Silver D. Spontaneous arterial perforation: The Ehlers-Danlos specter. J Vasc Surg 1987;5:248–255.

136. Neldner K. Pseudoxanthoma elasticum. In Royce P, Steinmann B, (eds). Connective Tissue and its Heritable Disorders. New York: Wiley-Liss, 1993, 425–436.

137. Yap E-Y, Gleaton M, Buettner H. Visual loss associated with pseudoxanthoma elasticum. Retina 1992;12:315–319.

138. Connor P, Juergens J, Perry H, et al. Pseudoxanthoma elasticum and angiod streaks: A review of 106 cases. Am J Med 1961;30: 537–543.

139. Raybould M, Birley A, Moss C, et al. Exclusion of an elastin gene (ELN) mutation as the cause of pseudoxanthoma elasticum (PXE) in one family. Clin Genet 1994;45:48–51.

140. Godfrey M, Cisler J, Geerts M, et al. Fibrillin immunofluorescence in pseudoxanthoma elasticum. J Am Acad Dermatol 1995;32: 589–594.

141. Iqbal A, Alter M, Lee S. Pseudoxanthoma elasticum: A review. Ann Neurol 1978;4: 18–20.

142. Carlborg U, Ejrup B, Gronblad E, et al. Vascular studies in pseudoxanthoma elasticum. Acta Med Scand 1959;350(suppl):1–84.

143. Mulvihill J, Parry D, Sherman J, et al. NIH conference: Neurofibromatosis I (Recklinghausen disease) and neurofibromatosis 2 (bilateral acoustic neurofibromatosis), an update. Ann Intern Med 1990;113:39–52.

144. Ballester R, Marchuk D, Boguski M, et al. The NF1 locus encodes a protein functionally related to mammalian GAP and yeast IRA proteins. Cell 1990;63:851–859.

145. Leone R, Schatzki S, Wolpow E. Neurofibromatosis with extensive intracranial arterial occlusive disease. AJNR 1982;3:572–576.

146. Tomsick T, Lukin R, Chambers A, et al.

Neurofibromatosis and intracranial arterial occlusive disease. Neuroradiology 1976;11: 229–234.

146a. Rizzo JF, Lessell S. Cerebrovascular abnormalities in neurofibromatosis type 1. Neurology 1994;44:1000–1002.

147. Parkinson D, Hay R. Neurofibromatosis. Surg Neurol 1986;25:109–113.

148. Deans W, Bloch S, Leibrock L, et al. Arteriovenous fistula in patients with neurofibromatosis. Radiology 1982;144:103–107.

149. Desnick RJ, Ioannou YA, Eng CH, et al. alpha-Galactosidase A deficiency: Fabry disease. In Scriver CR, Beaudet AL, Sly WS, et al (eds). The Metabolic and Molecular Basis of Inherited Disease. 7th ed. New York: McGraw Hill, 1995:2741–2784.

150. Mitsias P, Levine SR. Cerebrovascular complications of Fabry's disease. Ann Neurol 1996;40:8–17.

151. Desnick RJ, Dean KJ, Grabowski G, et al. Enzyme therapy in Fabry disease: Differential in vivo plasma clearance and metabolic effectiveness of plasma and splenic alpha-galactosidase A isozymes. Proc Natl Acad Sci USA 1979;76:5326–5330.

152. Sakuraba H, Igarashi T, Shibata T, et al. Effect of vitamin E and ticlopidine on platelet aggregation in Fabry's disease. Clin Genet 1987;31:349–354.

153. Jeffs B, Clark JS, Anderson NH, et al. Sensitivity to cerebral ischaemic insult in a rat model of stroke is determined by a single genetic locus. Nat Genet 1997;16:364–367.

154. Volpe M, Iaccarino G, Vecchioone C, et al. Association and cosegregation of stroke with impaired endothelium-dependent vasorelaxation in stroke prone, spontaneously hypertensive rats. J Clin Invest 1996;98: 256–261.

Part IV

Clinical Applications

Chapter 15

Evaluation of Stroke Patients for Inherited Causes of Stroke

Mark J. Alberts, MD, Jeffrey M. Stajich, PA-C

Introduction

This chapter will review some clinical, genetic, and laboratory guidelines for determining: (1) whether a stroke in a particular patient may be due to an inherited disease, and (2) the most appropriate manner by which to evaluate such patients. Because a myriad of genetic diseases can cause strokes, we have chosen to organize this chapter using a common classification scheme based on stroke subtypes. This should be familiar to most physicians (particularly neurologists and neurosurgeons). For the nonphysician researcher or health care provider, we hope the following sections will provide a useful system for classifying and evaluating stroke patients.

General Clinical Guidelines

As with any clinical problem, a thorough and complete history and physical is the first step in the evaluation. In addition to obtaining data on the onset and symptoms of the stroke, a complete family history will provide key information (and sometimes the only information) that will support a genetic etiology for the stroke. Because the family history is so important in the evaluation of patients with a suspected genetic disorder, we will re-

view below, in some detail, how to obtain a complete and thorough family history.

One of the most useful classifications of stroke is by stroke type and vessel involvement (Table 1). We will use this classification to organize some of the genetic disorders that can cause stroke. Later in this chapter, additional clinical and laboratory features will be included to complete this classification scheme.

A clinical "pearl" that we have used in evaluating stroke patients is that when we encounter "unusual strokes in unusual folks" we think of unusual stroke etiologies such as an inherited disease. The term "unusual strokes" refers to strokes with atypical or unusual clinical features (i.e. presenting symptoms, anatomic location, radiologic findings). The designation "unusual folks" indicates that the patient is atypical in terms of age (often younger than 45 years) or other clinical aspects (i.e., other systemic diseases, skin lesions, family history, past history). While this clinical paradigm certainly does not encompass all types of inherited stroke patients or syndromes, it will properly attract the clinician's attention in most cases.

The general history should include information on related symptomatology that may appear unrelated to the specific stroke, yet upon closer examination it may provide a

From Alberts MJ (ed). *Genetics of Cerebrovascular Disease.* Armonk, NY: Futura Publishing Company, Inc., © 1999.

Table 1
Classification of Stroke by
Type and Vessel Involvement

Ischemic Stroke	Hemorrhagic Stroke
Thrombotic	Intracerebral hemorrhage
large vessel	Subarachnoid hemorrhage
small vessel	Ischemic stroke with secondary
Embolic	hemorrhagic transformation
Venous thrombosis	Venous thrombosis

valuable clue about an underlying genetic etiology. For example, a strong history of migraine headaches and an abnormal MRI might suggest the CADASIL (cerebral autosomal dominant arteriopathy with subcortical infarcts and leukoencephalopathy) syndrome in some patients. A history of difficult to control hypertension can be seen in patients with renal artery fibromuscular dysplasia (FMD), which can also produce ischemic strokes and aneurysmal subarachnoid hemorrhage due to FMD of the carotid or vertebral arteries. Renal failure is a common complication of Fabry's disease. A past history of spontaneous abortions may increase suspicion of an antiphospholipid antibody syndrome.

Family History

Taking a detailed family history should be one of the primary components of a comprehensive patient evaluation. When there are atypical features present, as discussed above, suspicions are raised that a genetic etiology may be present. However, a complete family history should be obtained for all patients, even those with a typical presentation. A well-documented, complete family history can be of value by: (1) establishing that the disorder being evaluated is genetic; (2) aiding in establishing a diagnosis; (3) helping to determine the inheritance pattern; (4) providing natural history information of the disease as well as establishing the degree of variability in its expression; and (5) being valuable for further genetic counseling of other family members.

There are several important considerations that should be kept in mind prior to obtaining a family history. First, it is often helpful to identify an informant (besides the patient) within a family who knows or has access to extensive, complete, and accurate family data. While it is usual to take a family history from the presenting patient, the individual may (for any number of reasons) be uninformed about the extended family and thus able to provide only partial information. Since many stroke patients have language problems, obtaining a family history from other family members is particularly relevant for this disease. A follow-up phone call to the family historian who can provide the details that are needed to produce an accurate family history will be well worth the extra effort.

The second consideration is how questions are phrased in relation to the disease. In many hereditary diseases penetrance and expressivity are reduced and, therefore, full-blown disease states may be present in one or only a few individuals while premonitory symptoms may be present in others. This information may be missed if the informant is asked (in the case of SAH, for example) only "Who else in the family has had a hemorrhage or a stroke?". In addition to this question, other inquiries should be made, such as "Does anyone in the family have bad headaches?". "Has anyone died suddenly?" Although the differences may appear subtle, years of experience in ascertaining hundreds of families has shown the importance of complete questioning to avoid missing additional affecteds.

A comprehensive family history should ascertain information about the immediate and extended family. The presenting patient, from whom the family history information is being sought, is known as the proband, the propositus/proposita, or the consultant. Information for constructing a pedigree, or a diagrammatic representation of the family history, is needed from at least three generations including the generation prior to the proband. It may be necessary, depending on the disorder or the mode of inheritance, to expand the pedigree to include more than three generations. The specific information collected should include names, dates of birth and death, marriage(s), separation(s), divorce(s), children (including those out-of-wedlock), causes of death, medical conditions/disorders, and the data concerning the cerebrovascular disease in question. In addition consanguineous relationships must be recorded and ethnicity should be noted.

These data are summarized by the construction of the pedigree using a set of standardized symbols and relationships (Figures 1 and 2).[1,2] Individual data sheets for each family member should be generated if a genetic research project is contemplated. A typical series of questions might proceed as follows if a family history is taken from the proband:

> I would like to ask you some questions about your family history. First, let me start with your parents. Is your mother still living? What is her name? What is her date of birth? Was she related to your father in any way other than by marriage? When did she die (if deceased)? What was her cause of death? How many brothers and/or sisters does your mother have? Was your mother married more than once? How many children did she have by all marriages? Who were the fathers? Did any of the marriages end in separation, divorce, or death of the spouse? Did she have any children outside marriage? How many children, what were their sexes, and who was the father?

This series of questions is repeated for each individual in the three generations. An exception to this is that it is unnecessary to include information on spouses in regard to

their biological families unless the spouse also has the inherited disorder under study. In such cases, detailed genetic and medical information about the spouse and his/her family might be very important for both research and genetic counseling of the proband's children. Once the pedigree is completed, questions specific to the disorder under consideration should then be asked in regard to at-risk individuals. This is the time to inquire about premonitory signs, symptoms, and disease-associated risk factors in addition to determining who else in the family has had a stroke (or related condition).

If the pedigree structure appears to support a genetic etiology, as represented by the example in Figure 2, the next step would be to organize field studies for the collection of DNA samples and clinical information on the family. A family "organizer" first needs to be identified and this may also be the family historian. The organizer needs to understand the details about what will be asked of the family if they are to participate in research studies. With this knowledge he or she can next poll the family members who are necessary for the study, answer their questions, and determine whether there is sufficient family interest for participation. If there is, then a mutually agreeable time, date, and place can be established for the family members to meet with the researchers. Most often, field studies take place on a weekend and a family member's home is used. Because of size considerations, larger facilities such meeting rooms in houses of worship, schools, or hospitals may be necessary.

During the field study, medical histories are obtained, examinations are performed and recorded, and blood samples are drawn. Occasionally, because of poor venous access or an individual's refusal to have blood drawn, DNA samples can be obtained by utilizing buccal swabs. Using this method, endothelial cells are harvested by scraping the inside of the cheeks with a sterile brush. Although this process is painless (in contrast to venipuncture), the potential quantity of DNA yielded from this method is approximately 100 times less than that obtained from

STANDARDIZED PEDIGREE SYMBOLS
and RELATIONSHIPS

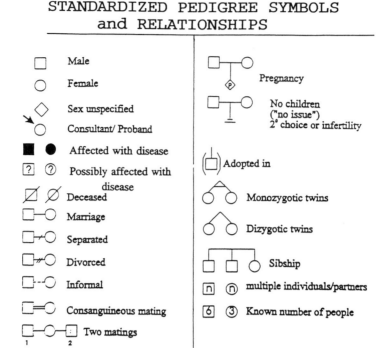

Figure 1. A depiction of the standard symbols and nomenclature used for drawing pedigrees. By using this standardized scheme, family data can be exchanged, with minimal ambiguity and inconsistencies, among centers and investigators from around the world.

venipuncture. This method is therefore used as a last resort when the researcher is not able to obtain a DNA sample by venipuncture from an important family member. The clinical information gathering and blood collection usuallly takes 15–20 minutes per person (depending on how extensive the data base is). If two or three researchers are evaluating family members at the same time, a family of the size represented by Figure 2 can be ascertained in less than 3 hours.

Physical Examination

Depending on the stroke type, there are several findings on physical examination that should be specifically investigated. Skin lesions are a common finding with many of the diseases causing ischemic or hemorrhagic stroke (Table 2). In cases of ischemic stroke, some common skin lesions to look for include the livedo-reticularis rash of Sneddon's syn-

drome, evidence of Raynaud's phenomenon, manifestations of neurofibromatosis (i.e., neurofibromas, axillary/inguinal freckling), changes that indicate peripheral embolization (suggesting heart disease or a systemic coagulation abnormality), or the unusual rash of pseudoxanthoma elasticum. The skeletal changes of Marfan's syndrome and the skin changes seen in Ehlers-Danlos type IV (EDS type IV) should be noted, since these diseases can produce ischemic as well as hemorrhagic strokes. Cutaneous angiokeratomas and corneal dystrophy are seen with Fabry's disease, and telangiectasias of the mucous membranes can be seen with hereditary hemorrhagic telangiectasia (also known as Rendu-Osler-Weber syndrome).

Although eye involvement, particularly retinal ischemia, is common in many stroke syndromes, there is a handful of diseases that specifically involve the eye and retina in unique ways. These findings may provide clues as to the underlying disease (Table 3).

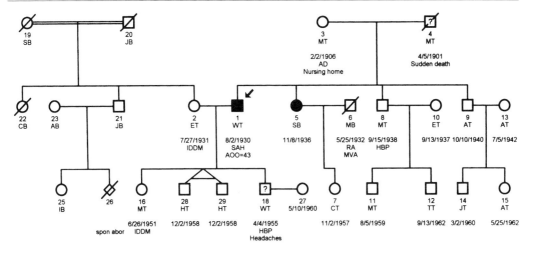

Figure 2. This fictitious pedigree is provided for the purpose of demonstrating some of the aspects of familial relationships and pedigree construction. The initials below each individual indicate names, and the numbers indicate the order in which the family members were ascertained. Different numbering systems can be used, depending upon local preferences. Each individual's date of birth is also below their symbol. Individual #1, WT, is the proband, as indicated by the arrow. He and individual #5, SB, are both affected, as indicated by the darkenend symbols. Individual #6, MB, is deceased, as are individuals #4, 19, 20, 22, and 26. Note the consanguineous mating between individuals #19 and 20. Individuals #28 and 29 are monozygotic twins. Important concomitant diseases are listed under some individuals, using standard abbreviations. The proband, individual #1, had an SAH (subarachnoid hemorrhage), and his wife, individual #2, has insulin-dependent diabetes (IDDM). (See Figure 1 for additional details about pedigree symbols and nomenclature.)

For example, retinal hemorrhage is common in cases of pseudoxanthoma elasticum, and cataracts are common in Fabry's disease.

Premature atherosclerosis is a common cause of ischemic stroke. A search for systemic atherosclerotic lesions (cardiac, peripheral vascular, etc.) may indicate an inherited lipid abnormality, homocystinuria or hyperhomocysteinemia, all of which can cause premature atherosclerosis. Many strokes are due to cardiogenic emboli, and many of the cardiac conditions can be inherited. A careful cardiac examination, with attention to rhythm disturbances and structural disease, is prudent in such cases. Aortic dissection and valvular disease are common with Marfan's syndrome, connective tissue diseases, and skin disorders.

Other findings on neurologic examination may provide valuable clues about an inherited disorder. Patients with an encephalopathy and/or myopathy may have the MELAS syndrome. Evidence of venous occlusive events should suggest a general-

ized hypercoagulable state due to a factor deficiency, an antiphospholipid antibody syndrome, activated protein C resistance/factor V Leiden mutation, or homocystinuria/hyperhomocysteinemia. Patients with anhydrosis and episodic painful crises may have Fabry's disease. Table 4 of *Chapter 13* lists some of the inherited disorders associated with moyamoya disease.

Systemic findings may suggest a particular underlying disorder. For example, evidence of lupus or other autoimmune diseases can be seen with the antiphospholipid antibody syndromes. Severe hypertension might suggest fibromuscular dysplasia as the cause of a stroke, since the renal artery is often involved in this disorder. Severe migraine headaches, though a nonspecific finding, are commonly found in patients with the CADASIL syndrome and the antiphospholipid antibody syndromes. Painful crises or a severe anemia might suggest a hemoglobinopathy.

For patients with intracerebral or sub-

Table 2
Inherited Stroke Syndromes Associated with Skin Abnormalities

Disease	Skin Lesion
Tuberous sclerosis	adenoma sebaceum (angiofibroma), hypopigmented areas, subungual fibromas, shagreen patches
Fabry's	angiokeratoma
Sneddon's syndrome	levido reticularis
Ehlers-Danlos type IV	loose, friable skin
Neurofibromatosis 1	neurofibromas, cafe-au-lait spots, axillary freckling
Protein C deficiency	purpura fulminans (neonetal) skin necrosis (adult)
Protein S deficiency	see protein C above
Hereditary hemorrhagic telangiectasia	telangiectasias
Wyburn-Mason	unilateral facial angioma

Table 3
Ophthalmologic Findings in Some Inherited Disorders Associated with Stroke

Disease	Ophthalmologic Finding
Fabry's	cataract/corneal opacity
Homocystinuria/ homocysteinemia	ectopic/dislocated lens
Marfan's	ectopic/dislocated lens
Neurofibromatosis 1	Lisch nodules
Pseudoxanthoma elasticum	retinal hemorrhage
Tuberous sclerosis	retinal hamartoma

arachnoid hemorrhage, a careful examination is required to look for peripheral manifestations of a systemic bleeding disorder (i.e. skin, fundi, gut), as well as for evidence of specific diseases (i.e., the lax skin seen in EDS type IV, dislocated lens seen with Marfan's). Gastrointestinal hemorrhages and paradoxical emboli due to pulmonary atriovenous malformations (AVMs) are seen in cases of hereditary hemorrhagic telangiectasia. Hemorrhagic strokes are a common complication of the inherited hemoglobinopathies, particularly sickle cell disease. Evidence of dementia in a patient with an ICH should raise suspicion of cerebral amyloid angiopathy as a possible etiology. Renal dysfunction in a patient with an SAH should raise concerns about polycystic kidney disease.

Laboratory Testing

When evaluating a young stroke patient, or a patient with a clear family history of stroke, there may be a temptation to order every test known in an effort to uncover the underlying etiology. However, a more rational process, using the historical and clinical clues outlined above, may lead to a more cost-effective approach for the testing of such patients.

Clearly, all stroke patients require a brain imaging study, preferably an MRI, to help localize and characterize the current stroke, as well as to detect prior or asymptomatic strokes and other brain lesions that may provide clues to the underlying disease (i.e. neurofibromas, small hemangiomas, AVMs). MRI is also superior to head computed tomography (CT) for visualizing small periventricular strokes that may be seen with CADASIL, as well as tiny cavernous hemangiomas and AVMs. The MRI can also be used to obtain a noninvasive image of the cerebral vasculature by performing an MR angiogram. This technique can detect large-vessel lesions such as atherosclerosis, some dissections and aneurysms, fistulae, etc. Conventional angiography is needed to detect small aneurysms, small vessel vasculitis, some AVMs, and moyamoya disease. CT angiography is a new, noninvasive technique that can detect cerebral aneurysms with a

Table 4
Suggested Special Laboratory Tests for Hereditary Ischemic Strokes*

Lupus anticoagulant (will usually have an elevated PTT or PT)
Anticardiolipin antibodies (IgG, IgM, IgA)
Serum lactic acid
Serum/urine homocysteine (may require a methionine challenge test)
Protein C and protein S levels and functional assays
Activated protein C resistance/factor V Leiden mutation
Plasminogen
Antithrombin III
Tissue Plasminogen activator
Collagen-vascular/autoimmune screen
   sedimentation rate
   anti-DNA antibody
   rheumatoid factor
   antineutrophil antibody
Hemoglobin electrophoresis
Fasting lipid profile/lipid electrophoresis
Holter monitor
Echocardiogram
Slit-lamp eye examination and fundiscopic examination
Leukocyte alpha-galactosidase assay
Skin biopsy

*See individual chapters for details of specific diseases, which will help determine the tests that should be ordered.

sensitivity equal to (or greater than) that of conventional angiography.

Routine blood tests for all stroke patients include a CBC, chemistries, PT/PTT, and sedimentation rate. Other routine tests that should be obtained (at least once) include an ECG, chest x-ray, and urinalysis. Once the clinician knows whether the stroke is ischemic or hemorrhagic, more directed laboratory testing can begin. For patients with an ischemic stroke, more detailed tests are listed in Table 4. A screen for hypercoagulable states is reasonable, as is a hemoglobin electrophoresis and tests for collagen vascular disease. A serum lactic acid is appropriate, as is a fasting lipid panel and serum homocysteine levels. Blood specimens can be sent to independent laboratories to screen for activated protein C resistance and the factor V Leiden mutation, which have been associated with some cases of premature thrombotic events.

In cases of hemorrhagic strokes, a further distinction can be made between parenchymatous (intracerebral, ICH) hemorrhages and subarachnoid hemorrhages (SAH). In patients with an ICH, a thorough search for vascular malformations and a vasculopathy should be performed using MRI and angiography. In cases of SAH, an angiogram to identify aneurysms and other vascular anomalies is mandatory. We recommend either a conventional cerebral angiogram or a high resolution CT angiogram. In all cases of cerebral hemorrhage, screening tests for a hypocoagulable state should be performed, including a PT, PTT, and bleeding time. If no structural abnormalities are found to explain an ICH or SAH, and the screening tests are normal, more detailed blood tests for a hypocoagulable state should be performed. These include assays for individual factor deficiences as well as *in vitro* tests of platelet function that examine ADP release and aggregation.

If further genetic testing is planned, or the family and/or disease is part of a research project, blood (or buccal swabs) for DNA extraction should be obtained (as outlined above in the family history section). If the patient has a severe stroke and may not survive, the family should be approached in a compassionate manner (after approval of the patient's attending physician). The legal representative (usually the next-of-kin) should be asked for permission for obtain blood or buccal samples for DNA extraction, with an explanation about the purpose and long-term goals of the study. Obtaining blood via venipuncture has a minimal risk of infection. If necessary, an informed consent should be signed by the patient or his or her legal representative.

Genetic counseling, if appropriate, should be performed by a qualified counselor who has knowledge about the particular disease affecting the patient and his/her family. Further detail regarding genetic counseling can be found in the next chapter.

Making a Diagnosis

As with most neurologic problems, the overall clinical paradigm begins with determining the anatomy and the etiology of a particular lesion and clinical syndrome. Then, based on the results of the evaluation and testing outlined above, a probable diagnosis, or a reasonable differential diagnosis, can be made for a patient or family with inherited stroke. One helpful clue is to separate those genetic diseases that affect only the brain from those with systemic manifestations. For example, sickle cell disease and the MELAS syndrome have systemic manifestations in most cases, while cerebral amyloid angiopathy does not. A listing of some of the more common inherited conditions likely to be encountered in clinical practice is in Table 5. (The reader is urged to refer to prior chapters for a more detailed description of these disorders.)

At this point, more detailed confirmatory testing may be needed. These tests might include specific blood tests, a tissue biopsy, testing DNA for a disease-specific mutation, or a second opinion from clinicians or researchers at another medical center. Every attempt should be made to establish a diagnosis, since it will help guide therapy and be central to any genetic counseling. There are several excellent textbooks that are valuable in making a specific diagnosis.[3,4] In addition, several on-line databases are available that can match symptom complexes and physical findings with thousands of potential genetic diseases. We recommend the Online Mendelian Inheritance in Man (OMIM, Internet address http://www3.ncbi.nlm.nih.gov/omim/), which has a wealth of valuable genetic information. For more detailed genetic mapping and sequence data, the Genome Database can be accessed online at http://gdbwww.gdb.org/

The possibility exists, however, that the clinician will be dealing with a disease that has not been previously described or characterized. In such cases, a thorough evaluation, including a complete and extensive family history, should be obtained. Referral of the patient and family to a tertiary care center or a specialized genetics center may be appropriate in such cases.

Once a diagnosis is made, treatment can begin. In some cases, specific and effective treatments for an inherited disease are well established. Examples include the use of coumadin to treat antiphospholipid antibody syndromes and protein C/protein S deficiency. In other cases, surgical treatment of an aneurysm or AVM may be needed. Therapies for specific diseases are discussed in previous chapters. However, in many cases there is no proven therapy; in such cases, there may be on-going research trials underway at major medical centers. An inquiry at a nearby center, or NIH, can be helpful in determining if any trials are underway. A search of the Internet may identify specific physicians, medical centers, or support groups with an interest and information about a specific disorder.

Conclusion

A systematic and comprehensive approach to patient evaluation is important when investigating patients and families for genetic causes of stroke. Although high-technology tools are available, such an eval-

Table 5
Common Inherited Causes of Stroke

Ischemic Stroke

Disease*	Mechanism**	Gene/Protein***
Hyperlipidemia	premature atherosclerosis	various
Homocystinuria	premature atherosclerosis	cystathione β-synthase
Hyperhomocysteinemia	premature atherosclerosis	MTHFR
Sickle cell disease	vasculopathy/sludging	globin genes
Antiphospholipid antibody syndrome	hypercoagulable state	unknown
Sneddon's syndrome	hypercoagulable state ? vasculopathy	unknown
Fibromuscular dysplasia	vasculopathy	unknown
MELAS	? vasculopathy/oxidative stress	mt-DNA mutations
CADASIL	small vessel occlusion	Notch 3
Marfan's syndrome	dissections	fibrillin
Protein C deficiency	hypercoagulable state	protein C gene
Protein S deficiency	hypercoagulable state	protein S gene
Activated protein C resistance	hypercoagulable state	mutation in factor V
Neurofibromatosis type 1	vessel occlusion	neurofibronin

Hemorrhagic Stroke**

Disease	Mechanism	Gene/Protein
HCHWA-D	amyloid angiopathy	APP
HCHWA-I	amyloid angiopathy	cystatin C
Cavernous hemangioma	malformation; ICH	CH7q11-q22
Polycystic kidney disease	aneurysmal SAH	polycystin 1 and 2 (autosomal dominant form)
Rendu-Osler-Weber	telangiectasias	endoglin and ALK1
Familial SAH	aneurysms	unknown
Ehlers-Danlos	aneurysms/fistulae	collagen type IIIA
Sickle cell disease	moyamoya rupture	globin gene
Marfan's syndrome	aneurysmal SAH	fibillin
Factor deficiencies	ICH and SAH	various genes
Neurofibromatosis type 1	fistulae, aneurysms	neurofibronin

* MELAS, mitochondrial encephalomyopathy, lactic acidosis, and stroke; CADASIL, cerebral autosomal dominant arteriopathy with subcortical infarcts and leukoencephalopthy; SAH, subarachnoid hemorrhage; HCHWA-D, hereditary cerebral hemorrhage with amyloidosis-Dutch type; HCHWA-I, hereditary cerebral hemorrhage with amyloidosis-Icelandic type.

**ICH, intracerebral hemorrhage.

***MTHFR, methylenetetrahydrofolate reductase; mt-DNA, mitochondrial DNA; APP, amyloid precursor protein; ALK1, activin receptor-like kinase 1. Chromosomal loci are given in cases where linkage studies have been performed, but a specific causative gene has not been identified.

uation should always begin with a thorough history and physical, including a detailed family history. Based on the results of the history and physical, laboratory screening tests and a brain imaging study can be ordered. These results will then guide the clinician in ordering additional studies so that a definitive diagnosis may be indicated. More detailed information related to testing for or diagnosing a particular disease can be found in previous chapters. In difficult cases, referral to a tertiary care center may be useful for making a diagnosis and commencing therapy.

References

1. Bennet RL, Steinhaus KA, Uhrich SB, et al. Recommendations for standardized human pedigree nomenclature. Am J Hum Genet 1995:745–751.
2. McGoldrick M, Gerson R. Genograms in Family Assessment. New York: WW Norton Co, 1995, 29–39.
3. Scriver CR, et al (eds). The Metabolic and Molecular Basis of Inherited Diseases. New York: McGraw-Hill, 1995.
4. Thompson MW, McInnes RR, Willard HF. Genetics in Medicine. 5th ed. Philadelphia: WB Saunders 1991, 8–11.

Chapter 16

Genetic Counseling

Shane M. Palmer, MS

Introduction

Medical genetics can be defined as a public health field as each of us carries three to eight harmful, lethal genes in recessive fashion, or dominant genes that affect our health and potentially the health of our children.[1] Genetic factors are important causes of common diseases in adults, including heart disease, cancer, diabetes, emphysema, strokes, high blood pressure, and psychiatric illness. In 1975, a committee of physicians and scientists defined the goals and objectives of genetic counseling:

Genetic counseling is a communication process which deals with the human problems associated with the occurrence or the risk of occurrence of a genetic disorder in a family. The process involves helping the individual or family:

1. comprehend the medical facts including the diagnosis, the probable course of the disorder and the available management;
2. appreciate the way heredity contributes to the disorder and the risk of occurrence in specified relatives;
3. understand the options for dealing with the risk of recurrence;
4. choose the course of action which seems appropriate to them in view of their risk, their family goals, and their ethical and religious standards, and act in accordance with that decision; and
5. make the best possible adjustment to the disorder in an affected family member and/or to the recurrence risk of that disorder. (Ad-hoc Committee on Genetic Counseling, 1975).[2]

The emphasis of this definition is on clear and thorough communication of the medical and genetic information while helping the individual or family to use the information provided in the most meaningful way for them. Now, genetic counselors may incorporate a more active counseling stance appropriate to the psychological and problem-solving needs of the genetic counselees.[3] Experienced counselors accomplish this through simple interview formulations, helping clients to think through their problems, and by reframing and role playing.[3] The goal of these procedures during the counseling session is to promote client autonomy and self-directedness, which are particularly valuable in circumstances requiring a nondirective stance.[3]

Primary prevention of genetic disease is not mentioned in the definition of genetic counseling, as the patient is presumed to already have the condition. However, other

From Alberts MJ (ed). *Genetics of Cerebrovascular Disease*. Armonk, NY: Futura Publishing Company, Inc., © 1999.

types of prevention are widely used in genetics, such as secondary prevention which is aimed at keeping the gene disorder from expressing itself as a significant clinical problem.[4] An example includes individuals with adult polycystic kidney disease (AKPD) at risk for a cerebral aneurysm, who may be given prophylatic medication to help prevent an aneurysm from developing. Another type of prevention is tertiary prevention, aimed at minimizing the problem caused by the disorder's expression.[4] An example includes heart bypass surgery for individuals with familial hypercholesterolemia.

The importance of genetic counseling has grown from the recognition of the complex emotional and social issues that may result from the diagnosis of a genetic disorder in a family. Many of the emotional reactions to genetic diseases are applicable to chronic illness in general. These reactions include denial, anger, blame, guilt, sorrow, stigma, alienation, and poor self-image. However, because of the permanence, intrinsic nature, and hereditary aspects of genetic disorders, these feelings (especially guilt) are intensified. By recognizing these reactions and exploring their impact, the counselor can help the family deal with the disorder more effectively.

The genetic, psychological, and social situations that arise in genetic counseling are not simple and straightforward, as reflected by Huntington's disease, breast, ovarian and other cancers, Alzheimer's disease, and mental disorders.[5] The genetic information continues to become more complex and the psychological context of such counseling is one in which the threat to the client's personal health, quality of life, and survivability is intensified. Unless skillful counseling procedures are employed, not only is communication likely to be ineffective but the potential for inflicting psychological harm increases.[5] This needs to be considered especially when dealing with cerebrovascular diseases that have or may have an underlying genetic component.

Who Should Be Referred for Genetic Counseling?

Individuals who are concerned about the potential occurrence of a disorder in themselves or their offspring should be referred for genetic counseling. In many situations, genetic counseling occurs after a condition has been diagnosed in a family. This is referred to as retrospective counseling. Many couples seek genetic counseling prior to having their first child to see if they are at risk of having a child with a birth defect. This is referred to as prospective counseling. The most common reasons for referrals to a genetic clinic are summarized in Table 1. Individuals and families who are seen in a genetic clinic usually have a variety of questions about the condition. However, all want to be informed about the prognosis, risk of recurrence, and available options.

Who Provides Genetic Counseling?

Genetic counseling is provided by a wide variety of professionals. Most often, a team approach is involved, which includes medical geneticists and genetic counselors. Many other health professionals such as nurses, social workers, psychologists, and nutritionists all partake in some aspect of the family's care. The genetic information itself is usually provided by individuals who are specifically trained in genetics.

Clinical medical geneticists are physicians who have completed a 2-year clinical genetic fellowship, after which they are eligible for board certification in this specialty. Recently, the American Board of Medical Specialities has recognized this board certification, which is administered by the American Board of Medical Genetics. In addition, the American Medical Association has recognized the newly formed American College of Medical Genetics.

Genetic counselors have master's-level training in clinical medical genetics. After

Table 1
The Most Common Reasons for Referrals to a Genetic Clinic

- single or multiple malformations
- metabolic disorder
- motor or mental regression without a known etiology
- undifferentiated mental retardation
- family history of genetic disease (i.e., familial hypercholesterolemia)
- consanguinous union
- ethnic background (Eastern European, or Russian Jews- Tay Sachs)
- reproductive losses
- prenatal diagnosis
- exposure to adverse agents during pregnancy
- multifactorial conditions (heart disease, high blood pressure, diabetes, etc.)
- sensory loss
- psychiatric illness

graduation from an accredited program they are eligible for board certification. The American Board of Genetic Counseling oversees the board certification procedure. Their nationwide organization is the National Society of Genetic Counselors (NSGC).

Genetic Counseling Process

Currently, there are a variety of approaches to genetic evaluation depending on the center, the special interests of the various team members, and the specific concerns of the family or individual. The session may be divided into four phases: intake, diagnostic evaluation, discussion, and follow-up. These phases may overlap and are not necessarily discrete. The genetic counseling process is outlined in Table 2.[6]

Intake

The intake phase may begin with a telephone contact with the family prior to the clinic session. This contact can serve to clarify both what the genetic service can offer and what the the family would like to receive. Expectations that are unrealistic can thus be uncovered. This early mutual understanding allows the genetic counselor to prepare for the family's needs and provides the family with a more realistic perspective

of what the session can accomplish. Initial contact can also save time if medical records, X-rays, further family history, or photographs need to be obtained for diagnostic purposes and risk assessment. If there is a specific genetic disorder or medical concern in question, a literature review prior to the family's visit to the clinic may be helpful. If possible, both parents of a child, or husband and wife or the individual's support person should be requested to attend the session to insure that each person receives the information firsthand.

The intake phase continues in the clinic where a family pedigree of up to three generations is compiled. The information requested on each individual is listed in Table 3, and general medical genetic information obtained on each family pedigree is included in Table 4. Any family member with symptoms or problems similar to the affected are noted and further information may be needed on these relatives. In cases of autosomal dominant, X-linked or consanquinity, a more extensive family history may be indicated. The family pedigree allows the calculation of risk figures, the identification of other persons at risk, and can sometimes help discern the mode of inheritance when the entity in question is undefined.

With knowledge of the ethnic background, religion, and educational status of the family, the genetic counselor may better

Table 2
Genetic Counseling Process

I. Intake
    A. Review genetic counseling process with prospective individuals
    B. Obtain further clinical information (i.e. medical records, x-rays, photographs, etc.)
    C. Family pedigree
    D. Medical history
    E. Social history

II. Diagnostic Evaluation
    A. Review histories and pedigree*
    B. Diagnostic testing*
    C. Establish diagnosis

III. Discussion
    A. Explain diagnosis
    B. Discuss treatment/management
    C. Clarify recurrence risks*
    D. Review options*
    E. Facilitate decision making, adjustment to diagnosis*
    F. Make appropriate referrals

IV. Follow-up
    A. Communicate with primary care providers*
    B. Provide written summary to family*
    C. Suggest extended family studies as needed*
    D. Inform families of new technologies available
    E. Schedule family for return visit

*From reference 6.

predict potential communication barriers and determine how best to approach technical explanations. A routine medical history is obtained with specific questions relating to the disease. The establishment of an empathetic rapport with the family during the initial meeting is invaluable. This type of environment, free of judgment, allows the family to feel comfortable expressing their concerns, to ask the questions they need to ask, and to stop the counselor when they do not understand. This atmosphere also allows the counselor to obtain all the necessary information.

Information is also obtained concerning the social and emotional impact of the disorder on the family. In most case scenarios of cerebrovascular disease, the affected individual is an adult who probably already has his/her desired number of children, and may in some cases be beyond the age of procreation. The late age of onset of these disorders poses unique psychological issues in genetic counseling. Individuals affected with cerebrovascular disease may feel guilty at having passed down to their offspring a predisposition for this condition. Profound emotional response to this information may be reflected by the entire family unit (including spouses) once a diagnosis has been made and the risk to other family becomes known.[7] In Lynch's study of families with cancer, he notes that emotional responses may be modified by the type of disease involved, its age of onset, clinical course, treatment, and the degree of financial stress the family sustains.[8] In addition, emotional conflicts are noted, especially since strong feelings can interfere with clear communication. This knowledge will help identify specific problem areas that need to be addressed. There may be a need for multiple or overly lengthy sessions.[5,9,] These sessions may

---

Table 3

Family Pedigree Information

- Names
- Dates of birth
- Health status
- Consanguineous unions
- Ethnic origin (birthplace, origination of ancestors)

Table 4

Genetic Family History Information

- Mental retardation and/or Down's syndrome
- Seizures
- Kidney problems
- Blindness/deafness
- Skeletal dysplasia
- Muscular dystrophy
- Early myocardial infarction
- Hemophilia
- Blood transfusions
- Cystic fibrosis
- Neural tube defects
- Miscarriages/stillbirths
- Cleft lip/palate
- Strokes

---

make the transferential phenomena more conspicuous.[5]

Diagnostic Evaluation and Discussion

An accurate diagnosis is fundamental for meaningful genetic counseling. Some of the problems inherent in reaching a diagnosis in genetics include heterogeneity, phenocopies, and variable expression. As cerebrovascular disease is a complex clinical entity, it is important to establish the etiology. Often, specialists in different fields are consulted for further confirmation of the clinical symptoms. On occasion, additional search of the literature, examination of other family members, or simply time for further development of the clinical "picture" may be required before a specific diagnosis can be established. In many cases where a diagnosis cannot be made, the lim-

iting factor may be the present state of knowledge or, as in cases of stillbirths or deaths, the appropriate genetic studies may not have been done and the family history is inconclusive. Once a diagnosis has been made, the prognosis, available management, and recurrence risks can be discussed.

The natural history of the condition should be reviewed thoroughly in terms the patient can comprehend so that the family can plan accordingly for the problems they may encounter. Misconceptions should be clarified and all issues regarding the condition should be addressed. Various options in the long-term care and management should be presented to give the family a broader perspective on the situation. Arranging contact with other families or support groups dealing with similar concerns can provide support and practical information for the family. Referrals can be made to special services in the community in order to help the family meet their needs.

In the communication of risk of recurrence, visual aids, or analogies to games of chance are often used to facilitate an understanding of probability, which can be a difficult concept to convey. In Mendelian disorders, families need to realize that the recurrence risk is the same for each pregnancy or for each of the children they may have already had. For example, in the case of autosomal recessive inheritance, after one affected child the couple will not necessarily have 3 subsequent unaffected offspring. Although the risk figures may be understood in a mathematical sense, the individuals may not perceive the personal implications of these figures.

The emotional response to disease in general plays a role in decision-making. One way to help the family work through the emotional "swell" that arises from the diagnosis of a genetic disorder is to explain how the disorder occurs and give them an opportunity to reevaluate the basis for their feelings. Patients and parents often blame someone, something, or themselves in the need to affix a tangible reason for their disease or their child's disease. Thus, it is important to point out that the genetic makeup of each of us is determined at the moment of conception. In some individuals, habits such as smoking and overeating could influence the underlying genetic predisposition to cerebrovascular disease. However, some patients may feel reassured knowing that each of us carries genes that may predispose us to certain conditions and that no one is genetically perfect.

How the family chooses a suitable course of action is believed to depend on how they perceive the severity of the disorder and its effect on their lives, the availability of treatment, and their previous life experiences. The genetic counselor does not directly participate in the family's decision-making process, but rather facilitates that process.

After each counseling session, a summary of the content should be sent to the family and referring physician. For the family, this summary provides a written document that can be reviewed when necessary. For the physician, it serves as an update of the management and reviews the genetic aspects of the condition.

Follow-Up

The extent of follow-up varies with different conditions and circumstances. Usually, follow-up visits are arranged according to the family's and counselor's perception of need. Initially, the family may not have grasped all the technical information presented, or because of situational stress the information was distorted. Thus, follow-up provides an opportunity to review the medical and genetic facts. Additional questions may have arisen with time, attitudes may have changed, and the family may now be encountering a different phase of the condition. During the follow-up visit the family's adjustment should be assessed. Sometimes unreconciled feelings cause the family to become

dysfuntional or to devote all of their energy and resources into the care of the affected individual. Follow-up is also important when new medical developments may change available options, or when new information or clinical developments enable a diagnosis to be established, or an uncertain diagnosis to be confirmed or revised. Because the follow-up phase helps to insure successful genetic counseling, its value cannot be overemphasized.

Goals of the Genetic Counseling Process

The goals of genetic counseling are twofold and involve: (1) providing the family with technical information concerning the medical and genetic nature of the disorder; and (2) helping the family to deal more effectively with the potential for or presence of the disorder in their lives. The methods for effective genetic counseling include: uncovering the family's concerns, establishing rapport so that effective communication can occur, and providing follow-up to insure that goals are met.

Genetic Screening

Genetic screening refers to the application of tests to groups of individuals for the purpose of detecting the carrier of deleterious genes.[10] The goals of genetic screening are: (1) to identify individuals with genetic disease (or genetic predisposition) so that they may receive treatment to obviate or decrease the effects of the phenotype; and (2) to expand our knowledge of these diseases through the collection of epidemiological data. The determination of the groups to be screened depends on the nature of the disease. With cerebrovascular disease, the group may be a family or it could be a subpopulation at risk for cerebrovascular disease. With respect to mass screening, various other factors play a role in choosing who should be screened. These factors include: disease frequency, disease burden, availability, efficacy of treatment, preventive measures, cost of testing, accuracy of diagnostic tests, and evidence that the program will be beneficial.

Nora et al.[10] state that optimal genetic screening programs must meet the following requirements: an adequate public education program should be administered prior to screening, with evidence of community support of and involvement in the program; those screened should be informed of the purpose of the test and consent to it; the results should be conveyed with appropriate nondirective counseling; and the results should be kept confidential. These authors also state that the screening tests should be accurate, simple, and inexpensive; evidence should be provided that the expected benefits justify the cost. In order to assure the success of the screening program, the necessary manpower and appropriate laboratory facilities need to be available. Finally, the screening program should also provide a means of assessing its effectiveness.[10]

It is not inconceivable that some day we will have genetic screening for cerebrovascular diseases. The importance of meeting all of the aforementioned requirements cannot be overstated. Screening programs have been instituted without much forethought and disastrous results have occurred. Screening for cerebrovascular diseases may eventually become an important element in the overall management of these conditions. Therefore, careful forethought and planning must be undertaken to assure success of the screening program.

Research Studies

As many of the studies regarding the genetics of cerebrovascular disease may be experimental, cooperation from family members is essential. Often, other family members are needed to pursue further testing, such as blood samples from other affected and nonaffected individuals to complete molecular genetics studies. The degree of rapport established between the patient and his/her own family and between the patient and physician will determine the individual's willingness to cooperate in data collection. Refusal to assist in information gathering, despite encouragement by other relatives, reflects a more general unwillingness to heed suggestions pertaining to possible treatment or monitoring of cerebrovascular disease.[8] Sometimes an indirect approach by the family's physician can be more successful in obtaining cooperation from the patient. Communicating that participation in research studies may generate more information about the condition for the patient and their children who may be at risk may facilitate cooperation.

References

1. North Carolina Medical Genetics Association, Inc. Fact Sheet. Senate Bill 903, 1987.
2. Ad-hoc Committee on Genetic Counseling. Genetic counseling. Am J Hum Genet 1975;27:240–242.
3. Kessler S. Psychological aspects of genetic counseling. X. Advanced counseling techniques. J Genet Counseling 1997;6:379–392.
4. Riccardi VM. The Genetic Approach to Human Disease. New York: Oxford University Press, 1977.
5. Kessler S. Psychological aspects of genetic counseling. XII. More on counseling skills. J Genet Counseling 1998;7:263–278.
6. Scott J. Genetic Counseling in Genetic Applications: A Health Perspective. University of Colorado Health Sciences Center, School of Nursing and School of Medicine, Genetic Unit, 1988 185–291.
7. Day E. The patient with cancer and the family. N Engl J Med 1966;274:883–886.
8. Lynch HT, Lynch PM, Lynch JF. Genetic counseling and cancer. In Kessler S (ed). Genetic Counseling Psychological Dimensions. San Diego, CA: Academic Press, 1979, 221–241.
9. Schneider KA. Counseling About Cancer: Strategies for Genetic Counselors. Boston, MA: Division of Cancer Epidemiology and Control, Dana-Farber Cancer Institute, 1994, 79.
10. Nora J, Clarke Fraser F. Medical Genetics: Principles and Practice. 3rd ed. Philadelphia, PA: Lea & Febiger, 1989, 355.

Chapter 17

Gene Therapy

Jay Lozier, MD, PhD

Introduction

Gene therapy may be defined as the transfer of genetic material to human cells or tissues for the purpose of curing or palliating disease.[1] Gene therapy was initially conceived of as a "cure" for genetic diseases by the addition of normal genetic material to affected cells in order to compensate for an abnormal or absent gene. Through this process normal function can be returned to affected cells. The initial success of gene therapy for adenosine deaminase deficiency[2–4] has demonstrated the feasibility of this approach in humans. Currently there is tremendous effort underway to transfer genetic material to human cells of all kinds by an ever-widening array of viral and nonviral vectors. Furthermore, it has become apparent that transfer of genetic material to normal tissues in patients with acquired or noncongenital diseases may represent a new avenue for treatment of these acquired disorders. At present, there are more than 250 proposals for gene therapy approved by the Recombinant DNA Advisory Committee in the United States or other regulatory agencies worldwide, and more than half are devoted to acquired noncongenital diseases such as cancer, AIDS, or other acquired disorders.[5]

Thus far, all proposals for human gene therapy have postulated transfer of genetic material into somatic cells of affected patients. Although cloning of higher mammals has recently been demonstrated[6] and gene transfer might someday be practiced at the level of the germ cell to prevent genetic disorders in the unborn, ethical considerations and current practical limitations preclude therapeutic manipulation of human germ line DNA.[7–10] Although most occurrences of genetic diseases are in families where the disease is already known, and correction of germ line DNA is a theoretical possibility, the continuing occurrence of de novo mutations means that manipulation of germ line DNA cannot prevent all genetic disorders in future generations. This is most clearly evident in X-linked disorders such as the hemophilias where perhaps one third of cases are due to de novo mutations.[11,12]

Strokes or cerebrovascular disorders are not typically due to a single genetic trait but may be due to one or more of many different genetic traits. Stroke may also be the result of diseases such as atherosclerosis which may be acquired on a genetic or nongenetic basis. For diseases with identified gene defects it may be possible to utilize a gene therapy approach in which the normal gene is introduced into affected tissue to correct the disorder (Table 1). In other situations, one or

From Alberts MJ (ed). *Genetics of Cerebrovascular Disease*. Armonk, NY: Futura Publishing Company, Inc., © 1999.

Table 1
Diseases Associated with Stroke Which May Be Candidates for Gene Therapy

Disease	Gene(s)	Chromosome	Animal Models	Human Trials
Atherosclerosis	LDL receptor (and others)	19	Watanabe heritable hyperlipidemic rabbits	Grossman, et al [119] (liver-directed, ex vivo gene therapy)
Homocystinuria	Cystathionine β-synthase (and others)	21	–	–
Protein C deficiency	Protein C	2	–	–
Protein S	Protein S	3	–	–
Antithrombin III	Antithrombin III	14	–	–
Hemophilia	Factor VIII (hemophilia A)	X	Hemophilia A canines, mice	
	Factor IX (hemophilia B)	X	Hemophilia B canines, mice	Hsueh, et al.[113] (skin fibroblast-directed ex vivo gene therapy)
Adult polycystic kidney disease	Adult polycystic kidney disease gene 1	16	–	–

more genes may be transferred to normal tissues in order to compensate for biochemical defects where the genetic basis is not yet known, e.g., overexpression of a modfied anticoagulant protein to compensate for an unknown biochemical/genetic defect associated with repeated cerebrovascular thrombosis. Some defects leading to cerebrovascular disease may not be correctable by gene therapy approaches in the forseeable future due to various technical limitations of gene transfer techniques. This chapter will review the basic concepts of gene transfer, highlight their potential for application to diseases leading to cerebrovascular disease, and address future issues in this area.

Methods for Gene Transfer

Transfer of genetic material to living cells may be accomplished *in vitro* using viral infection/transduction, physicochemical methods, or by methods that incorporate features of both approaches. Each method has certain inherent advantages and disadvantages and all may eventually be utilized for gene therapy of various human diseases (Table 2).

The terms transfection and transduction require definition in any discussion of gene transfer or gene therapy. Transfection is the process of introducing DNA into cells by physical or mechanical means, followed by expression of the gene product for a limited period of time by the modified cells. Usually the DNA is lost from the cell before it is integrated into the genomic DNA and gene expression is transient, and limited to the transfected cells (not transmitted to progeny or neighboring cells).

Transduction is a gene transfer process in which DNA is delivered into a cell, usually as part of a recombinant viral vector, after which the DNA is incorporated (integrated) into the genomic DNA of the target cell or persists for varying periods of time outside of the nucleus as an episome. Gene expression is observed in progeny derived from transduced cells. The term transduction has

Table 2
Methods for Gene Transfer

Method	Characteristics, Advantages, and Disadvantages
Viruses	
Retrovirus	RNA virus capable of transfer of up to 7 kb of nonviral genetic material. Capable of causing integration of transferred genetic material into genomic DNA of target cell. Requires division of target cell for integration to occur (except lentivirus vectors). Efficient gene transduction. Recombinant viral titers of $\sim 10^6$/ml or greater. Wide range of target cells
Adenovirus	DNA virus capable of transfer of up to 8 kb of nonviral genetic material. Does not cause integration of transferred genetic material into genomic DNA of target cell. Highly efficient gene transfer. Recombinant viral titers of $\sim 10^{11}$/ml or greater. Epithelial cells most efficiently modified by adenovirus gene transfer; other cells also susceptible.
Adeno- Associated Virus (AAV)	Defective single-stranded DNA (parvo) virus; requires adenovirus helper for production of virus. Capable of transfer of up to 4.5 kb of nonviral genetic material. Capable of causing integration of transferred genetic material into genomic DNA of target cell. Does not require division of target cell for integration to occur. Recombinant viral titers of $\sim 10^8$/ml. Wide range of target cells.
Herpes virus (Epstein-Barr and simplex groups)	DNA virus capable of transfer of large DNA fragments. Capable of persisting as episome (Epstein-Barr viruses). Capable of transferring genetic material to neuronal cells (simplex viruses). Recombinant viral titers of $\sim 10^9$/ml or greater.
Physicochemical Methods	
Calcium phosphate transfection	Precipitation of DNA with calcium phosphate permits transfer of DNA into cells. Low efficiency of gene transfer. Expression transient. Nontoxic. Applicable to wide range of cells.
Lipofection	DNA in lipid micelles transferred into cells by fusion with membranes. Variable efficiency of transfection. May be toxic to cells *in vitro* and *in vivo.* Expression transient. Applicable to wide range of cells.
Electroporation	Cells rendered transiently permeable to DNA by passage of an electric current through cells. Toxic to cells. Expression transient. Not likely to be useful for *in vivo* gene transfer.
Combined Physiochemical/Viral Methods	
Molecular Conjugates	DNA complexed with polylysine to which various molecules may be attached. According to molecule attached to polylysine, cell specificity may be conferred on conjugate. Adenovirus may be incorporated as part of conjugate to protect DNA from degradation after uptake. Efficient gene transfer. Applicable to wide range of cells.

been used to describe this type of viral-mediated gene transfer to distinguish it from viral infection.[13] In the former process, independent viral replication does not produce infectious particles capable of transferring the viral genome to other cells. In contrast, viral infection leads to repeated cycles of viral replication and release of more virus which infects other susceptible cells, often resulting in cell death. A key problem for recombinant viral vector design is the construction of recombinant viral vectors which can transfer genetic material without causing a virulent infectious process. Strategies for production of replication-defective viruses are outlined in the following description of various vectors currently under development.

Viral Vectors

It is fitting and perhaps ironic that certain pathogenic viruses with the ability to transfer their nucleic acid content to human cells can be modified and utilized for genetic modification for therapeutic purposes. Several viral vector systems are under development for gene therapy of various diseases at this time. These are described first as tools for gene transfer in general, then as vectors for gene therapy of cerebrovascular diseases.

Retroviruses

The most extensively studied viral vectors are the retroviruses. Retroviruses are double-stranded RNA viruses that are characterized by the ability to enter susceptible target cells and transcribe RNA into DNA, which is then capable of integration into the genomic DNA of the host. Transcription of DNA from RNA is catalyzed by the unique enzyme reverse transcriptase which is the product of the retrovirus *pol* gene.[14] The retrovirus genome consists of *gag*, *pol*, and *env* genes which produce the g antigen, reverse transcriptase enzyme, and envelope proteins, respectively. These three genes are flanked by the 5′ and 3′ long terminal repeats (LTRs) which are necessary for integration of proviral DNA and serve as promoters of gene expression.[15] During normal viral infection and replication viral RNA is enclosed with the reverse transcriptase enzyme within a coat derived from the retroviral *gag* and *env* proteins and lipids from the outer membranes from the host cell.[16] The ability of the retrovirus to fuse with a target cell is determined by the combination of proteins and lipids comprising the outer surface of the virus. Those capable of infecting a wide variety of species of target cells are termed *amphotropic*, while those capable of infecting only a single species of target cell are termed *ecotropic* viruses.[17–19] It has proved possible to remove the *gag*, *pol*, and *env* genes from the viral genome and replace this genetic material with up to 7 kilobases (kb) of nonviral nucleic acid.[1] Retroviral vectors with this configuration are incapable of independent virulent infection, and must therefore be produced by "helper" or "packaging" cell lines which provide the requisite *gag*, *pol*, and *env* proteins. It has been possible to produce retroviral packaging cell lines capable of serving this helper function.[20–22] Thus, replication-defective retroviral vectors can be produced which can enter a cell and transcribe DNA that integrates within the host cell genomic DNA. This schema is shown in Figure 1.

Recently, retroviral vectors based on the lentivirus family (which includes the human immunodeficiency virus) have been described as potential tools for gene therapy. The main advantage these vectors have over other retroviral vectors is their ability to enter the cell nucleus and insert their nucleic acid into the host cells' genomic DNA without cellular division.[23–25]

Adenoviruses

Adenoviruses are double-stranded DNA viruses which infect a wide range of cells, particularly epithelial cells of the lung and other organs. The 36 kb viral genome contains multiple regulatory genes as well as other genes which encode structural elements such as the hexons, pentons, and fiber proteins which constitute the outer surface of the virus. In a manner analogous to retro-

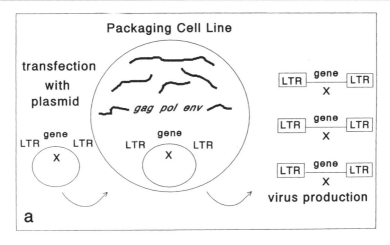

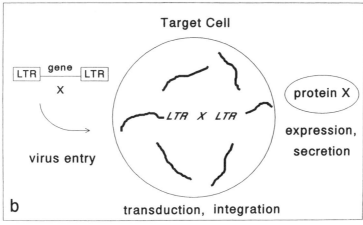

Figure 1. Retroviral vectors for gene therapy: **(a)** production of recombinant retrovirus containing gene X by transfection of a retroviral packaging cell line; **(b)** infection of a target cell by the recombinant retroviral vector results in integration of the provirus in genomic DNA, after which gene X is synthesized and secreted if appropriate signal sequences are present.

viruses, certain viral genes can be deleted from the adenovirus genome and replaced with genes encoding nonviral proteins of interest.[26] In particular, the E1a and E1b genes (which control the host cell cycle) and/or the E3 gene (which helps the infected cell evade host immunity) can be deleted and replaced with up to 8 kb of nonviral DNA as shown in Figure 2. Like the retroviral vectors, recombinant adenovirus vectors are incapable of independent infection and replication, and must be produced by helper cells designed to provide the missing viral gene products.[27] The design of cell lines for production of recombinant adenoviruses is analogous to that for retroviral packaging cell lines. Adenovirus helper cell lines must provide the genes for synthesis of recombinant adenoviruses which are replaced by the gene to be transferred by the recombinant virus, typically the E1 or E3 genes for first-generation vectors.[28–32] Recently, vectors devoid of the adenovirus E2 or E4 genes or even devoid of all adenovirus genes ("gutless" vectors) have been developed.[33–37] Unlike retroviruses, adenoviruses do not cause integration of their nucleic acid into the target cell DNA; rather, the adenovirus DNA exists for variable periods of time in the nucleus of the infected cell and is gradually lost over a period

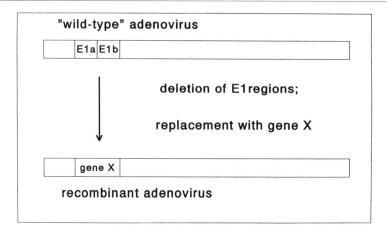

Figure 2. Recombinant adenovirus vectors can be produced by deletion of the viral E1a and E1b genes which are replaced with the gene X. The recombinant adenovirus vector is capable of infecting a target cell and causing gene X to be synthesized; the recombinant adenovirus is incapable of independent replication due to lack of E1a and E1b genes.

of days or weeks.[30–32,38–43] Accordingly, expression of foreign genes introduced by current adenoviral vectors is transient.

Adeno-Associated Viruses

The adeno-associated virus (AAV) is a single-stranded DNA virus which has been recognized as a potential vector for gene transfer. The wild-type virus was originally discovered during studies of adenoviruses, and is capable of latent or lytic infection of a wide variety of cell types in the presence of adenovirus which serves as a "helper" virus.[44] The AAV genome consists of *cap* (capsid) and *rep* (replication) genes between two terminal repeats (TRs) which form double-stranded DNA (hairpins) and serve as origins of viral replication.[45] The *cap* and *rep* genes contain multiple overlapping transcriptional reading frames which encode several proteins essential for viral replication.[46] The parvovirus family to which AAV belongs is associated with little pathology in humans; most humans have antibodies to the AAV2 serotype and can be shown to have latent AAV infection without apparent ill effect.[47] Wild-type AAV integrates preferentially at a specific site on chromosome 19,[48,49] and recombinant AAV vectors would also be expected to integrate at this site although integration at other sites in genomic DNA may be demonstrated as well.[50,51] Recombinant AAV vectors can be designed with nonviral DNA replacing the *cap* and *rep* genes between the two TRs.[52,53] The nonviral genetic material that can be inserted between the terminal repeats is limited to about 4.5 kb, due to packaging constraints of the virus. Recombinant AAV vectors can be produced by cotransfection of an appropriate cell line with a plasmid containing the desired genes flanked by the two TRs, and a "helper" plasmid with *cap* and *rep* genes. The transfected cells are then infected with adenovirus which effects rescue of the recombinant AAV vector, as shown in Figure 3. Recombinant AAV vectors are capable of transducing a wide variety of target cells and are able to effect transduction of nondividing cells,[54,55] like the lentiviral vectors.

Herpes Viruses

Herpes viruses are double-stranded DNA viruses capable of latent infection of various cell types. The simplex virus group is notable for the ability to cause latent infection of neural tissue or other terminally differentiated, nondividing cells. The Epstein-Barr

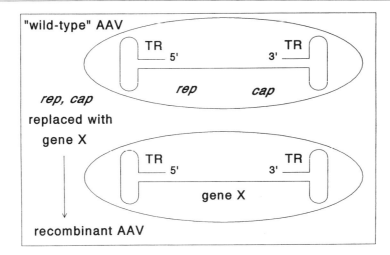

Figure 3. Adeno-associated virus (AAV) is a parvovirus containing a single strand of DNA with the two terminal repeats (TRs) flanking the AAV *cap* and *rep* genes; these genes can be replaced by a foreign gene to make a recombinant AAV vector for gene transduction.

virus group is capable of latent infection of B-lymphocytes as well as cells of epithelial derivation.[56,57] These viruses persist as non-integrated episomes which are passed to progeny of infected cells with each cell division. The herpes virus and Epstein-Barr genomes are among the largest of all known viruses (150–200 kb of DNA), and substantial portions of the viral DNA may be replaced with nonviral DNA for the purpose of gene transfer to susceptible cells.[58–61] Herpes virus and Epstein-Barr virus vectors have been used to tranduce neurons,[58,59,62] hepatocytes,[63] and B-cells [61,64] with a variety of transgenes. The full potential of these viruses as gene transfer vectors remains to be completely realized.

Physicochemical Methods

DNA can be introduced into cells by various nonviral means, including precipitation with calcium phosphate salts,[27] electroporation,[65] or incorporation into liposomes which can fuse with outer membranes and deliver DNA to the cytoplasm of the cell.[66,67] The first two of these methods are characterized by low efficiency of gene transfer and/or unacceptable toxicity, and probably will not be used for *in vivo* application in humans in the foreseeable future. Systemic or localized gene transfer *in vivo* using various liposome formulations has been demonstrated and appears to be nontoxic.[68–70] Localized delivery of vector DNA in liposomes to vascular tissue using double-balloon catheters has been reported, and may prove to be useful for gene transfer to vascular endothelium for therapeutic purposes.[71,72]

The main limitation of liposome-mediated gene transfer is the relatively low efficiency of transfection *in vivo*, which may be a result of deleterious interactions with host factors (e.g., negatively charged serum proteins)[73] and may be ameliorated by changes in the chemical composition and physical conformation of the liposome.[73,74] Liposomes have been combined with viruses to circumvent the need for cell surface receptors which normally mediate binding and uptake of viral vectors.[75,76]

Molecular Conjugates

One gene transfer approach is to couple DNA by noncovalent means to cationic polymers such as polylysine[77–79] or polyethylenimine[80–83] to which other molecules may be attached. The attached molecules may be ligands to cell surface receptors which result in cellular uptake. These ligands may be chosen for their specificity for particular cell types. One such example is the asialo-orosomucoid receptor molecule which is the ligand for hepatic uptake of asialoglycoproteins.[77] By linking DNA to polylysine coupled to this molecule, DNA can be transferred specifically to hepatocytes *in vitro* and *in vivo*.[78,79] One approach to increase the overall effectiveness of transfection has been to protect the transfected DNA from degradation after uptake. This can be accomplished by incorporating components that serve to prevent degradation of DNA in cytoplasmic endosomes after uptake. It has been discovered that certain external proteins from the surface of adenoviruses can serve to protect DNA from degradation in the cytoplasm of the infected cell, and incorporation of a nonvirulent adenovirus in conjugates may enhance the effectiveness of gene transfer.[84–90] Dissolution of adenovirus fiber protein under acidic conditions stabilizes lysosomes and prevents enzymatic degradation of the contents.[85–87] The transfected DNA is then transported to the nucleus where it is expressed for a variable period of time. One advantage of the molecular conjugate approach is that the size of DNA to be transferred is not subject to the limitations imposed by most viral vector systems, and conjugates containing DNA fragments as large as 48 kb have been used successfully for gene transfer.[87] This may allow the transfer of DNA containing complete genes and multiple genes to be coexpressed or large combinations of genes, promoters/enhancer elements and/or selectable markers too large for some viral vectors. Thus far, expression of genes as part of molecular conjugates has been associated only with transient expression which lasts for days or weeks.[91–93]

Ex Vivo and In Vivo Gene Therapy Approaches

Gene therapy may be performed by modification of cells outside the body or by modification of cells within the body of the patient. Each approach has certain advantages and disadvantages, and for certain diseases one approach may be preferred over the other for practical considerations.

Ex Vivo Approaches

Genetic modification can be performed by transducing or transfecting cells to produce the desired protein *in vitro* and then returning the functional cells to the patient. This *"ex vivo"* approach has the advantage that gene transduction or transfection conditions can be carefully controlled and adjusted for optimal gene transfer. Furthermore, elimination of uncorrected cells can be achieved through the use of a selectable marker gene and subjecting the treated cells to a toxic drug; this is particularly important when only a small fraction of cells are transduced by exposure to a vector.

The return of modified cells to the body is an important point to consider for *ex vivo* gene therapy, and may prove to be a limiting factor for this approach. A key problem is establishing a satisfactory blood supply to sustain the modified cells immediately upon implantation. This is especially difficult for cells which require adherence to a basement membrane or incorporation within a matrix to establish a proper relationship between the implanted cells and adjacent tissues. In contrast, some cells can be reinfused directly into the bloodstream, as is done routinely during blood transfusion or bone marrow transplantation. Furthermore, if the desired gene product is to be delivered from outside the bloodstream for distribution and activity elsewhere in the body, there must be opportunity for diffusion of the gene product into the circulation. The *ex vivo* approach to gene therapy might be contemplated for genetic modification of endothelial cells lining the lumen of the carotid arteries, for example. If the gene product can circulate in the bloodstream and reach the intended site of action, the positioning of the implanted cells may not be critical as long as the gene product can diffuse from the implanted cells to the bloodstream. Such is the case for coagulation factor

IX (absent in hemophilia B); however, coagulation factor VIII (absent in classic hemophilia A) is probably too large a molecule to diffuse into the bloodstream from extravascular sites.[94–97]

In Vivo Approaches

An alternative to the *ex vivo* approach to gene therapy is one in which a vector containing the genetic material is injected into the bloodstream or directly into tissues, and the genetic modification occurs *in vivo*. The advantages of this approach are its simplicity and the efficiency with which it might be applied to practical clinical treatment of patients, since tedious *in vitro* cultivation and transduction of cells from individuals would be eliminated. For such an approach to work efficiently via injections into the bloodstream, there must be a mechanism for targeting the vector to a desired cell type at a particular site. Mechanical approaches have been shown to result in gene delivery targeted to specific cells *in vivo*. Through the use of occlusive double-balloon catheters, it has proved possible to isolate segments of the vascular tree and perform genetic modifications of endothelial cells lining the segment. Such an approach has been demonstrated with retroviral-mediated transduction of the isolated endothelium *in situ*.[72] A limiting factor is likely to be the number of cells that can be genetically modified by a vector in an isolated vascular segment.

A key limitation to the *in vivo* genetic modification approach has been the limited number of dividing cells susceptible to transduction by certain viral vectors *in vivo*. Perhaps the adeno-associated virus vectors will permit transduction of these cells in the future.

Potential Applications of Gene Therapy to Cerebrovascular Disease

Gene therapy may be applicable to a number of genetic disorders which confer an increased risk for cerebrovascular hemorrhage or ischemia. Diseases associated with hemorrhagic stroke include hemophilia and related blood coagulation disorders, adult polycystic kidney disease, Ehlers-Danlos syndrome (associated with abnormal collagen), and other diseases such as familial amyloidosis which results in abnormal deposition of prealbumin around blood vessels. Diseases associated with ischemic stroke include atherosclerosis, homocystinuria, deficiencies of anticoagulant proteins (proteins C, S, and antithrombin III), and various other diseases. For application of gene therapy to prevent cerebrovascular disease a gene that encodes a therapeutic protein relevant to the underlying disease process must be transferred into an appropriate target cell. For certain diseases such as hemophilia, this may be a very straightforward proposition since the genes are well characterized and have been isolated in a form suitable for gene transfer experiments. In other diseases with no clear candidate gene (or multiple genetic defects), the approach may be directed towards providing genetic material to compensate for adverse effects of the disease without changing the underlying condition.

Genetic Diseases that May Cause Hemorrhagic Stroke

Several genetic diseases have consequences for the central nervous system which may directly or indirectly lead to hemorrhagic stroke. Typical of these diseases are the hemophilias and related coagulopathies that may result in fatal or debilitating CNS hemorrhage. Gene therapy may offer the prospect of cure at some time in the future, or may at least decrease the likelihood of hemorrhage in affected patients.

Hemophilia

Hemophilia A (factor VIII deficiency) and hemophilia B (factor IX deficiency) are ideal candidates for gene therapy since the genes have been identified and cloned,[98,99] and replacement of the deficient clotting factors to levels of even 1% of normal can lead to substantial improvement in the bleeding phenotype.[100] Patients with 5% or more of normal levels have very mild disease and are not prone to spontaneous CNS hemorrhage. Prior to the AIDS epidemic in the 1980s, the leading cause of death in patients with hemophilia was intracranial hemorrhage.[101] This was true despite the use of factor VIII and factor IX concentrates to stop bleeding, and it underscores the need for continuous replacement of factor VIII or factor IX to *prevent* rather than *react* to established bleeding (especially in the central nervous system). In this regard, gene therapy represents a potential avenue for a "cure" of hemophilia and could prevent much morbidity from hemorrhage in the CNS or elsewhere.

Animal models for hemophilia have been used to test various gene transfer methods. Due to the relatively large size of the factor VIII gene, cDNA, and protein compared to those for factor IX, efforts to transfer genetic material for factor VIII have lagged behind those for factor IX.[100,102] It has proved possible to transfer the factor IX cDNA into hepatocytes in hemophilia B canines using recombinant retroviruses and demonstrate low levels of factor IX production *in vivo* for more than 3 years,[103] suggesting that it should be possible to achieve permanent expression of recombinant factor IX with appropriate vectors. Although the levels of factor IX achieved were considerably less than the desired 1% level, it is encouraging that *in vitro* and *in vivo* blood coagulation parameters improved significantly.

It has also been possible to transfer genetic material for factors VIII or IX to the liver of normal rodents or hemophilic mice or canines *in vivo* using adenoviral vectors.[40-42] This approach has led to efficient gene transfer without resection of liver as required for retrovirus-mediated transduction *in vivo*. Very high levels of factor IX (> 3 times normal) were rapidly achieved by this approach in hemophilia B canines, but these levels declined to a zero baseline within 100 days due

to the transient nature of expression with adenoviruses.[41] The main obstacle to success for *in vivo* gene therapy of hemophilia has been sustaining high level gene expression for prolonged periods of time. Presumably this might one day be overcome by the use of improved vectors for gene transfer,[33-37] or by the concomitant use of immunosuppressive maneuvers to prevent elimination of adenoviral-transduced cells or to permit repeated administration of adenoviral vectors.[39,104,105]

The relatively large size of the factor VIII protein (280,000 daltons) precludes diffusion of the protein into the bloodstream from extravascular sites,[96,97] which is in contrast to factor IX, which has been delivered into the circulation from a variety of extravascular sites after *in vivo* or *ex vivo* gene transfer with retroviral vectors.[94,95,106-110] It is therefore likely that gene therapy for hemophilia A will require gene transfer to cells or tissues within the bloodstream, such as lymphocytes or endothelial cells, so that the expressed protein might be delivered directly into the bloodstream. Molecular conjugates have been used for *ex vivo* gene transfer of the factor VIII cDNA to mouse fibroblasts and myocytes followed by reimplantation in the liver or spleen; however, this approach has so far resulted only in transient expression of factor VIII.[91] Another possibility may be the use of vectors which transfer genes as episomes in lymphocytes as exemplified by the Epstein-Barr virus.

The most promising data for hemophilia gene therapy is seen with the use of AAV-factor IX vectors for *in vivo* gene transfer to correct hemophilia B. Sustained therapeutic levels of factor IX have been observed after gene transfer to liver[111] or to muscle[112] *in vivo* with AAV vectors in mice; it is likely that these vectors will be successfully employed in clinical trials in the near future if these results can be validated in larger animal models.

One trial of gene therapy for hemophilia B in humans has been reported by Hsueh et al.[113] Fibroblasts from patients with mild to moderate disease were transduced *ex vivo* with a factor IX-retroviral vector, and then reimplanted subcutaneously in these subjects. A slight increase in factor IX above baseline was detected and no adverse effects were noted. Although it is difficult to conclude that there was a lasting therapeutic effect from these experiments, it is encouraging that no problems were encountered with the protocol that was followed.

Deficiencies of other coagulation factors such as fibrinogen (the precursor to fibrin) or factor XIII, which cross-links and stabilizes fibrin, may lead to hemorrhagic stroke. Thus far, there has been little effort to correct either of these deficiencies, in part due to their relative rarity compared to hemophilia A and B. In addition, fibrinogen is produced by coordinate expression of three discrete genes, which complicates gene replacement efforts.[114] In the case of factor XIII deficiency, there are safe concentrates that can be used for replacement therapy; this method is much more convenient for factor XIII deficiency than for hemophilia A or B since the half-life of factor XIII is in the order of weeks compared to hours for factor VIII and IX.[114] It is therefore possible to treat factor XIII-deficient patients by administration of concentrates every few weeks and maintain a normal phenotype without much inconvenience. In theory, however, deficiency of factor XIII could be approached by gene therapy methods at some point in the future.

Other Genetic Diseases Associated with Hemorrhagic Stroke

Diseases such as adult polycystic kidney disease and Ehlers-Danlos syndrome are congenital diseases associated with hemorrhagic stroke. Both result from defects in connective tissue proteins which are abnormal due to congenital defects. It is theoretically possible that gene therapy could be applied to these diseases since the responsible genes have been identified. However, there are practical difficulties that make this unlikely in the near future. For example, it is likely to be very difficult with current technology to transfer genetic material efficiently into the media of arteries in the circle of Willis where aneurysms form and rupture. Although adenovirus-mediated gene transfer into blood vessels has been demonstrated,[115] undesirable results of neointimal hyperplasia and inflammation have been reported[116]; furthermore, adenovirus-mediated gene transfer typically results in transient expression of the desired gene product. Developments in gene transfer technology will be necessary to permit gene therapy of these diseases. It is encouraging that *in vivo* gene transfer into rat carotid arteries has been reported with AAV vectors, and transgene expression has lasted for more than 6 months in one report.[117]

Genetic Diseases That May Cause Ischemic Stroke

Atherosclerosis, homocystinuria, or deficiency of circulating anticoagulants such as proteins C or S, or antithrombin III may result in ischemic stroke, and all are diseases that may be treated by gene replacement therapy in the future. A trial of gene therapy in humans for the most severe form of atherosclerosis caused by congenital deficiency of the low-density lipoprotein (LDL) receptor has already taken place, with partial success.[118,119] Human gene therapy for homocystinuria or any of the anticoagulant protein deficiencies has not yet been reported. Each will require different approaches and strategies for success due to the nature of the biochemical defects involved. For instance, while familial hypercholesterolemia results from the absence of a cell-surface receptor,[120] homocystinuria is the result of a deficiency of an intracellular enzyme,[121] and the anticoagulant deficiencies result from lack of secreted extracellular proteins in the bloodstream.[122]

Atherosclerosis

Atherosclerosis is the end result of a series of events, including injury to the intimal endothelium in large and medium-sized arterial vessel, proliferation of vascular smooth muscle, and accumulation of oxidized cholesterol and related lipids at the site of vessel injury. Rupture of plaque from affected vessels (e.g., the carotid arteries) may result in embolization of fragments to the smaller arterial vessels of the brain which then causes stroke. Atherosclerosis may be the result of well-defined genetic disorders affecting the metabolism and transport of cholesterol (e.g., deficiency of the low-density lipoprotein receptor), consumption of diets rich in saturated fat and cholesterol, or the interplay between diabetes and/or hypertension and hypercholesterolemia.[123] It is likely that endogenous production of cholesterol (influenced by genetic traits) is as important as dietary intake of cholesterol and/or saturated fats.

Grossman et al. have reported the first liver-directed gene therapy trial in humans for treatment of familial hypercholesterolemia.[118,119] *Ex vivo* transduction was performed on human hepatocytes from five severely affected patients with disease not responsive to conventional treatment with bile sequestration resins and/or inhibitors of cholesterol biosynthesis. The human trial was based on *ex vivo* hepatocyte-directed experiments with the Watanabe heritable hyperlipidemic (WHHL) rabbit model in which LDL receptors are absent, resulting in hypercholesterolemia and simulating the human form of the disease.[124,125] These experiments demonstrated a reduction of elevated cholesterol after retroviral-transduced hepatocytes containing the LDL receptor were returned to the liver by portal infusion. When this approach was used in the human gene therapy trial for familial hypercholesterolemia, three of five patients had a prolonged and significant reduction of their LDL-cholesterol levels, one of which decreased 20% from pretreatment levels.[119] It remains to be seen whether this therapy will prevent further progression of atherosclerosis in those patients whose lipid profiles improved, but this trial represents an important first step in the field of gene replacement therapy.

Systemic reduction of cholesterol and lipids as exemplified by the liver-directed approach may have the benefit of reducing the risk of atherosclerosis in all arterial vessels, not only those of the carotid circulation. It may also prove possible to apply gene transfer techniques to isolated arterial segments and direct therapy to affected local areas only. This approach has been advanced in experimental animals using adenovirus vectors to transfer the thymidine kinase gene to proliferating vascular smooth muscle cells in porcine arterial segments after balloon-induced injury.[126] This approach kills target cells by rendering the genetically modifed smooth muscle cells susceptible to drugs such as gancyclovir given after gene transfer. Such an approach would be well suited to local therapy of atherosclerotic lesions which might be found in the carotid arteries, and would not entail the more extensive surgical risks associated with systemic gene therapy approaches directed to the liver. Other approaches may be directed at seeding grafts or stents with genetically modified endothelial cells which express tissue plasminogen activator to enhance fibrinolysis in the atheromatous plaque, especially after surgical therapy.[127] In these scenarios, therapy is directed not at correcting a specific gene defect but instead at compensating for a biochemical abnormality. Gene replacement therapy may be applied in the future to the treatment of local complications of the more common forms of atherosclerosis which are not due to well-defined gene defects. This may already be feasible for therapy of extracranial carotid artery disease but would likely be impractical with current balloon catheter technology as an approach to treat disease in smaller intracranial arteries. *In vivo* gene therapy in which vectors may be targeted to the vascular endothelium by the intrinsic specificity of the vector (rather than by mechanical means) would be critical to the development of this approach.

Homocystinuria

Elevated blood levels of homocysteine result in its accumulation in body fluids from which it is normally absent and leads to its excretion in the urine as the condensed form, homocystine (homocystinuria). Elevated homocysteine levels can be the result of various enzyme defects but is most commonly due to deficiency of cystathionine β-synthase enzyme activity. The clinical manifestations of homocystinuria include a predilection to thromboembolic events, particularly ischemic stroke and peripheral vascular occlusions due to accelerated atherosclerosis.[128] The clinical severity of homocystinuria varies and is correlated with the amount of residual cystathionine β-synthase activity in cells from affected patients and the resulting toxic effects of homocysteine in the circulation.[129,130] Retrospective surveys suggest that elevated homocysteine levels may be an independent risk factor for stroke and vascular disease, and that a significant number of patients with strokes or other vascular disease may be undetected heterozygous carriers for homocystinuria.[131–134] Patients with homocystinuria can be categorized into those whose homocysteine levels can be lowered by administration of pyridoxine and those who are unresponsive to pyridoxine.[128] Those who are unresponsive to pyridoxine have no residual enzyme activity, while those who respond to pyridoxine have low intracellular levels of cystathionine β-synthase in cultured skin fibroblasts.[121,128,135]

Restriction of dietary methionine and supplementation with pyridoxine, folic acid, and/or betaine to lower homocysteine levels may delay or prevent strokes or other thromboembolic complications, but are not completely effective, especially in the absence of residual cystathionine β-synthase activity. Replacement of absent cystathionine β-synthase enzyme activity by gene therapy may be a rational approach for lowering homocysteine levels in these patients. The hypothesis that the elevated level of homocysteine is the primary toxic factor responsible for thrombosis suggests that reduction of homocysteine levels in the bloodstream should prevent atherosclerosis and thrombosis in patients with homocystinuria. There is *in vitro* evidence for dose-related toxicity to endothelial cells, and some (though not all) animal studies show direct *in vivo* toxicity to endothelial cells.[136,137] The cDNA encoding cystathionine β-synthase is of a size (2.7 kb) that could be incorporated into most recombinant viral vectors easily, and the gene product has been expressed in cultured cells *in vitro*.[137] It is unclear whether placing the gene in endothelial cells would suffice to prevent local toxic effects and thrombosis and atherosclerosis, or whether replacement of the gene product in a major organ such as the liver might be needed to serve as a metabolic "sink" to lower circulating homocysteine levels and prevent toxicity to all tissues. Gene therapy approaches which decrease systemic levels of homocystine would presumably provide protection against complications and manifestations of homocystinuria other than thromboembolic disease, and might be preferred over therapy directed only to the vascular endothelium.

Deficiencies of Anticoagulant Proteins

Deficiency of the anticoagulant proteins such as proteins C, S, or antithrombin III are associated with predisposition to thromboembolic events, including stroke.[138–140] Antithrombin III exerts a negative regulatory effect on the serine proteases of the coagulation system, particularly (but not exclusively) against thrombin. Proteins C and S are vitamin K-dependent anticoagulant proteins whose joint function is to downregulate blood coagulation by inactivating coagulation factors V and VIII by cleavage of each protein.

Antithrombin III, protein C, or protein S deficiencies are clinically significant in heterozygotes who have 50% of normal levels of the anticoagulant protein. Homozygotes who completely lack any of these three proteins present with massive widespread thrombosis soon after birth, and these conditions are incompatible with life. Deficiency of these anticoagulant proteins has been associated with arterial thrombotic stroke and cerebral vein thrombosis, often in the superior sagittal sinus, which can be associated with infarction and secondary hemorrhage.[139,140] The peculiar tendency for cerebral vein thrombosis may derive from the high levels of tissue factor in the cerebral tissues which may appear in the venous outflow and predispose to thrombosis.

The complementary DNAs encoding antithrombin III, protein C, or protein S could be readily incorporated in most viral vectors used for gene transfer. The expression of proteins C and S would require that they be synthesized in cells capable of vitamin K-dependent post-translational carboxylation of glutamic acid side chains which are essential for calcium binding and full activity. This is identical to the situation for gene therapy of hemophilia B where factor IX biosynthesis requires the same vitamin K-dependent modifications. Gene therapy of the anticoagulant protein deficiencies will likely be directed toward genetic modification of hepatocytes, since these proteins are normally synthesized in the liver and the hepatocyte is capable of producing all of these proteins. It remains to be seen whether other cells can synthesize anticoagulant proteins with full activity.

There is uncertainty regarding the levels of protein synthesis that will be required to have a beneficial effect on heterozygotes with deficiencies of any of the anticoagulant proteins. Clearly, raising heterozygotes from 50% to 100% of normal levels of protein C or S or antithrombin III would completely prevent thrombosis. It is not clear whether lesser increments of improved protein synthesis could lead to a beneficial effect.

Issues of Safety and Efficacy

The examples cited for gene therapy of adenosine deaminase deficiency, hemophilia B, and familial hypercholesterolemia represent only the beginnings of human gene therapy. Alhough each approach has demonstrated varying degrees of success, the experiments also point to specific limitations of each therapeutic method. Before widespread application of these approaches, various issues relating to their safety and efficacy remain to be addressed.

Safety

Adverse effects could arise from direct toxicity of the viral vectors to the recipient, mutagenesis from insertion of a vector into genomic DNA (especially for retroviruses and adeno-associated viruses), or the unintended effects of expression of a gene in an unintended cell type, or at too high a level in the intended target tissue.

The direct toxicity of most viral vectors is related either to low level expression of viral gene products (despite substantial deletions of the viral DNA), or contamination by traces of endotoxins or eukaryotic cellular components from preparation of viral stocks.[27,141] These problems presumably can be overcome by improved vector design, improved methods for virus production, and meticulous adherence to good manufacturing practices during vector production. In addition, these problems may require adjustment of the dose of virus delivered to the recipient *in vivo;* this dose-dependent relationship is most clearly shown by the example of administration of recombinant adenoviruses to the lungs of animals or human patients with cystic fibrosis, where high doses of adenoviruses are associated with toxicity to the lung.[142,143]

Insertional mutagenesis has been considered as a hypothetical risk for *in vivo* administration of retroviral vectors due to their "random" integration into genomic DNA of target cells. These risks are difficult to calculate but probably are quite remote.[144,145] Perhaps more likely is the possible emergence of replication-competent retroviruses during vector manufacture. This has been encountered only once in all of the preparations of recombinant retroviruses for use in humans and was easily detected by relatively simple assay methods.[146] In this regard, the adeno-associated virus system may have some advantage over retroviruses since integration is preferentially targeted to human chromosome 19, and the virus is intrinsically incapable of replication in the absence of a helper, even in the wild-type form.[147] Vectors that maintain themselves as episomes (and integrate only at low frequencies) based on the Epstein-Barr virus system may diminish the risk of insertional mutagenesis to the lowest possible levels, although this system has not yet been tested as extensively as retroviral vectors.

The consequences of unintended over-expression of a desired gene product in the appropriate tissue, or unintended expression of a gene product in the wrong tissue(s), cannot be predicted and underscore the need for testing of gene transfer technologies in animal models wherever possible before human trials. For some diseases such as hemophilia where standard therapeutic replacement of coagulation factors occasionally simulates the effect of "overexpression," there would appear to be no adverse consequences in this regard; in fact, this has been shown during gene therapy of hemophilia B in canines using adenovirus vectors.[41] For other diseases, the consequences will remain unknown until the experiments are performed in animals or human subjects.

Efficacy

An important question to ask as human gene therapy trials proceed is "What are the goals of therapeutic gene replacement and how are these endpoints to be measured?". For some diseases such as hemophilia, the minimum therapeutic goal may be relatively low, since 1% of normal factor VIII or factor IX levels would serve to prevent most of the spontaneous hemorrhage associated with the hemophilias. For hemophilia and the anticoagulant protein deficiencies, the end product (factor VIII, factor IX, antithrombin III, protein C, or protein S) can be measured directly, and the measured levels correlate well with the intended goal of intervention (i.e., to prevent bleeding or thrombosis). For diseases such as familial hypercholesterolemia and homocystinuria, the measured serum levels serve as reasonably good indicators of expression of the primary gene products, i.e., cell surface LDL receptors and intracellular cystathionine β-synthase activity, respectively. For these diseases long-term clinical trials will be necessary to gauge the the efficacy of gene therapy protocols, since the undesired events to be prevented take years to develop.

Diseases such as adult polycystic kidney disease or Ehlers-Danlos syndrome present even greater difficulties in assessment of efficacy. The expression of the gene product will be difficult to measure, and the clinical effects of gene replacement therapy will be even more difficult to assess. It is likely that treatment of these diseases by gene therapy will lag considerably behind the diseases mentioned previously. Animal models, including transgenic mice which simulate these human diseases, will be extremely important for the testing of any gene therapy method to be tested in humans.

Suggested Reading

The reader may wish to consider the following readings which treat gene therapy methods and applications to particular diseases in greater detail.

Kay MA, Liu D, Hoogerbrugge PM. Gene therapy. Proc Natl Acad Sci USA 1997;94:12744–12746. A succinct review of the current status of gene therapy.

Carmen IH. Human gene therapy: A biopolitical overview and analysis. Hum Gene Ther 1993;4:187–193. Reviews social issues and their impact on the application of human gene therapy technology to clinical situations.

Rich DP, Couture LA, Cardoza LM, et al. Development and analysis of recombinant adenoviruses for gene therapy of cystic fibrosis. Hum Gene Ther 1993; 4:461–476. Provides a detailed description of adenovirus vectors in development for therapy of cystic fibrosis.

Michaels SI, Curiel DT. Strategies to achieve targeted gene delivery via the receptor-mediated endocytosis pathway. Gene Ther 1994;1:223–232. An excellent review of molecular conjugates for targeted gene transfer.

Caplen NJ. Lipid gene transfer: A story of simplicity and complexity. In Xanthopolous K (ed). Gene Therapy. NATO ASI Series Vol H 105. Berlin: Springer-Verlag, 1998, 193–202.

Lozier JN, Brinkhous KM. Gene therapy and the hemophilias. J Am Med Assoc 1994;271:47–51. A comprehensive review of gene therapy for hemophilia.

References

1. Miller AD. Progress toward human gene therapy. Blood 1990;76:271–278.
2. Blaese RM, Culver KW, Miller AD, et al. T lymphocyte-directed gene therapy for ADA-SCID: Initial trial results after 4 years. Science 1995;270:475–480.
3. Bordignon C, Notarangelo LD, Nobili N, et al. Gene therapy in peripheral blood lymphocytes and bone marrow for ADA-immunodeficient patients. Science 1995;270:470–474.
4. Kohn DB, Weinberg KI, Nolta JA, et al. Engraftment of gene-modified umbilical cord blood cells in neonates with adenosine deaminase deficiency. Nature Med 1995;1017–1023.
5. Anonymous. Human gene marker/therapy clinical protocols. Hum Gene Ther 1998;9:935–976.
6. Wilmut I, Schnieke AE, McWhir J, et al. Viable offspring derived from fetal and adult mammalian cells. Nature 1997;385:810–813.
7. Carmen, IH. Human gene therapy: A biopolitical overview and analysis. Hum Gene Therapy 1993;4:187–193.
8. Council for Responsible Genetics, Human Genetics Committee. Position paper on human germ line manipulation. Hum Gene Ther 1993;4:35–37.
9. Neel JV. Germ-line gene therapy: Another view. Hum Gene Ther 1993;4:127–128.
10. Fox JL. Germline gene therapy contemplated. Nature Biotechnol 1998;16:407.
11. Kemball-Cook G, Tuddenham EGD. The factor VIII structure and mutation resource site: HAMSTeRS version 4. Nucleic Acids Res 1998;26:216–219.
12. Giannelli F, Green PM, Sommer SS, et al. Haemophilia B: Database of point mutations and short additions and deletions. 8th ed. Nucleic Acids Res 1998;26:265–268.
13. Miller AD. Human gene therapy comes of age. Nature 1992;357:455–460.
14. Brown PO, Bowerman B, Varmus HE, Bishop JM. Retroviral integration: Structure of the initial covalent product and its precursor, and a role for the viral IN protein. Proc Natl Acad Sci USA. 1989;89:2225–2229.
15. Verma IM. Gene therapy. Sci Am 1990;63:68–84.
16. Hoeben RC, Valerio D, van der Erb AJ, et al. Gene therapy for human inherited disorders: Techniques and status. Crit Rev Oncol/Hematol 1992;13:33–54.
17. Miller AD, Law M-F, Verma IM. Generation of helper-free amphotropic retroviruses that transduce a dominant-acting, methotrexate-resistant dihydrofolate reductase gene. Mol Cell Biol 1985;5:431–437.
18. Cone RD, Mulligan RC. High-efficiency gene transfer into mammalian cells: Generation of helper-free recombinant retrovirus with broad mammalian host range. Proc Natl Acad Sci USA 1984;81:6349–6353.
19. Sorge J, Wright D, Erdman VD, et al. Amphotropic retrovirus vector system for human cell gene transfer. Mol Cell Biol 1984;4:1730–1737.
20. Miller AD, Trauber DR, Buttimore C. Factors involved in production of helper virus-free retrovirus vectors. Som Cell Mol Genet 1986;12:175–183.
21. Miller AD. Retrovirus packaging cells. Hum Gene Ther 1990;1:5–14.
22. Salmons B, Gunzburg WH. Targeting of retroviral vectors for gene therapy. Hum Gene Ther 1993;4:129–141.
23. Naldini L, Blomer U, Gallay P, et al. In vivo gene delivery and stable transduction of nondividing cells by a lentiviral vector. Science 1996;272:263–267.
24. Zufferey R, Nagy D, Mandel RJ, et al. Multiply attenuated lentiviral vector achieves efficient gene delivery in vivo. Salk Institute, La Jolla, CA 92037–1099, USA. Nat Biotechnol 1997;15(9):871–875.
24a. Naldini L, Blomer U, Gage FH, et al. Efficient transfer, integration, and sustained long-term expression of the transgene in adult rat brains injected with a lentiviral vector. Proc Natl Acad Sci USA 1996;93:11382–11388.
25. Kim VN, Mitrophanous K, Kingsman SM, Kingsman AJ. Minimal requirement for a lentivirus vector based on human immunodeficiency virus type 1. J Virol 1998;72: 811–816.
26. Berkner KL. Development of adenovirus vectors for the expression of heterologous genes. Biotechniques 1988;6:616–628.
27. Graham FL, van der Eb AJ. A new technique for the assay of infectivity of human adenovirus 5 DNA. Virology 1973;52:456–467.
28. Li Q, Kay MA, Finegold M, et al. Assessment of recombinant adenoviral vectors for hepatic gene therapy. Hum Gene Ther 1993;4: 403–409.

29. Rich DP, Couture LA, Cardoza LM, et al. Development and analysis of recombinant adenoviruses for gene therapy of cystic fibrosis. Hum Gene Ther 1993;4:461–476.

30. Rosenfeld MA, Siegfried W, Yoshimura K, et al. Adenovirus-mediated transfer of a recombinant alpha-1 antitrypsin gene to the lung epithelium *in vivo*. Science 1991;259:431–434.

31. Rosenfeld MA, Yoshimura K, Trapnell BC, et al. In vivo transfer of the human cystic fibrosis transmembrane conductance regulator gene to the airway epithelium. Cell 1992;68:143–155.

32. Engelhardt JF, Simon RH, Yang Y, et al. Adenovirus-mediated transfer of the CFTR gene to lung of nonhuman primates: Biologic efficacy study. Hum Gene Ther 1993;4:759–769.

33. Gao GP, Yang Y, Wilson JM. Biology of adenovirus vectors with E1 and E4 deletions for liver-directed gene therapy. J Virol 1996;70:8934–8943.

34. Wang Q, Greenburg G, Bunch D, et al. Persistent transgene expression in mouse liver following in vivo gene transfer with a delE1/delE4 adenovirus vector. Gene Ther 1997;4:393–400.

35. Mitani K, Graham FL, Caskey CT, Kochanek S. Rescue, propagation, and partial purification of a helper virus-dependent adenovirus vector. Proc Natl Acad Sci USA 1995;92:3854–3858.

36. Lieber A, He CY, Kirillova I, Kay MA. Recombinant adenoviruses with large deletions generated by Cre-mediated excision exhibit different biological properties compared with first-generation vectors in vitro and in vivo. J Virol 1996;70:8944–8960.

37. Chen H-H, Mack LM, Kelly R, et al. Persistence in muscle of an adenoviral vector that lacks all viral genes. Proc Natl Acad Sci USA 1997;94:1645–1650.

38. Schowalter DB, Tubb JC, Liu M, et al. Heterologous expression of adenovirus E3-gp19K in an E1a-deleted adenovirus vector inhibits MHC I expression in vitro, but does not prolong transgene expression in vivo. Gene Ther 1997;4:351–360.

39. Yang Y, Nunes F, Berencski K, et al. Cellular immunity to viral antigens limits E1-deleted adenovirus for gene therapy. Proc Natl Acad Sci USA 1994;91:4407–4411.

40. Smith TAG, Mehaffey MG, Kayda DB, et al. Adenovirus mediated expression of therapeutic plasma levels of human factor IX in mice. Nat Genet 1993;5:397–402.

41. Kay MA, Landen CN, Rothenberg SR, et al. *In vivo* hepatic gene therapy: Complete albeit transient correction of factor IX deficiency in hemophilia B dogs. Proc Natl Acad Sci USA 1994;91:2353–2357.

42. Connelly S, Andrews JL, Gallo AM, et al. Sustained phenotypic correction of murine hemophilia A by in vivo gene therapy. Blood 1998;91:3273–3281.

43. Ye X, Robinson MB, Pabin C, et al. Adenovirus-mediated in vivo gene transfer rapidly protects ornithine transcarbamylase-deficient mice from an ammonium challenge. Pediatr Res 1997;41:527–534.

44. Berns KI. Parvovirus replication. Microbiol Rev 1990;54:316–329.

45. Samulski RJ, Srivastava A, Berns KI, et al. Rescue of adeno-associated virus from recombinant plasmids: Gene correction within the terminal repeats of AAV. Cell 1983;33:135–143.

46. Samulski RJ, Berns KI, Tan M, et al. Cloning of adeno-associated virus into pBR322: Rescue of intact virus from the recombinant plasmid in human cells. Proc Natl Acad Sci USA 1982;79:2077–2081.

47. Grossman Z, Mendelson E, Brok-Simoni F, et al. Detection of adeno-associated virus type 2 in human peripheral blood cells. J Gen Virol 1992;73:961–966.

48. Kotin RM, Linden RM, Berns KI. Characterization of a preferred site on human chromosome 19q for integration of adeno-associated virus DNA by non-homologous recombination. EMBO J 1992;11:5071–5078.

49. Samulski RJ, Zhu X, Xiao X, et al. Targeted integration of adeno-associated virus (AAV) into human chromosome 19. EMBO J 1991;10:3941–3950.

50. Flotte TR, Solow R, Owens RA, et al. Gene expression from adeno-associated virus vectors in airway epithelial cells. Am J Respir Cell Mol Biol 1992;7:349–356.

51. Ponnazhagan S, Erikson D, Kearns WG, et al. Lack of site-specific integration of the recombinant adeno-associated virus 2 genomes in human cells. Hum Gen Ther 1997;8:275–284.

52. Walsh CEJ, Liu JM, Xiao X, et al. Regulated high-level expression of a human gamma globin gene introduced into erythroid cells by an adeno-associated virus vector. Proc Natl Acad Sci USA 1992;89:7257–7261.

53. Nahreini P, Larsen SH, Srivastava A. Cloning and integration of DNA fragments in human cells via the inverted terminal repeats of the adeno-associated virus 2 genome. Gene 1992;119:265–272.

54. Podsakoff G, Wong KK, Chatterjee S. Efficient gene transfer into nondividing cells by adeno-associated virus vectors. J Virol 1994;6:5656–5666.

55. Koeberl DD, Alexander IE, Halbert CL, et al. Persistent expression of human clotting factor IX from mouse liver after intravenous injection of adeno-associated virus vector. Proc Natl Acad Sci USA 1997;94:1426–1431.

56. Thorley-Lawson DA. Basic virological aspects of Epstein-Barr virus infection. Semin Hematol 1988;25:247.

57. Straus SE, Cohen JI, Tosato G, et al. Epstein-Barr virus infections: Biology, pathogenesis, and management. Ann Int Med 1993;118: 45–58.

58. Fink, DJ, Sternberg LR, Weber PC. *In vivo* expression of galactosidase in hippocampal neurons by HSV mediated gene transfer. Hum Gene Ther 1992;3:11–19.

59. Ho DY, Mocarski ES, Sapolsky RM. Altering central nervous system physiology with a defective herpes simplex virus vector expressing the glucose transporter gene. Proc Natl Acad Sci USA 1993;90:3655–3659.

60. Sun T-Q, Livanos E, Vos J-MH. Engineering a mini-herpesvirus as a general strategy to transduce up to 180 kb of functional self-replicating human minichromosomes. Gene Ther 1996;3:1081–1088.

61. Sun T-Q, Fenstermacher DA, Vos J-MH. Human artificial episomal chromosomes for cloning large DNA fragments in human cells. Nat Genet 1994;8:33–41.

62. Glorioso JC, Bender MA, Goins WF, et al. Herpes simplex virus as a gene-delivery vector for the central nervous system. In Kaplitt MG, Loewy AD, (eds). Viral Vectors: Gene Therapy and Neuroscience Applications. New York: Academic Press, 1995, 1–23.

63. Miyonahara A, Johnson PA, Elan RL, et al. Direct gene transfer to the liver with herpes simplex virus type 1 vectors: Transient production of physiologically relevant levels of circulating factor IX. New Biologist 1992; 4:238–246.

64. Banerjee S, Livanos E, Vos J-MH. Therapeutic gene delivery in human B-lymphoblastoid cells by engineered non-transforming infectious Epstein-Barr virus. Nat Med 1995;1:1303–1308.

65. Potter H. Electroporation in biology: Methods, applications, and instrumentation. Anal Biochem 1988;174:361–373.

66. Felgner PL, Gadek TR, Holm M, et al. Lipofection: A highly efficient, lipid-mediated DNA transfection procedure. Proc Natl Acad Sci USA 1987;84:7413–7417.

67. Nicolau C, Legrand A, Grosse E. Liposomes as carriers for *in vivo* gene transfer and expression. Methods Enzymol 1987;149:157–176.

68. San H, Yang Z-Y, Pompili VJ, et al. Safety and short-term toxicity of a novel cationic lipid formulation for human gene therapy. Hum Gene Ther 1993;4:781–788.

69. Caplen NJ, Alton EWFW, Middleton PG, et al. Liposome-mediated CFTR gene transfer to the nasal epithelium of patients with cystic fibrosis. Nat Med 1995;1:39–46.

70. Oudrhiri N, Vigneron J-P, Peuchmaur M, et al. Gene transfer by guanidinium-cholesterol cationic lipids into airway epithelial cells *in vitro* and *in vivo*. Proc Natl Acad Sci USA 1997;94:1651–1656.

71. Lim CS, Chapman GD, Gammon RS. Direct *in vivo* gene transfer into the coronary and peripheral vasculatures of the intact dog. Circulation 1991;83:2007–2011.

72. Nabel EG, Plautz G, Nabel GJ. Site-specific gene expression *in vivo* by direct gene transfer into the arterial wall. Science 1990;249: 1285–1288.

73. Yang J-P, Huang L. Overcoming the inhibitory effect of serum on lipofection by increasing the charge ratio of cationic liposome to DNA. Gene Ther 1997;4:950–960.

74. Liu Y, Mounkes LC, Liggitt HD, et al. Factors influencing the efficiency of cationic liposome-mediated intravenous gene delivery. Nat Biotechnol 1997;15:167–173.

75. Qiu C, De Young MB, Finn A, Dichek DA. Cationic liposomes enhance adenovirus entry via a pathway independent of the fiber receptor and alpha$_v$-integrins. Hum Gene Ther 1998;9:507–520.

76. Hodgson CP, Solaiman F. Virosomes: Cationic liposomes enhance retroviral transduction. Nat Biotechnol 1996;14:339–342.

77. Wu GY, Wu CH. Receptor-mediated *in vitro* gene transformation by a soluble DNA carrier system. J Biol Chem 1987;262:4429–4432.

78. Wu CH, Wilson JM, Wu GY. Targeting genes: Delivery and persistent expression of a foreign gene driven by mammalian regulatory elements in vivo. J Biol Chem 1989;264: 16985–16987.

79. Wu GY, Wilson JM, Shalaby F, et al. Receptor-mediated gene delivery *in vivo*: Partial correction of genetic analbuminemia in Nagase rats. J Biol Chem 1991;266:14338–14342.

80. Boussif O, Lezoualch'h F, Mergny MD, et al. A versatile vector for gene and oligonucleotide transfer into cells in culture and in vivo: Polyethylenimine. Proc Natl Acad Sci USA 1995;92:7297–7301.

81. Abdallah B, Hassan A, Benoist C, et al. A powerful nonviral vector for in vivo gene transfer into the adult mammalian brain: Polyethylenimine. Hum Gene Ther 1996;7:1947–1954.

82. Kircheis R, Kichler A, Kursa M, et al. Coupling of cell-binding ligands to polyethylenimine for targeted gene delivery. Gene Ther 1997;4:409–418.

83. Baker A, Saltik M, Lehrmann H, et al. Polyethylenimine (PEI) is a simple, inexpensive and effective reagent for condensing and linking plasmid DNA to adenovirus for gene delivery. Gene Ther 1997;4:773–782.

84. Curiel DT, Agarwal S, Wagner E, et al. Ade-

novirus enhancement of transferrin-polyly-sine-mediated gene delivery. Proc Natl Acad Sci USA 1991;88:8850–8854.

85. Yoshimura K, Rosenfeld MA, Seth P, et al. Adenovirus-mediated augmentation of cell transfection with unmodified plasmid vectors. J Biol Chem 1993;268:2300–2303.

86. Seth P, Fitzgerald D, Ginsberg H, et al. Evidence that the penton base of adenovirus is involved in potentiation of toxicity of *Pseudomonas* exotoxin conjugated to epidermal growth factor. Mol Cell Biol 1984;4: 1528–1533.

87. Cotten M, Wagner E, Zatloukal K, et al. High-efficiency receptor-mediated delivery of small and large (48 kilobase) gene constructs using the endosome-disruption activity of defective or chemically inactivated adenovirus particles. Proc Natl Acad Sci USA 1992;89:6094–6098.

88. Cristiano RJ, Smith LC, Kay MA, et al. Hepatic gene therapy: Efficient gene delivery and expression in primary hepatocytes utilizing a conjugated adenovirus-DNA complex. Proc Natl Acad Sci USA 1993;90:11548–11552.

89. Lozier JN, Thompson AR, Hu P-C, et al. Efficient transfection of primary cells in a canine hemophilia B model using adenovirus-polylysine-DNA complexes. Hum Gen Ther 1994;5:313–322.

90. Meunier-Durmort C, Grimal H, Sachs LM, et al. Adenovirus enhancement of polyethyleneimine-mediated transfer of regulated genes in differentiated cells. Gene Ther 1997; 4:808–814.

91. Zatloukal K, Cotten M, Berger M. *In vivo* production of human factor VIII in mice after intrasplenic implantation of primary fibroblasts transfected by receptor-mediated, adenovirus-augmented gene delivery. Proc Natl Acad Sci USA 1994;91:5148–5152.

92. Michaels SI, Curiel DT. Strategies to achieve targeted gene delivery via the receptor-mediated endocytosis pathway. Gene Therapy. 1994;1:223–232.

93. Cristiano RJ, Roth JA. Molecular conjugates: A targeted gene delivery vector for molecular medicine. J Mol Med 1995;73:479–486.

94. St. Louis D, Verma IM. An alternative approach to somatic cell gene therapy. Proc Natl Acad Sci USA 1988;85:3150–3154.

95. Palmer TD, Thompson AR, Miller AD. Production of human factor IX in animals by genetically modified skin fibroblasts: Potential therapy for hemophilia B. Blood 1989;73: 438–445.

96. Hoeben RC, van der Jagt CM, Schoute F, et al. Expression of functional factor VIII in primary human skin fibroblasts after retro

97. Hoeben RC, Fallaux FJ, Van Tilburg NH, et al. Toward gene therapy for hemophilia A: Long-term persistence of factor VIII-secreting fibroblasts after transplantation into immunodeficient mice. Hum Gene Ther 1993; 4:179–186.

98. Gitschier J, Wood WI, Goralka TM, et al. Characterization of the human factor VIII gene. Nature 1984;312:326–330.

99. Yoshitake S, Schach BG, Foster DC, et al. Nucleotide sequence of the gene for human factor IX (antihemophilic factor B). Biochemistry 1985;24:3736–3750.

100. Lozier JN, Kessler CM. Clinical aspects and therapy for hemophilia. In Hoffman R, Benz EJ, Shattil SJ, et al (eds). Hematology Basic Principles and Practice. 3rd ed. New York: Churchill Livingstone. 1998. In press.

101. Eyster ME, Gill FM, Blatt PM, et al. Central nervous system bleeding in hemophiliacs. Blood 1979;51:1179.

102. Lozier JN, Brinkhous KM. Gene therapy and the hemophilias. J Am Med Assoc 1994;271: 47–51.

103. Kay MA, Rothenberg S, Landen CN, et al. *In vivo* gene therapy of hemophilia B: Sustained partial correction in factor IX-deficient dogs. Science 1993;262:117–119.

104. Kay MA, Holterman AX, Meuse L, et al. Long-term hepatic adenovirus-mediated gene expression in mice following CTLA4Ig administration. Nat Genet 1995;11:191–197.

105. Ilan Y, Sauter B, Chowdhury NR, et al. Oral tolerization to adenoviral proteins permits repeated adenovirus-mediated gene therapy in rats with pre-existing immunity to adenoviruses. Hepatology 1998;27:1368–1376.

106. Scarfmann R, Axelrod JH, Verma IM. Long-term in vivo expression of retrovirus-mediated gene transfer in mouse fibroblast implants. Proc Natl Acad Sci USA 1991;88: 4626–4630.

107. Gerrard AJ, Hudson DL, Brownlee GG, et al. Towards gene therapy for haemophilia B using primary human keratinocytes. Nature Genet 1993;3:180–183.

108. Yao SN, Kurachi K. Expression of human factor IX in mice after injection of genetically modified myoblasts. Proc Natl Acad Sci USA 1992;89:3357–3361.

109. Roman M, Axelrod JH, Dai Y, et al. Circulating human or canine factor IX from retrovirally transduced primary myoblasts and established myoblast cell lines grafted into murine skeletal muscle. Somat Cell Mol Genet 1992;18:247–258.

110. Dai Y, Roman M, Naviaux RK, et al. Gene therapy via primary myoblasts: Long-term

expression of factor IX protein following transplantation in vivo. Proc Natl Acad Sci USA 1992;89:10892–10895.

111. Snyder RO, Miao CH, Patijn GA, et al. Persistent and therapeutic concentrations of human factor IX in mice after hepatic gene transfer of recombinant AAV vectors. Nat Genet 1997;16:270–276.

112. Herzog RW, Hagstrom JN, Kung S-H, et al. Stable gene transfer and expression of human blood coagulation factor IX after intramuscular injection of recombinant adeno-associated virus. Proc Natl Acad Sci USA 1997;94:5804.

113. Hsueh JL, Lu DR, Zhou JM, et al. Gene therapy for hemophilia B using fibroblasts. Part I. Sci China (Ser B). 1993;23:53–60.

114. Roberts HR, Lozier JN. Other clotting factor deficiencies. In Hoffman R, Benz EJ, Shattil SJ, et al (eds). Hematology Basic Principles and Practice. New York: Churchill Livingstone, 1991, 1332–1342.

115. Lemarchand P, Jones M, Yamada I, Crystal RG. In vivo gene transfer and expression in normal uninjured blood vessels using replication-deficient recombinant adenovirus vectors. Circ Res 1993;72:1132–1138.

116. Newman KD, Dunn PF, Owens JW, et al. Adenovirus-mediated gene transfer into normal rabbit arteries results in prolonged vascular cell activation, inflammation, and neointimal hyperplasia. J Clin Invest 1995; 96:2955–2965.

117. Rolling F, Nong Z, Pisvin S, Collen D. Adeno-associated virus-mediated gene transfer into rat carotid arteries. Gene Ther 1997;4:757–761.

118. Grossman M, Raper SE, Kozarsky K, et al. Successful ex vivo gene therapy directed to liver in a patient with familial hypercholesterolaemia. Nat Genet 1994;6:335–341.

119. Grossman M, Rader DJ, Muller DWM, et al. A pilot study of ex vivo gene therapy for homozygous familial hypercholesterolaemia. Nat Med 1995;1:1148–1154.

120. Brown MS, Goldstein JL. A receptor-mediated pathway for cholesterol homeostasis. Science 1986;232:34–37.

121. Hu FL, Gu Z, Kozich V, et al. Molecular basis of cystathionine β-synthase deficiency in pyridoxine responsive and nonresponsive homocystinuria. Hum Mol Genet 1993;2: 1857–1860.

122. Rodgers GM, Chandler WL. Laboratory and clinical aspects of inherited thrombotic disorders. Am J Hematol 1992;41:113–122.

123. Havel RJ, Kane JP. Introduction: Structure and metabolism of plasma lipoproteins. In Scriver CR, Beaudet AL, Sly WS, Valle D (eds). The Metabolic Basis of Inherited Disease. 6th ed. New York: McGraw-Hill, 1989, 1129–1138.

124. Wilson JM, Chowdhury JR, Grossman M, et al. Temporary amelioration of hyperlipidemia in low density lipoprotein receptor-deficient rabbits transplanted with genetically modified hepatocytes. Proc Natl Acad Sci USA 1990;87:8437–8441.

125. Chowdhury JR, et al. Long-term improvement of hypercholesterolemia after ex vivo gene therapy in LDLR deficient rabbits. Science 1991;254:1802–1805.

126. Ohno T, Gordon D, San H, et al. Gene therapy for vascular smooth muscle cell proliferation after arterial injury. Science 1994; 265:781–784.

127. Dichek DA, Neville RF, Zwiebel JA, et al. Seeding of intravascular stents with genetically engineered endothelial cells. Circulation 1989;80:1347–1353.

128. Cacciari E, Salardi S. Clinical and laboratory features of homocystinuria. Haemostasis 1991;19(suppl 1):10–13.

129. Mudd SH, Levy HL, Skovby F. Disorders of transulfuration. In Scriver CR, Beaudet AL, Sly WS, Valle D (eds). The Metabolic Basis of Inherited Disease. 6th ed. New York: McGraw-Hill, 1989, 693–734.

130. Mudd SH, Skovby F, Levy HL, et al. The natural history of homocystinuria due to cystathionine β-synthase deficiency. Am J Hum Genet 1985;37:1–31.

131. Boers GHJ, Smals AGH, Trijbels FJM, et al. Heterozygosity for homocystinuria in premature peripheral and cerebral occlusive arterial disease. N Engl J Med 1985;313: 709–715.

132. Malinow MR. Hyperhomocyst(e)inemia: A common and easily reversible risk factor for occlusive atherosclerosis. Circulation 1990; 81:2004–2006.

133. Clarke R, Daly L, Robinson K, et al. Hyperhomocysteinemia: An independent risk factor for vascular disease. N Engl J Med 1991;324:1149–1155.

134. Clarke R, Fitzgerald D, O'Brien C, et al. Hyperhomocysteinaemia: A risk factor for extracranial carotid artery atherosclerosis. Ir J Med Sci 1992;161:61–65.

135. Kraus JP, Le K, Swaroop M, et al. Human cystathionine β-synthase cDNA: Sequence, alternative splicing and expression in cultured cells. Hum Mol Genet 1993;2:1633–1638.

136. Harker LA, Ross R, Schlichter SJ, et al. Homocysteine-induced arteriosclerosis: The role of endothelial cell injury and platelet response in its genesis. J Clin Invest 1976;58: 731–741.

137. Donahue S, Sturman JA, Gaull G. Arteriosclerosis due to homocyst(e)inemia: Fail-

ure to reproduce the model in weanling rabbits. Am J Pathol 1974;77:167–174.

138. Camerlingo M, Finazzi G, Casto L, et al. Inherited protein C deficiency and nonhemorrhagic arterial stroke in young adults. Neurology 1991;41:1371–1373.

139. Roos KL, Pascuzzi RM, Kuharik MA, et al. Post-partum intracranial venous thrombosis associated with dysfunctional protein C and deficiency of protein S. Obstet Gyn 1990;76:492–494.

140. Muramatsu S, Mizuno Y, Murayama H, et al. Hereditary antithrombin III deficiency with a superior sagittal sinus thrombosis: Evidence for a possible mutation starting in the mother of the propositus. Thromb Res 1990;57:593–600.

141. Cotten M, Baker A, Saltik M, et al. Lipopolysaccharide is a frequent contaminant of plasmid DNA preparations and can be toxic to primary human cells in the presence of adenovirus. Gene Ther 1994;1: 239–246.

142. Simon RH, Engelhardt JF, Yang Y, et al. Adenovirus mediated transfer of the CFTR gene to lung of nonhuman primates: Toxicity studies. Hum Gene Ther 1993;4:771–780.

143. Crystal RG, McElvaney NG, Rosenfeld MA, et al. Administration of an adenovirus containing the human CFTR cDNA to the respiratory tract of individuals with cystic fibrosis. Nat Genet 1994;8:42–51.

144. Anderson WF. What about those monkeys that got T-cell lymphomas? Hum Gene Ther 1993;4:1–2.

145. Anderson WF. Was it just stupid or are we poor educators? Hum Gene Ther 1994;5: 791–792.

146. Otto E, Jones-Trower A, Vanin EF, et al. Characterization of a replication-competent retrovirus resulting from recombination of packaging and vector sequences. Hum Gene Ther 1994;5:567–575.

147. Kotin RM. Prospects for the use of adeno-associated virus as a vector for human gene therapy. Hum Gene Ther 1994;5:793–801.

Index

Hypoalphalipoproteinemia, 88
Hypobetalipoproteinemia, 90
Hypoplasia of anterior cerebral artery, 246–247

ICH; *see* Intracerebral hemorrhage
Identical by descent, 35–36
Identical by state, 36
In situ hybridization, 13
In vivo approach in gene therapy, 361
Inflammation in atherosclerosis, 120–123
Inflammatory cytokine expression, 120–121
Informative linkage information, 23–25
Inherited disease, 313–332
    analysis of mutations, 49
    cardiac, 183–193; *see also* Heart disease
    coagulation disorders, 316–319
    complex diseases, 50–52
    connective tissue disorders, 324–326
    evaluation for causes of stroke, 335–344
    gene expression, 49–50
    genetic counseling, 345–351
    hemoglobinopathies, 313–316
    homocysteine metabolism disorders,
        321–323
    mitochondrial disorders, 319–320
    pediatric stroke, 261–311; *see also* Pediatric
        stroke
    protein analysis, 42–43
    reverse genetics, 43–48
Insulin, hypertension and, 61
Insulin receptor, 61
Insulin resistance, 73–74
Intercellular adhesion molecule, 121–122
Interleukin-1, 120–121
Intermediate density lipoprotein, 83, 84
Internal carotid artery dissection, 202–204
Internet, 342
Intracerebral hemorrhage, 209–220
    amyloid angiopathies, 212–220
    spontaneous/hypertensive, 209–212
Intracranial aneurysm, 237–259
    see also Subarachnoid hemorrhage
    candidate genes for, 247–251
    clinical evaluation and genetic counseling,
        251–253
    coarctation of aorta, 243–244
    in Ehlers-Danlos syndrome, 242–243
    in fibromuscular dysplasia, 244–245
    genetic epidemiology of, 237–240
    human leukocyte antigen and, 245–246
    hypoplasia of anterior cerebral artery,
        246–247
    in Marfan's syndrome, 247
    in polycystic kidney disease, 241–242

    in pseudoxanthoma elasticum, 247
    risk factors for, 240
Intracranial hemorrhage
    see also Intracerebral hemorrhage; Subarach-
        noid hemorrhage
    in Ehlers-Danlos syndrome, 294
    genetic epidemiology in, 167–168
    inherited coagulation factor abnormalities
        in, 134, 286
    sickle cell disease and, 266
Intracranial thromboembolism, 141
Intron, 5
Iron-binding proteins, 123–124
Ischemic cardiomyopathy, 184
Ischemic stroke
    genetic epidemiology in, 166–167
    inherited causes of, 343, 363–366
    pediatric, 264
Isovaleric acidemia, 289

Kallikrein, 61
Kawasaki disease, 292
Kininogen, 139
Kohlmeier-Degos disease, 277–278
Kringle, 93

Laboratory testing, evaluation for inherited
    causes of stroke, 340–342
LCAT; *see* Lecithin:cholesterol acyltransferase
LDL-receptor-like protein, 83
Lecithin:cholesterol acyltransferase, 82, 85, 97
Leigh's disease, 289
Leukoencephalopathy
    in CADASIL syndrome, 195–199
    in familial arteriosclerotic leukoen-
        cephalopathy with alopecia and skeletal
        disorders, 199–200
Ligases, 9
Linkage analysis, 19–40
    in arrhythmias, 188
    complications in, 30–34
    in familial dilated cardiomyopathy, 186–187
    genetic markers, 22–25
    LOD score approach, 27–30
    measures of linkage information, 25–27
    nonparametric methods of, 34–36
    planning of, 37
    recombination and, 20–22
    reverse genetics in, 44–48
    in von Hippel-Lindau disease, 229–230
Linkage phase, 24
Lipid metabolism abnormalities, 81–115
    apolipoproteins and, 86–94
    atherosclerosis and, 119
    candidate genes in, 98